Handbook of
Quality Assurance
in Mental Health

Handbook of Quality Assurance in Mental Health

EDITED BY

GEORGE STRICKER

Institute of Advanced Psychological Studies
Adelphi University
Garden City, New York

AND

ALEX R. RODRIGUEZ

Preferred Health Care, Ltd.
Wilton, Connecticut

Springer Science+Business Media, LLC

Library of Congress Cataloging in Publication Data

Handbook of quality assurance in mental health.

Includes bibliographies and index.
1. Mental health services—Standards—United States. 2. Quality assurance—United States. I. Stricker, George. II. Rodríguez, Alex R. [DNLM: 1. Mental Health Services—United States. 2. Quality Assurance, Health Care—United States. WM 30 H2358]
RA790.5.H3   1987   362.2'068   87-29101

ISBN 978-1-4684-5238-9   ISBN 978-1-4684-5236-5 (eBook)
DOI 10.1007/978-1-4684-5236-5

© 1988 Springer Science+Business Media New York

Originally published by Plenum Press, New York 1988

Softcover reprint of the hardcover 1st edition 1988

Contributors

MICHAEL J. BENNETT   Harvard Community Health Plan, Harvard Medical School, Boston, Massachusetts 02215

RUSSELL J. BENT   School of Professional Psychology, Wright State University, Dayton, Ohio 45435

DONALD N. BERSOFF   Ennis, Friedman, & Bersoff, 1200 17th Street, N.W., Washington, D.C. 20036

ROBERT H. BROOK   Rand Corporation, 1700 Main Street, Santa Monica, California 90406–2138

WILLIAM J. CHESTNUT   Indiana University, Student Health Center, Bloomington, Indiana 47405

J. JARRETT CLINTON   Office of Health Affairs, The Pentagon, Department of Defense, Washington, D.C. 20310

LAWRENCE H. COHEN   Department of Psychology, University of Delaware, Newark, Delaware 19716

JOSEPH N. CRESS   4645 Brady Street, Davenport, Iowa 52806

PATRICK H. DeLEON   Office of Senator Daniel K. Inouye, 722 Hart Senate Office Building, Washington, D.C. 20510

ROBERT W. GIBSON   Sheppard and Enoch Pratt Hospital, 6501 North Charles Street, Baltimore, Maryland 21285–6815

PETER G. GOLDSCHMIDT   World Development Group, 5101 River Road, Bethesda, Maryland 20816.

JUDY E. HALL   New York State Education Department, Cultural Education Center, Albany, New York 12230

AGNES B. HATFIELD   Department of Education, University of Maryland, College Park, Maryland 20742

KIT KINPORTS   Ennis, Friedman, & Bersoff, 1200 17th Street, N.W., Washington, D.C. 20036

NANCY LANE-PALES   3112 Westover Drive, S.E., Washington, D.C. 20020

ROBERT S. LONG   Mutual of Omaha Insurance Company, Omaha, Nebraska 68175

LORRAINE L. LUFT   Peninsula Hospital Community Mental Health Center, 1783 El Camino Real, Burlingame, California 94010

MYRENE McANINCH   Accreditation Program for Psychiatric Facilities, Joint Commission on Accreditation of Hospitals, 875 North Michigan Avenue, Chicago, Illinois 60611

ELIZABETH MEID   10 Minot Avenue, Auburn, Maine 04210

DONALD E. NEWMAN   Peninsula Hospital Community Mental Health Center, 1783 El Camino Real, Burlingame, California 94010

BETH EGAN O'KEEFE   2100 52nd Avenue, Moline, Illinois 61265

NORMAN R. PENNER   American Psychiatric Association, 1400 K Street, N.W., Washington, D.C. 20005

ALEX R. RODRIGUEZ   Preferred Health Care, Ltd., 15 River Road, Wilton, Connecticut 06897

ABRAM ROSENBLATT   Department of Psychology, University of Arizona, Tucson, Arizona 85721

LEE SECHREST   Department of Psychology, University of Arizona, Tucson, Arizona 85721

SHARON A. SHUEMAN   Shueman Troy and Associates, 246 North Orange Grove Boulevard, Pasadena, California 91103

H. BERNARD SMITH   Private practice, Washington, D.C. 20005

MILTON THEAMAN   565 West End Avenue, New York, New York 10024

WARWICK G. TROY   California School of Professional Psychology, 2235 Beverly Boulevard, Los Angeles, California 90057

GARY R. VANDENBOS   American Psychological Association, 1200 17th Street, N.W., Washington, D.C. 20036

KENNETH B. WELLS   Rand Corporation, 1700 Main Street, Santa Monica, California 90406–2138

JOAN G. WILLENS   Private practice, Beverly Hills, California 90212

JACK A. WOLFORD   Western Psychiatric Institute and Clinic, University of Pittsburgh, School of Medicine, Pittsburgh, Pennsylvania 15213

Contents

PART I INTRODUCTION

PART II GENERAL ISSUES

Chapter 6

The Effects of Contemporary Economic Conditions on Availability and
Quality of Mental Health Services 137

Alex R. Rodriguez

Chapter 7

The Role of the Consumer in Quality Assurance 169

Agnes B. Hatfield and H. Bernard Smith

Chapter 8

Alternative Futures for Assuring the Quality of Mental Health Services ... 185

Peter G. Goldschmidt

Chapter 9

Milton Theaman

PART III   LEVEL OF CARE

Chapter 10

Jack A. Wolford

I

INTRODUCTION

1

An Introduction to Quality Assurance in Mental Health

ALEX R. RODRIGUEZ

INTRODUCTION

Among the most frequently encountered terms in the health services' literature and at meetings of various health professional disciplines is *quality assurance*. Since the early 1970s, the term has spread well beyond the health sciences to other fields, so that now it serves as a common term of reference for both principles and activities that reflect professional attention to quality services. In this context, quality assurance has come to represent the profession's attention to its innate concern about quality services reflected in professional ethics, identity, science, and training. It has thus become contiguous with professionalism and serves as a cornerstone and touchstone for professional commitment to quality and service.

As the term has gained such common parlance that it is now often associated with numerous nonprofessional activities, its initial and continuing denotations have primarily been connected with health care. The health professions literature is rich in its range and depth of focus on quality assurance. In particular, mental health services have been a frequent source of attention because the nature and diversity of care are such that the assurance of quality requires special modes of assessment and delivery of services.

Although much has been developed and written about quality assurance in mental health over the years, those activities have tended to be focused on various conceptual, operational, or research-related aspects of the subject. This book is a first attempt to pull together this diverse knowledge into an overview of the primary fields of study and development to date. Although the authors and editors have not attempted to construct a complete or comprehensive review, they have set out to develop a compendium that is both broadly defined and informative. The usefulness of this book will be not in providing a programmatic "how to" approach to quality assurance in mental health, but in its conceptual and descriptive summaries of research, programs, and other developments.

This introductory chapter should serve to provide an overview of the field and

ALEX R. RODRIGUEZ • Preferred Health Care, Ltd., 15 River Road, Wilton, Connecticut 06897.

to invite the reader to explore further both the many references available on specific subjects in the chapter and the highly thoughtful summations and insights in subsequent chapters. The authors hope that each chapter will stimulate the reader to ask new questions and to make new contributions to this once and future field.

DEFINITIONS OF QUALITY ASSURANCE

The editors have approached this book with no single or simple concept or definition of quality assurance because they perceived it as not being simply defined. The authors have been asked to "beat the bushes" in an attempt to explore previous definitions promoted by others within their areas and to expand both their own views and their own limitations in defining this term. The reader will find that there is a common area regarding quality mental-health services that is more readily agreed on by most persons concerned with such issues: patients, professional providers, and payers. Outside this core consensus area, there is a broader area that is characterized by a polyglot of qualifying definers, reflecting the concepts and requirements of each category and the individuals within those categories. The diversity in opinion has provided, and will continue to provide, both fruitful professional consensus development and troubling disagreements, until some eventual hypothetical point is reached where there can be a broader societal consensus.

It should be remembered throughout this book that quality mental-health care *can* be an absolute state but is almost always provided and evaluated in a relative state. Thus, quality may be ideal, adequate, or normative, depending on who is assessing it—and what values and consensus are used in evaluation—and by what implicit or explicit standards or gauges it is being objectively or subjectively evaluated. Therefore, quality, like most social realities, is relative to a host of variables that modify the scientific-objective realities.

A "pure" definition of quality assurance would be oriented toward means of assuring the highest level of optimal care for all patients. Thus, such care would be available, appropriate, and efficacious and would be provided without consideration of costs. This is the care generally desired by patients and attempted by providers. The veritable "greed for life and health" that Morris Abram (1984) of the President's Commission for the Study of Ethical Problems in Medicine and Biomedical and Behavioral Research consideres a primary contributor to the costs of high-cost and high-quality health care will continue to be a market determiner of quality provided. Parallel to the consumer demand promoting quality and prodding professional considerations of quality assurance, consumer protection and legal trends in recent years have extended the views that quality is both an entitlement and part of the "contract" between the health care system and the beneficiary. Thus, pure quality is an intrapsychic, professional-scientific, and "demand side" ideal, which serves as an ethical, legal, and personal touchstone in defining quality. Of course, the road to heaven is not paved in gold, but in dollars. For this reason, the term *quality assurance* was born, as will be discussed in the next section.

The federal government has provided one view of quality assurance (Department of Health, Education, and Welfare, 1975) in its attempt to delineate the role of professional-standards-review organizations (PSRO) in defining quality of care for the Medicare program; it is a "shared responsibility of health professionals and government to provide a reasonable basis for confidence that action will be taken, both to assess whether services meet professionally recognized standards and to correct any deficiencies that may be found" (p. 14). Similar pronouncements have been made for the quality assurance activities of the Department of Defense's CHAMPUS program and of the 1980s successor to the PSROs, the federally designated peer-review organizations (PROs), established to ensure quality and utilization-efficient care for Medicare. Links between the federal and state governments and between professional associations and private review entities have been developed to make this "shared responsibility" manifest in the delivery and reimbursement of health services. This responsibility is seen in light of both professional and legal accountability, a view noted by Gibson and Singhas (1978) and Alger (1980). Accountability, then, becomes a concentric concept that elaborates on the pure view of quality and reflects the federal government's consumer protection activities during the 1970s.

The Joint Commission on Accreditation of Hospitals (JCAH), which has provided another primary historical leadership role in defining quality assurance, has promoted the evolution of the concept of resource limitations as a part of the definition of quality assurance. That body sees qualtiy care as "the greatest achievable health benefit, with minimal unnecessary risk and use of resources, and in a manner satisfactory to the patient" (Joint Commission of Accreditation of Hospitals, 1985, p. 76), and quality assurance as those professional and hospital activities that lead to quality care. This more recent JCAH definition further elaborates on the pure concept of quality care by underscoring the role of resource conservation. In doing so, it reflects the growing social value of resource conservation as it pertains to broader social concerns about world recessionary and national inflationary trends in recent years and specifically to the significant escalation in health care costs overall, particularly those related to high-cost, high-technology and low-yield care. In this definition, utilization, efficacy, and cost effectiveness are blended into the concept of quality care and reflect the increasing utilization-review focus of third-party payers and, in turn, professional entities.

The more layering that occurs in refining the definitions of quality care and quality assurance, and the more social, economic, and legal values that are brought into qualifying the "outlier" scientific-professional perspectives, the more difficult become the tasks of separating the objective from the subjective components of such a multifaceted definition; of weighing the relative values of providers, beneficiaries, and payers and developing a consensus for multiple treatments in multiple situations; and of developing objective measures, standards, and criteria that can define quality across structural, process, and outcome parameters of care with sensitivity, validity, and reliability. These tasks become both the creative laboratory and the battleground for the field of quality assurance, and they become more difficult to accomplish when services that are less readily objectively assessed and provided—such as mental health services—are being considered.

The special considerations in mental health services quality assurance require some flexibility and subjective therapeutic judgment in both treatment and review of that treatment, rather than what Menninger (1977) considers "reductionistic ... simple or explicit measure of quality" (p. 479). He believes that quality mental-health services must be viewed within the unique needs of each patient, which the treatment program is ethically bound to thoroughly and conscientiously assess and treat. He believes that quality mental health care "might be defined as the goodness of fit between the problem requiring therapeutic attention; the desired outcome (the goal or purpose of treatment); the treatment used, as sensed or experienced by the patient, as judged by the physician and his colleagues, and as verified by outcome studies (Menninger, 1977, p. 480). This "goodness-of-fit" concept of quality care must be taken into account when defining and assessing mental health services, but it does not preclude the development and use of normative and objective standards and criteria.

Thus, contemporary definitions of quality assurance and quality care must take into account the professional-scientific emphasis of providers, the legal-protectionist emphasis of patients and their advocates, and the economic emphasis of payers for health services. In this respect, the definitions are not ideal or pure but reflect the "real world" of health care services. The definitions do not set the parameters of the services; on the contrary, the evolving systems in which services occur establish evolving definitions. This is an important distinction in approaching the subject areas of this book, which will repeatedly demonstrate that quality care and quality assurance are more social systems issues than solely professional ones. In this sense, the focus on quality will demonstrate that mental health care occurs in evolving multiaxial contexts, which require a continuing developmental orientation to definitions of quality.

THE DRIVING FORCES IN QUALITY OF CARE

Although subsequent chapters will elaborate on specific elements of quality assurance, some note should be made of the evolutions and current considerations in quality assurance programming. A number of general references (Bausell, 1983a, b; Donabedian, 1980, 1982; Greenspan, 1980; Jessee, 1983; LoGerfo & Brook, 1979; Mattson, 1984; Sanazaro, 1980; Ware, Johnston, Davies-Avery, & Brook, 1979; Williams & Brook, 1978; Williamson, Hudson, & Nevins, 1982b) are noted that reflect the evolution of the field conceptually and operationally. Recommended topical references related to quality of care and quality assurance are also noted for the reader (Brook, 1977; Brook, Kamberg, & Lohr, 1982; Donabedian, 1978; Ginzberg, 1975; Jessee, 1977; Luke & Boss, 1981), as they underscore the inherent difficulties in defining and assessing quality care, particularly mental health services. Such difficulties have been confronting and confounding professionals concerned with quality-of-care standards and assessment since the beginnings of the quality assurance movement in the United States (Bowman, 1920; Wetherill, 1915) and will be elaborated on in several chapters. This chapter highlights the four primary conditions and groups that have an impact on contemporary quality of care in the United States and thus reflect the essential modifying parameters of quality assurance programs: (1) economic conditions and cost of care; (2) the professional and institutional providers of care; (3) the payer for services; and (4) the consumer of care and his or her representatives. The confluence of these variables within the environment of current health care, and their dynamic interactions, set the pace and scope of quality assurance activities.

ECONOMIC CONSIDERATIONS

Had this book been written in the early-1970s, economic considerations and costs of mental health care would very likely not have been considered a leading variable in defining quality or quality assurance, or in developing policy and programs related to them. Yet the consequences of world recessionary and inflationary trends since the mid-1970s have been particularly demonstrable in health care costs (Chassin, 1983; Fuchs, 1982; General Accounting Office, 1983; Ginzberg, 1984; Goldfarb, Mintz, & Yeager, 1982; "Hospital Costs," 1983; Lindsay, 1982; "Medical Care," 1983; Moore, Martin, & Richardson, 1983; Relman, 1983c; Rubin, 1982; Sloan & Schwartz, 1983), which have constantly increased over the years so that they now represent over 10% of the gross national product (GNP). Although other Western nations variably spend somewhat more or less of their GNP on health care, the 10% threshold has been generally acknowledged by economists over the years as the point where health care costs represent a disproportionately high segment of the economy. This progressive rise in the cost of health care has led to the expression of many opinions and proposals for controlling such escalations, so that "cost containment" initiatives and their consequences have become part and parcel of health care benefits and services (Bedrosian, 1983; Belndon & Rogers, 1983; Council on Medical Service, 1983; Davis, 1983; Ginsburg & Sloan, 1984; Ginzberg, 1983; Greenfield, 1983; Hart & Hart, 1980; Kinzer, 1983; "More M.D.'s," 1984; "Name of New Game," 1983; Platt, 1983; Pollard, 1983; Rappleye, 1983; Schorr, 1983; Sloan, 1983; Stuart & Stockton, 1973) and likewise have become an integral part of quality assurance activities (Luke & Modrow, 1982).

The alarm about rising health costs was sounded further when, in 1981, wide concern was expressed in the United States about the impending bankruptcy of the Medicare fund, which was then considered inevitable if health costs were not brought down for that program and Medicaid. Similarly, several Blue Cross and Blue Shield plans reported significant losses in the early 1980s. Other third-party payers, such as commercial insurance companies and the Civilian Health and Medical Program of the Uniformed Services (CHAMPUS), were also experiencing serious rises in benefit payouts. Thus, such third-party payers, for the first time in 30 years, began considering reversing the trend toward consistent benefit expansions by limiting and capping benefits, increasing beneficiary financial liabilities for costs, and incorporating alternative health-care financing and delivery mechanisms in coverage agreements ("New Trends," 1983; Rodriguez, 1985).

This complex of changes has stimulated competition and the emergence of capital-supported, investor-owned hospitals (Levenson, 1982; Pattison & Katz, 1983; Relman, 1983b). However, it has also signaled an uncertain future for mental health services ("Mental Health," 1981; Rodriguez, 1985; Sharfstein & Taube, 1982), which are frequently the first health services sacrificed during lean times because of traditional stigmatization of mental illness, limited advocacy, limited data related to the efficacy of psychotherapy, and the relatively high costs per patient for care over time. These developments have raised further concerns about the consequences of budget and program costs, as they might result in impediments to quality of care, containment of quality assurance activities, and reversals in positive health status because of limited access to care, limited seeking of services, and inadequate provision of care. These situations, naturally, have numerous potential professional responses, professional and societal ethical implications, and political reverberations (Bayer, Callahan, Fletcher, Hodgson, Jennings, Monsees, Sieverts, & Veatch, 1983; Evans, 1983; Kohlmeier, 1983; Leaf, 1984; Lundberg, 1983; McGuire & Montgomery, 1982; Ricardo-Campbell, 1982; Rubin, 1980; Sammons, 1983; Snoke, 1982).

Although these costs issues have become more pronounced in more recent years, they were demonstrably evident as far back as the late 1960s and the early 1970s. The earlier recognition of the need to regulate potentially unrestricted care and charges to the Medicare program led to implementation of professional-standards-review organizations, which became the focus of development of the field of utilization review (UR). UR activities were similarly developed by other third-party payers as a means of assessing the documented medical necessity of care. Over time, utilization review became synonymous with medical auditing (in some instances a non-health-professional function), and quality assurance began to be associated by some with health care protectionism and apologists for expenses related to quality medical services. Thus, an unfortunate adversarial division has tended to alienate the fields and agents of utilization review and quality assurance. In more recent years, because of efforts by the JCAH and other organizations, such as the American College of Utilization Review Physicians, that recognize the common areas of focus of these fields, there has been a greater acknowledgment of the need to ensure quality through utilization review activities and, likewise, to ensure appropriate utilization of services and limited resources through quality assurance activities. This contemporary theme and focus will be reiterated consistently in this book and should be understood by the reader to reflect a basic consensus about the interactions and interdependence of such functions.

THE HEALTH CARE PROVIDER

In an environment where health care resources and the economic system become the primary defining parameters of social policy about health care, the roles of the health care practitioner and facility become critical in different ways from when economic considerations are secondary or incidental. During the period of geyserlike growth of modern Western medicine since World War II, research and quality care have developed exponentially with the financial supports afforded through both the federal government and health insurance. Thus, health-professional training programs and practices have been steered by principles related to providing the best possible care that contemporary knowledge could afford. In more recent years, however, with cost concerns and containment becoming the defining principles, practitioners have been directed to provide the best possible care that contemporary finances can afford. This evolving shift has been a very difficult one for practitioners, who have tended to see quality care as a professionally absolute end that is essentially independent of costs. It has also been difficult for consumers of care, who see quality of care as such a common and available commodity—through their own or third-party payment—that it is to be expected (i.e., an inherent right) rather than an optional commodity to be purchased in a market economy.

Thus, the contemporary practitioner is caught up in a multisided struggle among internalized standards of practice, patient care expectations, and the emerging influence of external forces that attempt to change his or her practice behavior as it relates to health care costs. The concern about the practitioner's decisions has now resulted in a new system of shared decision-making, increasingly involving the consumer and the payer. Although this system is seen by many professionals as intruding in the time-honored privileged and confidential relationship between practitioner and patient, nevertheless incursions into the provider's "territory" are continuing, based on assumptions about the need to protect the interests both of the payer and society and of the patient. This assumption becomes paradoxical when the patient doesn't seek such protection from the payer. Nevertheless, quality does continue as an important commodity in the purchase of health care from the position of both the payer and the patient.

For these reasons, practitioners are progressively coming under the scrutiny of third-party payers who are requiring increasingly more evidence that the care they provide meets the often variable requirements of payers to justify quality and cost effectiveness. Providers of care are held to a general requirement of establishing the "medical necessity" of care, which is commonly defined as care that is

- in keeping with quality and utilization standards of practice, as generally provided in the United States;
- adequate and essential for the evaluation and/or treatment of a defined medical (psychological) disease, condition, or illness.

Under this general benefit policy requirement, payers develop variably explicit and implicit criteria for determining medical (psychological) necessity. Although payers have traditionally paid for most services billed either because of a partnership attitude engendered by participation agreements (e.g., Blue Cross and Blue Shield) or because they lacked the incentive or ability to identify unnecessary care, in more recent years utilization review activities have dramatically increased. They have led to increasing levels of involvement by health professionals in developing standards criteria for review and in actually conducting review for the payer. In some instances, payers have simply hired individual reviewers, and in other instances, local or state professional associations have been consulted. With the rapid evolution of such activities, national professional associations such as the American Medical Association have become more active in calling for peer review of services. Their concerns about nonprofessional or nonpeer judgments in determining the appropriateness of care, on which reimbursements would be based, have led to a number of public statements and advocacy for peer review in determining quality and appropriate utilization (Ertel & Alredge, 1977).

The most notable exercise of the call for peer review occurred in the mid-1970's by the American Psychiatric Association (APA) and the American Psychological Association (APA). They were reasonably concerned about the often arbitrary decisions made by third-party payers about benefits that were beginning to have serious consequences for mental health practitioners and patients. Psychotherapies in general were suspect by payers because they tended to be long-term and thus, expensive over time. A lack of understanding about the value of psychotherapy merged with incipient sociocultural biases against the mentally ill and the often poor or guarded justifications of care by practitioners to result in increasing problems in reimbursement for psychotherapy. Fortuitously, the Department of Defense's Civilian Health and Medical Program for the Uniformed Services (CHAMPUS) was experiencing a crisis with Congress over quality problems identified in CHAMPUS-authorized psychiatric residential treatment centers. By 1981, the first national peer review programs were instituted by the APAs with CHAMPUS (Hamilton, 1985; Rodriguez, 1983b). This landmark project set the stage for further professional incursions into the rapidly evolving marketplace of health care, and it established important roles for professional associations in setting up standards and criteria for both professional practice and reimbursement and in becoming operationally involved in the business of insurance.

Despite these generally successful initial ventures, the economic storm clouds within health care have cast vast shadows over providers' abilities to make consistent and definable decisions about necessary patient care, especially as they relate to the current *sturm und drang* about cost and quality. In the quest for answers to rising health-care costs, much attention has been focused on providers—both individual practitioners and facilities—as a primary cause of the rise in these costs. An economic perspective on the healthcare market would look at both supply and demand variables as they affect costs and quality. Unlike the common trend in free markets, in which increased numbers of service providers tend to reduce costs through competitive pricing, in health care it seems that increased numbers of providers stimulate both demand and price escalation. One reason appears to be the very nature of professional care, which is based on the presumption of quality, and on American market phenomena in which consumer demand for quality is associated with acceptance of higher prices. As insurance benefits have supported such practices by not putting the patient-beneficiary at financial risk, costs have naturally risen. In mental health, the proliferation of facilities has been supported both by government policies and programs (e.g., Hill-Burton funds) and by local consumer demand. Similarly, increases in the numbers of practitioners have been supported by factors as diverse as government training grants, greater public acceptability of mental illness and status for mental health professionals, creative developments that have stimulated diverse professional disciplines, and increasing income levels.

All of this has led to increasing concerns about costs as they relate to provider-induced demand (Wilensky & Rossiter, 1983), especially where those costs are associated with variable intensity and comprehensiveness of services—and related charges—for the treatment of similar conditions. These variations in professional and hospital practice patterns are related to geographic and specialty variables, and they seem most pronounced in mental health care. They raise questions not only about variable quality but also about variable costs ("Dramatic Variations," 1984). Mental health care receives extra scrutiny because of the long-standing cultural stigma about and biases against the mentally ill, the questionable scientific foundations of many psychological theories and treatments, and the difficulties that non-mental-health professionals generally have in understanding and assessing the medical (psychological) necessity of treatment, which is often longterm and hence expensive. Although a case can be made that mental health providers have a relatively limited control over the rise in mental health costs (Biegel & Sharfstein, 1984) and that modern quality health care can have a positive effect on the economy that is not often calculated with the actual costs of care (A. R. Nelson, 1984), nevertheless it appears a national consensus has been formed:

- Health care costs both are too high and are escalating at an unacceptable rate.
- Health care providers are the main cause of this rise, notwithstanding the influences of inflation and recession.
- Higher cost care is not directly correlated with quality outcomes, vis-à-vis improved health status, productivity, and other desired personal and social goals.

The great concerns about the costs of care have focused an unprecedented amount of inquiry on the many factors that influence the provider's evaluation and treatment decisions, especially those that result in expensive or relatively inexpensive care (Anderson & Shields, 1982; Cuningham, 1981; J.M. Eisenberg, 1979; Glick, 1985; Hetherington, 1982; Hubbard, Levit, Schumacher, & Schnabel, 1965; Jackson, 1969; Jessee, 1981; Linn, 1970; Liston & Yager, 1982; Luke, Kreuger, & Modrow, 1983; Newble, 1983; Payne, Lyons, & Neuhaus, 1984; Sanazaro & Williamson, 1970; Shortell & LoGerfo, 1981; Weinstein & Fineberg, 1980; Wendorf, 1982; Wilder & Rosenblatt, 1977). The confluence of individual competence, professional ethics and standards, organizational structure of the practice environment, socioeconomic conditions, and quality and cost management systems forms the contemporary boundaries of both the quality and the costs of care. Ironically, professional interests in achieving high-quality treatment outcomes have tended to be driven by the same forces that now worry about the costs of that care: regulatory bodies, third-party payers, and consumers of care. Professional training is intensive and expensive and contributes significantly to health care costs, but it also contributes to the quality of care that is demanded. Professional standards are shaped by societal standards for "the best" in care, which often means it will be both of high quality and expensive. Thus, the provider is generally as much the passive recipient of socioeconomic forces as the active driver of treatment decisions and costs. It should come as no surprise, then, that providers frequently are confused about what they should be doing or not doing, in conflict between what they've been professionally prepared to deliver and are increasingly inhibited from providing, and frustrated by the general confusion that characterizes national health care goals. In no segment of health services are these cost-generated problems more evident for providers, payers and patients than in mental health care. This progressively intensifying situation is leading to focused efforts by payers to especially affect mental-health-care providers' evaluation and treatment decisions. It is also leading to an evident schism between "medically necessary" quality, as defined by professional associations and individual practitioners, and allowable benefits as generally determined by utilization normative data. As these conflicting circumstances are well known to contemporary mental-health practitioners, so the consequences in terms of steering practitioner decisions toward possible adverse health outcomes are increasingly more evident in clinical practice. This situation has created the major current professional crisis in quality assurance in mental health.

The Payer

The course of health services in general and of mental health care in particular is increasingly being determined by payers for care. Chapter 6 details many of the cost containment initiatives of payers through the initiation of alternative health-care financing and delivery systems. This section will serve to highlight the increasingly active role of payers (such as private employers and public health benefits programs such as Medicaid and CHAMPUS) and their agents (such as insurance companies and benefits management companies) in understanding how they may affect providers and patients through changes in health services reimbursement and delivery. Payers are becoming much more sophisticated in their understanding of the complex interactions of reimbursement policies and such variables as benefits utilization and health status. Much thinking is currently being given to benefit plan designs that will maximize quality and related goals such as employee productivity and satisfaction, while minimizing costs, that is, benefits with optimal value (Fielding, 1984; Wolfson & Levin, 1985).

Mental health benefits are posing special challenges to payers because of traditional concerns—often expressed by employees through unions and by legal protections—about confidentiality in mental health benefits. The tendency of mental health professionals to limit clinical information, in medical records related to privileged communications, makes it difficult for payers and their agents to understand what "value" they and beneficiaries are receiving in mental health care because records are often limited in clinical details. Additional problems in assessing psychological necessity of care ensue from frequently significant variations in utilization patterns among providers for the treatment of similar conditions and from the difficulty payers frequently have in either understanding or believing clinical documents. Many of these problems can be attributed to poor documentation, lack of qualified mental-health personnel evaluating care for payers, and inherent difficulties in analyzing the benefit value from claims data. Nevertheless, problems in assessing care for reimbursement purposes remain the single greatest dilemma for payers, as they daily balance a mountain of claims against their legal requirements to cover "medically necessary" care. Meanwhile, payers are becoming increasingly more uncomfortable with psychiatric benefits and utilization, as they see increasingly unacceptable budget levels being allocated for health care that they may or may not understand. The tension between payers and providers is currently dramatized and accelerated in the plethora of cost-containment programs that payers have implemented to restrain costs. These various approaches will be commented on in a number of chapters, especially as they are seen as affecting access to care, quality, health status, and overall costs in mental health care. Operating in the context of governmental and professional regulation of health professionals (Gaumer, 1984), cost-containment programs establish another layer of regulation of provider and patient. Of especial interest to each party is the effect of such programs on quality of care (Donabedian, 1984; Mechanic 1985; Wyszewianski, 1982), especially, in mental health services (Schulz, Greenley, & Peterson, 1983). Pardoxically, the threat some payers see in quality monitoring as part of cost containment is the potential increase in costs of care related to increasing provider quality performance and positive health outcome in the treatment of certain conditions. There is widespread and growing concern by payers that an outcome of quality standards and monitoring, as well as such other programs as health promotion and disease prevention, could be longer life and the potential for both more health encounters and costs over time, and also the potential for very costly care associated with aging and end-stage diseases. Similarly, for mental health services, quality monitoring is questioned because of its administrative costs, where cost and overall health consequences may not be known. Thus, payers are frequently in conflict about quality monitoring as a major facet of cost containment efforts, but they continue to place some priority on it as a commodity that is demanded by patients, providers, social standards, and the law, and as a potential effective component of overall cost management of benefits.

As reimbursement for health services has effects on providers' decisions about necessary care and their fees (National Center for Health Services Research, 1983; Sloan, 1982), so cost-containment initiatives have been shown to have significant effects on providers' practice behaviors (J.M. Eisenberg & Williams, 1981). Moreover, it has been shown that reimbursement policies can be structured to ensure quality (Shaughnessy & Kurowski, 1982) while restraining costs. These and other findings reinforce the resolve and ensure the legal safety zone for payers in establishing cost-containment programs, even though some gatekeeper approaches are drawing fire for their threat to quality ("Gate Keeper Payment," 1984) and their possible boomerang effect on longer term costs, as necessary early care and preventive services are denied and result in health deterioration and ultimately greater costs.

Unfortunately and all too commonly, most cost-containment programs rarely have carefully developed evaluation components established for utilization management activities that project longer term costs or health consequences or look at broader cost or social impact of generous or limited mental-health benefits, such as on disability, productivity loss, or related macroeconomic effects. Most are born of a crisis intervention attempt(s) to reduce costs in the short run and tend to be shotgun approaches that differentially affect patients with different conditions. The federal government's implementation of a prospective payment system using Diagnosis-Related Groups (DRGs) is a manifest example of a "successful" cost-containment program that rewards the treatment of certain conditions and defaults others, while raising numerous notable questions about quality assurance in medical care (B.S. Eisenberg, 1984; Grimaldi & Micheletti, 1984; Horn, Sharkey, Chambers, & Horn, 1985; Kapp, 1984; Kreitzer, Loebner, & Rovetti, 1982; Mariner, 1984; "Prospective Payment," 1983; Schumacher, Clopton, & Bertram, 1982; Ziegenfuss, 1985). A DRG approach to reimbursement for psychiatric institutional or professional services poses special problems because of the nature of psychiatric illness and treatments (Jencks, Goldman, & McGuire, 1985; Mezzich & Sharfstein, 1985). Other reimbursement methodologies will need to be examined more thoroughly (Brewster, Jacobs, & Bradbury, 1984; R.G. Frank & Lave, 1985; Gonnella, Hornbrook, & Louis, 1984; Gordon, Jardiolin, & Gordon, 1985a; Horn, Sharkey, & Bertram, 1983b; National Association of Private Psychiatric Hospitals, 1985), unless psychiatric quality of care and practices shaped by reimbursement be adversely affected. Such approaches to reimbursement for health-care services have raised questions about the rationing of care and the tiering of quality services through reimbursement maneuvers that have serious social, if not economic, implications. This sort of approach has implications not only for short-term acute conditions, but also for catastrophic and long-term care.

Quality assurance for costly long-term care, especially for individuals who are often unemployable and thereby don't contribute to the payer's economic resources, is a particular problem. Although cost-containment efforts are generally highly active in such care situations, quality assurance efforts are limited by the payer's often lessened investment in the future of the patient. Although quality assurance efforts are certainly possible (Kane, 1981; Munich, Carsky, & Appelbaum, 1985; Rzaha, 1983), they frequently result in improved quality of life and, not uncommonly, allow for some return to economic productivity; nevertheless, the emotional investment of the payer seems to dwindle for the patient with chronic or hopeless conditions. Nowhere is this more evident than in end-stage "custodial" care and for the chronically mentally ill. Cost-containment efforts are thus frequently both successful and a moral failure for such persons.

Hence, the drive to reduce the costs of care seems to be resulting in other prices to pay and may ultimately result in sizable social and other debts that will have to be settled. Payers are over a log in these respects because they have both current and future budgets to account for and are increasingly alienating providers and patients in their efforts to reduce current costs. Although the relationship between payer and provider has become progressively more adversarial, there is a building mutual understanding of the need to provide incentives for patients and providers to seek reimbursement only for "truly necessary" care. Both parties are beginning to acknowledge that health costs are escalating in ways that are ultimately not in their own or collective societal interests. In doing so, they are also acknowledging the efforts of the payer to do something to stabilize costs. Educating providers and patients about costs has some impact on cost containment (Atchinson & Brooks, 1981; General Accounting Office, 1982; Jessee & Brenner, 1982; Thompson, 1981; Williamson & Associates, 1982a; Youel, 1983), as do reimbursement systems that reward prudent benefit utilization and health behavior.

It does appear that some balances will eventually be struck for cost restraint and quality among payers, providers, and patients. Whether this will occur through consensus or regulation, or through economic or social pressures, remains to be seen. What does seem already evident is the emerging power of payers in determining future professional practice, whether it is through selective reimbursement mechanisms, the organizing of alternative financing-delivery systems, or the promotion of quality assurance systems. This power over the traditional compact of the patient and the provider to seek and provide health services will have dramatic consequences for quality in all health care, especially mental health.

THE PATIENT-CLIENT

It is a paradox that the most important person in the health services covenant has consistently the least direct influence on the structure and process of care. Yet, it is the recipient of that care for whom the health care system is supposedly established and who will ultimately express his or her needs, either directly and appropriately or in some misdirected or pathological manner. When the expression of distress or disease is negatively influenced by the structure or process of the health care system, then the system must be considered inefficient or incompetent. That there has been relatively little formal or coordinated inquiry into the forces that drive this person and his or her family to effectively use or influence the care system is, then, ironic. The apparent conflict that the payers and the deliverers of health care services have in appreciating this person is expressed many ways:

- The tendency to make assumptions about what the person "really needs" in the way of health benefits and services.
- The relative lack of organized research into factors that the recipient brings into the care setting that influence the detection and treatment of medical and psychological conditions, as well as health.
- The general lack of training for most administrative and health professionals in the communication skills necessary for providing services and in healing.

Moreover, there is a problem in developing a consensus over what to call this person: *patient, client, consumer, customer, recipient, dependent,* or *beneficiary.* All of these manifestations, among others, of a lack of real appreciation of this person have *significant* implications for both the quality and the costs of care. The documented (Taube & Barrett, 1985) and undocumented prevalence rates of psychiatric disorders in the community and in general medical settings (Houpt, Orleans, George, & Brodier, 1979) which are inadequately evaluated and treated at each level exemplify the problems that the emotionally disrupted person experiences in understanding how and where to seek relief. Additionally, these situations have considerable impact on the quality assurance benefit analysis, and other management systems established to deliver optimal care. Thus, the patient-client is not passive but exerts a profound, if often misunderstood influence on individual and collective health services.

Certainly, one of the reasons that health costs are as high as they are today is because of the high expectations of the patient-client as beneficiary, consumer, and customer. The growth of health benefits emerged in the years following World War II with a surging economy, the predominance of motivational-human-relations and competitive-benefits practices by employers, the emergence of the entitlement concept in health care, the influence of unions and the consumer movement, the growth of the insurance industry, and the realization that ever-improving levels of health were possible with advances in technology and infusions of captial into the health care system. The patient-client was explicitly or implicitly at the foundation of all of these developments as the federal and state governments and employers both initiated what they thought was needed and responded to consumer influences and demands. Enlightenment and technical capability merged with financial plenty to build the most sophisticated health care system in the world and to reduce morbidity and mortality from most diseases.

Yet, all of this either hasn't been enough at times or too often has resulted in a waste of resources. Although, infant mortality in the United States has steadily decreased in recent years, it remains unacceptably high in certain impoverished populations. And although many have paid exorbitant amounts for mind expansion provided by mental health professionals or "pop-psych" agents, many others have suffered without available or required mental-health services (Regler, Shapiro, Kessler, & Taube, 1984). These persons have exacted serious societal tolls through impaired productivity, disability, alcoholism and substance abuse, family and community violence, and other manifestations of disrupted emotional states.

Unfortunately, the focus of the health care systems on treating disease rather than preventing it, and on isolated and specialized care rather than on care that is systems-oriented and effectively managed, has resulted in a self-limiting system. Further, it has led to the current situation, in which there is a general perception that health benefits must be cut or capped and costs must be "contained." Unfortunately, it is the insurance customer turned psychiatric patient who must pay for this situation, as decreasing insurance coverage and increasing costs are contributing to diminished access to needed specialty psychiatric services. This situation is compounded by the lack of awareness of the emotionally distressed person in seeking the appropriate evaluation and treatments and the lack of effective evaluation and referral systems for persons expressing their distress through physical illness, antisocial behavior, or other nondirect manifestations. It should be of real concern that approximately $6 billion is expended on mental health costs in the general medical sector annually and that $3 billion is spent in the human services sector (R. A. Frank & Kamlet, 1985). This care is too often uncoordinated, uneven in delivery, and poorly justified.

What does the consumer think of the quality and availability of health care services? Opinion polls indicate a general level of satisfaction with the coverage, the availability and the perceived quality of medical and psychological care (Englehart, 1984) although this satisfaction varies somewhat with population, locale, health status, and system of delivery of care (Smith & Metzner, 1970). In general, most see health care practitioners as empathetic and capable, even though too many professionals, especially nonpsychiatric physicians, are seen as being deficient in communication and interpersonal skills (Cousins, 1985). No doubt, patients' changing doctors or therapists is due as much to problems in communication and establishing a healing alliance as it is to such causes as psychological transference, the higher costs of some providers, or displeasure with appointment or office administrative practices. The great concern about this phenomenon is both the inherent cost effects due to duplication in evaluations and treatments and the potentially adverse impact on quality outcome in some patients associated with their seeking treatments from a series of providers. An unfortunate aspect of the current predominant system of health benefits, based on beneficiary freedom of choice in selecting health professionals, is the lack of a coordinative function that would steer persons to appropriate health-care and social-services resources and restrict inappropriate provider charges for services. The triage and case-management functions being built into health systems are increasingly popular approaches to dealing with the adverse consequences of patients' randomly tracking through the health care system.

A notable amount of attention to matching patient needs with services is already being reflected in both internal and external quality-assurance systems. Effective hospital and clinic quality-assurance and risk-management programs are attentive to the interests of persons using their services, through monitoring and follow-up on complaints, opinion surveys (Morrison, Rehr, Rosenberg, & Davis, 1982), and patient care assessments by quality assurance professionals. More formalized approaches to understanding patient-client needs and concerns are becoming a more standard aspect of clinical and administrative management of care. The importance of such approaches is being driven home not only by evidence that such activities can result in an improved quality of care outcome (Thompson & Rodrick, 1982), but also because such activities improve business effectiveness, through reduction in adverse legal actions, more prompt payment of bills by satisfied patients, and improved community reputation and referrals. Business-oriented approaches to improving patient comforts are often successfully marketed to individuals who appreciate service amenities in their care. However such popular current activities may be justified as necessary in a competitive health field, they most definitely result in increased costs of care, without predictable improvements in quality outcome. Nevertheless, they further demonstrate the influence that the consumer-patient has on the shape of clinical services and their related costs.

Another impact that the patient-client has on the organization and resource allocation of the health care system is through organized advocacy. Such groups as the National Mental Health Association, the National Alliance for the Mentally Ill, and the National Association of Retired Persons are demonstrable examples of the power of organizations to influence public opinion, budgetary allocations, and health benefits legislation at the national and state level. Likewise, veterans' groups have stimulated the growth of the significant health care system that is the Veterans Administration and have helped create support for such special programs as mental health outreach programs for Vietnam Veterans. And military active duty and retiree organizations have had a profound effect on the progressive expansion of the Department of Defense's Civilian Health and Medical Program of the Uniformed Services (CHAMPUS).

At every turn, these groups have acted like unions in demanding assurances of inexpensive, available, timely, accommodating, and high-quality health services. When quality has been questioned, they have been quick to use political pressure, the media, and legal action to make sure that quality is not compromised. Thus, they have frequently functioned as an *ad hoc* external quality-assurance monitor, supplementing various external operational quality-assurance programs established by governmental agencies, professional associations, and private entities such as the Joint Commission on the Accreditation of Hospitals. Of course, while demanding and securing high-quality care, they have also further contributed to health care and related costs. Under any circumstances, whether acting as part of an organized group or as an individual, the patient-client exerts a profound influence on the costs and quality of health care. Unfortuantely, these influences are uneven in application and unpredictable in outcome, and they too commonly result in situations that do not improve health status or societal goals.

QUALITY ASSURANCE AND RELATED SYSTEMS

Quality assurance (QA), utilization review (UR) and risk management (RM) systems are a standard part of the interactions between the payer, the provider, and the patient in contemporary American health care. In the covenant relationship that they share, these systems are established both by internally guided values, such a professional and personal ethics and standards, and by legal provisions, such as through governmental regulations and insurance contracts. Not uncommonly, these value systems and their associated expectations come into conflict with one another. The results are well known and are exemplified by such common oc-currences as malpractice suits, denied claims for health services, and audits. However, the formal and informal systems established to monitor and influence quality, to reduce risk, and to restrain unnecessary costs of care can also be seen in a positive light. They have begun to reshape health care into a range of sophisticated professional services that are increasingly better managed, accounted for, and mutually agreed on.

INTERNAL SYSTEMS

Internal systems refer to those formal QA-UR-RM programs that are developed and provided within organized care settings. Although many are instituted as the result of some external requirements, such as JCAH accreditation or state licensing specifications, they are also internally provided as a part of the clinical and adminis-trative operations of health care programs. Although the most elaborately de-veloped programs are in institutional settings, well-structured programs are evident in clinic settings (e.g., ambulatory surgical settings and partial psychiatric hospitals), outpatient practices (e.g.. clinical and financial records, structured patient educa-tion, and informed consent procedures and forms), and community residential-care settings. Larger group practices and most facilities and clinics now have identified persons dedicated to establishing and monitoring QA-UR-RM functions and have developed both written policies and procedures and education programs for staff and patients to inform them of the parameters of these provisions (Restuccia & Holloway, 1982; Scott, 1982).

Planned and structured activities are a routine part of the daily activities in such settings. The most effective programs are those that anticipate and provide for quality, risk, and utilization management interventions rather than reacting to iden-tified problems (Long, 1984). They include medical care evaluation studies and con-ferences and various clinically oriented audits and assessments, for example, morbidity–mortality data analyses, employee and patient surveys, assessments of discharge abstract data, and analyses of incident reports. Such activities allow clini-cal program planning and interventions that will better ensure quality, low-risk, and utilization-efficient care. Timely and accurate analyses are essential for a treatment setting to be financially stable in a competitive environment, by avoidance of pay-ment delays, litigation, and a negative reputation for quality management.

This market-driven aspect of quality assurance programming is becoming an essential component of the structure and marketing of alternative delivery systems, such as health maintenance organizations (HMOs), preferred and exclusive pro-vider organizations (PPOs and EPOs), and other organized and managed care plans. These newer market-driven systems recognize that a competitive approach to health care requires administrative and professional internal systems to ensure the "product": quality. Thus, the current and likely future environments will con-tinue to drive internal systems to be more administratively sophisticated, pro-fessionally resource-intensive, and market-oriented. They will also, of necessity, have to be consonant with external QA, UR, and RM systems.

EXTERNAL SYSTEMS

External QA, UR, and RM systems are those that are directed and provided by an entity not affiliated with the institutional or the individual provider. They include review programs for licensing, certification or accreditation, credentials and health care services. Professional associations, governmental agencies, and private con-sultation and education enterprises are the agents of these diverse activities. The in-dividual purpose and collective action of these external systems establish structural, process, and outcome parameters for quality, safe, and utilization-efficient health services. Increasingly, they are playing roles in a competitive, market-driven health environment for which they were not originally established. For instance, such pro-grams as CHAMPUS require specialized mental health services accreditation and adherence to additional program standards for certification. Organized and man-aged care programs and government certification now compel providers to sign contracts holding them to audit reporting requirements and adherence to and review by admission and treatment criteria.

The cumulative weight of these programs is both qualitatively and quan-titatively beyond the scope of proposals by Wetherill (1915), Bowman (1920), Flex-ner (1910), and Codman (1916) for establishing the foundations of modern American quality assurance. By anyone's appraisal, the now modest propositions of these authors for building a professionally based system of quality and account-ability through training and hospital standards are now dwarfed by the multi million-dollar, multitiered and labyrinthine business that comprises external qual-ity-utilization management programs. These programs may be categorized by their legal status:

- Private—not for profit: Review companies or certification agencies, either established as offshoots of professional associations (e.g., medical care foun-dations, American Psychiatric Association, and American Psychological Association peer review programs) or as a consortium of professional organizations (e.g., the JCAH)
- Public: Governmental agencies or their agents (e.g., peer review organ-izations for Medicare review, and the CHAMPUS Office of Quality Assur-ance certification and review activities)

- Private—proprietary: Privately or publicly owned companies engaged in the business of health services administration, management, and payment (e.g., insurance review activities, benefits management programs, and hospital management and consultation services).

The diversity of these programs is well known to providers. Some have multi-service functions and some are highly specialized. Some use qualified health professionals and professional standards, and some are primarily fiscally centered or benefit-(legally)-centered. All criss-cross through the doors, phone lines, and mailrooms of providers as they attempt to establish and monitor the quality, risks, and costs of health services. Providers are routinely beseiged by auditors, surveyors, and reviewers, all of whom are engaged in defining and defending professional, business, and consumer interests—through external review. There is little privacy, and there is a common professional, legal, and consumer concern about confidentiality (Borenstein, 1985; Hiller & Seidel, 1982; Schuchman, Nye, Foster, & Lanman, 1980).

Despite concerns about intrusion into the practitioner–patient relationship and the time-honored authority of the professional to make treatment decisions, payers are now asserting their rights to know what they are paying for (Hunt, 1983; "Industries Use," 1982; Sandrick, 1982). In doing so, they are using an array of private consultants and agents to establish the product and the value of their health benefit payments. This exercise is inherently a strained one, because practitioners basically do not like to have their decisions reviewed—even by peers—and because health care administrators' primary focus is on the financial stability of their facility, clinic, or group practice. Conflicts over the bequeathing and withholding of dollars for health care services pervade the current health-care landscape, affecting both public and private institutions as they struggle in a market environment.

Despite these basic difficulties and the risk of a dog-eat-dog situation adversely affecting quality, there is both some altrusim and a fear of legal or professional reprisals that allows a focus on a more than the dollar bottom line. A number of creative and effective approaches to establishing quality-focused assessment and management programs have shown that QA programs can be cost-effective and supportive of therapeutical efforts (Bausell, 1983a,b; Donabedian, 1980, 1982; Greenspan, 1980; LoGerfo & Brook, 1979; Mattson, 1984; Reeder, 1981; Sanazaro, 1980; Williams & Brook, 1978). These many approaches and activities serve to underscore the power of momentum that has occurred in the fields of QA, UR, and RM, driven by the accountability invective. Moreover, professional review activities have witnessed two notable changes, motivated by the increasing business-management requirements of payers:

- A shift of review timing from retrospective to concurrent and prospective, exemplified by the popularity of preadmission certification, the second surgical opinion, and concurrent (continuing-care) review services, including those for psychiatric care (Rodriguez & Maher, 1986).
- A shift from case-review to case-management activities, in which an agent of the payer, generally a health professional, interacts with the provider(s) to ensure timely, focused, and quality treatments, discharge planning, and an

aftercare course of care, thus reducing the costs associated with health services and disability

Hence, external review has begun to be transformed into a more active process of external cooperative or directed case management. This practice will become a standard payer operation in the quality-utilization management of health benefit resources in the future. The leverage of payment is likely to allow further incursions into the health-services decision-making freedoms conventionally allowed practitioners and patients. This should create some anxiety in all three parties, relative to the potential impact on quality of care.

STANDARDS OF PRACTICE AND REVIEW

To evaluate health care, one must have clinical and economic frames of reference within which to define quality, safety, efficacy, and costs. "Standard" clinical care is interpreted in terms of both professional definitions and delineations of standards of practice and statistical analyses of normative and variant treatment practices and outcomes. Professional definitions of standard care have tended to be *implicit* in recent years and have been reflected in "standard" textbooks on the clinical subject area. In more recent years, government agencies, payers, and accreditation or licensing agencies have engaged professional associations in developing *explicit* (written) standards to define the acceptable boundaries of clinical practice.

On the basis of these standards, and related specific admission and treatment criteria, developed by professional associations or other professional bodies, these payers and agencies have established policies and procedures that recognize acceptable professional practices for certification and reimbursement purposes. This professional *consensus-development* approach draws on a composite body of scientific evidence and common or historical professional practices as a basis of establishing standards and criteria. Examples of this approach include the Diagnostic and Therapeutic Technology Assessment (DATTA) Project of the American Medical Association and the peer review manuals developed by the American Psychological Association (1980) and the American Psychiatric Association (1985). The advantage of such an approach is a sharpening of professional consensus on certain evaluations and treatments. The disadvantages are primarily related to those areas where there are a range of "acceptable" professional practices and to the inherently guarded positions professionals in groups may take in establishing norms and ranges for themselves and their colleagues, especially when they are aware that their judgments will be tied to reimbursement decisions.

Another approach to defining standards of practice and review is rapidly gaining favor with both payers and providers: the *empirical* approach. Data analyses of various clinical, administrative, and cost elements are increasingly being arrayed into standardized scales, criteria sets, and other evaluation systems (see the next section). An example of this approach is the research-diagnostic criteria method of the third edition (revised) *Diagnostic and Statistical Manual* (American Psychiatric Association, 1987). The upsurge of such methodologies concerns some practitioners because of the fear of "cookbook" approaches to care, the loss of the "art" of treatment in a scientific-statistical approach, and the potential ramifications for research, teaching, and reimbursement. An alternative view would claim that such approaches (1) affirm the scientific validity of and the ethical requirements for indicated treatments, (2) establish the indications for "hassle-free" reimbursement, and (3) allow the establishment of further research parameters and priorities that better define the scientifically derived efficacy and safety of evaluations and treatments.

Standards and criteria are a hot item currently, largely because of the crisis in health care costs and reimbursement. Without defined standards and criteria, health care services cannot be objectively assessed for individual case reimbursement, benefit-trend analysis, insurance risk-rating, and budgeting for health care. Without standards and criteria, benefit determinations run the risk of being arbitrary, with significant therapeutic and legal ramifications. Standards related to mental health services are particularly needed because of criticisms about the limited specificity, reliability, and validity of current psychiatric clinical standards and utilization norms (S.H. Nelson, 1979; Richman & Barry, 1985) and of psychotherapies (London & Klerman, 1982).

Psychotherapy standards are not the only specialty service area being questioned. The significant geographical variations in clinical practices for surgical procedures, hospitalization rates, and lengths of stay for medical-surgical-obstetrical/hospitalizations are well known (Commission on Professional and Hospital Activities, 1985; Davidson, 1986). The safety and efficacy of numerous medical procedures and practices have likewise not been scientifically validated. Further, the significant geographical variations in charges for similar procedures, even where the locus of service is not significantly distant from areas of divergent charges, raise great concerns about the credibility of charges (Wennberg, 1986), as well as about benefits to the patient and payer. Such local variations in practice are becoming increasingly difficult to justify. It is clear that *national* standards of practice and review will be demanded, obtained, and used by payers ("National Utilization," 1983; Rodriguez, 1983a; "Test Programs," 1981).

Payers are now embarking on ambitious claims analyses, using "small-area analysis" and other utilization–cost profiling approaches to better understand the value and wisdom of their reimbursement policies. Unfortunately, claims analysis data are a relatively inadequate source of understanding the need or effectiveness of psychotherapy services. Future data systems that will collate normative clinical information from professional services review should shed more light on the normative aspects of standards of practice. Research on efficacy and safety will need to be linked into such data analysis systems through the use of standardized instruments and criteria. Although such approaches seem forbidding to some, they are seen as promising, necessary, and inevitable by others.

EVALUATION MODES

Evaluating quality, risks, efficacy, and costs by means of scientific-statistical approaches has been an area of largely academic interest in past years, relegated to obscure journals and highly specialized professional meetings. However, most health professionals acknowledge the emergence and growing status of such research in standard national publications, both professional and lay. The interest in more standardized approaches to evaluation and paying for care goes beyond those evaluation and reimbursement methods that are prominently related to the variable costs and charges of variably delivered services. It is increasingly more difficult for practitioners to ethically or scientifically justify variations in evaluation and treatment decisions. Legal actions, through legislation and litigation, are also driving practitioners and professional associations to support a more standardized approach to treatment and review.

Evaluation research, then, is quickly becoming "in," especially for employee benefit managers and insurance companies. Standard approaches to health-program and individual treatment are expanding from an already defined field of action (Guttentag & Struening, 1975; Schulberg & Baker, 1979). Structural, process, and outcome studies of health services are generating a high level of research interest and support, largely because they allow a better analysis of the highly complex links between the quality, the benefits, and the costs of care (Donabedian, 1968; McAuliff, 1979; Newman, 1982; Nutting, Shorr, & Burkhalter, 1981; Rutstein, Berenberg, Chalmers, Child, Fishman, & Perrin, 1976). Such research may allow professionals, payers, and regulators to better understand how such variables as professional training, certification, and facility accreditation affect the process of care and treatment outcome. The medium of communication in evaluation (direct, phone-based, or document-based) is an area of important inquiry, as researchers and payers attempt to find the means of obtaining core information on which a justifiable reimbursement can be made. This emerging and more sophisticated system of evaluation of course, is all part of a much larger and refined quality-assurance system as it helps steer health care evaluation and treatment decisions toward reliable, valid, and specific standards.

The intermediary for such health care evaluations has tended to be human in the past, guided by general consensus-developed professional standards and criteria. An example of such an expert human system is the CHAMPUS peer-review program established with the American Psychological Association and the American Psychiatric Association, 1980; Hamilton, 1985; Rodriguez, 1983b). Although programs such as this have been assumed to be effective in ensuring quality and in reducing unjustified costs, the basis for such conclusions is mostly anecdotal and has not been empirically established. Thus, after 6 years of the operation of national psychiatric and psychological reviews for a number of insurance companies and other corporations, the APAs are finding that payers want better validation of the care they are reimbursing, that is, more objective data. The need for better cost–benefits data has caused payers to turn to a number of proprietary data consultants to define costs, benefits, and standards. This trend has concerned the American Psychiatric Association and many private review companies enough so that they have essentially abandoned marketing document-based retrospective review, in favor of concurrent phone-based review. This market demand underscores the need of payers for timely and focused case evaluations.

Yet, the issue of more objective standards remains. Payers are uncomfortable with the limited objective criteria for mental health services, because they are used to using the more reliable and objective criteria sets developed for medical-surgical reviews. Moreover, the emergence of prospective payment systems has spurred interest in objective evaluation instruments that will allow reimbursement based on a more scientifically reliable and quality-assuring basis than case mix by DRGs. Numerous studies are now being conducted to develop alternatives to the DRG approach. They include disease staging (Gonnella, 1983; Gonnella et al., 1984), the Appropriateness Evaluation Protocol ("Acceptance Growing," 1982; Walter, 1983), the Non-Acute Profile ("Acceptance Growing," 1982; Borchardt, 1981) the standardized Medreview Instrument ("SMI Presents," 1983) and the Severity of Illness Index (Horn, Chachich, & Clopton, 1983a; Wagner, Knaus, & Draper, 1983). Similar approaches to linking reimbursement to standardized psychiatric instruments are stimulating interest in the Psychiatric Status Schedule (Spitzer, Endicott, Fleis, & Cohen, 1970), the Global Assessment Scale (Spitzer, Gibbon, & Endicott, 1973), and the Strain Ratio (Gordon, Vijay, Sloate, 1985b). Other psychiatric rating scales—both developed (Spitzer & Endicott, 1975) and not developed—will drive the direction of the future evaluation of psychiatric treatments to more standardized objective instruments. The attractiveness of such scales should not supplant the development of other research-based evaluation criteria or obviate the role of consensus development in shaping evaluation and review parameters; rather they should supplement them. The evolution of psychiatric evaluation in the 1970s (J.D. Frank, 1975) and the 1980s (Schacht & Strupp, 1985) will underscore, by the 1990s how fertile this field currently is and how fervent is the desire of payers to have some objective answers to their psychiatric benefit policy questions.

PROBLEMS IN EVALUATIONS AND ASSURING QUALITY

The growth of quality assurance systems has been characterized by troubling and challenging problems, as well as by professionally exciting developments. The initiation of such systems has been characterized by professional resistance because, among other reasons, they are new, are interdisciplinary, are perceived to intrude on the patient-practitioner relationship, are annoyingly focused on practitioner inadequacies, and are actually or potentially tied to reimbursement, professional self-image and reputation among peers and patients (Luke & Boss, 1981). Effective implementation has taken these concerns into account and has allowed empathy, time, education, and indirect pressure to take their course in correcting practitioner aberrancies or impairments as they affect the quality and the unnecessary costs of care (Farrington, Flech, & Hare, 1984; Forquer & Anderson, 1982). Administrators have occasionally been reluctant to support such programs because they are costly or focus on deficiencies that one may not want to deal with (Warner, 1983). However, both health professionals and administrators have been eased into accepting such programs by payers' and regulators' insistence that they be put into place, by the leadership of professional associations such as the APAs, and by direct and indirect peer pressure. At this point, most practitioners and administrators support such activities because they are required, are professionally sanctioned, and are tied to the overall financial stability of the facility or practice.

Yet, there are lingering problems that require a higher level of resolution. In addition to the problems previously cited related to the need for more sophisticated and standardized evaluation systems and better data on the relative efficacy of psychotherapies, three major areas will require further attention: the development of better information systems, the resolution of confidentiality conflicts, and ensuring that economic considerations affecting quality assurance do not result in a lower quality of care for any individual or category of individuals.

INFORMATION SYSTEMS

Quality assurance, utilization review-management, and risk management programs are inherently dependent on accurate and timely information flowing between the patient, the provider, and the payer. Claims data are inadequate for most purposes in understanding the quality, the efficacy, or the cost benefits of health services. Likewise, clinical records are often inadequate for evaluating care because they tend to be disorganized, unfocused, illegible, and circumspect. Additionally, psychiatric records are difficult to evaluate because of (1) practioners' concerns about confidential information which lead to limited chart entries; (2) the specialized nature and the tendency toward vagueness of psychiatric terminology; and (3) the problems practitioners have in connecting specific treatment goals to defined signs, symptoms, and behaviors, psychological processes, severity of dysfunctions(s), and need for the identified level of care. These common deficiencies often lead to problems in assessment and reimbursement, effective clinical and administrative management, and therapuetic outcome (Henisz, Levine, & Etkin, 1981; Perlman et al., 1982).

In an age in which so many decisons are based on the quality of information, this general inadequacy of conventional clinical records cannot be sustained or justified. With the explosion of developments within the computer field, it was inevitable that one of the major potential users would be the health care field. Although the therapeutic benefits of such systems may not have seemed obvious to many practitioners in recent years, the business applications have seemed very obvious, given the woefully inadequate capacities of reimbursement systems based on paper claims.

Thus, most hospitals, clinics, and practices are converting at least some levels of billings to computers, and many are venturing into highly efficient, automated clinical-records systems (Barnett, Winickoff, Dorsey, Morgan, & Lurie, 1978; Eden & Eden, 1979; National Center for Health Sciences Research, 1979; Oulton, 1981; "Revolutionary Computer," 1984), including those in mental health services (Hedlund, Viowig, Evenson, 1979; Looney, Claman, Moir, Costello, Henderson, & Steffek, 1984). Some paperless-claims-reimbursement systems are proving to be effective and popular, allowing reduced time lags in payment through more rapid and detailed communication of the data on which a benefit determination can be made and amended. This will surely be an area of significant future developments, with import in internal and external review systems, research, and training. However, a current and looming problem area is the protection of confidential clinical information in highly automated and integrated treatment and reimbursement systems (Hiller & Beyda, 1981).

CONFIDENTIALITY

Perhaps no single area raises such basic ethical and legal problems for providers, and personal problems for patients, as the threatened loss of privacy historically protected in the patient–practitioner relationship. This is particularly vexing for the mental health professions, given the often sensitive nature of patient-client revelations in psychotherapy and the traditional pact of secrecy binding the patient-client and psychotherapist, which allows trustful uncovering. Mutual concerns about adverse occupational or other outcomes related to the public disclosure of therapeutic material are reinforced by lingering reminders of public stigmatization of the mentally ill.

Thus, the threat of disclosure of clinical information has been, and continues to be a problem for health professionals (Borenstein, 1985; Hiller & Seidel, 1982; Schuchman et al., 1980), especially with the growth of internal and external quality-assurance and utilization-review programs demanding and gaining access to such information. Although the considerable experience of the APAs and CHAMPUS has not uncovered any evidence of breaches of confidentiality (Hamilton, 1985; Rodriguez, 1983b), the situation does seem vulnerable to slip-ups, especially considering the volumes of information exchanged, the lack of security systems in some review programs, and the increasing use of computers. This is an area that will require professional monitoring and will no doubt invite legal challenges and definitions (Taub, 1983).

QUALITY OF CARE

Too often, efficiency can result in unanticipated inefficient or painful outcomes. One of the paradoxical results of modern QA, UR, and RM systems is the "sorting-out effect." Such systems identify cost-efficient and cost-effective (Doubilet, Weinstein, & McNeil, 1986) care, which, in turn, allows more efficient benefit determinations and analyses. In a competitive, market-driven environment with limited resources, the threat to quality should be readily apparent. Review systems, then, sort out the patients who are "winners" and those who are "losers." Winners generate uncontested income from reimbursement, require relatively low overhead (e.g., staffing intensity) to serve, and otherwise do not create significant disruptions in the treatment program. Losers do just the opposite. The logical business strategy is to obtain winners and to unload losers. An example of the potential negative repercussions for some individuals with the wrong diagnoses is the current situation with DRGs (Kapp, 1984; Mariner, 1984; "Prospective Payment," 1983; Ziegenfuss, 1985). Widespread implementation of a system of psychiatric DRGs would create serious disruptions in quality of care that could not be ensured by QA programs. Likewise, utilization review programs have no doubt led to many mentally ill persons, being declared losers in the high-stakes gamble that is contemporary health insurance and entitlements to be denied necessary treatment.

The limitations in quality and availability of health services resulting from decreased health care dollars pose vexing problems for ethical practitioners, payers that either have run out of funding or face a loss in profits, and patients and families

that both suffer and often have limited financial, informational, or emotional resources to deal with catastrophic or long-term mental illness. We live in times of
hard decisions: Who will care for the chronically mentally ill, many of whom are
homeless (Bassuk, Rubin, & Lauriat, 1984; Freedman & Moran, 1984; Lamb, 1984;
Public Health Service, 1980)? Where is quality assurance for the children, many of
whom are living in impoverished settings (American Academy of Child Psychiatry,
1983; Public Health Service, 1981; Saxe, Cross, Siverman, 1985) or with an ill or
abusive family member? Who accounts for the mentally ill in prisons (Cold, 1984),
in medical settings, and in nursing homes (Shadish & Bootzin, 1984)?

In sum, what will happen to all of those whom the President's Commission on
Mental Health (1978) declared eight years ago to be underserved and unserved?
Too many people are funneling into private and public programs that claim they
have too few dollars. Will there, in fact, be a rationing of care? The bad news is that
there has been rationing for a long time—and it's getting worse, especially for mental health services (Mollica, 1983; Rodriguez, 1985). In these times, then, is quality
assurance in mental health a misnomer for something else, a pipe dream, or the best
hope for securing necessary and equitable benefits? It may be all of these. Yet, one
thing is clear. Without the *substantial* involvement of mental health practitioners—
individually and organized—coordinating and influencing QA, UR, and RM systems, there will be serious cutbacks in the quality of the availability of, and the
rational payment for psychiatric care. This involvement or lack thereof will affect
research, training, and delivery in ways that could have profound effects not only on
the mental health professions, but on the productivity and health of many communities, if not the entire nation. A tragic chance now exists that the great advances
in mental health may too soon enter a "back-to-the-future" scenario that could not
be adequately recovered except for extraordinary efforts at a national level.

THE FUTURE

Mental health professionals are broadly informed and experienced about the
pitfalls in predicting future human behavior. The same could be said about societal
trends, yet there is a substantial body of evidence and speculation that points to future trends in health services (Goldschmidt, 1982), medical and other professional
practice (Relman, 1983a; Tarlov, 1983), psychotherapy (Furrow, 1983; R.W. Gibson,
1985), public mental-health policy (Merwin & Ochberg, 1983), and quality assurance (Brook & Lohr, 1981; Brook, Williams, & Davies-Avery, 1976; Duncan, 1980).
The consensus is that the enterprises of medicine and health care are in the first
phases of a significant evolution that will result in a period of several years of major
changes in the organization, financing, and delivery of health care. Prognosticators
foresee a limited number of major business entities entering into joint ventures or
purchasing various specialized services that will allow them to compete for the ap-
proximaely 80% of Americans who have some form of health insurance or public
program benefits, including Medicare, Medicaid, and CHAMPUS. Similar organizational changes directed toward capitated-managed health plans for publicly provided health care services are possible. The variables of world and national

economic, social, and political events make any forecast speculative (Gold-schmidt, 1982).

Nevertheless, the next changes are already under way. Managed mental-health plans are currently being developed and marketed by a number of companies and have been started for several companies (Flagg, 1985). The Department of Defense is establishing a major demonstration project in the Tidewater, Virginia, area that will evaluate the effectiveness of a capitated purchasing arrangement for mental health services during 1986–1989, and that will provide model approaches to evaluating, managing, and reimbursing all health care services under CHAMPUS in the future. These changes should forebode real concerns and provide a positive challenge to the mental health professions. The agenda of the mental health pro-fessions will be a full one and will need to accomodate to others having a greater say in the process of decision making about the future directions in mental health care (Talbott, 1984). Yet, the professions must not either play a passive or resistant role or continue to work at odds with one another *if* professionally oriented and directed mental health services are to continue within the changing structures of financing and delivery. Nothing less than a unified strategy is required as the business and nonprofessional interests consolidate to reorganize health care within the "free market." These interests are occasionally sensitive to quality assurance issues in mental health but are more sensitive to competitive and profit-oriented business priorities. For reasons previously noted, this competitive market driven process should create a further crisis in mental health coverage for the unserved and the un-derserved, especially the chronically mentally ill, the impoverished, children, and others who have limited resources and capacities for organized advocacy.

There will be certain roles for professional associations and individual mental-health professionals in these changing environments. Professional roles will con-tinue but change in management-oriented direct care, training, and clinical re-search. Increased professional roles will be required in professional licensing and institutional certification and accreditation activities; quality-assurance and util-ization-risk management; the development of objective (research-based) standards and criteria; and the support of legal, ethical, and scientific needs in patient care. Data on the efficacy of psychotherapies and related objective indicators of required care will need to be the highest priority of a consolidated mental-health professional effort. This consortium approach to developing a consensus about reliable in-dications for mental health services will need to break with the historical trends of interprofessional sniping. If the mental health professions develop an antagonistic intercompetitive strategy, relying on marketing and lobbying strength, the Federal Trade Commission and free market theorists may be satisfied, but patients, payers, and providers will ultimately suffer. There is no safety net in such an internecine ap-proach to determining what's the "best" in mental health care.

Ultimately, the democratic process will prevail in all of these changes, so it's hoped that equity and conscientiousness will prevail in the balancing of multiple and conflicting needs. Few areas will raise the level of passion in legal, legislative, and professional deliberations about equity in health care more than who will receive what shares of quality mental-health services. The abandonment of the nation's commitment to equal protection under the law for mental health care will reverberate against communities' struggles to comprehend the plight of the homeless, to cope with stress in the workplace, and to understand an increasingly troubled and large population of adolescents. The mental health professions will be called on to provide creative and collaborative solutions to these and other problems. The extent of their wisdom, leadership, and success will become the future legacy of quality assurance in mental health.

REFERENCES

Abram, M. (1984, September). Paper presented at the meeting of the Blue Cross and Blue Shield Association, Annual Medical and Dental Directors' Conference, Chicago.

Acceptance growing for new utilization tools—AEP and NAP. (1982, October). *Hospital Peer Review*, pp. 117–121.

Alger, I. (1980). Accountability: Human and political dimensions. *American Journal of Orthopsychiatry, 50*, 388–393.

American Academy of Child Psychiatry. (1983). *Child psychiatry: A plan for the coming decades.* Washington, DC: Author.

American Psychiatric Association. (1987). *Diagnostic and statistical manual of mental disorders* (3rd ed., revised). Washington, DC: Author.

American Psychiatric Association, (1985). *Manual of psychiatric peer review* (3rd ed.). Washington, DC: Author.

American Psychological Association. (1980). *APA/CHAMPUS outpatient psychological peer review manual.* Washington, DC: Author.

Anderson, D.W., & Shields, M.S. (1982). Quality measurement and control in physician decision-making: State of the art. *Health Services Research, 17,* 124–156.

Atchinson, T.A., & Brooks, E.T. (1981). Centralized continuing education programs. *Quality Review Bulletin, 7*(1), 13–16.

Barnett, G.O., Winickoff, R., Dorsey, J.L., Morgan, M.M., & Lurie, R.S. (1978). Quality assurance through automated monitoring and concurrent feedback using computer-based medical information system. *Medical Care, 16,* 962–970.

Bassuk, E.L., Rubin, L., & Lauriat, A. (1984). Is homelessness a mental health problem? *American Journal of Psychiatry, 141,* 1546–1550.

Bausell, R.B. (1983a). Quality assurance: An overview. *Evaluation and the Heath Professions, 6,* 139–255.

Bausell, R.B. (1983b). Quality assurance methods. *Evaluation and the Health Professions, 6,* 259–375.

Bayer, R., Callahan, D., Fletcher, J., Hodgson, T., Jennings, B., Monsees, D., Sieverts, S., & Veatch, R. (1983). The care of the terminally ill: Morality and economics. *New England Journal of Medicine, 309,* 1490–1494.

Bedrosian, J.C. (1983, December). Market forces creating change, absent a national health policy. *Financier,* pp. 51–54.

Biegel, A., & Sharfstein, S.S. (1984). Mental health care provides: Not the only cause or only cure for rising costs. *American Journal of Psychiatry, 141,* 668–672.

Blendon, R.J., & Rogers, D.E. (1983). Cutting medical care costs. *Journal of the American Medical Association, 250,* 1880–1885.

Borchardt, P.J. (1981). Nonacute profiles: Evaluation of physicians' nonacute utilization of hospital resources. *Quality Review Bulletin, 7*(11), 21–26.

Borenstein, D.B. (1985). Confidentiality. In J. Hamilton (Ed.)., *Psychiatric peer review.* Washington, DC: American Psychiatric Press.

Bowman, J.G. (1920). Hospital standardization series—General hospitals of 100 or more beds: Report for 1919. *Bulletin of the American College of Surgeons, 4,* 3–36.

Brewster, A.C., Jacobs, C.M., & Bradbury, R.C. (1984). Classifying severity of illness by using clinical findings. *Health Care Financing Review, 6,* 107–108.

Brook, R.H. (1977). Quality: Can we measure it? *New England Journal of Medicine, 296,* 170–172.

Brook, R.H., Kamberg, C., & Lohr, K.N. (1982). Quality assessment in mental health. *Professional Psychology, 13,* 34–39.

Brook, R.H., & Lohr, K.N. (1981). Quality of care assessment: Its role in the 1980s. *American Journal of Public Health, 71,* 681–682.

Brook, R.H., Williams, K.N., & Davies-Avery, A. (1976). Quality assurance today and tomorrow: Forecast for the future. *Annals of Internal Medicine, 85,* 809–817.

Chassin, M.R. (1983). *Variations in hospital length of stay: Their relationships to health outcomes* (Office of Technology Assessment, Technology Case Study No. 24). Washington, DC: U.S. Government Printing Office.

Codman, E.A. (1916). *A study in hospital efficiency.* Boston: Thomas Todd.

Cold, J. (1984). How many psychiatric patients in prison? *British Journal of Psychiatry, 145,* 78–86.

Commission on Professional and Hospital Activities. (1985). *Length of stay by diagnosis, United States, 1984 (north central region, northeastern region, southern region, western region).* Ann Arbor, MI: Author.

Council on Medical Service. (1983). Effects of competition in medicine. *Journal of the American Medical Association, 249,* 1864–1868.

Cousins, N. (1985). How patients appraise physicians. *New England Journal of Medicine 313,* 1422–1424.

Cunningham, L.S. (1981). Early assessment for discharge planning. *Quality Review Bulletin, 7*(10), 11–13.

Davidson, J. (1986, March 5). Research mystery: Use of surgery, hospitals vary greatly by region. *Wall Street Journal,* p. 35.

Davis, E.C. (1983, December). Local coalitions best hope to control delivery costs. *Financier,* pp. 27–30.

Department of Health, Education and Welfare. (1975). *Forward plan for health: 1977–1981.* Washington, DC: U.S. Government Printing Office.

Donabedian, A. (1968). Promoting quality through evaluating the process of patient care. *Medical Care, 6,* 181–202.

Donabedian, A. 91978). The quality of medical care. *Science, 200,* 856–864.

Donabedian, A. (1980). *The definition of quality and approaches to its assessment.* Ann Arbor, MI: Health Administration Press.

Donabedian, A. (1982). *Explorations in quality assessment and monitoring: The criteria and standards of quality.* Ann Arbor, MI: Health Administration Press.

Donabedian, A. (1984). Quality, costs and cost containment. *Nursing Outlook, 32,* 142–145.

Doubilet, P., Weinstein, M.C., & McNeil, B.J. (1986). Use the misuse of the term "cost effective" in medicine. *New England Journal of Medicine, 314,* 253–255.

Dramatic variations in medical practice blamed for misspent health dollars. (1984, August 13). *Medical World News,* pp. 18–19.

Duncan, A. (1980). Quality assurance: What now and where next? *British Medical Journal, 180* 300–302.

Eden, H.S., & Eden, M. (1979). *Microprocessor-based intelligent machines in patient care.* Washington, DC: National Institutes of Health.

Eisenberg, B.S. (1984). Diagnosis-related groups, severity of illness, and equitable reimbursement under Medicare. *Journal of the American Medical Association, 252,* 645–646.

Eisenberg, J.M. (1979). Sociologic influences on decision-making by clinicians. *Annals of Internal Medicine, 90,* 957–964.

Eisenberg, J.M., & Williams S. V. (1981). Cost containment and changing physician's practice behavior. *Journal of the American Medical Association, 246,* 2195–2201.

Englehart, J.K. (1984). Opinion polls on health care. *New England Journal of Medicine, 310,* 1616–1620.

Ertel, P.Y., & Alredge, M.G., (1977). *Medical peer review: Theory and practice.* St. Louis: C. V. Mosby.

Evans, R.W. (1983). Health care technology and the inevitability of resource allocation and rationing decisions. *Journal of the American Medical Association, 249* 2208–2219.

Farrington, J., Flech, W., & Hare, R. (1984). Sounding board: Quality assurance. *New England Journal of Medicine, 303,* 154–156.

Fielding, J.E. (1984). *Corporate health management.* Menlo Park, CA: Addison-Wesley Publishing.

Flagg, D.C. (1985, October). A business perspective on the future of health care. *American College of Utilization Review Physicians Newsletter,* pp. 1–2, 7–8.

Flexner, A. (1910). *Medical education in the United States and Canada*. New York: Carnegie Foundation for the Advancement of Teaching.

Forquer, S.L., & Anderson, T.B. (1982). A concerns-based approach to the implementation of quality assurance programs. *Quality Review Bulletin, 8*(4), 14–19.

Frank, J.D. (1975). Evaluation of psychiatric treatment. In A.M. Freedman, H.I. Kaplan, & B.J. Sadock (Eds.), *Comprehensive textbook of psychiatry/II* (2nd ed., Vol. 2, pp. 2010–2014). Baltimore, MD: Williams & Wilkins.

Frank, R.G., & Kamlet, M.S. (1985). Direct costs and expenditures for mental health care in the United States in 1980. *Hospital and Community Psychiatry, 3,* 165–168.

Frank, R.G., & Lave, J.R. (1985). A plan for prospective payment for inpatient psychiatric care. *Hospital and Community Psychiatry, 36,* 775–776.

Freedman, R.I., & Moran, A. (1984). Wanderers in a promised land: The chronically mentally ill and reinstitutionalization. *Medical Care, 22*(12, Suppl.), 1–56.

Fuchs, V. (1982). The battle for control of health care. *Medical Director, 8*(6), 5–9.

Furrow, B.R. (1983, Fall). Will psychotherapy be transformed in the 1980s? *Law Medicine and Health Care,* p. 96.

Gatekeeper payment system could result in lower quality of care. (1984, May). *Hospital Peer Review,* p. 61.

Gaumer, G.L. (1984). Regulating health professionals: A review of the empirical literature. *Milbank Memorial Fund Quarterly—Health and Society, 62,* 380–416.

General Accounting Office. (1982). *Physician cost containment training can reduce medical costs* (GAO/HRD Publication No. 82-36). Washington, DC: U.S. Government Printing Office.

General Accounting Office. (1983). *GAO finds many ancillary services unnecessary* (GAO/HRD Publication No. 83-74). Washington, DC: U.S. Government Printing Office.

Gibson, R. (1985). The future of the practice of psychotherapy. *Psychiatric Hospital, 16,* 155–159.

Gibson, R. & Singhas, P. (1978). Professional accountability and peer review. In G. U. Balis (Ed.), *The behavioral and social sciences and the practice of medicine*. Woburn, MA: Butterworth.

Ginsburg, P.B., & Sloan, F.A. (1984). Hospital cost shifting. *New England Journal of Medicine, 310,* 893–898.

Ginzberg, E. (1975). Notes on evaluating the quality of medical care. *New England Journal of Medicine, 292,* 366–368.

Ginzberg, E. (1983). The grand illusion of competition in health care. *Journal of the American Medical Association, 249,* 1857–1859.

Ginzberg, E. (1984). The monetarization of medical care. *New England Journal of Medicine, 310,* 1162–1165.

Glick, I.D. (1985, May). *Quality treatment and practitioner competence*. Paper presented at the meeting of the American Psychiatric Association, Dallas.

Goldfarb, D.L., Mintz, R., & Yeager, M.S. (1982). Why did hospital costs increase in 1981? *Hospitals, 16,* 109–114.

Goldschmidt, P.G. (1982). *Health 2000*. Baltimore: Policy Research.

Gonnella, J.S. (1983). *Clinical criteria for disease staging*. Santa Barbara, CA: SysteMetrics.

Gonnella, J.S., Hornbrook, M.C., & Louis, D.Z. (1984). Staging of disease: A case mix measurement. *Journal of the American Medical Association, 251,* 637–544.

Gordon, R.E., Jardiolin, P., & Gordon, K.K. (1985a). Predicting length of hospital stay of psychiatric patients. *American Journal of Psychiatry, 142,* 235–237.

Gordon, R.E., Vijay, J., Sloate, S.G., et al. (1985b). Aggravating stress and functional level as predictors of length of psychiatric hospitalization. *Hospital and Community Psychiatry, 36,* 773–774.

Greenfield, W.M. (1983). New approaches to using the determination-of-need process to contain hospital costs. *New England Journal of Medicine, 309,* 372–374.

Greenspan, J. (1980). *Accountability and quality assurance in health care*. Baltimore: Charles Press.

Grimaldi, P.L., & Micheletti, J.A. (1984). Utilization and quality review under the prospective rate system. *Quality Review Bulletin, 10*(2), 30–37.

Guttentag, M., & Struening, E.L. (Eds.). (1975). *Handbook of evaluation research* (Vol. 2). Beverly Hills, CA: Sage.

Hamilton, J. (Eds.). (1985). *Psychiatric peer review: Prelude and promise*. Washington, DC: American Psychiatric Press.

Hart, D.K., & Hart, J.D. (1980). Hospitals and economic turbulence in the 1980s. *Journal of Contemporary Business, 9,* 97–110.

Hedlund, J., Viowig, B., Evenson, C., *et al.* (1979). *Mental health infomation systems: A state of the art report.* Columbia: University of Missouri.

Henisz, J.E., Levine, M.S., & Etkin, K. (1981). The psychiatric record and quality review. In C. Siegel & S.K. Fischer (Eds.), *Psychiatric records in mental health care.* New York: Brunner/Mazel.

Hetherington, R. (1982). Quality assurance and organizational effectiveness in hospitals. *Health Services Research, 17,* 185–201.

Hiller, M.D., & Beyda, V. (1981). Computers, medical records and the right to privacy. *Journal of Health Politics, Policy and the Law, 6,* 463–487.

Hiller, M.D., & Seidel, L.F. (1982). Patient care management systems, medical records and privacy: A balancing act. *Public Health Reports, 97,* 332–345.

Horn, S.D., Chachich, B., & Clopton, C. (1983a). Measuring severity of illness: A reliability study. *Medical Care, 21,* 705–714.

Horn, S.D., Sharkey, P., & Bertram, D.A. (1983b). Measuring severity of illness: Homogeneous case mix group. *Medical Care, 21,* 14–25.

Horn, S.D., Sharkey, .D., Chambers, A.F., & Horn, R.A. (1985). Severity of illness within DRGs: Impact on prospective payment. *American Journal of Public Health, 75,* 1195–1197.

Hospital costs for psychiatric care up sharply. (1983, June 6). *American Medical News,* p. 7.

Houpt, J.L., Orleans, C.S., George, L.K., & Brodie, H.K. (1979). *The importance of mental health services to general health care.* Cambridge, MA: Ballinger.

Hubbard, J.P., Levit, E.J., Schumacher, C.F., & Schnabel, R.G. (1965). An objective evaluation of clinical competence. *New England Journal of Medicine, 272,* 1321–1328.

Hunt, K. (1983, June 27). Who'll decide which patients you hospitalize? *Medical Economics,* pp. 153–159.

Industries use preadmission screening to cut hospitalization. (1982, January). *Hospital Peer Review,* pp. 1–2.

Jackson, J. (1969). Factors affecting the treatment environment. *Archives of General Psychiatry, 21,* 39–45.

Jencks, S.F., Goldman, H.H., & McGuire, T.G. (1985). Challenges in bringing exempt psychiatric services under a prospective payment system. *Hospital and Community Psychiatry, 36,* 764–769.

Jessee, W.F. (1977). Quality assurance systems: Why aren't there any? *Quality Review Bulletin, 3*(11), 16–26.

Jessee, W.F. (1981). Approaches to improving the quality of health care: Organizational change. *Quality Review Bulletin, 7*(7), 13–18.

Jessee, W.F. (1983). Assuring the quality of health care: Policy and perspectives. In C. Sager & L.C. Jain (Eds.), *Policy issues in personal health services* Rockville, MD: Aspen.

Jessee, W.F., & Brenner, L.H. (1982). Approaches to improving health care: Dealing with the problem physician. *Quality Review Bulletin, 8*(1), 11–14.

Joint Commission on Accreditation of Hospitals. (1985). *Accreditation manual for hospitals.* Chicago: Author.

Kane, R.A. (1981). Assuring quality of care and quality of life in long term care. *Quality Review Bulletin, 7*(10), 3–10.

Kapp, M.B. (1984, December). Legal and ethical implications of health care reimbursement by diagnosis-related groups. *Law Medicine and Health Care,* pp. 245–253.

Kinzer, D.M. (1983). Massachusetts and California—Two kinds of hospital cost control. *New England Journal of Medicine, 308,* 838–841.

Kohlmeier, L.M. (1983, December). Question of morality, economics now also confront U.S. medicine *Financier,* pp. 9–13.

Kreitzer, S.L., Loebner, E.S., & Roveti, G.C. (1982). Severity of illness: The DRGs' missing link? *Quality Review Bulletin, 8*(5), 21–33.

Lamb, H.R. (1984). Deinstitutionalization and the homeless mentally ill. *Hospital and Community Psychiatry, 35,* 899–907.

Leaf, A. (1984). The doctor's dilemma—and society's too. *New England Journal of Medicine, 310,* 718–720.

Levenson, A.I. (1982). The growth of investor-owned psychiatric hospitals. *American Journal of Psychiatry*, *139*, 902–907.

Lindsay, C.M. (1982, May). Is there really a crisis in the cost of health care? *Colloquium*, *3*, 1, 2, 8.

Linn, L. (1970). Measuring the effectiveness of mental hospitals. *Hospital and Community Psychiatry*, *21*, 17–22.

Liston, E., & Yager, J. (1982). Assessment of clinical skills in psychiatry. In *Teaching psychiatry and behavioral science*. New York: Grune & Stratton.

LoGerfo, J.P., & Brook, R.H. (1979). Evaluation of health services and quality of care. In S.J. Williams & P.R. Torrens (Eds.), *Introduction to health services*. New York: Wiley.

London, P., & Klerman, G.L. (1982). Evaluating psychotherapy. *American Journal of Psychiatry*, *139*, 709–717.

Long, D.A. (1984). Prospective quality assurance. *Quality Review Bulletin*, *10*(5), 143–145.

Looney, J.G., Claman, L., Moir, R., Costello, A.J., Henderson, P.B., & Steffek, J.C. (1984). Data systems in child psychiatry. *Journal of the American Academy of Child Psychiatry*, *23*, 99–104.

Luke, R.D., & Boss, R. (1981). Barriers limiting the implementation of quality assurance programs. *Health Services Research*, *16*, 305–314.

Luke, R.D., & Modrow, R. (1982). Professionalism, accountability, and peer review. *Health Services Research*, *17*, 113–123.

Luke, R.D., Kreuger, J.C., & Modrow, R.E. (Eds.). (1983). *Organization and change in health care quality assurance*. Rockville, MD: Aspen.

Lundberg, G.D. (1983). Rationing human life. *Journal of the American Medical Association*, *249*, 2208–2219.

Mariner, W.K. (1984, December). Diagnosis related groups: Evading social responsibility? *Law Medicine and Health Care*, pp. 243–244.

Mattson, M.A. (1984). Quality assurance: A literature review of a changing field. *Hospital and Community Psychiatry*, *35*, 605–616.

McAuliff, W.E. (1979). Measuring the quality of medical care: Process versus outcome. *Milbank Memorial Fund Quarterly-Health and Society*, *57*, 118–131.

McGuire, T.G., & Montgomery, J.T. (1982). Mandated mental health benefits in private health insurance. *Journal of Health Politics, Policy and the Law*, *7*, 380–406.

Mechanic, D. (1985). Cost containment and the quality of medical care: Rationing stategies in an era of constrained resources. *Milbank Memorial Fund Quarterly-Health and Society*, *63*, 453–475.

Medical care costs rose twice as fast as other prices. (1983, April 18). *American Medical News*, p. 4.

Menninger, R.W. (1977). What is quality care? *American Journal of Orthopsychiatry*, *47*, 476–483.

Mental health benefits: Shrinking reimbursement. (1981, November 9). *Washington Report on Medicine and Health*, pp. 1–4.

Merwin, M.R., & Ochberg, F.M. (1983). The long voyage: Policies for progress in mental health. *Health Affairs*, *2*, 96–127.

Mezzich, J.E., & Sharfstein, S.S. (1985). Severity of illness and diagnostic formulation: Classifying patients for prospective payment systems. *Hospital and Community Psychiatry*, *36*, 770–772.

Mollica, R.F. (1983). From asylum to community: The threatened disintegration of public psychiatry. *New England Journal of Medicine*, *308*, 367–373.

Moore, S.H., Martin, D.P., & Richardson W.C. (1983). Does the primary care gatekeeper control the costs of health care? *New England Journal of Medicine*, *309*, 1400–1404.

More, M.D.'s support proposals to control costs. (1984, January 20). *American Medical News*, p. 13.

Morrison, B.J., Rehr, H., Rosenberg, G., & Davis, S. (1982). Consumer opinion surveys. *Quality Review Bulletin*, *8*(2), 19–24.

Munich, R.L., Carsky, M., & Appelbaum, A. (1985). The role and structure of long-term hospitalization: Chronic schizophrenia. *Psychiatric Hospital*, *16*, 161–169.

Name of new game: Allocation of resources. (1983, January 7). *American Medical News*, pp. 1, 7, 8.

National Association of Private Psychiatric Hospitals. (1985). *A proposal on prospective reimbursement for psychiatric hospitals*. Washington, DC: Author.

National Center for Health Sciences Research. (1979). *Computer applications in health care* (DHEW Publication No PH5 79–3251). Washington, DC: Author.

National Center for Health Services Research. (1983). *Economic incentive and physician practice: An examination of Medicare participation decisions and physician-induced demand* (Publication No PB-83-208132). Washington, DC:Author.

National utilization standard urged. (1983, July). *Hospital Peer Review*, p. 96.

Nelson, A.R. (1984, April). *Health care and its effect on the economy.* Statement of the American Medical Association to the Joint Economic Committee, United States Congress, Washington, D.C.

Nelson, S.H. (1979). Standards affecting mental health care: A review and commentary. *American Journal of Psychiatry, 136,* 303–307.

New trends in care seen as reimbursements cut. (1983, January 7). *American Medical News*, pp. 1, 24.

Newble, D.I. (1983). The critical incident technique: A new approach to the assessment of clinical performance. *Medical Education, 17,* 401–403.

Newman, F. (1982). Outcome evaluation and quality assurance in mental health. *Quality Review Bulletin, 8*(4), 27–31.

Nutting, P.A., Shorr, G.I., & Kurkhalter, B. (1981). Assessing the performance of medical care systems: A method and its application. *Medical Care, 19,* 281–296.

Oulton, R. (1981). Use of incident report data in a system-wide quality assurance/risk management program. *Quality Review Bulletin, 7*(6), 2–7.

Pattison, R.V., & Katz, H.M. (1983). Investor-owned and not-for-profit hospitals. *New England Journal of Medicine, 309,* 347–353.

Payne, B.C., Lyons, T.G., & Neuhaus, E. (1984). Relationships of physician characteristics to performance quality and improvement. *Health Services Research, 19,* 307–332.

Perlman, B.B., Schwartz, A.H., Paris, M., Thronton, J.C., Smith H., & Weber, R. (1982). Psychiatric records: Variations based on discipline and patient characteristics, with implications for quality of care. *American Journal of Psychiatry, 139,* 1154–1157.

Platt, R. (1983). Cost containment—Another view. *New England Journal of Medicine, 309,* 726–730.

Pollard, M.R. (1983). Competition or regulation. *Journal of the American Medical Association, 249,* 1860–1863.

President's Commission on Mental Health. (1978). *Report to the President.* Washington, DC: U.S. Government Printing Office.

Prospective payment presents challenges for UR, QA. (1983, October). *Hospital Peer Review*, pp. 57–60.

Public Health Service. (1980). *Report to the Secretary: Toward a national plan for the chronically mentally ill* Washington, DC: U.S. Government Printing Office.

Public Health Service. (1981). *Better health for our children—A national strategy* (DHHS Publication No. 79-55071). Washington, DC: U.S. Government Printing Office.

Rappleye, W.C. (1983, December). System for rational choices. *Financier*, pp. 19–29.

Reeder, M.P. (1981). The benefits of areawide quality review. *Quality Review Bulletin, 7*(12), 2–3.

Regier, D.A., Shapiro, S., Kessler, L.G., & Taube, C.A. (1984). Epidemiology and health services resource allocation policy for alcohol, drug abuse and mental disorders. *Public Health Reports, 99,* 483–492.

Relman, A.S. (1983a, July). The future of medical practice. *Health Affairs*, pp. 5–19.

Relman, A.S. (1983b). Investor-owned hospitals and health care costs. *New England Journal of Medicine, 309,* 370–372.

Relman, A.S. (1983c, December). Powerful new policy force: The medical industial complex. *Financier*, pp. 40–44.

Restuccia, J.D., & Holloway, D.C. (1982). Methods of control for hospital quality assurance systems. *Health Services Research, 17,* 182–197.

Revolutionary computer system helps assure quality. (1984, May). *Hospital Peer Review*, pp. 57–61.

Ricardo-Campbell, R. (1982). *The economics and politics of health.* Chapel Hill, NC: University of North Carolina Press.

Richman, A., & Barry, A. (1985). More and more is less and less. *British Journal of Psychiatry, 146,* 164–168.

Rodriguez, A.R. (1983a, October). *National standards of review.* Paper presented at the meeting of the American College of Utilization Review Physicians, Denver.

Rodriguez, A.R. (1983b). Psychological and psychiatric peer review at CHAMPUS. *American Psychologist, 38,* 941–947.

Rodriguez, A.R. (1985). Current and future directions in reimbursement for psychiatric services. *General Hospital Psychiatry, 7,* 341–348.

Rodriguez, A.R., & Maher, J.J. (1986). Psychiatric case management at CIBA-GEIGY. *Business and Health,* 3(3), 14–17.

Rubin, J. (1980). A survey of mental health policy options. *Journal of Health Politics, Policy and the Law, 5,* 210–232.

Rubin, J. (1982). Cost measurement and cost data in mental health setting. *Hospital and Community Psychiatry, 33,* 750–754.

Rutstein, D.D., Berenberg, W., Chalmers, T.C., Child, C.G., Fishman, A.P., & Perrin, E.B. (1976). Measuring the quality of medical care. *New England Journal of Medicine, 294,* 582–588.

Rzaha, C.B. (1983). Quality assurance in long term care. *Quality Review Bulletin,* 9(8), 229–232.

Sammons, J.H. (1983, December). Doctors' active role in shaping tough policy choices. *Financier,* pp. 46–50.

Sanazaro, P. J. (1980). Quality assessment and quality assurance in medical care. *Annual Review of Public Health, 1,* 31–48.

Sanazaro, P.J., & Williamson J.W. (1970). Physician performance and its effects on patients. *Medical Care, 8,* 299–308.

Sandrick, K.M. (1982). Private review—Its impact on health care and physicians. *Quality Review Bulletin,* 8,(4), 5–6.

Saxe, L., Cross, T., Siverman, N., Batchelor, W.F., (1986). *Children's mental health: Problems and services—A background paper.* (Office of Technology Assessment, OTA-BP-H-33) Washington, DC: U.S. Government Printing Office.

Schacht, T.E., & Strupp, H.H. (1985). Evaluation of psychotherapy. In H.I. Kaplan, & B. J. Sadock (Eds.), *Comprehensive textbook of psychiatry* (4th ed., Vol. 2). Baltimore: Williams & Wilkins.

Schorr, B. (1983, December 2). Hospitals scramble to track costs as insurers limit reimbursements. *Wall Street Journal,* p. 2.

Schuchman, H., Nye, S.G., Foster, L.M., & Lanman, R.B. (1980). Confidentiality of health records: A symposium on current issues, opportunities and unfinished business. *American Journal of Orthopsychiatry, 50,* 639–677.

Schulberg, H.C., & Baker, F. (Eds.). (1979). *Program evaluation in the health fields* (Vol. 2). New York: Human Science Press.

Schulz, R.I., Greenley, J.R., & Peterson R. W. (1983). Management, cost and quality of acute inpatient psychiatric services. *Medical Care, 21,* 911–928.

Schumacher, D.N., Clopton, C.J., & Bertram, D.A. (1982). Measuring hospital case mix. *Quality Review Bulletin,* 8(4), 20–26.

Scott, W.R. (1982). Managing professional work: Three models of control for health organizations. *Health Services Research, 17,* 198–211.

Shadish, W.R., & Bootzin, R.R. (1984). Social integration of psychiatric patients in nursing homes. *American Journal of Psychiatry, 141,* 1203–1207.

Sharfstein, S.S., & Taube, C.A. (1982). Reductions in insurance for mental disorders: Adverse situation, moral hazard and consumer demand. *American Journal of Psychiatry, 139,* 1425–1430.

Shaughnessy, P.W., & Kurowski B. (1982). Quality assurance through reimbursement. *Health Services Research, 17,* 157–182.

Shortell, S.M., & LoGerfo, J.P. (1981). Hospital medical staff organization and quality of care. *Medical Care, 19,* 1041–1055.

Sloan, F.A. (1982). Effects of health insurance on physicians fees. *Journal of Human Resources, 17,* 533–557.

Sloan, F.A. (1983). Rate regulation as a strategy for hospital cost control. *Milbank Memorial Fund Quarterly—Health and Society, 61,* 195–221.

Sloan, F.A., & Schwartz, W.B. (1983). More doctors: What will they cost? *Journal of the American Medical Association, 249,* 766–769.

SMI presents alternative to current UR tools. (1983, April). *Hospital Peer Review,* pp. 41–43.

Smith, D., & Metzner, C. (1970). Differential perceptions of health care quality in a prepaid group practice. *Medical Care, 8,* 264–275.

Snoke, A.W. (1982). The need for a comprehensive approach to health and welfare. *American Journal of Public Health, 72,* 1028–1034.

Spitzer, R.L., & Endicott, J. (1975). Psychiatric rating scales. In A.M. Freedman, H.I. Kaplan, & B.J. Sadock (Eds.), *Comprehensive textbook of psychiatry/II* (2nd ed., Vol. 2). Baltimore: Williams & Wilkins.

Spitzer, R.L., Endicott, J., Fleis, J., & Cohen, J. (1970). The psychiatric status schedule: A technique for evaluating psychopathology and impairment in role functioning. *Archives of General Psychiatry, 23,* 41–55.

Spitzer, R.L., Gibbon, M., & Endicott, J. (1973). *Global assessment scale.* New York: New York State Department of Mental Hygiene.

Stuart, B., & Stockton, R. (1973). Control over the utilization of medical services. *Milbank Memorial Fund Quarterly—Health and Society, 51,* 341–394.

Talbott, J.A. (1984). Psychiatry's agenda for the 80s. *Journal of the American Medical Association, 251,* 2250.

Tarlov, A.R. (1983). The increasing supply of physicians, the changing structure of the health services system, and the future practice of medicine. *New England Journal of Medicine, 308,* 1235–1244.

Taub, S. (1983, June). Psychiatric malpractice in the 1980's: A look at some areas of concern. *Law Medicine and Health Care,* pp. 97–103.

Taube, C.E., & Barrett, S.A. (Eds.). (1985). *Mental health, United States 1985* (DHHS Publication No. ADM 85-1378). Washington, DC: U.S. Government Printing Office.

Test program to assess physicians in Maryland, Virginia, Washington, D.C. (1981, May). *Hospital Week,* p. 3.

Thompson, R.E. (1981). Relating continuing education and quality assurance activities. *Quality Review Bulletin, 7*(1), 3–6.

Thompson, R.E., & Rodrick, A.G. (1982). Integrating patient concerns into quality assurance activities. *Quality Review Bulletin, 8*(2), 16–18.

Wagner, D.P., Knaus, W.A., & Draper, E.A. (1983). Statistical validation of a severity of illness measure. *American Journal of Public Health, 73,* 878–884.

Walter, R.S. (1983, February). The hottest new thing in utilization review. *Medical Economics,* pp. 177, 180–181.

Ware, J.E., Johnston, S.A., Davies-Avery, A., & Brook, R.H. (1979). *Conceptualization and measurement of health for adults in the health insurance study: Vol. III. Mental health* (R-1987/3-HEW). Santa Monica, CA: Rand Corporation.

Warner, A.M. (1983). Thoughts about the cost of quality assurance. *Quality Review Bulletin, 9*(2), 39–41.

Weinstein, M.C., & Fineberg, H.V. (1980). *Clinical decision analysis.* Philadelphia: W.B. Saunders.

Wendorf, B. (1982). Ethical decision-making in quality assurance. *Quality Review Bulletin, 8*(1), 4–6.

Wennberg, J. (1986). Which rate is right? *New England Journal of Medicine, 314,* 310–311.

Wetherill, H.G. (1915). A plea for higher hospital efficiency and standardization. *Surgery, Gynecology, Obstetrics, 20,* 705–707.

Wilder, J.F., & Rosenblatt, A. (1977). An assessment of accountability areas by psychiatrists, psychologists and social workers. *American Journal of Orthopsychiatry, 47,* 336–340.

Wilensky, G.R., & Rossiter, L.F. (1983). The relative importance of physician-induced demand in the demand for medical care. *Milbank Memorial Fund Quarterly—Health and Society, 61,* 252–277.

Williams, K.N., & Brook, R.H. (1978). Quality measurment and assurance: A review of the recent literature. *Health and Medical Care Services Review, 1,* 3–15.

Williamson, J.W., & Associates. (1982a). *Teaching quality assurance and cost containment in health care.* San Francisco: Jossey-Bass.

Williamson, J.W., Hudson, J.I., & Nevins, M.M. (1982b). *Principles of quality assurance and cost containment in health care.* San Francisco: Jossey-Bass.

Wolfson, J., & Levin, P.J. (1985). *Managing employee health benefits.* Homewood, IL: Dow Jones-Irwin.

Wyszewianski, L. (1982). Market-oriented cost containment strategies and quality of care. *Milbank Memorial Fund Quarterly—Health and Society, 60,* 518–550.

Youel, D.B. (1983, February). Patient care, continuing health professions education, and quality assurance. *American College of Utilization Review Physicians Newsletter,* pp. 1–2.

Ziegenfuss, J.T. (1985). *DRGs and hospital impact.* New York: McGraw-Hill.

II

GENERAL ISSUES

2

Historical Trends in Quality Assurance for Mental Health Services

KENNETH B. WELLS AND ROBERT H. BROOK

INTRODUCTION

Concerns about the quality of health services are as old as the healing professions. The modern concept of quality assurance, however, evolved in the twentieth century, chiefly in the United States,[1] and has had lasting effects on the mental-health-care delivery system. For example, the demands for accountability that quality assurance entails have stimulated a greater concern with the reliability and validity of mental health diagnoses and of measures of outcome. Similarly, the concept of peer review has stimulated reconsideration of the therapist- client (doctor–patient) relationship and the limits of confidentiality. Further, the difficulties inherent in instituting effective peer review have been cited as a partial justification for the current trends toward decreasing the insurance coverage of mental health services. And finally, quality assurance legislation has played a significant role in the developing conflict between mental health provider groups.

How did these developments occur? The answers to this question are of more than historical interest. Despite doubts about the effectiveness of current quality-assurance legislation and programs, quality assurance is likely to remain a prominent feature of the U.S. health-care delivery system in the foreseeable future. New strategies to ensure quality and to contain costs continue to be developed. A historical approach can suggest strategies for enhancing the effectiveness of current or future quality-assurance activities.

---

[1]Quality assurance activities can be broadly defined as those designed to evaluate, certify, or improve the quality of medical care in order to enhance the health of consumers and the cost effectiveness of care.

KENNETH B. WELLS AND ROBERT H. BROOK • Rand Corporation, 1700 Main Street, Santa Monica, California 90406–2138. This project has been funded, at least in part, with federal funds from the U.S. Department of Health and Human Services under Grant No. 016B80. The contents of this publication do not necessarily reflect the views or policies of the U.S. Department of Health and Human Services nor does mention of trade names, commercial products, or organizations imply endorsements by the U.S. government.

This chapter summarizes the historical events leading to the development of quality assurance programs, the general historical trends enhancing this development, and the specific historical trends relating to the development of quality assurance programs for mental health services. The chapter also provides a brief summary of the literature on mental health quality assurance programs and concludes by attempting to relate this historical review to current and future needs for quality assurance programs in mental health services.

HISTORICAL EVENTS IN MENTAL HEALTH QUALITY ASSURANCE

Attempts to standardize or assess the quality of health care date to ancient times. As early as 1500 B.C. autonomous schools of medicine existed in Egypt, Assyro-Babylonia, Persia, India, China, Japan, and Mexico. Professional codes of physician conduct can be dated to Greek prescientific medicine and the Hippocratic oath (fourth century B.C.).

Physicians have tried to classify and treat mentally disturbed persons at least since Graeco-Roman times (e.g., Galen, A.D. 138–201). Until the sixteenth century A.D., however, religious orders, benefactors, and penal institutions were primarily responsible for caring for the "insane." Consequently, concerns about the quality of such care evolved chiefly outside medicine.

The history of quality assurance in health can possibly be traced to Juan Cuidad Duarte, a sixteenth-century Spanish merchant who suffered a transient psychotic episode (Ellenberger, 1974). After treatment by floggings, Duarte recovered and went on to found a hospital dedicated to the humane treatment of the mentally ill. Following his death, he was canonized as Saint John of God, and the Order of Charity for the Service of the Sick was established to implement his practices throughout Europe. This order is believed to have influenced Phillippe Pinel and Jean-Etienne Esquirol, who are credited with establishing the modern psychiatric hospital. Pinel, a physician, developed clear and logical standards of humane practice, based on a scientific method as evidenced in his *Medico-Philosophical Treatise on Insanity* (1801). His work was all the more influential because it occurred just as the discipline of psychiatry was emerging. With the development of university psychiatry, especially by Wilhelm Griesinger (1817–1868) in Germany, the study of mental disorders eventually achieved the status of a legitimate academic discipline. Furthermore, the rise of autonomous medical schools in Europe and in the United States and the credentialing of graduates represent additional steps in defining or controlling standards of care.

Toward the end of the nineteenth century, concern over the adequacy of U.S. medical schools led to the creation of the American Medical Association's Council on Education. This council cosponsored an evaluation of U.S. medical schools with the Carnegie Foundation in 1908. The result was the Flexner Report (1910), which sharply criticized medical school organization, facilities, and curricula and suggested uniform standards based on a scientific approach. This report has had a lasting and profound impact on medical education in the United States.

Partly in response to this report, the American College of Surgeons, founded in

1913, sponsored several major quality-assurance programs. It developed the Hospital Standardization Program in 1917, which required medical staffs to use clinical records to review their performance (Affeldt, Roberts, & Walczak, 1983). The college also assessed compliance with its standards and pioneered an evaluation of surgical competence based on an assessment of outcomes (Langsley, 1980). The activities of this organization represent the main efforts of organized medicine to ensure quality before the establishment of the Joint Commission on accreditation of Hospitals (JCAH).

The JCAH was formed in 1951 by the American College of Physicians, the American College of Surgeons, the American Hospital Association, and the American Medical Association. It offers voluntary accreditation to health care facilities and has worked with professional medical organizations and, to a much smaller extent, nonmedical professional organizations to develop standards for health facilities and services. The JCAH now evaluates roughly three fourths of the nation's hospitals (including virtually all hospitals with more than 50 beds), and it has served as an important model for other quality-assurance programs. Until 1980, standards for mental health services were developed by the JCAH accreditation councils for long-term facilities (1966), the mentally retarded (1969), psychiatric inpatient facilities (1970), and child and adolescent psychiatric facilities (1974). Advisory councils subsequently replaced these councils (Affeldt et al., 1983).

The Medicare and Medicaid programs (Public Law 89-94, 1965) profoundly affected the growth of the JCAH, as of all other quality-assurance activities. This law required quality assurance programs for facilities seeking reimbursement for Medicare (Title XVIII) and Medicaid (Title XVIV). Retrospective utilization reviews, conducted by a physician staff committee at each facility, were required to determine whether hospitalizations were medically necessary and whether care was appropriate (for selected short-term and all long-term stays). The mandatory review system was unpopular and was considered ineffective by providers and policymakers (Greenblatt, 1975). Medicare and Medicaid programs became increasingly costly, and the retrospective reviews did little to contain costs. Providers perceived the reviews as arbitrary and the standards as rigid.

Sweeping reforms of the review process were mandated in 1972 by PL92-603, which established professional standard review organizations (PSROs). The Secretary of Health, Education, and Welfare (HEW) designated PSRO regions, each of which was to develop a PSRO through provider groups, primarily physicians. If the region did not do so by a specific date, then HEW could so designate the PSRO members. Each PSRO was responsible for performing the quality assurance activities required for reimbursing facilities and individual providers for services rendered to recipients of Medicare, Medicaid, or Maternal and Child Health program benefits. The initial review process was developed for short-term inpatient care, but the law required it eventually to include extended care and outpatient practice.

The mandated quality-assurance programs included an internal review process subject to external monitoring. Both concurrent and retrospective reviews were required. The concurrent review activities included admission certification and periodic rereview of hospital stays. The retrospective reviews included "medical-care-evaluation studies" and "profile analysis." Medical-care-evaluation studies concern the quality of individual episodes of care, and profile analysis concerns identifying patterns of patient care problems based on data aggregated across patients (e.g., the identification of high users of services). PSROs were expected to establish their own regional criteria with the aid of national guidelines called "model criteria sets" (Fowler, 1977).

The initial reactions to PSRO and mandated review of care were strong, especially in organized medicine. Physicians were concerned abut external control of medical practice. All health providers were concerned about threats to the confidentiality of treatment.

Despite strong objections to mandated review, the American Medical Association and its specialty affiliates developed model criteria sets and peer review systems for other third-party insurers to retain some control over the process. For example, the American Psychiatric Association proposed its model criteria in 1975. The nonmedical professions, notably psychology, also developed quality assurance programs, but largely in response to specific third-party insurers (see later).

Since PL93-60, quality assurance programs have grown rapidly in number and variety. Quality assurance programs have been mandated for other federal-and state-supported programs. Important examples are health maintenance organizations by PL93-222 (1973) and community mental health centers by PL94-63 (1975). The mandated evaluations for community mental health centers had two important effects on the development of quality assurance for mental health services. First, several provider groups (e.g., psychologists and social workers) became involved in developing the mandated evaluation. Second, these programs pioneered the coupling of quality assurance with continuing education to improve care.

The JCAH has attempted to develop general programs as models for PSROs and other quality-assurance programs. In 1975, for instance, it developed the Performance Evaluation Procedure for Auditing and Improving Patient Care (PEP), a form of retrospective audit of medical records that served as a standard for many PSROs. However, the JCAH found that system to be costly and ineffective and, in 1979, introduced a quality assurance system based on identifying significant medical problems and developing specific programs to address these problems (Affeldt et al. 1983). In 1980, a similar system was developed for psychiatric care. The JCAH has also worked to develop and evaluation system for community mental-health centers (Towery & Windle, 1978).

Private third-party carriers also developed quality assurance programs. Although some private insurance carriers, notably Blue Cross programs, used claims review before 1965, the legislated Medicare and Medicaid programs ushered in such programs in most insurance plans. The federal government had hoped that other third-party carriers would use the PSROs to coordinate quality assurance activities. This was not to be. Many insurance carriers, such as the Civilian Health and Medical Program of the Uniformed Services (CHAMPUS), developed independent peer-review systems. In 1975, CHAMPUS contracted with the American Psychiatric and Psychological Associations to develops systems of peer review (described in the "Specific Issues" section).

Despite the impact of PSRO and related legislation on the development of quality assurance programs, the future of the mandated quality-assurance program is uncertain for several reasons. First, PSRO activities themselves appear to be costly, even though the program evaluation of PSROs demonstrated that they save about as much money as they cost (*Health Care Financing Administration*, 1979). Second, the effectiveness of the programs has been difficult to ascertain, and the changes they recommend for correcting deficiencies in care have often been difficult to institute (Donaldson & Keith, 1983). Third, PSRO activities have met persistent resistance from some provider groups.

Currently, the PSRO legislation has been revised, and PSROs have been recreated as professional review organizations (PROs) through the Peer Improvement Act of 1982. To date, Congress has continued to support the role of provider groups in determining medical necessity and the appropriateness of care.

TRENDS AFFECTING THE DEVELOPMENT OF QUALITY ASSURANCE

•Five general societal trends have influenced the development of quality assurance programs. The first is the explosion of medical knowledge and technology during the twentieth century. It has become virtually impossible for one physician to master all medical knowledge and skills completely. As a result, specialization in medicine has increased, and consumers have demanded evidence (e.g., certification) of competence in specific fields or technologies. Because medical advances such as cancer chemotherapy have often been associated with risks, consumers have demanded some assurance of appropriate care and protection from unnecessary harm.

•The second trend affecting quality assurance is third-party coverage of health services. Before third-party coverage, the quality of health care was primarily the concern of the individual provider, the patient, and the patient's family. With third-party insurance coverage, insurance companies required information to determine eligibility for reimbursement.

•The third trend is an increasing sense of social responsibility, as reflected in the development of public work employment programs and federal entitlement programs (Medicaid and Medicare). Private attempts to provide more comprehensive health to broad populations are reflected as early as the 1930s in the development of Kaiser programs.

•The fourth trend is the rising cost of medical care, which has had two effects on quality assurance. First, consumers and third parties have required assurance that their money is well spent. Second, rising costs have stimulated the development of cost-containment programs, such as diagnosis-related groups (DRGs). By limiting the amount or type of care delivered, such programs may accentuate the importance of quality assurance.

• The fifth general trend is the consumer movement (Chodoff, 1978; Zano & Kilberg, 1982). Greater accountability of the health professions has paralleled a trend toward greater accountability of all marketed services. Increased malpractice suits and legislation supporting consumer's rights attest to the increasing role of the consumer in health transactions.

These five trends can be understood in part as a natural consequence of market principles. As Getzen (1983) noted, market theory predicts that high quality is ensured for many commodities because quality attracts consumers. However, the health-care-delivery system is not a normal market. There are "quality differences that affect health outcomes, which are highly uncertain and measured much more readily by knowledgeable professionals in large-scale research efforts" than by the individual consumer (Getzen, 1983, p. 309). The consumer does not have sufficient information to evaluate quality, and market principles alone cannot ensure quality.

SPECIFIC ISSUES IN MENTAL HEALTH QUALITY ASSURANCE

In addition to factors influencing the development of quality assurance programs in general, three others have influenced quality assurance in mental health: the diversity of provider groups, concerns about confidentiality, and insurance coverage of mental health services.

The Role of the Professions

Providers that deliver mental health services include both physicians (psychiatrists and nonpsychiatrists, e.g., internists and general practitioners) and nonphysicians (psychologists, psychiatric social workers, psychiatric nurses, and marriage and family counselors). With such a diversity of provider groups, the history of quality assurance is naturally complex. To clarify this history, we first identify the quality assurance activities common to all the professions, and then, we consider four professional groups: psychiatrist, clinical psychologists, clinical social workers, and nonpsychiatrist physicians.[2]

All mental health professions share the selection of candidates, the administration of professional degrees, licensure, the development of ethical standards, and continuing education as quality-assurance-related activities. The selection of candidates and the administration of degrees occur through professional schools, which may or may not be accredited by the relevant professional organizations. Licensing in the United States occurs through state regulations, including standardized examinations. Most professional groups have recommended standards for licensing, but the requirements vary considerably from state to state. Ethical standards are developed through professional organizations, court decisions, and legislation. Continuing-education activities are offered by a variety of organizations, including professional organizations and universities.

Concomitant with the development of PSROs, most states developed more stringent requirements for state licensure, including minimum requirements for continuing education. Continuing education originated as a voluntary, even recreational, activity to improve clinical knowledge and skill. Since 1960, an increasing number of states have required continuing education for relicensure, and some pro-

---

[2]A subsequent chapter describes the role of the nursing profession in mental-health quality-assurance activities.

fessional societies have specified continuing education requirements for membership. As a result, continuing education has become a vast enterprise. The Continuing Education Training Branch of the Division of Manpower and Training Programs of the National Institute of Mental Health (NIMH) was established in 1966 to coordinate these activities for all the mental health professions (NIMH, 1974).

Psychiatry

Organized psychiatry in the United States is an outgrowth of the Association of Medical Superintendents of American Institutions, founded in 1844. This organization was the first national society for physicians, and it was the precursor of the American Psychiatric Association, founded in 1889. Following World War I and an increased acceptance of psychiatry, the American Board of Psychiatry was established to develop standards for training and certification. The American Psychiatric Association has requred continuing education for membership since 1975.

In the early 1950s and 1960s, two developments occurred that were to become characteristics of quality assurance in psychiatry. First, as a result of the prominence of psychoanalysis in American psychiatry, clinical supervision was established as one of the most salient components of psychiatric education. Although supervision has ancient origins (the apprenticeship model), psychoanalysis established a conceptual model for supervision and gave it scientific credibility. Although supervision represents a very intensive form of "peer review," clinical supervision has not been proposed as an alternative to peer review for reimbursement purposes, and supervision has largely retained the quality of an educational activity.

Second, since World War II, psychiatry, through the American Psychiatric Association, has played an increasingly important role in the quality assurance programs—especially JCAH councils and PSROs. For example, in the 1950s, JCAH used individual psychiatrist to apply general JCAH standards to psychiatric facilities. During the 1960s, the American Psychiatric Association encouraged the JCAH to develop separate standards for psychiatric facilities. During the 1960s, the American Psychiatric Association encouraged the JCAH to develop separate standards for psychiatric facilities, and in 1970, the Council on Accreditation of Psychiatric Facilities was established. In 1971, in response to the threat of external review, the American Psychiatric Association appointed a Task Force on Peer Review to prepare guidelines for psychiatric peer review (Clayton, 1976; Sullivan, 1977a). After the PSRO legislation, the American Psychiatric Association worked with 30 other specialty branches of the American Medical Association to develop model criteria sets for PSROs. Since then, the American Psychiatric Association has continued to revise or develop its peer review criteria. Despite working with the PSRO system, psychiatry has repeatedly voiced considerable objection to mandated review (Dorsey, 1974), especially to a mandated external-review system (e.g., claims review by nonprofessionals). Through its Position Statement on Peer Review in Psychiatry (1973), the American Psychiatric Association advocates peer review as the appropriate mechanism to ensure quality (Sullivan, 1977a).

Clinical Psychology

The origins of the field of clinical psychology date to about the same time as the development of university psychiatry (the mid-nineteenth century). The American Psychological Association was founded in 1892, and licensing in clinical psychology began in 1946. About the same time, the American Psychological Association developed standards for the accreditation of schools of professional psychology. As of September 1986, 15 states required continuing education for the relicensure of clinical psychologists (American Psychological Association, 1986).

Clinical psychology is a discipline that, historically, is closely affiliated with the psychology of perception, cognition, and behavior. As a result of this scientific heritage, the concept of defining the goals, process, and outcomes of treatment are central to the professional identity of psychologists.

Nevertheless, the history of development of peer review standards for psychology is very different from that for psychiatry. Because clinical psychology developed outside the medical profession, it has not had the support of the medical profession as a whole. As a result, it has not enjoyed as powerful a political position as has psychiatry in the development of quality assurance.

For example, until recently, psychology has had little representation in the JCAH and other medically oriented accreditation organizations. Although PSROs were mandated to develop standards for care provided by nonphysician health professionals, physicians have had the predominant voice in these organizations. Most legislated quality-assurance programs have required psychologists to be under the direction of physicians, despite increasing trends in state legislation and private insurance companies to establish psychologists as independent practitioners (Dorken & Morrison, 1976).

The relationship of the American Psychological Association to the JCAH illustrates the point. The American Psychological Association became a member of the advisory Committee for Psychiatric Facilities (which replaced the accreditation council) only in 1979. The JCAH standards for such facilities allow psychologists to be members of the clinical staff, but the senior position must be held by a physician (Crosby, 1982). Psychologists have an status equal to that of physicians on only one accreditation council: the Council for Services for the Mentally Retarded and Other Developmentally Disabled Persons (Zano & Kilber, 1982).

Although the American Psychological Association has had little formal control over JCAH activities, it has been developing standards for insurance review since the early 1970s. The first such standards were developed in 1968 by the Committee on Health Insurance. A task force on peer review, eventually called the Committee on Professional Standards Review, developed the *Procedures Manual for Professional Standards Review Committees* to establish such committees in state organizations as a service to psychologists, clients, and third-party payers (Albee & Kessler, 1977).

Probably the most significant overtures to psychology as a profession for quality assurance activities occurred in 1977 through a contract between the American Psychological Association and CHAMPUS to develop peer review of outpatient services (Claiborn & Stricker, 1979; Rodriguez, 1983). The system differs markedly from that of the American Psychiatric Association, which is based on criteria of care for specific diagnostic groups. By contrast, the American Psychological Association encourages both psychologist and client to jointly identify clinical problems, set evaluation goals, and attain these goals within a specified time period. In association with CHAMPUS, the association has also begun empirical evaluations of this system (Cohen & Pizzirusso, 1982).

Clinical Social Work

Although social work is one of the oldest professions, clinical social work has its origins in the 1930s and the work of Florence Hallis. From its beginnings, clinical social work has been closely affiliated with psychiatry and has used psychoanalytic principles and psychiatric terminology. Further, clinical social work grew out of a tradition of close clinical supervison by psychiatrists. The National Association of Social Workers recognized the private practice of psychotherapy as a legitimate activity of social workers in 1958. In 1963, the same organization developed standards of private practice (Phillips, 1975). Most states have had licensing requirements for clinical social workers for at least 10–15 years. As of 1982, 19 states required continuing education for the relicensure of clinical social workers. Two national organizations of social workers have been involved in the development of quality assurance programs for social work: the National Association of Social Workers (NASW), founded in 1955, and the National Federation of Societies for Clinical Social Work (NFSCSW), founded in 1971.

Clinical social workers have gained recognition as independent mental-health providers only recently. For example, in California, effective in 1977, freedom-of-choice legislation (AB 23–74) required California-based health insurers to reimburse social workers, if referral was by physician (Fields, 1977). AB 11-60, effective in 1984, requires all third-party carriers (whether based in California or not) to reimburse services delivered by clinical social workers in California, without physician referral. However, as of 1987, similar freedom-of-choice legislation had been passed in only a handful of states.

Although clinical social work is just being recognized as an independent profession, peer review activities have been performed almost entirely by psychiatrists. In 1982, CHAMPUS contracted with the National Association of Clinical Social Work to develop a program to train social workers in peer review. Consistent with the history of clinical social workers, their peer review system uses psychiatric diagnostic categories.

If current trends continue, clinical social work is likely to become increasingly recognized as an independent health-care profession, with a significant voice in the review of care delivered by social workers.

Nonpsychiatrist Physicians

For centuries, general physicians have probably been attempting to treat some mental disorders and psychosocial aspects of physical disorders. However, few studies focused on this phenomenon until about the mid-1970s. Epidemiological work has suggested that such providers deliver mental health services to a substantial proportion of the population (Goldberg, Kay, & Thompson, 1976; Hankin & Oktay, 1979; Hoeper, Nycz, Cleary, *et al.*, 1979; Liptzin, Regier, & Goldberg, 1980; Shepherd, Cooper, Brown, & Kalton, 1966). For example, Regier *et al.* (1979a) estimated that 54% of all persons with mental disorders receive all of their health care, including mental health care, from general medical physicians. By comparison, 21% use formally trained mental-health-care professionals.

Despite the relatively high use of general medical physicians for mental health services, little is known about the quality of these services (Hankin, 1980). General medical providers may underreport mental disorders because they want to protect patient confidentiality or because they believe insurance reimbursement will be more generous for physical disorders. Patients may have both physical and mental disorders, and care for both types of disorder may occur during the same visit. As a result, it is difficult to completely distinguish mental health care from other care delivered by these providers. Finally, because of negative societal attitudes, patients may be unwilling to provide information on the mental health care they have received.

Despite the importance of nonpsychiatrist physicians as providers of mental health services, many insurance companies do not reimburse such providers for mental health services. Further, as standards for peer review have developed through the JCAH and PSROs, separate standards have not been developed for care from these providers (Berg & Kelly, 1980).

From a cost perspective, it seems understandable that early quality-assurance efforts focused on mental health specialists. Wells, Manning, Duan, Ware, and Newhouse (1982) reported that, although one half of those who receive mental health services visit general medical providers only, these providers account for only 5% of total ambulatory mental health care expense. That is, concerns about containing the cost of mental health care apply almost exclusively to care delivered by mental health specialists. By contrast, concerns about the *quality* of care apply to both provider groups.

Currently, the importance of general medical physicians in the delivery of mental health care is becoming more widely appreciated. Quality assurance activities that reflect this awareness include continuing medical-education courses in interviewing skills and in the managment of specific mental disorders, the use of questions about mental disorders in licensing exams, and required undergraduate courses in human behavior and psychiatry.

The future is likely to bring a sharper delineation of the role of nonpsychiatrist physicians in providing mental health care. Necessary steps in achieving this goal include the development of typologies of disorders and services appropriate for general medical settings. In addition, deficiencies in care need to be identified, and strategies for improving them need to be formulated and tested.

CONFIDENTIALITY

Protecting confidentiality between the provider and the patient or client is central to the ethical codes of all health professionals (Bent, 1982; Plaut, 1974;

Reynolds, 1976; Sullivan, 1976). It is an even greater concern in mental health care because of the highly personal nature of the information communicated during psychotherapy sessions and the social stigma attached to mental disorders.

Quality assurance programs have always raised concern about confidentiality (Bent, 1982; Plaut, 1974; Schwed, Kuvin, & Baliga, 1979). With the growth of third-party coverage, insurance companies required information on diagnosis and treatment; not surprisingly, the intrusion of a third party into the therapist-patient relationship, as well as the manner in which some third parties handle personal information, has caused considerable controversy. Bent (1982) outlined some of the problems that have occurred: (a) the nonprofessionals who are often used to process claims information are inadequately trained to handle confidential information; (b) quality assurance programs may use storage systems that are inadequate for protecting confidentiality; (c) programs may fail to convey to providers and patients the limitations on access to stored information; and (d) insurance carriers, when establishing administrative policies, may fail to develop standards of confidentiality that are compatible with professional ethics. Further, Sullivan (1977) noted that most insurance carriers and peer review boards provide inadequate information to consumers about the nature of the review process.

Despite concerns about the confidentiality of third-party review, few studies have provided data on the extent or the consequences of the problem. As a result, it is not clear what changes are needed in quality assurance progams. Further, some of the problems in communication between the third party, the patients, and the providers may be due to the providers themselves. For example, in the early days of third party coverage, many providers were reluctant to provide the information requested by third-party carriers, or to serve as reviewers. Insurance carriers consequently tended to develop review systems without feedback from professional organizations.

A noted exception is the peer review system developed by the American Psychological Association for CHAMPUS (described above). This system keeps the detailed personal information needed for peer review purposes separate from the identifying information needed for insurance purposes. The rationale and description of the system is given in Bent (1982). CHAMPUS has developed similar systems with other mental-health professionals to protect confidentiality (Rodriguez, 1983).

The development of similar data-management systems in other quality assurance programs may encourage providers to be more cooperative with third parties. Further, empirical studies of the existing systems would lead to a realistic understanding of the confidentiality of data in quality assurance programs.

INSURANCE FOR MENTAL HEALTH SERVICES

Developing effective peer-review systems has become increasingly important as private insurance carriers have reduced their coverage of these services. Difficulty in implementing effective peer review has been cited as one of the main reasons for reduced coverage. As a context for understanding this issue, we provide a brief review of the history of insurance coverage for mental health services, based in part on previous reviews (Cooper, 1979; McGuire & Weisbrod, 1981).

Major insurance coverage for ambulatory mental-health services began in the 1950s. Coverage was initially generous (e.g., unrestricted visits and low coinsurance rates) and comparable to coverage for medical services. Because intensive treatment (e.g., psychoanalysis) was a common mode of treatment, insurance carriers faced high costs for relatively few patients (Reed, Myers, & Scheidemandel, 1972). This pattern of expenditures for such services was not entirely anticipated by insurance carriers. Most carriers responded to this initial experience by providing coverage that was less generous for ambulatory mental-health care than for medical services (Health Insurance Institute, 1979; Reed, 1975).

Recent events in the federal employee plans have kept this issue prominent. In the past, Aetna and Blue Cross and Blue Shield federal employee plans offered high-option policies with generous coverage of mental health services, but concern about the costs of such coverage has led to less generous insurance coverage. Although expenditures for all health services covered by the Blue Cross and Blue Shield plans roughly tripled between 1966 and 1975, expenditures for mental health services increased by over six times. Aetna limited its coverage of mental health services to 20 visits per enrollee per year in the federal employee plan in the mid-1970s; in 1981, the Blue Cross and Blue Shield plans increased coinsurance and restricted mental health benefits. Costs were not the only concern, however. Other factors contributed to this curtailment of benefits, including difficulties in instituting peer review.

Because of decreasing benefits, the mental health professions have been under continuing pressure to develop effective peer-review systems. The nature of mental disorders and of most psychotherapy procedures, however, has made the establishment of standards for care of mental health disorders more difficult than for that of physical disorders. These difficulties are discussed in the next section.

RESEARCH IN QUALITY ASSURANCE

The growth of quality assurance activities described in previous sections has been accompanied by a growth in research on the quality of care. This section provides a brief synopsis of the research on quality assurance for mental health services, identifies components of quality assurance programs that are generic to evaluations of the quality of health care, and reviews future needs for quality-assurance program development and research.

THE LITERATURE ON QUALITY ASSURANCE

Several literature reviews and bibliographies attest to the large number of studies that have emerged in quality assurance for medical care in general (Demlo, 1983; Donabedian, 1969, 1978, 1980; Kahn & Berger, 1980; Maguire, 1978; Palmer & Nesson, 1982; Racusin & Krell, 1980; Towery, Seidenberg, & Santero, 1979; Williams & Brook, 1978; Williamson, 1977) and for mental health services in particular (Brook, Kamberg, & Lohr, 1982; Cohen & Stricker, 1983). A complete review of this literature is beyond the scope of this chapter. Instead we distinguish among specific types of quality assurance studies, and we highlight literature on the effectiveness of quality assurance.

Many studies have used the classic formulation of Donabedian (1969, 1980), which separates assessments of the quality of care into studies of structure, process, or outcomes. *Structure* refers to physical facilities, the organization and financing of care, and personnel. *Process* refers to the transactions between health care personnel and patients during episodes of care. The term *outcomes* refers to the costs of care and to patient health status and attitudes as affected by care. Demlo (1983) used Donabedian's distinctions to describe the assessment techniques that have been developed. Palmer and Nesson (1982) reviewed techniques that are applicable to ambulatory care.

The bulk of the literature on quality of care deals with the assessment of care, not the assurance of standards of care. For most policy purposes, it is important to demonstrate that quality assurance programs actually improve the quality of care (Ciarlo, 1982). Although most studies of specific programs report that they consider their quality assurance program valuable, relatively few link quality assurance activities to changes in process or outcome (Brown & Uhl, 1970; Fleisher, Brown, Zeleznik, Escovitz, & Omdal, 1976). The relative lack of studies in this area and the results obtained have been considered disappointing (Anderson & Shields, 1982; Sinclair & Frankel, 1982).

The literature on quality assurance programs for mental health services largely describes specific peer-review programs associated with specific hospitals (e.g., Edelstein, 1976; Longabaugh, Fowler, Hostetler, McMahon, & Sullivan, 1978; Morrison, 1977; Richman & Pinsker, 1974), state or national professional organizations (Allen, 1979; Miller, Black, Entel, & Ogram, 1974; Stricker, 1980; Wilson, 1982), or community mental health centers (Diamond, Tislow, Snyder, & Rickels, 1976; Goldblatt, Bauer, Garrison, Henisz, & Malcolm-Laves, 1973; Henisz, Goldblatt, Flynn, & Garrison, 1974; Luft, Sampson, & Newman, 1976; Tischler & Riedel, 1973). Only a few studies have compared the cost or yield of different quality-assurance programs. There are few empirical data to guide mental-health-program developers in their selection of a program (Henisz, *et al.*, 1974; Seidenberg & Johnson, 1979). We found only one study that evaluated the effectiveness of quality assurance activities using a control group and a prospective design (Sinclair & Frankel, 1982). These authors evaluated the process of care of community mental-health-treatment teams that did or did not have a quality assurance program. The program consisted of a peer review system and education of clinical supervisors and case workers in delivering high-quality care. One year after implementing the program these authors found significant improvements in the process of care for the experimental groups, but not the control groups. They also determined the cost per reviewed case of the quality assurance program. However, because of their design, it is impossible to separate the effects of the education program and the peer review.

This study is consistent with other studies that report positive effects of quality assurance programs in community mental health centers. These programs combine peer review, clinical supervision, and continuing education in a supportive professional environment.

Other mental health quality assurance studies have focused on the appropriateness of prescribing practices for psychotropic medications. Several such studies have used hospital survey data to evaluate psychiatrists' use of psychotropic drugs in inpatient settings. These studies have concluded that psychiatrists may make excessive use of antipsychotics (i.e., may use too many drugs for the same patient), and may use antiparkinsonian drugs unnecessarily (Altman, Evansen, Sletten, et al., 1972; Eastaugh, 1980; Hartmann, Allison, & Hartig, 1979; Michel & Kolakowska, 1981; Schroeder, Caffey, & Lorei, 1977; Winstead, Blackwell, Eilens, & Anderson, 1976). Prien, Balter, and Caffey (1978), however, discussed the need for caution in inferring inappropriate use of psychotropic medication from hospital survey data. In many of these studies, data were not available to show fully the appropriateness of the medications prescribed for a particular patient. In fact, some of the studies that emphasized the inappropriateness of psychiatrists' prescribing habits also provided data suggesting that psychiatrists may use these drugs appropriately.

Michel and Kolakowska (1981) offered an example of this tendency. They examined the care delivered to 511 inpatients in two British psychiatric hospitals and concluded that psychiatrists tended to prescribe too many psychoactive drugs for each patient and tended to prescribe antidepressants for some patients who did not have a diagnosis of affective disorder. Nevertheless, every patient with a diagnosis of depression was prescribed lithium, an antidepressant, or a neuroleptic. Further, when antidepressants were prescribed, drugs of known efficacy were most often selected (i.e., amitriptyline or imipramine). Only one third of the patients who were prescribed antidepressants received daily doses lower than the equivalent of 100 mg of imipramine. Those with lower dosages either were elderly or had a diagnosis other than an affective disorder. Lower dosages may have been appropriate for these patients.

Studies have also examined the practices of nonpsychiatrist physicians in prescribing psychotropic drugs (Gullick & King, 1979; Hasday & Karch, 1981; Johnson, 1973, 1974). These studies suggest that nonpsychiatrist physicians tend to rely on sedative-hypnotic agents for managing anxiety and depression; antidepressants are often prescribed in subclinical dosages. But because these studies provide few data on the case mix of patients, it is difficult to draw firm conclusions about the appropriateness of the care.

CURRENT RESEARCH NEEDS

Brook et al, (1982) described seven components of quality assurance programs that are generic to quality assurance programs. Below we discuss each component in light of our historical review of quality assurance for mental health services.

Problem Identification

The first step in developing a quality assurance program is the selection of a diagnosis or a patient-management problem. Theoretically, the problems selected for study should be common and of clinical importance, and there should be the potential for improvement in the process and outcomes of care. For mental health, the selection of appropriate clinical problems is further complicated by the different conceptual models of the major provider groups. Whereas psychiatry has concentrated on different diagnostic groups, psychology has focused on specific clinical problems as defined by the client and the psychologist. Social work uses a model based on the interaction of diagnostic problems and environmental stresses. For general medical providers, a conceptual framework has yet to be elucidated.

This diversity means that the professional groups face different issues in identifying clinical problems for study. Psychiatric quality-assurance programs should probably initially concentrate on the more common clinical diagnosis, such as the depressive disorder, the phobias, and alcoholism. For each disorder, meaningful differences in the care that are related to health outcomes should be identified. Appropriate use of antidepressant medication is a prime candidate for studies of the quality of care for depression. Indeed, some guidelines for peer review of prescriptions for antidepressants have been developed (Dorsey, Ayd, Cale, Klein, Simpson, Turin, & DiMascio, 1979). One limitation, however, of the quality assurance programs that focus on specific diagnostic groups is that treatment modalities, notably psychotherapy, are used to manage patient problems that occur *across* diagnostic groups. To determine whether psychotherapy is indicated for a particular patient, researchers or reviewers would therefore need information on such patient problems, as well as on diagnostic categories.

The peer program developed by psychologists for CHAMPUS appears to be especially useful for evaluating the appropriateness of psychotherapy in relation to the needs of the particular patient. As we discussed above, this approach is based on the identification of the salient clinical problems by the client and the psychologist together. Treatment response is monitored by improvement in these problem areas. But such an approach also has at least two limitations: it is not clear whether information on quality could be aggregated *across* patients or providers and although psychotherapy is designed to address the particular needs manifested by each patient, few data exist that link specific problems to more general outcomes of care, such as reduced psychological distress. Thus, an important research need for psychologists is to obtain data on the prevalence of specific patient problems and the relation of specific treatment objectives to outcomes.

Although clinical social workers tend to use the diagnostic classifications developed by psychiatrists, they tend to place greater emphasis on family and environmental influences than do psychiatrists. Quality assurance programs in mental health have not focused on these problems; future studies of the quality of care delivered by these providers should.

Like clinical social workers, nonpsychiatrist physicians are also trained in the diagnostic scheme used by the psychiatrists. However, the context of mental health care is different for nonpsychiatrist physicians and mental health specialists. The former provide mental health care as part of an office visit that includes physical health evaluations and procedures (Verbrugge, 1984). The latter usually provide only mental health care. Nonpsychiatrist physicians may be less oriented than psychiatrists toward identifying specific diagnostic classifications, but they may be more aware of emotional problems or life stresses that complicate the management of physical disorders (Baker & Tversky, 1977; Regier, Kessler, Burns, & Goldberg, 1979b).

Thus, direct applications of quality asssurance programs developed for psychiatrists are unlikely to be useful for nonpsychiatrist physicians. Rather, such programs should take into account the particular objectives of care and the practice constraints of nonphychiatrist physicians.

Although, each mental-health-provider group is identified with a unique theoretical perspective, individual providers probably use several of these paradigms in working with patients. Each provider type may assign a diagnosis, isolate salient problems or symptoms, and evaluate the patient's family. Further, providers from different professional groups may work together on the same cases. Consequently, we believe that the quality assurance programs for each provider group would benefit by incorporating some features of the quality assurance approach used by other provider groups. Psychiatric peer-review systems, for instance, should include major diagnostic categories and salient symptoms or functional limitations.

Targeting the Assessment

After a clinical problem is identified, quality assurance programs are targeted for a particular patient group, depending on the pattern of deficiencies in care for a particular clinical problem. If most patients receive poor or questionable care for a given condition, then the quality assurance program should be directed to the *average* patient. If only a small proportion of patients are thought to receive deficient care, the program is targeted for a specific outlier group. The literature contains examples of both types of studies of mental health services. Large peer-review efforts, such as programs of state professional societies (e.g., Wilson, 1982), are usually designed to identify outlier cases. In the relatively intimate setting of community mental-health centers, quality assurance programs are designed to enhance the general level of care (e.g., Luft, Sampson, & Newman, 1976).

One problem that program developers face in selecting a target population is the current lack of knowledge of the distribution of deficiencies in care. Several factors have made this knowledge difficult to obtain. First, there is a wide variety of psychotherapies. Studies of one type may not apply to another. Second, some providers of mental health services are not members of a licensed professional group. There are a few data collected on these providers. Third, some forms of treatment are either controversial or controlled by legislation (e.g., shock therapy). Because of social stigma and legal restrictions, these treatments may be difficult to study. Fourth, mental health specialists have been somewhat "on the defensive" for decades, especially since the efficacy of psychotherapy in general has been seriously questioned. In such an environment, it is not surprising to find little research by mental health specialties in "deficiencies" in mental health care. With the publication of several major reviews supporting the efficiency of psychotherapy, we may find more data forthcoming on deficiencies (American Psychiatric Advisory Council on Psychotherapies, 1980; Garfield & Bergin, 1978). This should be a high research priority.

For patients who receive mental health care from general medical physicians, the issue of targeting quality assurance studies is somewhat different. Evidence is accumulating that suggests that, although nonpsychiatrist physicians see many patients with mental health problems, they provide such patients with very little care— at least in terms of care as defined by mental health specialists (Hankin, 1980; Liptzin *et al.*, 1980; Wells *et al.*, 1982). Thus, studies or programs should be directed to the *average* patient receiving mental health care from general medical physicians.

Measuring Process and Outcome

Mental-health-care studies have a rich tradition of focusing on patient outcomes (Linn & Linn, 1975). A wide range of measures have been used to evaluate personal affects, psychological well-being and distress, physical health, patient satisfaction, social health, occupational and role functioning, family relationships, and so on (Garfield & Bergin, 1978). Although, it can be said that there is little consensus on the most adequate measure for a given construct, the breadth of outcomes and measures available is quite impressive. The nursing and social work professions have an especially strong history of using outcomes as indicators of the quality of care (Lang & Clinton, 1983).

A key problem in program development and research is the selection of outcomes that are appropriate to the clinical problems identified. Several self-report and observer instruments have been developed and validated as indicators of general psychopathology. Some of these, such as the Hopkins Symptoms Check List (HSCL), were specifically designed to measure change in response to psychotherapy (Kelman & Parloff, 1957). Yet, relatively few studies have used the HSCL to evaluate the outcome of treatment for patients with a particular diagnosis or clinical management problem. For some specific diagnoses, a variety of outcome measures have been widely used and validated. The Hamilton Depression Rating Scale (Hamilton, 1967) and the Raskin Clinical Rating Scale (Raskin, 1972) are observer assessments of depression that have been widely used as outcome measures in controlled trials of antidepressants. In selecting outcome measures for particular target diagnoses or clinical problems, program developers should first enumerate the specific outcomes that are relevant to the target condition. They should then carefully evaluate the available outcome measures for reliability and validity (discussed below) and, finally, match those measures to specific clinical outcomes.

Quality assurance studies also require measures of the process of care, which can refer broad patterns of care, such as the selection of a treatment modality, or to specific behaviors for a particular visit, such as interpretations or emotional support. Quality assurance programs should focus on components of process that are related to treatment outcome and that providers or patients can alter.

For many mental health diagnoses, quality assurance programs can productively focus on more general aspects of process, such as the selection of a treatment modality and the appropriateness of its administration. The efficacy of tricyclic antidepressants, electroconvulsive therapy, and monoamine oxidase inhibitors for the treatment of depressive disorders has been clearly established (Paykel, 1982). In addition, several studies suggest that specific short-term psychotherapies are effective in treating these disorders (Klerman & Schecter, 1982), and that antidepressants and psychotherapy have different effects on depressive symptoms (DiMascio, Weissman, Prusoff, *et al.*, 1979). Quality assurance studies for depression can focus on whether an effective treatment is administered.

There is a large literature on the process of psychotherapy at a more detailed level (Garfield & Bergin, 1978). Studies have examined the efficacy of specific psychotherapeutic activities (e.g., interpretations) and the affective nature of the doctor–patient relationship. However, measures of process at this level of detail are probably not useful for large-scale quality-assurance programs. Further, the link to outcome of such components of care is more uncertain. Few studies have evaluated the effect of specific psychotherapeutic activities on outcome while controlling for the pretreatment characteristics of patients, such as mental health status, that could also affect outcome (Rounsaville, Weissman, & Prusoff, 1981).

Future studies should describe the distribution of components of the process of care, as evidenced in simple, concrete behaviors that are clinically meaningful. Likely candidates, such as the appropriate use of psychotropic drugs or the frequency of sessions, should be related to carefully selected outcomes.

Reliable and Valid Data

Studies of quality assurance should use reliable and valid measures of the target condition, and of the process and outcome of care. Medical records and insurance claims, however, are notoriously unreliable sources of data on mental health diagnosis and care (Baker, 1983), partly bacause providers underreport mental health care to protect patient confidentiality. Further, the diversity of the available conceptual models of mental disorders has led to a variety of uses of the same diagnostic labels (e.g., depression). Consequently, the reliability of diagnostic information may be low even when it is present on the record. The third edition of *The Diagnostic and Statistical Manual* (Committee on Nomenclature and Statistics, 1980) was developed, in part, to standardize psychiatric diagnoses, but there have been few studies of its reliability and validity.

Most of the more widely used mental health outcome-measures of acceptable reliability have a level of reliability that is appropriate for group comparisons, but not for individual comparisons. The internal consistency of the General Health Questionnaire, for instance, is between .80 and .90 (Vieweg & Hedlund, 1982). Conclusions about the outcomes of care using this measure are more appropriate for quality assurance studies of patient groups (e.g., profile analysis) than for studies of the care of individual patients (e.g., peer review). Nonetheless, a few mental-health outcome-measures, such as the HSCL, may be appropriate for peer review purposes. Its total score has a reliability (.95) that may allow individual comparisons (Murphy, 1980).

Establishing Criteria for Quality Care

The hallmark of evaluations of the quality of care is the establishment of standards for care. Such standards may be developed by experts or may be derived from empirical studies of the relation of process to outcome. Disagreement over the definition of mental health problems and treatments and concerns about confidentiality have made it extremely difficult to establish standards of care for mental-health-care services. For some concrete aspects of care, standards have been proposed or are currently being proposed, such as the guidelines of the American Psychiatric Association for claims review of psychotropic drug prescriptions (Dorsey et al., 1979). We adapted these criteria to examine prescribing practices of general internists (Wells, Goldberg, Brook, & Leake, 1986).

In the absence of clearly defined standards for high-quality care, quality assurance programs have tended to emphasize either the identification of "outlier" cases or the assessment of the structural components of care (e.g., provider licensing). The lack of specific standards has contributed to pessimism about the insurability of mental health care (Simon, 1976). It should be a high priority of all professional groups to develop and empirically test standards for specific conditions (e.g., diagnostic groups or client problems).

Validating the Criteria

Standards of quality care are useless unless they are valid, that is, unless they reflect true differences in care that lead to different outcomes. The validation of criteria for mental health care may be especially complex for two reasons. First, mental health problems are believed to be influenced by a host of factors, (e.g., biological disposition, environment, and intrapsychic conflict). To demonstrate that differences in care *cause* different outcomes, it is necessary to control for these factors. Such control is accomplished in experimental studies largely through random assignment to treatment groups. In nonexperimental studies, the relevant factors must be identified and included as covariates in the analyses. Most quality-assurance studies are naturalistic, that is, they use nonexperimental designs. Because we have an incomplete understanding of the factors that cause mental health problems, we may be unable to identify the relevant control variables.

Second, few measures of the severity of illness for particular mental disorders have been developed. However, to link patient improvement to the process of mental health care, measures of the severity of specific mental-health problems are needed. The available measures of severity largely measure variation in symptoms or functioning across specific diagnostic categories. The Beck Depression Inventory (BDI; Beck, Ward, Mendison, Mock, & Erbaugh, 1961), for example, is a measure of the severity of depressive symptoms. It has been validated against clinical assessments of the global severity of depression and observer measures of the severity of the depressive symptoms. The validity of the BDI as a measure of the severity or specific depressive disorders (e.g., major depressive disorder or dysthymic disorder) is uncertain, however.

Assuring Quality

Quality assurance programs should be oriented toward improving deficiencies in care through the feedback or the education of providers and/or patients (clients).

Sinclair and Fankel (1982) demonstrated some improvement in the process of mental health treatment as a result of a peer review system. They provided feedback to the providers through educational seminars and clinical supervision.

In an excellent review of the effectiveness of continuing medical education, Sanazaro (1983) noted that continuing education has been demonstrated to improve knowledge of appropriate care (competency), but that this knowledge is not necessarily reflected in clinical practice. He identified programmatic factors associated with real change in clinical behavior. Among these are the presence of recognized experts and opportunities to practice recommended changes under individual observation. These are the essential components of clinical supervision in the mental health professions. Clinical supervision may represent an untapped source of quality assurance for the mental health professions. Once deficiencies in care are detected, insurance companies could require clinical supervision at the therapist's expense as an alternative to stopping reimbursement of services.

THE FUTURE OF QUALITY ASSURANCE

The future of quality assurance, as we know it, is uncertain. The programs implemented in response to PSRO legislation are still in place, but their motivation (the PSRO legislation) has changed. Current proposals to contain costs emphasize reorganizing medical practice in such a way as to provide incentives to restrict the use of some health services. Examples of such systems are health maintenance organization, preferred provider organizations, and diagnosis-related groups. By curtailing use, these systems may enhance the need for programs to evaluate or ensure the quality of care.

Developers of these programs may profit from a review of the problems that have attended past quality-assurance efforts. Mandated programs required providers to organize and develop quality assurance programs, but these programs were instituted on the basis of an incomplete understanding of the relation among process and outcome variables. Future program developers and researchers should carefully test the assumptions on which past programs were based and should incorporate new research findings into their programs. For example, standardized definitions of psychiatric disorders are now available through the third edition of *The Diagnostic and Statistical Manual* (Committee on Statistics and Nomenclature, 1980). For some diagnoses, such as major depression, validated measures of severity and outcome exist. Further, standards for the use of psychotropic medications are curently being developed. These measures and standards should be used consistently in quality assurance studies of care for specific depressive disorders.

Interspecialty rivalry, especially between psychiatrists and psychologists, has also contributed to the development of separate and largely polarized quality-assurance systems (i.e., diagnosis versus patient problems). Although the variety of systems may be seen as reflecting a conceptual richness, the degree of polarization makes it less likely that provider groups will borrow from one another to enhance *both* types of quality assurance systems. Creative cooperation among mental health professionals is especially important because some patients use more than one type of mental health care provider.

Finally, many mental health professionals have been reluctant to provide information to fiscal third parties, in part because of concern about confidentiality. Providers need to cooperate to some extent, however, if mental health services are to be insured. To facilitate their cooperation, the adequacy of current data-management systems to protect confidentiality should be evaluated.

To implement an effective quality-assurance program for mental health services, we need an interdisciplinary effort to obtain data on the epidemiology of the quality of care. First, we need studies to determine the prevalence of the functional impact of specific patient-care problems or psychiatric diagnoses. Second, we need to describe the distribution of care for these problems in various health-care settings. Third, we should link the components of the process of care to salient outcomes. And fourth, we should undertake studies to evaluate whether education of providers and patients improves the process or the outcome of care. While this information is being collected, we need quality assurance systems that identify the "outliers," that is the patients who receive markedly deficient care and the providers who are delivering that care.

Quality assurance studies should focus initially on specific clinical problems where there is a rich history of research and where the efficacy of specific treatments is established. In these studies, classification schemes and measures of severity and outcome of known validity should be used. Such studies would enhance the credibility of mental health quality assurance efforts and would provide a model for future research.

REFERENCES

Advisory Panel on Psychotherapy. (1980). *The implications of cost effectiveness of medical technology: Backround paper 3—The efficacy and cost effectiveness of psychotherapy.* Washington, DC: Office of Technology Assessment.

Affeldt, J.E., Roberts, J.S., & Walczak, R.M. (1983). Quality assurance—Its origin, status, and future direction—A JCAH perspective. *Evaluation and Health Professions, 7,* 45–255.

Albee, G. W., & Kessler, M. (1977). Evaluating individual deliverers: Private practice and professional standards review organizations. *Professional Psychology, 8,* 502–515.

Allen, M.G. (1979). Peer review of group therapy: Washington, D.C., 1972–1977. *American Journal of Psychiatry, 136,* 444–447.

Altman, H., Evansen, R.C., Sletten, A., *et al.* (1972). Patterns of psychotropic drug prescribing in four midwestern state hospitals. *Current Therapeutic Research, 14,* 667–672.

American Psychiatric Association Commission on Psychotherapies (1982). *Psychiatric research—Methodological and efficacy issues, Washington, DC:* American Psychiatric Association.

American Psychological Association. (1986, January). *A summary of state laws regulating psychological practice through licensure or certification.* Washington, DC: Author.

Anderson, O.W., & Shields, M.C. (1982) Quality measurement and control in physician decision making: State of the art. *Health Services Research, 17,* 125–155.

Baker, F. (1983) Data sources for the health care quality evaluation. *Evaluation and the Health Profession, 6,* 263–281.

Baker, R., & Tversky, R.K. (1977). Classification and coding of psychosocial problems in family medicine, *Journal of Family Practice, 4,* 88–89.

Beck, A.T., Ward, C.H., Mendelson, M., Moek, J., & Erbaugh, J. (1961). An inventory for measuring depression. *Archives of General Psychiatry, 4*, 561–571.

Bent, R.J. (1982) Multidimensional model for control of private information. *Professional Psychology, 13*, 27–33.

Berg, J.K., & Kelly, J.T. (1980). Psychosocial health care and quality assurance activities. *Journal of Family Practice, 11*, 641–643.

Brook, R.H., Kamberg, C.J., & Lohr, K.N. (1982). Quality assessment in mental health. *Professional Psychology, 13*, 34–39.

Brown, C., Jr., & Uhl, H. (1970). Mandatory continuing education, sense or nonsense? *Journal of American Medical Association, 213*, 1660–1668.

Chodoff, P. (1978). Psychiatry and the fiscal third party. *American Journal of Psychiatry, 135*, 114–1147.

Ciarlo, J.A. (1982). Accountability revisited—The arrival of client outcome evaluation. *Evaluation and Program Planning, 5*, 31–36.

Claiborn, W.L., & Stricker, G. (1979). Professional standards review organizations, peer review, and CHAMPUS. *Professional Psychology, 10*, 631–639.

Clayton, T. (1976). Peer review: A progress report. *Hospital and Community Psychiatry, 27*, 660–663.

Cohen, L.H., & Pizzirusso, D. (1982). Peer review of psychodynamic psychotherapy: Experimental studies of the American Psychological Association/CHAMPUS program. *Evaluation and the Health Professions, 5*, 415–439.

Cohen, L.H., & Stricker, G. (1983). Mental health quality assurance: Development of the American Psychological Association/CHAMPUS program. *Evaluation and the Health Professions, 6*, 327–338.

Committee on Nomenclature and Statistics. (1980). *Diagnostic and statistical manual of mental disorders*, (3rd ed.) Washington, DC: American Psychiatric Association.

Cooper, M.L. (1979). *Private health insurance benefits for alchoholism, drug abuse, and mental illness.* Washington, DC: Intergovernmental Health Policy Project, George Washington University.

Crosby, K.G. (1982). Accreditation and associated quality assurance efforts. *Professional Psychology, 13*, 132–140.

Demlo, L.K. (1983). Assessing quality of health care—An overview. *Evaluation and the Health Professions, 6*, 161–196.

Diamond, H., Tislow, R., Snyder, T., Jr., & Rickels, K. (1976). Peer review of prescribing patterns in a CMHC. *American Journal Psychiatry, 133*, 697–699.

DiMascio, A., Weissman, M.M., Prusoff, B.A., *et al.* (1979). Differential symptom reduction by drugs and psychotherapy in acute depression. *Archives of General Psychiatry, 36*, 1450–1456.

Donabedian, A. (1969) *A guide to medical care administration, medical care appraisal.* New York: American Public Health Association.

Donabedian, A. (1978). *Needed research in the assessment and monitoring of the quality of medical care,* (Research Report Series). Hyattsville, MD: U.S. Department of Health, Education, and Welfare, National Center for Health Services Research.

Donabedian, A. (1980). *Explorations in quality assessment and monitoring: The definition of quality and approaches to its assessment.* Ann Arbor, MI: Health Administration Press.

Donaldson, M.J., & Keith, K.J. (1983). Planning for program effectiveness in quality assurance. *Evaluation and the Health Professions, 6*, 233–244.

Dorken, H., & Morrison, D. (1976). JCAH standards for accreditation of psychiatric facilities—Implications for the practice of clinical psychology. *American Psychologist, 31*, 774–784.

Dorsey, R. (1974). PSROs: Salvation or suicide for psychiatry? *Psychiatric Opinion, 11*, 6–12.

Dorsey, R., Ayd, F.J., Cale, J., Klein, D., Simpson, G., Turin, J., & DiMascio, A. (1979). Psychopharmacological screening criteria development project. *Journal of American Medical Association, 241*, 1021–1031.

Eastaugh, S.R. (1980). Limitations on quality assurance effectiveness—Improving psychiatric inpatient drug prescribing habits of physicians. *Journal of Medical Systems, 4*, 27–43.

Edelstein, M.G. (1976). Psychiatric peer review: A working model. *Hospital and Community Psychiatry, 27*, 656–659.

Ellenberger, H.F. (1974). Psychiatry from ancient to modern times. In A.Arieti (Ed.), *American handbook of psychiatry* (2nd Ed.): Vol. 1. The foundation of psychiatry. New York: Basic Books.

Fleisher, D.S., Brown, C.R., Jr., Zeleznik, C., Escovitz, G.H., & Omdal, C. (1976). The mandate project: In-
stitutionalizing a system of patient care quality assurance. *Pediatrics, 57*, 775–782.

Flexner, A. (1910). *Medical education in the United States and Canada.* New York: Carnegie Foundation.

Fowler, D.R. (1977). Current practice in psychiatric utilization review. *International Journal of Mental
Health, 5*, 49–57.

Garfield, S.L., & Bergin, A.E. (Eds.). (1978). *Handbook of psychotherapy and behavior change: An empirical
analysis* (2nd ed.). New York: Wiley.

Getzen, T.E. (1983). Market and evaluation roles in quality assurance. *Evaluation and the Health Professions, 6*, 299–310.

Goldberg, D., Kay, C., & Thompson, L. (1976). Psychiatric morbidity in general practice and the community. *Psychological Medicine, 6*, 565–567.

Goldblatt, P.B., Bauer, L.P., Garrison, V., Heinsz, J.E., & Malcolm-Laves, M. (1973). A chart-review
checklist for utilization review in a community mental health center. *Hospital and Community Psychiatry, 24*, 753–756.

Greenblatt, M. (1975). PSRO and peer review: Problems and opportunities. *Hospital and Community Psychiatry, 26*, 354–358.

Gullick, E.L., & King, L.J. (1979). Appropriateness of drugs prescribed by primary care physicians for
depressed outpatients. *Journal of Affective Disorders, 1*, 55–58.

Hamilton, M. (1967). Development of a rating scale for primary depressive illness. *British Journal of Social
and Clinical Psychology, 6*, 278–296.

Hankin, J.R. (1980). Management of emotionally disturbed patients in primary care settings: A review of
the north american literature. In *Mental health services in primary care settings* (Mental Health Services
Reports, U.S. Department of HHS, Series DN No. 2., DHHS Publication No. ADM 80-995).
Washington, DC: U.S. Government Printing Office.

Hankin, J., & Oktay, J.S. (1979). *Mental disorder and primary medical care and analytical review of the literature.*
(National Institute of Mental Health, Series D, No. 5., DHEW Publication No. ADM 78-661). Washington, DC: U.S. Governmental Printing Office.

Hartmann, K., Allison, J., & Hartig, P. (1979). Measuring prescribing practices in a state hospital. *Hospital
and Community Psychiatry, 30*, 467–469.

Hasday, J.D., & Karch, F.E. (1981). Benzodiazepine prescibing practices in a family medicine center. *Journal of the American Medical Association, 246*, 1321–1325.

Health care financing administration professional standards review organizatiohs—1978 program evaluation.
(1979). Washington, DC: Office of Policy Planning and Research, Department of Health, Education,
and Welfare.

Health Insurance Institute. (1979). *Source book of health insurance data, 1978–1979.* Washington, DC:
Author.

Henisz, J., Goldblatt, P.B., Flynn, H.R., & Garrison, V. (1974). A comparison of three approaches to
patient care appraisal based on chart review. *American Journal of Psychiatry, 131*, 1142–1144.

Hoeper, E.W., Nycz, G.R., Cleary, P.D., *et al.* (1979). Estimated prevalence of RDC mental disorder in
primary medical care. *International Journal of Mental Health, 8*, 6–15.

Johnson, D.A.W. (1973). Treatment of depression in general practice. *British Medical Journal, 2*, 18–
20.

Johnson, D.A.W. (1974). A study of the use of antidepressant medication in general practice. *British
Journal of Psychiatry, 125*, 186–192.

Kahn, T.J., & Berger, S.S. (1980). *Quality assurance for alcohol, drug abuse, and mental health services: An annotated bibligraphy of recent literature.* Washington, DC: Department of Health, Education, and
Welfare.

Kelman, H.C., & Parloff, M.B. (1957). Interrelations among three criteria of improvement in group
therapy: Comfort, effectiveness and self-awareness. *Journal of Abnormal and Social Psychology, 54*, 281–
288.

Klerman, G.L., & Schecter, G. (1982). Drugs and psychotherapy. In E.S. Paykel (Ed.), *Handbook of affective
disorders.* New York: Guilford Press.

Lang, N.M., & Clinton, J.F. (1983). Assessment and assurance of the quality of nursing care—A selected
overview. *Evaluation and the Health Professions, 6*, 211–231.

Langsley, D.G. (1980). Quality assurance in psychiatric treatment. *National Association of Private Psychiatric Hospitals Journal, 11,* 13–17.

Linn, M.W., & Linn, B.S. (1975). Narrowing the gap between medical and mental health evaluation. *Medical Care, 13,* 607–614.

Liptzin, B., Regier, D.A., & Goldberg, I.D. (1980). Utilization of health and mental health services in a large insured population. *American Journal of Psychiatry, 137,* 553–558.

Longabaugh, R., Fowler, D.R., Hostetler, M., McMahon, C., & Sullivan, C. (1978). Use of the problem-oriented record in a proposed evaluation study of social isolation. *Quality Review Bulletin, 4,* 4–7.

Luft, L.L., Sampson, L.M., & Newman, D. (1976). Effects of peer review on outpatient psychotherapy: Therapist and patient follow-up survey. *American Journal of Psychiatry, 133,* 891–895.

Maguire, L. (1978). Peer review in community mental health. *Community Mental Work Journal, 14,* 190–198.

McGuire, T.G., & Weisbrod, B.A. (Eds.). (1981). *Economics and mental health* (National Institute of Mental Health, Series EN No. 1., DHHS Publications No. ADM 81-1114). Washington, DC: U.S. Government Printing Office.

Michel, K., & Kolakowska, T. (1981). A survey of prescribing psychotropic drugs in two psychiatric hospitals. *British Journal of Psychiatry, 138,* 217–221.

Miller, R.R., Black, G.C., Entel, P.Y., & Ogram, G.F. (1974). Psychiatric peer review: The Ohio system. *American Journal of Psychiatry, 131,* 1367–1370.

Morrison, S.D. (1977). Retrospective audit: Depressive neurosis. *American Journal of Psychiatry, 134,* 299–301.

Murphy, J.J. (1980). *Psychiatric instrument development for primary care research: Patient self-report questionnaire* (Final Report for Contract No. 80MOI4280I OID). Rockville, MD: National Institute of Mental Health.

National Institute of Mental Health. (1974). *Continuing education in mental health: Project summaries* (DHEW Publication No. HSM 73-9126).

Palmer, R.H., & Nelson, H.R. (1982). A review of methods for ambulatory medical care evaluations. *Medical Care, 20,* 758–781.

Paykel, E.S. (Ed). *Handbook of affective disorders.* New York: Guilford Press.

Phillips, D.G. (1975). The swing toward clinical practice. *Social Work, 20,* 61–63.

Plaut, E.A. (1974). A perspective on confidentiality. *American Journal of Psychiatry, 131,* 1021–1024.

Prien, R.F., Balter, M.B., & Caffey, E.M. (1978). Hospital surveys of prescribing practices with psychotropic drugs—A critical examination. *Archives of General Psychiatry, 34,* 1271–1275.

Racusin, R., & Krell, H. (1980). Quality assurance in community mental health centers. *Administration in Mental Health, 1,* 292–303.

Raskin, A. (1972). The NIMH collaborative depression studies—A progress report. *Psychopharmacology Bulletin, 8,* 55–59.

Reed, L. (1975). *Coverage and utilization of care for mental health conditions under health insurance, various studies, 1973–1974.* Washington, DC: American Psychiatric Association.

Reed, L., Myers, E., & Scheidemandel, P. (1972). *Health insurance and psychiatric care: Utilization and cost.* Washington, DC: American Psychiatric Association.

Regier, D. A., Goldberg, I. D., & Taube, C. A. (1979a). The de facto U.S. mental health services system. *Archives of General Psychiatry, 35,* 685–693.

Regier, D. A., Kessler, L. G., Burns, B.J. & Goldberg, I. D. (1979b). The need for a psychosocial classification system in primary care settings. *International Journal of Health, 8,* 16–29.

Reynolds, M.M. (1976). Threats to confidentiality. *Social Work, 21,* 108–113.

Rodriguez, A. R. (1983). Psychological and psychiatric peer review at CHAMPUS. *American Psychologist, 38,* 941–947.

Rounsaville, B. J., Weissman, M. M., & Prusoff, B. A. (1981). Psychotherapy with depressed outpatients: Patient and process variables as predicators of outcome. *British Journal of Psychiatry, 138,* 67–74.

Sanazaro, P. J. (1983). Determining physician's performance—Continuing medical education and other interacting variables. *Evaluation and the Health Professions, 6,* 197–210.

Schroeder, N. H., Caffey, E. M., Lorei, T. M. (1977). Antipsychotic drug use: Physician prescribing practices in relation to current recommendations. *Diseases of the Nervous System, 38,* 114–116.

Schwed, H. J., Kuvin, S. F., & Baliga, R. K. (1979). Medicaid audit crisis in confidentiality and the patient-psychiatrist relationship. *American Journal of Psychiatry, 136,* 447–450.

Seidenberg, G., & Johnson, F. S. (1979). A case study in defining developmental costs for quality assurance in mental health center programs in mental health center programs. *Evaluation and Program Planning, 2,* 143–152.

Shepherd, M., Cooper, A. G., Brown, A. C., & Kalton, G. (1966). *Psychiatric illness in general practice.* London: Oxford University Press.

Simon, G. (1976). NIMH funds study of mental health utilization and costs. *APA Monitor, 7*(7), 26, 55.

Sinclair, C., & Frankel, M. (1982). The effect of quality assurance activities on the quality of mental health services. *Quality Review Bulletin, 8*(7), 7–15.

State comparison of laws regulating social work. (1983). Silver Spring, MD: National Association of Social Workers.

Stricker, G. (1980). Peer review of outpatient psychological services. In A. G. Awad, H. B. Durost, & W. O. McCormich (Eds.), *Evaluation of quality care in psychiatry.* New York: Pergamon Press.

Sullivan, F. W. (1976). Peer review and PSRO: An update. *American Journal of Psychiatry, 133,* 51–55.

Sullivan, F. W. (1977a). Peer review and PSRO in American psychiatry. *Psychiatric Quarterly, 49,* 331–337.

Sullivan, F. W. (1977b). Peer review and professional ethics. *American Journal of Psychiatry, 134,* 186–188.

Tischler, G. L., & Riedel, D. C. (1973). A criterion approach to patient care evaluation. *American Journal of Psychiatry, 130,* 913–916.

Towery, O. B., Seidenberg, G. R., & Santero, V. (1979). *Quality assurance for alcohol, drug abuse, and mental health services: An annotated bibliography.* Washington DC: Department of Health, Education, and Welfare.

Verbrugge, L. M. (1984). How physicians treat mentally distressed men and women. *Social Science and Medicine, 18,* 1–9.

Vieweg, B. W., & Hedlung, S. L. (1982). The General Health Questionnaire (GHQ): A comprehensive review. *Journal of Operational Psychiatry, 14,*8–81.

Wells, K. B., Goldberg, G., Brook, R. H., & Leake, B. (1986, November). Quality of care for psychotropic drug use in internal medicine group practices. *Western Journal of Medicine, 145,* 710–714.

Wells, K. B., Manning, W. G., Jr., Duan, N., Ware, J. E., Jr., & Newhouse, J. P. (1982). *Cost sharing and the demand for ambulatory mental health services* (R-2960-HHS). Santa Monica, CA: Rand Corporation.

Williams, K. N., & Brook, R. H. (1978). Quality measurement and assurance. *Health and Medical Care Services Review, 1,* 1–5.

Williamson, G. W. (1977). *A bibliographic guide to information in quality assurance and continuing education.* Cambridge MA: Ballinger.

Wilson, S. (1982). Peer review in California: Summary findings in 40 cases. *Professional Psychology, 13,* 517–521.

Winstead, D. K., Blackwell, B., Eilens, M. II., & Anderson A. (1976). Psychotropic drug use in five city hospitals. *Research of the Nervous System, 37,* 504–509.

Zano, J. S., & Kilberg, R. R. (1982). The role of the APA in the development of quality assurance in psychological practice. *Professional Psychology, 13,* 112–118.

3

Research on Mental Health Quality Assurance

A program of mental health quality assurance is designed to (1) evaluate the quality of mental health services delivered to the public and (2) provide corrective or educational feedback to remedy detected deficiencies in service quality. The first objective involves quality assessment and therefore shares with traditional psychotherapy research and program evaluation the complex conceptual and methodological issues inherent in the measurement of clinical effectiveness. The second objective, remediation, involves strategies for behavior change, on the part of either individual clinicians or a delivery system as a whole.

Most writers in the field distinguish between quality assurance and utilization review, the latter being concerned primarily with cost control and resource distribution. However, it should be understood that, in practice, quality assurance programs are often guided by financial considerations, with reduction in unnecessary and inefficient care seen as servicing both utilization review and quality assurance purposes (Sechrest & Hoffman, 1982).

The purpose of this chapter is to review research on programs of mental health quality assurance. It should be stated at the outset that the research literature on mental-health quality assurance is extremely sparse. For this reason, the ensuing review also describes relevant research on medical quality assurance and includes an emphasis on conceptual and methodological issues. The organization of the review follows the major assumptions and the formal components of existing mental health quality-assurance programs.

EVALUATION OF TREATMENT PROCESS

In a classic paper, Donabedian (1966) categorized quality assurance (in the medical field) into three interdependent types: (1) structural appraisal—an evaluation of the facilities and resources necessary for service provision; (2) process evaluation—a comparison of actual service provision with explicit standards or normative criteria; and (3) outcome quality assessment—an evaluation of the effects of

LAWRENCE H. COHEN • Department of Psychology, University of Delaware, Newark, Delaware 19716.

treatment on actual patient outcome. Although quality assurance efforts by the Joint Commission on Accreditation of Hospitals (JCAH) have emphasized structural appraisal, most mental health (and medical) programs have emphasized the evaluation of process. Therefore, quality assurance has, for the most part, concerned the development and evaluation of criteria for judging the adequacy of clinical *service delivery*.

In addition, most large-scale quality-assurance programs rely on tiers or levels of review: (a) an initial clerical review of form or record completion; (b) a second-level review that compares documented treatment process with explicit process criteria; and (c) a third level of professional peer review, which is usually based on implicit (subjective) process criteria and which is applied to those cases that deviate from the second-level criteria and occasionally to other cases that are chosen randomly or that exceed some specified treatment length.

For example, the quality assurance program developed at the Connecticut Mental Health Center represents a three-tiered system designed to evaluate clinical process (Riedel, Tischler, & Myers, 1974). The Psychiatric Utilization Review and Evaluation (PURE) project included a first-level, clerical review of the completeness of information found in medical charts and a second-level chart review based on a checklist of acceptable services organized by clinical problems and treatment goals. Those cases that deviated from the above criteria, as well as some cases chosen randomly, were channeled to an internal peer-review committee, which relied on implicit criteria to evaluate treatment process.

The program of the American Psychological Association/Civilian Health and Medical Program of the Uniformed Services (APA/CHAMPUS) similarly evaluated treatment process in a three-tiered system (Claiborn, Biskin, & Friedman, 1982). This program is described in Chapter 20, but because research on the program is reviewed in the present chapter, it is necessary to highlight here its major characteristics.

In 1977, the APA entered into a contractual agreement with the Office of CHAMPUS to develop a program of psychological peer review of outpatient mental health claims. CHAMPUS covers approximately 8 million beneficiaries, primarily retired members of the armed services and dependents of active-duty personnel. Level 1 review was primarily technical, in that CHAMPUS evaluated the eligibility of the beneficiary and the provider and determined if the service was a covered benefit. In general, claims representing 8, 24, and 40 sessions of outpatient psychotherapy were automatically referred to Level 2 reviewers, usually nurses, who relied on explicit process criteria (developed by the APA). At these Level 2 review points, CHAMPUS providers completed a treatment report that contained a description of the patient's problems, the goals of treatment, the treatment procedures, and the patient's progress toward these goals.

After 60 sessions of treatment with an adult (or 40 sessions with a child), the provider again completed a treatment report. When this point had been reached, or when the treatment report described a process that deviated from the Level 2 criteria, the claim was sent to peer review. For each claim, three psychologist peer reviewers from the provider's region independently evaluated the treatment reports and made advisory recommendations concerning the reimbursement for previous (already provided) and proposed (future) care. In general, then, the APA/ CHAMPUS program, like the PURE project, provided for evaluation of the clinical process, with a reliance on explicit screening criteria and professional peer review.

It is important to note that, beginning in 1985, the administration of CHAM-PUS peer review was awarded to the American Psychiatric Association. Today, the American Psychological Association contracts with private insurance companies to establish peer review programs, programs that differ markedly from the original CHAMPUS model. For example, the reviewers' sole responsibility is to judge the usualness and customariness, or reasonableness, of treatment, and it is the client's responsibility, not the provider's, to provide information for the treatment report (Joseph Avellar, personal communication, December 1985).

EXPLICIT SCREENING CRITERIA

RELIABILITY

The first research issue concerns the reliability of the explicit screening criteria used in a program of mental health quality assurance. Broadly speaking, the study of reliability entails measurement of the interrater and temporal consistency with which these criteria are applied in the evaluation of a documented process.

A few studies have shown that explicit process criteria can be reliably applied in the evaluation of clinical records. Specifically, Hays (1977) compared peer review systems developed at three state mental hospitals in Texas. Explicit admission criteria were formulated on psychiatric diagnoses based on the second edition of *The Diagnostic and Statistical Manual of Mental Disorders* (DSM-II; American Psychiatric Association, 1968). Judgments of admission appropriateness were generally consistent when judgments by in-house reviewers, local professionals, and out-of-house reviewers were compared. Sinclair and Frankel (1982) reported the use of a reliable checklist to evaluate the degree to which child clinical services are consistent with facility goals. There are also a number of medical studies that describe reliable screening checklists for the evaluation of medical records (e.g., Hastings, Sonneborn, Lee, Vick, & Sasmor, 1980). From a conceptual standpoint, it would seem that the reliability of explicit process criteria can be ensured if (a) the criteria are specific; (b) the clinical records or treatment reports contain sufficiently specific process information; and (c) reviewers are trained in the application of the explicit criteria.

VALIDITY

The next research issue concerns the validity of explicit process criteria used in a program of mental health quality assurance. Stated simply, the study of validity requires measurement of the relationship between explicit process criteria and actual clinical outcome. This issue is extremely important, because an audit based on explicit criteria is unlikely to lead to meaningful corrective action if the true usefulness of care processes is unknown (Brook, Kamberg, & Lohr, 1982).

There are, unfortunately, no data that bear directly on the relationship between normative or empirical process criteria and actual mental health outcome. The PURE project (Riedel et al., 1974) intended to compare normative process criteria for specific clinical problems with empirical data from targeted outcome and follow-up studies, but these efforts appear to have been unsystematic, and it is unclear if and how these studies confirmed the validity of the process criteria or contributed to criteria modification. Similarly, the APA/CHAMPUS screening criteria were intended to reflect both normatively and empirically derived practice standards, and to be responsive to empirical data on psychotherapy effectiveness (Stricker & Sechrest, 1982). To date, however, there have been no attempts to relate these criteria to outcome data, in part because of the inchoate state of psychotherapy research. However, a number of medical studies have shown that carefully derived, explicit process criteria can be predictive of medical outcome (e.g., Hastings et al., 1980; Mates & Sidel, 1981).

IMPLICIT PROCESS JUDGMENTS

RELIABILITY

Another research issue concerns the reliability of implicit process evaluations of mental health records. As was mentioned previously, most mental health quality-assurance programs include a third (final) level of professional peer review, based on implicit (subjective) criteria for evaluating treatment process. A number of studies conducted on the APA/CHAMPUS program bear indirectly on the issue of interreviewer reliability but bear directly on the factors that affect psychological reviewers' evaluations of clinical treatment reports.

When APA/CHAMPUS reviewers' recommendations were categorized as representing reimbursement denial, partial reimbursement, or full reimbursement, interreviewer agreement was relatively high. Specifically, for fiscal year 1982, peer reviewers (working independently in panels of three persons) made past-care *majority* recommendations for 98% of all cases sent to peer review ($N = 1,494$ cases) and proposed-care *majority* recommendations for 88% of these cases (S. Shueman, personal communication, October 1983). Similarly, Dall and Claiborn (1982) analyzed review recommendations received by Aetna claims and found that 98% of these claims had received past-care *majority* recommendations, and that about 80% had received proposed-care *majority* recommendations. Overall, the probability of any one reviewer's matching another reviewer's recommendation (for a particular case) was .54, compared with a chance agreement probability of .33.

Data by Cohen and Nelson (1982), however, indicated that precise interreviewer agreement was low when the specific number of past and proposed sessions recommended for reimbursement was analyzed. Because of the design of their study (different groups of peer reviewers each evaluating a different claim), a reliability index could not be directly computed. As an indirect estimate of interreviewer agreement, intraclass correlations were computed on APA/CHAMPUS reviewers' previous and total (previous plus proposed) session-reimbursement recommendations. These correlations did not exceed .09, an outcome indicating that reviewers' recommendations were characterized by a large variability within claims relative to the variability between claims.

Finally, in a somewhat related study, Smits, Feder, and Scanlon (1982) found that PSRO (professional standards review organization) officials' judgments of Medicare-covered patient conditions for nursing home treatment varied greatly among the PSROs sampled. The researchers specifically selected cases that were not clearly defined by Medicare regulations and that were dependent on PSRO reviewers' interpretation of the patient's condition.

FACTORS AFFECTING IMPLICIT PROCESS JUDGMENTS

Although not specifically concerned with the reliability of peer review evaluations, a number of studies of the APA/CHAMPUS program examined the factors that affect peer reviewers' implicit process evaluations of clinical treatment reports. Specifically, three factorial experiments examined the effects of APA/CHAMPUS reviewers' theoretical orientation on the peer review of long-term (60 previous sessions) treatment reports describing the psychodynamic care of adult outpatients (Cohen, 1981; Cohen & Oyster-Nelson, 1981; Cohen & Pizzirusso, 1982). The Cohen (1981) and Cohen and Pizzirusso (1982) experiments also tested the effects of documented treatment progress. In each experiment, self-described psychodynamic, behavioral, and eclectic reviewers were mailed treatment reports and asked to complete review reports. The review-report-dependent measures included reimbursement recommendations for previous and proposed care.

In general, psychodynamic reviewers provided the most generous previous and future-care reimbursement recommendations. All reviewer groups were very generous in their past-care recommendations (a finding confirmed by Dall & Claiborn, 1982, and by S. Shueman, personal communication, October 1983), with psychodynamic reviewers recommending reimbursement for close to 100% of past care, compared with 80%–90% reimbursement of past care recommended by nonpsychodynamic reviewers. The future-care recommendations were more variable: psychodynamic reviewers were consistently very generous, recommending reimbursement for about 90% of proposed care, but behavioral and eclectic reviewers' recommendations ranged from very restrictive to moderately generous.

Cohen (1981) found virtually no effects for a treatment progress manipulation. This study, however, had a number of methodological limitations (e.g., the treatment reports were very brief and superficial). In a methodologically superior experiment, Cohen and Pizzirusso (1982) found consistent treatment progress effects on reviewers' reimbursement recommendations. In general, their data suggest that APA/CHAMPUS peer review was sensitive to reported treatment progress, and that the pattern of previous and future-care recommendations appropriately reflected treatment quality, the prospects for success in continued treatment, and the outpatient's need for additional care.

The studies previously described found that psychodynamic reviewers compared with nonpsychodynamic reviewers were more positive in their evaluations of long-term psychodynamic psychotherapy. These reviewer-orientation effects are troubling and suggest a potential bias inherent in a peer review system based on implicit process criteria. However, because only psychodynamic treatment reports were evaluated in these studies, it is impossible to distinguish between the conclusion that psychodynamic reviewers are more lenient in their approach to insurance reimbursement, and the alternate conclusion that reviewers, in general, are more lenient when the reviewed treatment is congruent with their own professional practice.

In an analogue study of APA/CHAMPUS review, employing staffs at university counseling centers as reviewers, Biskin (1985) compared reviews of 24-session psychodynamic and cognitive-behavioral treatment as a function of reviewers' theoretical orientation. He found no main or interaction effects for reviewer orientation. Although this study stands in sharp contrast to the previous findings of Cohen (e.g., Cohen & Oyster-Nelson, 1981), it should be noted that Biskin evaluated 24-session rather than 60-session review, these reviews being provided by primarily counseling psychologists in institutional settings rather than private practicing clinical psychologists.

The issue of reviewer orientation effects was quite controversial and, unfortunately, has not been empirically resolved. The current American Psychological Association review program attempts to match the theoretical orientations of provider and peer reviewer.

The Cohen and Pizzirusso (1982) study found that APA/CHAMPUS reviewers were appropriately sensitive to documented treatment progress. This finding was encouraging, given that one of the major purposes of the APA/CHAMPUS program was to monitor the *effectiveness* and *efficiency* of psychological services rendered to CHAMPUS beneficiaries. However, as will be discussed in a later section of this chapter, these data do not address the relationship between documented treatment progress and actual treatment progress and outcome, that is, the *validity* of documented treatment progress.

VALIDITY

The validity of implicit process evaluations refers to (a) the relationship between subjective peer-review judgments and actual clinical outcome and (b) the degree to which clinical outcomes are enhanced if peer review recommendations are followed. There are, unfortunately, no empirical studies of the validity of implicit process evaluations of mental-health treatment reports. In the medical field, Brook and Appel (1973) found that implicit process judgments were not predictive of the health outcomes obtained by patients with various conditions. These authors were quite critical of implicit process evaluations, where reviewers rely on criteria more reflective of conventional wisdom than of empirical data that link care processes to health status.

OUTCOME-BASED QUALITY ASSURANCE IN MENTAL HEALTH

It is evident that most mental health (and medical) quality assurance programs rely on process assessment. The reader is directed to the work of Williamson

(1978), who developed the health accounting approach to quality assurance, an approach based on *outcome* rather than process assessment of medical care. He has argued that the identification of deficient outcomes is the most direct focus of assessment for isolating deficient care processes that require improvement. Williamson developed the concept of "achievable benefits not achieved," determined by a comparison of actual outcome data with consensual judgments of achievable benefits of care.

This model, however, does not appear to be applicable to the field of mental health. Measurement of outcomes in psychotherapy is still in a primitive state, as quantitative indices of psychosocial functioning are not yet widely used or even agreed on (Brook *et al.*, 1982). Furthermore, establishing links between health outcomes and care processes depends on how well the natural history of a health problem is known (Williamson, 1978); unfortunately, the natural history of most mental health problems is poorly understood. Similarly, an outcome-based quality-assurance program requires knowledge of the most valid point in time that outcomes should be assessed. Specifically, in the medical field, Mates and Sidel (1981) and Rubenstein, Mates, and Sidel (1977) demonstrated that the relationship between explicit process criteria and actual medical outcome depended on the temporal lag between service delivery and outcome measurement for specific medical conditions. In mental health, it is unclear when outcome should be measured for a large number of clinical problems.

TREATMENT DOCUMENTATION

There are a number of potential data sources for the evaluation of mental-health service delivery, for example, (a) direct observation of care; (b) face-to-face meetings between service providers and reviewers; (c) clinical records; (d) abstracts of clinical records; (e) clinical and administrative staff surveys; (f) patient surveys; (g) significant other surveys; and (h) population surveys (Baker, 1983). The Peninsula Hospital program (see Chapter 21) requires face-to-face meetings between clinicians and a peer review committee (Newman & Luft, 1974), whereas the Boulder County (Colorado) Mental Health Center program (to be described in a later section of this chapter) included a reinterview of a client by independent clinicians (Tash, Stahler, & Rappaport, 1980). However, these are relatively small-scale programs, and for obvious reasons of efficiency, most large-scale quality-assurance programs rely on *document-based* peer review (Cohen & Stricker, 1983). In the medical field, quality assessment has been based primarily on the evaluation of medical charts, whereas in the APA/CHAMPUS system evaluations were based on information provided in clinical treatment reports.

VALIDITY

There are, however, no data bearing on the *validity* of clinicians' treatment documentation. A program of document-based peer review presupposes that a provider's documentation of treatment accurately reflects his or her actual provision of clinical service. Inaccuracies in treatment reports may reflect a provider's willing misrepresentation of treatment procedures. The prevalence of this problem is unknown and is probably impossible to determine adequately, and a document-based quality-assurance program must rely on ethical adherence on the part of providers. Interestingly, Sharfstein, Towery, and Milowe (1980) found that diagnostic information on insurance claim forms submitted by psychiatrists was often inaccurate, most probably because of the clinicians' concerns about the confidentiality of the treatment reports.

Inaccuracies in treatment reports may also reflect a provider's inability to describe adequately the rationale and components of a treatment plan. This second problem is seemingly more amenable to research, perhaps involving interview studies of clinicians or more unobtrusive investigations in clinical facilities, examining the relationship between treatment documentation and actual practice as determined by observational data (Stricker & Cohen, 1984). Providers of various theoretical orientations may differ in their ability to provide sufficiently detailed information in treatment reports. For example, Dall and Claiborn (1982) found that treatment reports prepared by psychoanalytic psychologists were rated as significantly less adequate by APA/CHAMPUS staff than reports prepared by psychodynamic psychologists.

In the current American Psychological Association review program, it is the client's *responsibility*, not the provider's, to furnish treatment report information. It is unknown, however, whether this *legal* change will affect providers' actual role in the process of treatment documentation on treatment reports.

SELECTION OF PEER REVIEWERS

In order to enhance acceptance of mental-health (and medical) peer-review programs, reviewers have generally been chosen on the basis of outstanding credentials, allowing for expert or blue-ribbon review panels (Sechrest & Hoffman, 1982). Utilization of highly trained and competent providers as reviewers may better serve the goal of increasing overall quality of care and may make the review process more educative, as providers would receive feedback and recommendations from experts (Shueman, 1982). For example, the APA/CHAMPUS program used "expert" reviewers (N = about 400) who were nominated by the president of their state psychological association or by the chairperson of a professional standards review committee and who met certain selection criteria. Comparison of APA/CHAMPUS reviewers with a random sample of psychologist providers affirms that the reviewers were more experienced and credentialed (Cohen & Holstein, 1982). The current American Psychological Association review program still draws on the pool of reviewers selected for the APA/CHAMPUS system.

Interestingly, the findings of Pizzirusso and Cohen (1983) indicated that "expert" APA/CHAMPUS reviewers and randomly selected psychologist providers did not differ significantly in their evaluations of a psychodynamic treatment report. *National Register* psychologists of either a psychodynamic or an eclectic theoretical orientation evaluated one of three long-term psychodynamic-treatment reports that varied on reported progress level. This experiment represented a replication of the Cohen and Pizzirusso (1982) study of APA/CHAMPUS peer review. Like the APA/CHAMPUS data, the *National Register* data revealed significant reviewer theoretical orientation and documented progress effects in the expected direction. Direct comparison of the Cohen and Pizzirusso (1982) and Pizzirusso and Cohen (1983) data revealed only trivial effects for respondent group membership (APA/CHAMPUS versus *National Register* sample). The Pizzirusso and Cohen findings suggest that psychologist providers in general represent a source of competent personnel for peer review activities.

ATTITUDES TOWARD PEER REVIEW

There are a few studies that have surveyed mental health practitioners' attitudes toward peer review. In small-scale surveys of practitioners from diverse disciplines, Block (1975) and Gottesman (1974) found that a sizable minority had relatively negative attitudes toward specific programs of peer review.

SURVEY OF APA/CHAMPUS REVIEWERS AND PSYCHOLOGIST PROVIDERS

Cohen and Holstein (1982) surveyed APA/CHAMPUS reviewers and a random sample of psychologists listed in the *National Register*. The return rates for the two samples were about 61% (270/443) and 55% (551/1,000), respectively. The peer reviewer and provider questionnaires included a number of questions concerning the respondents' attitudes toward the APA/CHAMPUS peer-review program. It should be noted that, although the reviewers had not yet received cases for review at the time of the survey, they had received a detailed written description of the review system. The providers, on the other hand, were presumably unaware of the mechanics of the review system and were viewing peer review from a more general perspective.

Cohen and Holstein (1982) found that, although most respondents in each group reported that peer review would enhance the quality of psychological services, peer reviewers were more positive in their responses. Compared with the providers, the reviewers rated as less likely the possibility of a review program's being unfair to either providers or clients. Both groups of respondents were uncertain about the effect of peer review on the development of innovative treatment, although again, the reviewers were more favorable toward peer review. When asked whether peer review in general could be a workable method for improving the quality and cost effectiveness of care, the reviewers were much more enthusiastic in their endorsement of such a quality assurance program.

In general, Cohen and Holstein (1982) found that APA/CHAMPUS peer reviewers and psychologist providers differed rather consistently in their attitudes toward peer review. The general pattern of the survey data indicates that attitudes toward peer review are quite consistent with the respective vested interests and roles of peer reviewers and providers. Unfortunately, these data were collected before the APA/CHAMPUS program had been implemented, and it is necessary to conduct a follow-up survey to measure reviewers' and providers' attitudes after experience with the quality assurance process.

CORRECTIVE FEEDBACK TO CLINICIANS

As was mentioned previously, one of the primary objectives of quality assurance is the provision of corrective or educational feedback in order to reduce deficiencies in service quality. The assumption has been that peer review feedback will influence the subsequent professional practice of clinicians, although there are no convincing empirical data to support this assumption (see the next major section of this chapter) and no data whatsoever that bear directly on the relative effectiveness of feedback strategies (Donaldson & Keith, 1983; Woy, Lund, & Attkisson, 1978).

Research on psychotherapy supervision is similarly lacking, with few studies generalizable to the quality assurance process. Lambert's review (1980) of this literature indicates that, of the relatively few empirical studies of supervision, most have focused on supervision of novices or paraprofessionals, with measurement of the acquisition of elementary interviewing skills. He concluded that "research needs to be done on the outcome of supervision with the more experienced student or practicing psychotherapist" (p. 442).

CONTINUING MEDICAL EDUCATION

There are, however, a number of studies of the effects of continuing medical education (CME) on the subsequent performance of physicians. In general, the majority of these studies have shown that CME participation often results in increased *competence*, that is, new knowledge and skills (Sanazaro, 1983). However, increased competence does not necessarily result in improved medical *performance* in the care of patients. Sanazaro emphasized that exposure to medical knowledge is often insufficient to affect physicians' medical practice. He outlined the numerous prerequisites for effecting behavior change on the part of physicians, for example, (a) prompt individual feedback on performance; (b) a professional environment conducive to improved performance; (c) adequate resources; and (d) direct interventions with physicians (e.g., supervision and consultation). Sanazaro's review highlights the various individual and organizational behavior-change strategies that may promote improved performance by physicians. Given that peer review has an explicit educational and consultative objective, the lack of research on corrective feedback represents a serious gap in the quality assurance literature.

EFFECTS OF QUALITY ASSURANCE ON QUALITY OF CARE

The final, and perhaps most important, research issue concerns the ultimate effects of a quality assurance program on the quality of mental health services, especially its effects on clinical outcome. The focus of effect assessment, of course, would depend on whether the program was designed to improve the performance of all practitioners or just those identified as outliers (Brook *et al.*, 1982). There are, however, only a few empirical studies relevant to this issue.

For example, the Peninsula Hospital program requires professional peer review of outpatient cases that have reached six sessions of care. The therapist presents the case to an internal review committee, which then considers the need for additional care (Newman & Luft, 1974; Chapter 21 in this book).

After the Peninsula Hospital program had been operational for a number of years, the clinical staff and a sample of current patients were interviewed regarding the perceived effects of peer review (Luft & Newman, 1977; Luft, Sampson, & Newman, 1976). In general, the therapists had mixed reactions to the program. Specifically, (a) more than 90% believed that the program was educational; (b) more than 50% stated that they would present a case to peer review even if not required to do so; (c) about 33% believed that the program had resulted in more directive history-taking and concrete goal-setting with reviewed cases; but (d) about 80% reported that peer review had had no effect on their evaluations of reviewed cases (i.e., the consultation confirmed their clinical impressions); and (e) only about 10% stated that the program had influenced their care of nonreviewed patients. Of the patients surveyed, only about one third reported negative feelings about having their care presented to a peer review committee. About two thirds of the patients and their therapists agreed that peer review did not have a negative effect on the therapeutic relationship.

In general, then, these interview data suggest that the Peninsula Hospital program influenced clinicians' history-taking and goal-setting with reviewed cases. Although the clinicians believed that the program was educational, it did not seem to affect their evaluations of reviewed cases nor their care of nonreviewed cases. In any case, because of the lack of an experimental methodology and of measurement of patient outcome, these data do not specifically address the effects of peer review on therapy effectiveness.

Sinclair and Frankel (1982) reported an experimental study of the effects of quality assurance of child clinical care. Their quality assurance program included a record audit, an interview of staff by an auditing clerk, and face-to-face meetings between clinical staff and an internal review committee. Cases were randomly selected for review. All outpatient units received the program during the pretest and posttest, but only some (randomly assigned) received the program during the pretest–posttest interval. The dependent measures reflected clinical services' conformity to facility goals, based on record audits and the auditors' interviews with service providers. The results demonstrated significantly greater pretest–posttest improvement in service quality in the experimental compared with the control units.

Sinclair and Frankel's data (1982), however, are merely suggestive. Although the implementation of a true experimental design is noteworthy, the quality of their dependent measures is suspect, especially their assumed relationship to actual clinical effectiveness.

In an ambitious experimental study, Tash *et al.* (1980) tested the effects of an outcome-based quality-assurance program on clients' level of functioning. Their program represents one of the very few mental health programs that selects cases for peer review on the basis of clinical outcome criteria. Outpatients in a large mental health center (Boulder County MHC) were randomly assigned to the quality assurance program or a control condition. The patients received level-of-functioning (LOF) ratings by their therapists at intake, 2 months later (interim), and at termination. Each month, those 20 experimental cases that evidenced the least LOF improvement from intake to interim were peer-reviewed by an internal review committee. Of these cases, about 12 were interviewed at that time by independent clinicians. A comparable number of control patients were also selected (randomly) for an interview.

The results of the Tash et al. (1980) study were notably weak. Specifically, they found that (a) experimental compared with control patients did not evidence significant change in LOF ratings from intake to termination; (b) interviewed experimental cases, compared with experimental cases selected for an interview who were no-shows, did not show significant change in LOF ratings from interim to termination; and (c) interviewed control cases, compared with control cases not selected for an interview, did not show significant change in LOF ratings from interim to termination.

The Tash et al. (1980) experiment was characterized by a number of methodological limitations. For example, (a) there was no control group of poor-progress patients who were not peer-reviewed; (b) it appears that those peer-reviewed cases with the worst intake-to-interim progress were more likely to be selected for an interview; (c) in fact, only a very small number of experimental cases selected for an interview appeared for the interview, and it is possible that the no-shows represented the worst functioning experimental cases; and (d) the dependent measure was the therapists' level-of-functioning ratings, but the therapists were obviously not blind to the experimental condition of their patients. Only the experimental patients completed a clinical questionnaire at intake, interim, and posttest, and it is unclear why this questionnaire was not similarly administered to the control patients for comparative purposes.

Obviously, there are insufficient data that directly address the effects of mental-health quality assurance on clinical care. In any case, these effects may be extraordinarily difficult to measure. Some writers have hypothesized that programs of quality assurance may have primarily indirect effects on practitioners' treatment documentation and goal-oriented service delivery (e.g., Block, 1975; Demlo, 1983; Donaldson & Keith, 1983; Stricker & Cohen, 1984).

MEDICAL STUDIES

Data on medical quality assurance are relevant to this discussion. Evaluations of medical quality assurance, especially the PSRO program, have been inconclusive. The 1979 evaluation conducted by the Health Care Financing Administration (as described in Lohr & Brook, 1984) revealed that PSRO review positively affected some hospitals' "variation rate" between initial MCE (medical care evaluation) audits and reaudits. A variation rate represents the proportion of medical records that did not meet a specific standard in some quality-of-care area. The greatest improvements were in problem conditions with initially high variation rates. In general, however, Lohr and Brook concluded that the PSRO program was inconsistently operationalized over time and across geographical regions, a condition precluding an adequate large-scale evaluation. With respect to its cost–benefit effects, the consensus seems to be that the PSRO program probably consumed as many resources as it saved (Lohr & Brook, 1984).

MENTAL HEALTH QUALITY ASSURANCE AND COST CONTROL

In practice, quality assurance often has cost control as a major, if only implicit, objective (Sechrest & Hoffman, 1982). There are, however, no empirical data that address the potential cost savings of programs of mental health quality assurance. Some writers believe that quality assurance may, in some instances, result in *increased* rather than decreased costs, as high-quality care may be more expensive than low-quality care (e.g., Donabedian, 1983; Simon & Rosenberg, 1982). Furthermore, if a mental-health program places restrictions on some forms of mental-health service, patients may begin to overuse more costly medical care.

CONCLUSIONS

The publication of this book attests to the emergence of systems of accountability and the proliferation of programs of mental health quality assurance. The quality assurance literature has grown dramatically in recent years, with valuable contributions to conceptualization and design. As is evident from the present chapter, the establishment of mental health quality assurance programs has not been accompanied by systematic empirical research. The agenda for future research, discussed throughout this chapter, will not only guide efforts of quality assurance but also contribute more generally to the measurement of clinical effectiveness and the study of individual and organizational behavior change.

REFERENCES

Baker, F. (1983). Data sources for health care quality evaluation. *Evaluation and the Health Professions, 6,* 263–281.

Biskin, B. (1985). Peer reviewer evaluations and evaluations of peer reviewers: Effects of theoretical orientation. *Professional Psychology: Research and Practice, 16,* 671–680.

Block, W. (1975). Applying utilization review procedures in a community mental health center. *Hospital and Community Psychiatry, 26,* 358–362.

Brook, R., & Appel, B. (1973). Quality-of-care assessment: Choosing a method for peer review. *New England Journal of Medicine, 288,* 1323–1329.

Brook, R., Kamberg, C., & Lohr, K. (1982). Quality assessment in mental health. *Professional Psychology, 13,* 34–39.

Claiborn, W., Biskin, B., & Friedman, L. (1982). CHAMPUS and quality assurance. *Professional Psychology, 13,* 40–49.

Cohen, L. (1981). Peer review of psychodynamic psychotherapy: An experimental study of the APA/CHAMPUS program. *Professional Psychology, 12,* 776–784.

Cohen, L., & Holstein, C. (1982). Characteristics and attitudes of peer reviewers and providers in psychology. *Professional Psychology, 13,* 66–73.

Cohen, L., & Nelson, D. (1982). Peer review of psychodynamic psychotherapy: Generous versus restrictive reviewers. *Evaluation and the Health Professions, 5,* 130–144.

Cohen, L., & Oyster-Nelson, C. (1981). Clinicians' evaluations of psychodynamic psychotherapy: Experimental data on psychological peer review. *Journal of Consulting and Clinical Psychology, 49,* 583–589.

Cohen, L., & Pizzirusso, D. (1982). Document-based peer review of psychodynamic psychotherapy: Experimental studies of the APA/CHAMPUS program. *Evaluation and the Health Professions, 5,* 415–436.

Cohen, L., & Stricker, G. (1983). Mental health quality assurance: Development and evaluation of the APA/CHAMPUS program. *Evaluation and the Health Professions, 6,* 327–338.

Dall, O., & Claiborn, W. (1982). An evaluation of the Aetna pilot peer review project. *Psychotherapy: Theory, Research and Practice, 19,* 3–8.

Demlo, L. (1983). Assuring quality of health care. *Evaluation and the Health Professions, 6,* 161–196.

Donabedian, A. (1966). Evaluating the quality of medical care. *Milbank Memorial Fund Quarterly, 4,* 166–206.

Donabedian, A. (1983). Quality assessment and monitoring. *Evaluation and the Health Professions, 6,* 363–375.

Donaldson, M., & Keith, K. (1983). Planning for program effectiveness in quality assurance. *Evaluation and the Health Professions, 6,* 233–244.

Gottesman, D. (1974). Measuring attitudes about peer review in a university department of psychiatry. *Hospital and Community Psychiatry, 25,* 39–41.

Hastings, G., Sonneborn, R., Lee, G., Vick, L., & Sasmor, L. (1980). Peer review checklist: Reproducibility and validity of a method for evaluating the quality of ambulatory care. *American Journal of Public Health, 70,* 222–228.

Hays, J. (1977). Three methods of peer review in a state mental hospital system. *Psychological Reports, 41,* 519–525.

Lambert, M. (1980). Research and the supervisory process. In A. Hess (Ed.), *Psychotherapy supervision.* New York: Wiley.

Lohr, K., & Brook, R. (1984). Quality assurance in medicine. *American Behavioral Scientist, 27,* 583–607.

Luft, L., & Newman, D. (1977). Therapists' acceptance of peer review in a community mental health center. *Hospital and Community Psychiatry, 28,* 889–894.

Luft, L., Sampson, L., & Newman, D. (1976). Effects of peer review on outpatient psychotherapy: Therapist and patient follow-up survey. *American Journal of Psychiatry, 133,* 891–895.

Mates, S., & Sidel, V. (1981). Quality assessment by process and outcome methods: Evaluation of emergency room care of asthmatic adults. *American Journal of Public Health, 71,*687–693.

Newman, D., & Luft, L. (1974). The peer review process: Education versus control. *American Journal of Psychiatry, 131,* 1363–1366.

Pizzirusso, D., & Cohen, L. (1983). Psychologists' evaluations of psychodynamic psychotherapy: A peer review analogue experiment. *Professional Psychology, 14,* 57–66.

Riedel, D., Tischler, G., & Myers, J. (Eds.). (1974). *Patient care evaluation in mental health programs.* Cambridge, MA: Ballinger.

Rubenstein, L., Mates, S., & Sidel, V. (1977). Quality of care assessment by process and outcome scoring: Use of weighted algorithmic assessment criteria for evaluation of emergency room care of women with symptoms of urinary tract infection. *Annals of Internal Medicine, 86,* 617–625.

Sanazaro, P. (1983). Determining physicians' performance: Continuing medical education and other interacting variable. *Evaluation and the Health Professions, 6,* 197–210.

Sechrest, L., & Hoffman, P. (1982). The philosophical underpinnings of peer review. *Professional Psychology, 13,* 14–18.

Sharfstein, S., Towery, O., & Milowe, I. (1980). Accuracy of diagnostic information submitted to an insurance company. *American Journal of Psychiatry, 137,* 70–73.

Shueman, S. (1982). A model system of peer review. *Professional Psychology, 13,* 50–57.

Simon, G., & Rosenberg, A. (1982). Reviewing professional practice: Implications for the mental health consumer. *Professional Psychology, 13,* 159–166.

Sinclair, C., & Frankel, M. (1982). The effect of quality assurance activities on the quality of mental health services. *Quality Review Bulletin, 8,* 7–15.

Smits, H., Feder, J., & Scanlon, W. (1982). Medicare's nursing-home benefit: Variations in interpretation. *New England Journal of Medicine, 307,* 855–862.

Stricker, G., & Cohen, L. (1984). The APA/CHAMPUS peer review project: Implications for research and practice. *Professional Psychology, 15,* 96–108.

Stricker, G., & Sechrest, L. (1982). The role of research in criteria construction. *Professional Psychology, 13,* 19–22.

Tash, W., Stahler, G., & Rappaport, H. (1980). *Test of an outcome-based quality assurance system.* (Final report of NIMH Contract 278-78-0064). Rockville, MD: Horizon Institute.

Williamson, J. (1978). *Assessing and improving health care outcomes: The health accounting approach to quality assurance.* Cambridge, MA: Ballinger.

Woy, J., Lund, D., & Attkisson, C. (1978). Quality asssurance in human service evaluation. In C. Attkisson, W. Hargreaves, M. Horowitz, & J. Sorenson (Eds.), *Evaluation of human service programs.* New York: Academic Press.

4

Evaluating Peer Review

LEE SECHREST AND ABRAM ROSENBLATT

To be effective, any quality assurance program in psychology must be based on systematic research designed to inform both the development and the overall effectiveness of that program. These programs must be built on systematic data collection, experimental research, and program evaluation. Once built, the completed program must be judged by these same methods. Although it is debatable and perhaps even unlikely that the quality of mental health care can ever be assured (*enhanced* might be the preferable term), any given program will make major progress toward assuring quality of care only if it is based on quality scientific information.

Evaluating quality assurance programs seems, at first blush, a complex and expensive task. This may be owing, in part, to the multitude of mechanisms subsumed under the general label *quality assurance.* Such mechanisms include, but are certainly not limited to, selection for graduate programs; accreditation of training programs; socialization of students into the profession; licensure; continuing education; special competency exams; claims review by third-party payers; and peer review (Sechrest, 1984). Thus, to simplify matters and make the task manageable, this chapter focuses on one mechanism: peer review. This mechanism is of considerable current interest, is controversial, and is possessed of a small, but growing, empirical data base. Though our task here is to focus on the specific conceptual and methodological issues requiring consideration in evaluating a peer review program, we do not suggest that such issues are germane only to peer review, nor that peer review itself is any more important a mechanism in quality asurance than any of the mechanisms we do not cover. Indeed, we hope that some of this discussion will prove useful to researchers wishing to examine any of the other existing or potential components of an overall quality assurance program.

Before we go into specific dimensions and methodologies in evaluating peer review, it is important to address the distinction between evaluating the process of program development and evaluating the outcome of an already developed program. As we shall see, this distinction is of primary import in conducting an effective evaluation.

LEE SECHREST AND ABRAM ROSENBLATT • Department of Psychology, University of Arizona, Tucson, Arizona 85721.

PROCESS AND OUTCOME

Complex programs are not easy to translate fom conceptualization into implementation, and implementation is, itself, a hazardous process. Programs develop, grow, and change before ever reaching a steady state endpoint. Traditionally, studies that evaluate the developmental phase of a program have been called *process* (after Cronbach, 1964) or *formative evaluations* (Scriven, 1967). Studies that evaluate the steady state endpoint have been called *outcome* (after Cronbach, 1964) or *summative* (Scriven, 1967) *evaluations*. More recently, Tharp and Gallimore (1979) applied ecology terminology to social programs. Basically, they described a program as moving through a set of "seral" stages, composed of relatively transitory associations of elements, that culminate in a stable condition called "climax." They noted that not all programs reach a climax stage, described as "an association of program elements, organized for and producing a defined social benefit, which will continue to exist, and in which there will not be a replacement by other element types, so long as social values, goals and supporting resources remain constant." (p. 43).

They further described four external conditions necessary for a program to reach climax: (a) longevity—the climax condition cannot be reached quickly; (b) stability of values and goals—a program must be designed to meet a stable value; (c) stability of funding—a program obviously must remain stable in its funding if it expects to survive; and (d) power of the evaluator—the influence of the evaluator must be maintained in a way that ensures the integrity of research and development. The distinction between sere and climax (or formative and summative/ process and outcome) evaluation is crucial when examining peer review systems. As Sechrest (1984) noted, before arriving at any conclusions about the value of peer review systems, the processes involved in the system must be scrutinized with the greatest care. The evaluator who performs a summative evaluation on a program still in its initial stages of development is putting the cart before the horse; he or she is likely to use incorrect methodologies and is as likely as not to confirm the null hypothesis because of the weakness of his or her independent variable (the program).

This chapter, therefore, examines separately the research issues involved in effectively evaluating the processes and the outcomes of peer review. Although the distinction between process and outcome will necessarily be idealized (often these distinctions are far more blurred in the real world), separating the two will at least strongly encourage the reader to address this issue before conducting the evaluation.

PROCESS EVALUATION AND PEER REVIEW

The major evaluation task needing to be done on peer review at present is process evaluation. Peer review programs in psychology have simply not reached the point at which outcome evaluations should be considered crucial to further decisions about such programs. For example, the CHAMPUS peer-review project— the most developed, best publicized, and largest peer-review program in psychology—first contracted with the American Psychological Association (APA) in July 1977 (Clairborn, Biskin, & Friedman, 1982). Considering that Tharp and Gallimore described a project that took 9 years to reach what they considered *partial* climax conditions, we can see that peer review in psychology is clearly in its formative stages. If we combine this early stage of program development with the instability in funding that faces APA peer-review programs (see Fisher, 1985) as well as administrative instability (CHAMPUS, for example, was turned over to the American Psychiatric Association in the spring of 1985) we see that practically speaking, evaluating peer review largely means evaluating the processes of peer review systems.

Tharp and Gallimore (1979) described an approach called "evaluation succes sion" for evaluating program development. This is a systematic process that involves testing each program element at each stage of its development, from conceptualization to formulation to implemantation. The results from each test at each step are used to guide further development of the program element. An implemented element may then be tested in association with other elements of the program to determine how the elements work together in achieving program goals.

THE INDEPENDENT VARIABLES

Each element of a program may be considered an independent variable. Peer review programs are complex and contain many elements suitable for development and evaluation, including, but certainly not limited to, the problems for which services are offered; the nature of the services provided; the purposes for which the review is conducted; the criteria for peer review; the nature of the written records used in the peer review; and the qualifications and training of the peer reviewers. For puposes of illustration, we will examine two of these in more depth: the criteria for peer review and the nature of the written records used in the peer review.

Records and Peer Review

Peer review obviously requires some type of information base. Currently, this information consists of paper records prepared to meet peer review requirements. The question arises whether these documents are adequate for such a review. Reports may be biased in a variety of ways. Clinicians may distort matters by trying to convey a positive impression in these documents, or they may conceal important information out of a concern for patient confidentiality. Documents may simply be inadequate in conveying the true value of a complex treatment. We can begin to address some of these issues by systematically evaluating these documents. For example, we could relatively easily devise and test different forms of treatment reports. Other ways of obtaining needed information, such as telephone interviews, might well be explored. A major advantage of evaluation succession is that it provides for early detection of nonfruitful methods.

Criteria in Peer Review

Current versions of peer review rely on explicit criteria used to determine whether cases should be referred for peer review. These criteria may be based on expert opinion, patterns of practice, or experimental findings (Stricker & Sechrest, 1982). The exact role of criteria in peer review nonetheless remains uncertain. We do not even know if different types of criteria yield different types of decisions. It is possible, for example, that these decisions are based on a set of implicit criteria by which reviewers operate. We need to test different types of criteria to determine their effects (or lack of effects) on reviewers' decisions.

Evaluating the Independent Variable

Identifying, conceptualizing, and refining the appropriate independent variables is a crucial step in conducting any evaluation. In addition, too little attention is often paid to the strength and integrity of the independent variable (Sechrest & Redner, 1979). The extent to which an element is delivered as planned determines the integrity of treatment. Steps need to be taken (e.g., monitoring the implementation of the element; clearly describing the treatment or element; and insuring staff commitment to implementing the element) to ensure the integrity of the independant variable. The term *strength of treatment* refers to the *a priori* likelihood that a treatment could have the intended outcome (Yeaton & Sechrest, 1981), or to the amount of treatment provided (Sechrest, West, Phillips, Redner, & Yeaton, 1977; Yeaton & Sechrest, 1981). The evaluator needs to be aware of the need to evaluate the strength of the independent variables (for specific methods, see Sechrest, Ametrano, & Ametrano 1982; Sechrest & Redner, 1979; Sechrest *et al.*, 1977).

Thus, for example, a peer review program could be planned in such a way that there would be no likelihood of its having any impact on the quality of services. Some current proposals to limit reporting requirements (purportedly to protect confidentiality) would result in such inadequate information that peer reviewers would be in a position to do little more than say, "Pay the claim." That would be a weak form of peer review or, at least, a weak information base. On the other hand, an otherwise promising mechanism for peer review might be poorly implemented; that is, it might have low integrity. An instance would be failure to train peer reviewers adequately.

THE DEPENDENT VARIABLE

Assessment of the dependent variable has received far more attention than that of the independent variable, yet we still lack precision and basic knowledge in too many cases. Outcome variables must be chosen that will both adequately represent the domain expected to be affected by the intervention and be sensitive to the intervention.

We are unaccustomed to thinking in psychometric terms about outcome variables. The psychometric properties of dependent variables are, however, important. As Sutcliffe (1980) and Bejar (1980) have shown, very misleading results may ensue from studies using unreliable outcome measures or outcome measures that are differentially reliable across groups.

Finding the dependent variables appropriate to a process evaluation of peer review is not entirely simple. Although the conceptual dependent variables of an outcome evaluation of an entire peer-review program fundamentally consist of enhancing quality of care and cost containment, single elements of a peer review system are difficult to evaluate on these bases. A look at which dimensions of a peer review system are critical to its functioning is in order. Sechrest (1984) identified the critical dimensions of a peer review system as reliability, fairness, validity, and credibility. These dimensions are certainly not exhaustive, but they do provide a starting point for conceptualizing the dependent variable in peer review.

Reliability

A basic problem with the means by which grant proposals are evaluated is that two reviewers may read a proposal or an article and arrive at quite different conclusions and recommendations. This is sure to be the case in the peer review of clinical services as well (e.g., Cohen & Oyster-Nelson, 1981). It is difficult to justify a peer review system that does not result in substantial agreement from one reviewer to another. Research to date suggests that the theoretical orientation of the reviewer is an important potential source of disagreement (Cohen & Holstein, 1982a; Cohen & Oyster-Nelson, 1981; Pizzirusso & Cohen, 1983). Other variables that contribute to disagreement are not now known. Research needs to evaluate this situation to determine how and why disagreement occurs. For example, why should reviewers of even the same theoretical persuasion disagree about such factors as necessity of treatment and appropriateness of the approach taken by a therapist? If the problem stems from different assumptions growing out of inadequate information, then program elements can be designed and tested to eliminate this problem, perhaps by making the information adequate. Of course, if psychologists as a whole cannot produce reliable assessments and treatment plans, then achieving reliability in a peer review system becomes unlikely, and an evaluation of the profession itself is in order!

Fairness

A peer review system needs to be free of unwarranted, unsupportable bias. For example, as noted, theoretical orientation appears important in judgments concerning the appropriateness of treatment strategies (Cohen & Holstein, 1982b; Cohen & Oyster-Nelson, 1981; Pizzirusso & Cohen, 1983). In general, these studies have shown that psychodynamically oriented therapists are, not surprisingly, more in favor of long term therapy than behavior therapists with the difference being more pronounced in evaluating cases after 60 sessions than after only 40. Behaviorally oriented therapists are also not in full sympathy with the goals of psychodynamic therapy and are likely to want the therapy restructured. A system might, therefore, be reliable but nonetheless biased for or against specific treatment strategies. Behaviorally oriented therapists may advocate shorter term therapy than psychodynamically oriented therapists and would be likely to see psychodynamic therapy as inefficient. The psychodynamic therapists might see behavioral therapies as superficial and ineffective. This age-old issue in psychology presents serious problems for peer review because third-party payers are not likely to be impressed with a system that produces drastically different recommendations for seemingly similar cases. Designing program elements to reduce this bias would be a beneficial (although difficult) task. In any case, any program element or set of elements that resulted in less potential for bias (e.g., prearranged lengths of treatment for different problems) would be useful.

Validity

The judgments and recommendations of peer reviewers ought also to be demonstrably "valid"; that is, the recommendations should, if followed, lead to near-optimal outcomes. If peer reviewers recommend terminating treatment on the grounds that the maximum benefits have already been attained, then it should in some manner be possible, at least in principle, to show that such a judgment is warranted. If, on the other hand, reviewers recommend a continuation of treatment in expectation of important further gains, it ought to be demonstrable that such gains would at least on the average, be attained. As yet, we have no evidence at all that the judgments of peer reviewers are valid in the sense just described. One set of findings of interest in this context is that therapists and patients within a system providing for review agreed substantially on the amount of improvement achieved in therapy and on the need for further treatment. Nonetheless, although the therapists believed that only about 30% of their patients had had an adequate amount of therapy, 60% of the patients did not use all the time allowed by the peer review committee (Luft, Sampson, & Newman, 1976). The median number of sessions that had actually been approved was 15, a number probably much smaller than would be thought necessary by psychodynamic therapists. On the other hand, there was no relationship for either therapists or patients between amount of treatment and judged improvement.

Credibility

Any peer-review system must be credible to be accepted by all the parties involved: patients, therapists, and third-party payers. In the only research on the topic to date, Luft *et al.* (1976) reported that about one third of the patients informed about the requirements for peer review disliked the idea of having their treatment discussed with outsiders, about one third were indifferent, and about one third were positive, feeling that they could only benefit from the experience and recommendations of the review committee. Those with unfavorable reactions were primarily concerned about confidentiality, but some were afraid that their therapists might not be effective in presenting their cases and needs, and some were afraid that the committee might set a higher fee than they were paying.

About one third of the therapists reported that they would be less inclined to accept a patient whose case would be peer-reviewed. Of interest is the fact that psychologists were most resistant (50%) and that psychiatrists (35%) and social workers (11%) displayed less concern. Cohen and Holstein (1982a) reported that a sample of psychologists at risk for peer review were generally supportive of the CHAMPUS peer-review program. There were some differences in attitudes toward specific features of the program that were associated with recency of degree (Cohen & Holstein, 1982b). More senior therapists were less positive toward the features of the system that required greater patient involvement in treatment planning and assessment. The credibility of any program element must therefore be assessed. If therapists do not believe that they can adequately convey the nature of their services by means of a treatment report, then the review process will be jeopardized. The solution, one to be explored empirically, is either to improve the treatment report form or to attempt to induce greater confidence in it.

OUTCOME MEASURES IN STUDIES OF PEER REVIEW

As one mechanism in the array of quality assurance activities, peer review may have diverse effects, some intended and some not. It is important to be precise about what effects one is looking for and how those effects may be brought about.

Those who must pay for services may be most interested in cost containment and in the expectation that peer review will at least keep costs from rising as fast as they might otherwise rise and will perhaps even reduce them. Effects or costs may be achieved through "surveillance" that tends to eliminate unnecessary or questionable services. Surveillance effects should be detected at the system level and, perhaps, at the individual clinician level. The other side of the coin is, of course, the possibility of a chilling effect that reduces desirable services in the short run with longer run increases in costs. Effects on costs may also occur through a limitation on services that results when some are deleted from lists of acceptable services. For example, cases may be limited to a certain dollar amount, or certain procedures (e.g., tests) may be proscribed. Peer review may enter in if provision is made for exceptions to limitations with proper review. Again, effects may be detected at either system or individual provider levels, and risks to quality of care may be entailed.

The second aim of peer review is to improve quality of care. Improvements in quality may occur through a policing function that eliminates incompetent or otherwise undesirable providers. Quality improvements may also result from standard setting, that is, making evident the standards of good care. Quality of care may be assessed at individual care levels and may be detected at either provider or system levels.

The specification of outcomes and of the mechanisms for producing them is important, for such specification influences the research process, the design of the study, and the interpretation of the data. If, for example, costs were to be affected through a surveillance effect, it is important to know to what extent providers were aware that their practices were being watched. A policing function, on the other hand, might be carried out without most providers being aware of anything unusual. A study focusing on costs ought still to take quality issues into account.

DESIGN ISSUES

A wide variety of methodologies are available for linking these independent and dependent variables in the evaluation of a developing peer-review program.

Selection of the appropriate methodology should be guided by several factors. The first set of factors have to do with reducing threats to internal validity. A detailed description of these threats has been provided elsewhere (Campbell & Stanley, 1966; Cook & Campbell, 1979), and a summary of these threats with respect to program evaluation can be found in Sechrest et al. (1982). Therefore, we will not focus on these specific issues here; instead, we will examine causality and plausibility in program evaluation.

Causality and internal validity go hand in hand, the goal of internal validity being to increase confidence in causal inferences. Plausibility, on the other hand, largely concerns issues of external validity (Cook & Campbell, 1979) and conclusiveness of findings (Sechrest et al., 1982). The final factor influencing the researcher's choice of methodology concerns issues of ethics and practicality. When working within an action setting, ethics and practicality often clash with principles of research methodology. We will examine each of these guiding factors more closely before moving on to a discussion of some specific methodologies.

CAUSALITY

We basically take for granted the likelihood of establishing a dependable empirical relationship (correlation) between the phenomena of interest. Problems arise in evaluation, or just about any field, when we attempt to insist that a *causal* relationship is involved (see Sechrest, 1984, for examples of causal and not-so-causal relationships in the health-care arena). The problem of inferring causality is a long-standing one in the social sciences (see Babbie, 1983, for a brief summary; Hirschi & Selvin, 1973, for an example of causation within the context of a particular research finding). Unfortunately, the issue cannot be ignored in evaluating peer review programs. The most important questions are causal in nature. For example, does peer review result in (cause) decreased costs or increased quality of care?

Traditionally, the most straightforward and compelling evidence for causality is achieved by means of a randomized experimentation. Cook and Campbell (1979) provided a cogent discussion of the problems of inferring causality and thinking in causal terms. Many of these problems result from the presence of a third variable that may plausibly explain why the variables of interest are related. A well-done experiment can resolve these problems. The advantage of the randomized experiment, if properly done, is that research cases assigned to treatment conditions can be assumed to be equivalent in all ways save exposure to the experimental variable (treatment), with the consequence that any final differences would be unequivocally attributable to the treatment itself.

Unfortunately, things are not that simple. True randomization is rarely achieved even in the laboratory. The moment any single case is lost from a group, the groups are no longer strictly equivalent. Generally, we do not concern ourselves with these relatively small perturbations in protocol. However, in an action setting such as peer review system, random assignment to groups may be well-nigh impossible. It may be ethically or practically difficult to withhold treatment from certain groups. The "subjects" are in the real world, patients or employees who are not captive in a laboratory situation. Thus, the dropout rate is apt to be high. Moreover, mental-health-care providers and peer reviewers alike are not research assistants and may have reactions to having to withhold certain treatments (e.g., Borgatta, 1955) or to having their effectiveness questioned.

In short, a wide variety of factors (see Weiss, 1972, for a summary) make the experiment an often difficult option in evaluating a living, breathing program. Thus, a great deal of our discussion will involve a variety of quasi-experimental methods that may, with the proper cautions, produce useful, if not completely persuasive, causal data.

PLAUSIBILITY

One would assume that the more causal the results of a project, the more plausible they would be. Yet, this connection remains an unknown empirical question. It is important to keep in mind that the audiences for a study evaluating a peer review system differ from editors and referees of scientific journals. Members of Congress, businesspeople, program heads, and peer reviewers are not necessarily as likely as editors and journal referees to be swayed by methodological rigor. Weiss and Bucuvales (1980) did find that the judged utility of research was related to its perceived quality, but the quality judgments included variables beyond methodological rigor (e.g., lack of bias in conclusions). Furthermore, Holland (1984) found that a group of mental-health-service providers did not distinguish between the "usefulness" and the "truthfulness" of findings concerning the effectiveness of programs. The first question that the researcher needs to ask in considering the plausibility of his or her reasearch is: To whom does this need to be plausible? In terms of process evaluations, the likely answers include program participants, directors, and perhaps funding sources or third-party payers. The second question becomes: Given one's audience, how does one make research plausible as a basis for policy?

In this area, the researcher may do well to attend to issues of external validity. The term *External validity* refers to the generalizability of the study. It has been clearly demonstrated that different methods, done properly and addressing the same issue, can lead to very different results (Konecni & Ebbesen, 1979). As an illustration of the case, consider the evaluator who is trying to assess how different criteria affect peer reviewer decisions. So the evaluator constructs a true (randomized), internally valid causal experiment in which college students serve as the peer reviewers, cases are randomly assigned to groups, the experimental group has one set of criteria, and the control group has another or no set of criteria. The results may be causally interpretable, but they would not be plausible. Peer reviewers have different backgrounds from those of college students, have different pressures, and so on. These results would be unlikely to influence any real-world peer-review system. On the other hand, a well-done quasiexperiment conducted in the system might be far more convincing.

Although this example is certainly exaggerated, we want to consider the possibility that, under some circumstances, quasi experimental research may be at least equal to randomized experimentation in its overall impact.

Much of the variance in the impact of research findings may depend on how the findings are couched. There is only one universal code for expressing research findings, and that is in terms of statistical significance. Unfortunately, for a great many reasons, statistical significance does not convey much useful information about a research finding, particularly information that might be useful to a policymaker (Sechrest & Yeaton, 1981b). Policymakers need to know how large an effect is produced by some intervention or how much difference it makes when one is considering one group or another. There is simply not a good, generally agreed-upon metric for expressing effect sizes that has much practical utility.

Plausible Rival Hypotheses

For any relationship between two variables, there is always at least one alternative, or rival, hypothesis to any that might be put forth. That means that, in any instance in which we may have a favored causal hypothesis, there is at least one other possible explanation. Some rival hypotheses we are likely to consider implausible. For example, supernatural explanations are always possible, but we reject them out of hand as implausible. Other rival hypotheses may be considerably more plausible, and much of the art of research design is in creating arrangements that render most or all rival hypotheses implausible. To begin with, for example, chance is always a rival hypothesis, and we must do our research in such a way that the operations of chance come to seem implausible as an explanation for our findings.

The issue of plausible rival hypotheses arises most often in relation to internal validity, although it can arise with respect to any other feature of a research project. By their very nature, most plausible rival hypotheses have limits on their plausibility. One limit is Bayesian in nature. That is, our inclination to accept a research outcome as real will depend, in part, on our prior expectations about the likelihood of finding it. Therefore, we will—and should—have a bias in favor of research findings that make theoretical sense.

Plausible rival hypotheses also very often involve rather subtle effects—and presumably, nuances—in findings. We might believe that, if an intervention is likely to produce large effects, even a relatively weak design might be impervious to most rival hypotheses. Unhappily, those occasions when we might trust weak designs to reveal truth are not frequent in research, but the possibility should not be ignored.

Ethics and Practicality

Ethical issues may present a threat to or a conflict in research design. One common ethical dilemma raised in evaluation studies on treatment programs involves the possibility of withholding potentially beneficial treatment elements from some of the patients of the program. This dilemma is tempered by the possibility that untested treatments may not prove beneficial or may even prove harmful. Under such circumstances, withholding the treatment may actually be the more ethical option. Borgatta (1955) argued, for example, that providing an ineffective treatment may prevent a client from receiving an effective treatment.

In the case of peer review, ethical problems involving the withholding of treatment are likely to be minor. It may be useful to assign patients to a treatment reviewed by one type of program or element as opposed to a similar treatment reviewed by another type of program or element, but it is unlikely that the evaluator will need to assign patients to a no-treatment condition. More germane to evaluating peer review systems are ethical issues revolving around patient confidentiality. Peterson, Andrews, Spain, and Greenberg (1956), for example, observed physicians with actual cases in studying quality of care. Therefore, videotapes of therapy sessions might prove useful as a basis of peer review when compared to paper records as a basis of peer review. Confidentiality, however, becomes an issue (though such tapes are frequently used in supervising therapists) requiring careful consideration by the researcher. Careful informed consent is one way around this problem, though the possibility that patients may withdraw from therapy after reading the consent form plays havoc with random selection.

Ethical problems are likely to be decided as much by politics as by any absolute ethical standards. Practitioners and peer reviewers may disagree with the ethics of many interventions. In this area, as well as many other areas of the evaluation, researcher and practitioner or peer reviewer are likely to clash. Such problems are common in evaluation research (Aronson & Sherwood, 1967; Weiss, 1971, 1972) and are to be expected in evaluating peer review and implementing program elements. In the case of peer review, for example, the evaluator can be assured that even the most confident and competent practitioner will be a bit wary of having his or her treatment strategies scrutinized. Solutions to these problems include getting the necessary support from administrators; involving practitioners in the evaluation; minimizing disruption; emphasizing theory and benefits; providing information useful to the program; and clearly defining roles and authority structures (Weiss, 1972).

Careful selection of research and design may also serve to minimize ethical and practical problems in evaluation. A delayed treatment design (Weiss, 1972) may, for example, get around problems of withholding treatments from certain groups. This design, similar to that used for waiting-list-control studies, could be especially useful in circumstances in which services were limited and only a part of the population could be served at any one time. For example, if funds and personnel were available to review only a portion of the cases in a system, that fact could be exploited by a researcher. Scriven (1967) proposed another option for dealing with this problem by offering controls weaker versions of the treatment rather than placebos or no treatment (e.g., as might be accomplished by limiting the extent of review). Such a strategy does, however, reduce the chances of finding any advantages in the experimental condition (Baum, Amish, Chalmers, Sacks, Smith, & Fagerstrom, 1981). The researcher must, as in any setting, judge the value of the work against any potential harm that the work may do. Valueless, poorly conducted work cannot justify putting subjects through even minimal discomfort.

EXPERIMENTATION

When properly conducted, experimentation leads to the least causally ambiguous results of any method. The true experiment is the only design that allows one to eliminate alternate explanations of the results (Campbell & Stanley, 1966). To achieve this, this method relies on (a) the random assignment of cases to treatment groups and (b) the inclusion of a control group assumed to be equal to the experimental groups in every way except for receiving the treatment. All other designs allow for one or more rival hypotheses to remain plausible and thus fail to ensure internal validity. True experimental designs have the additional advantage of being very efficient (more powerful) and thus more able to detect small effects (Gilbert, Light, & Mosteller, 1977). The advantages of well-done, true experiments are such that they should be used whenever possible.

A word or two of caution about randomized experiments is, however, in order. Conducting a true, tightly controlled experiment is a tricky task, and things are likely to go wrong, all of which will very likely serve to weaken the causal interpretation of any results. For example, clinicians assigned to different groups may become aware of the existence of the other groups and may talk to clinicians in those groups or may attempt to switch groups. Randomization can break down in many places. For example, patients may self-select for different types of systems; providers may preferentially select patients for certain treatments; and patient attrition may create systematic error.

In addition, experiments provide only a very specific service in the formative evaluation of this program. They can evaluate the probable value of isolated elements in a program. An experiment (or quasi experiment), however, cannot accurately predict the value of the element when it is implemented in the program and combined with other elements (Tharp & Gallimore, 1979). For example, an experiment may prove that videotape is a more fair, reliable, and credible basis for peer review than written records. Such a finding would not, however, ensure that videotape will be effective when implemented in the program as a whole or that the program based on videotapes will prove effective as a whole.

QUASI-EXPERIMENTAL DESIGNS

Because experimental designs are not always feasible in an action setting such as a peer review system, the evaluator must possess full knowledge of various "quasi-experimental" designs (Campbell & Stanley, 1966; Cook & Campbell, 1979). In choosing one of the following designs, the researcher must carefully consider which design will provide the strongest evidence of causality given the circumstances. When conducted properly, many of these alternatives provide data that approach those from experiments in permitting inference in causality and in plausibility.

Nonequivalent Comparison-Group Designs

Nonequivalent comparison-group designs are among the most frequently used quasi-experimental designs. They vary from experimental designs in that cases are assigned to the comparison groups on a nonrandom basis. Nonrandom assignment creates a variety of methodological problems (summarized by Judd & Kenny, 1981; Reichardt, 1979) owing to the possibility that cases ending up in different groups may be different in ways that will affect the outcome measures. Thus, if differences are found between the experimental (treatment) group and the comparison group, the researcher cannot be sure whether the outcome differences are due to the manipulation or to preexisting differences in the two groups. For this reason, it is usually advisable to have the two groups be as similar as possible. Finding groups that weaken potentially plausible rival hypotheses constitutes much of the work in nonequivalent comparison-group designs. The researcher needs to identify plausible rival hypotheses and to select the comparison groups in such a way as to eliminate these sources of error. If multiple plausible confounding hypotheses exist, it might be possible to select more than one comparison group, each designed to eliminate an alternate hypothesis and thus to strengthen the conclusiveness of the results.

Two other ways of equating controls seem intuitively useful but must be viewed with caution. Matching would seem to reduce the variance between the groups, but serious problems with this tactic make it almost always undesirable. Matching may have serious negative consequences for making causal inferences. All the relevant variables cannot be matched for; matching for some variables will result in opposite mismatching for others; and the likelihood of regression effects, which usually do not pose a threat in nonequivalent comparison-group designs, is greatly increased. Studies that use matching must be looked on with skepticism (Campbell & Stanley, 1966; Cook & Campbell; 1979; Neale & Libert, 1973). A second approach to dealing with nonequivalence between groups is through statistical means (e.g., analysis of covariance). Much as is the case with matching, it is likely that allowances for initial differences are wrong in unknown, or even unknowable, degrees (Sechrest, 1984). Thus, nonequivalence remains a problem in these designs, especially when the differences in outcomes between groups are small.

Patching Up Designs

Institutional cycle designs or patched-up designs (Campbell & Stanley, 1966) usually involve ongoing modifications to the design as the evaluation progresses. These designs may prove useful in formative evaluations of programs because they allow the experimenter to account for threats to internal validity that may arise as the program changes by adding comparison groups or series as necessary. An example of such a research opportunity might be the periodic (or aperiodic) changes in personnel within an organization that often result in systematic changes in policy and practice. These designs can be as varied as each individual program or set of circumstances and can be equally varied in their usefulness.

Time Series

Interrupted-time-series designs are a fairly attactive quasi-experimental technique capable of yielding high-quality data and defensible conclusions (Judd & Kenny, 1981). These designs involve a series of measures taken at periodic intervals before an intervention begins (a baseline phase), followed by a series of measures taken after the intervention starts. These designs are similar to the designs fre-

quently used in assessing the efficacy of behavioral interventions in individual subjects (see Kazdin, 1974, for a summary). They are different, however, in that they
may be used with a large sample size and are formulated around complex statistical
tests (Gottman & Glass, 1978) that ensure that changes in data will be due to the
treatment and not to some general trend. A time series may show that the mean
levels of some phenomenon are different before and after treatment, that the slope
of change over time is different, or that there is a change at the time of treatment that
is of an abrupt nature. For example, researchers may decide that they wish to examine the effects of some new set of criteria for peer review on the reliability of the
recommendations of the peer review panel. The researcher may take a series of
measures assessing reliability, may introduce the new system, and may take a
further series of measures assessing the reliability. The subsequent analysis might
show that reliability increased after the new criteria were introduced. Or it might
show that reliability was generally increasing but increased more rapidly after the
criteria were introduced. Or it might show that reliability quickly increased after the
introduction of the criteria but eventually declined to former levels.

Other factors, such as history (e.g., a new record system may have been installed at the same time as the new criteria were introduced) or maturation (e.g.,
reliability may have simply increased as peer review members became more
friendly after working together for a period of time), can explain the results of a simple time-series design. To control for these problems, a researcher can use an elaborated version of a time series design called a multiple-time-series design. These
designs usually combine a time series design with a nonequivalent comparison-
group design. In the example provided above, a second, similar peer-review board
could have served as a nonequivalent control group, receiving the same measures
as the treatment group over the same period of time without receiving the new set
of peer review criteria. Much like nonequivalent control-group designs, multiple
comparison groups may be used to eliminate confounding hypotheses. When conducted properly, multiple-time-series designs can provide information that approaches that of an experiment in terms of causality and plausibility.

There are, however, some practical problems with these designs, the main one
being that they often take considerable time. They require a long series of observations, which may mean a considerable delay before the results of the intervention
are known. This delay is an ethical problem if there are potential negative outcomes
of the intervention (e.g., if patients are at risk for termination of services). Delay in
availability of results may also provide considerable practical problems if a program
element is unlikely to be implemented without immediate data about its effectiveness. In our criterion example, peer reviewers would very likely not respond
favorably to having to try a new, more difficult set of criteria for a long period of
time without evidence of its effectiveness.

The Regression–Discontinuity Design

It often happens that decisions about persons or cases must be or may be made
on the basis of some numerical indicator, a cutting score, so that persons with a
score at some level or above are to be treated one way, whereas persons below the

critical score are to be treated another way. For example, service providers whose caseloads include at least 20 cases per year from a given third-party payer may be reviewed by that payer, whereas clinicians with fewer cases may not be required to submit to peer review. Or cases that run over 24 or more sessions may be required to submit to peer review, whereas cases that run 23 sessions or fewer may not. Or a set of criteria may be used to "score" cases for potential review, with all cases above a certain score automatically going to review. We assume that, in the absence of intervention, there is a smooth function relating the score on the decision measure to any outcome measure. Thus, we would probably assume that, in the absence of any intervention, the relationship between the number of cases from a third party's panel and the cost to that third party would be linear, with a positive slope, as in Figure 1. In the absence of any intervention, the relationship between initial review criteria and some measure of quality of treatment would be linear, perhaps with a zero slope, as in Figure 2.

If, however, there is an intervention at some particular point along the range of scores or other numerical indicators, if that intervention had an effect it should result in a discontinuity in the relationship (regression) between the two variables. Let us suppose, for example, that cases that undergo peer review will show greater subsequent quality of care than cases that do not undergo peer review, presumably because of increased attention resulting from the peer review process. In such a case, rather than the relationship portrayed in Figure 2, we would expect something like Figure 3 (p. 97).

In the case of number of cases from a panel, if peer review has the desired effect, something like Figure 4 (p. 98) should occur.

The phenomena sketched here fall into the general category of regression–discontinuity effects (Trochim, 1984). When decisions about persons are made on

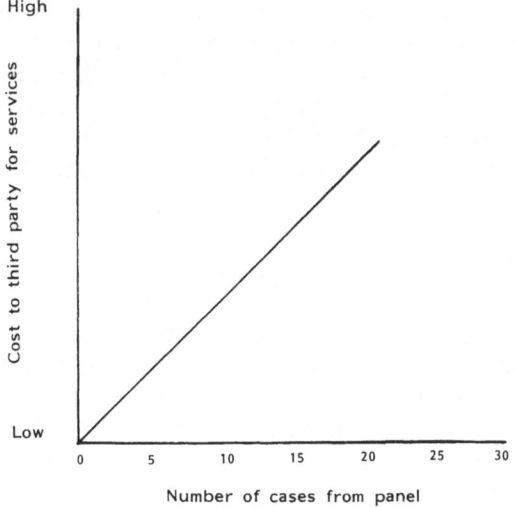

FIGURE 1. The relationship between the number of cases from a third party's panel and the cost to that third party.

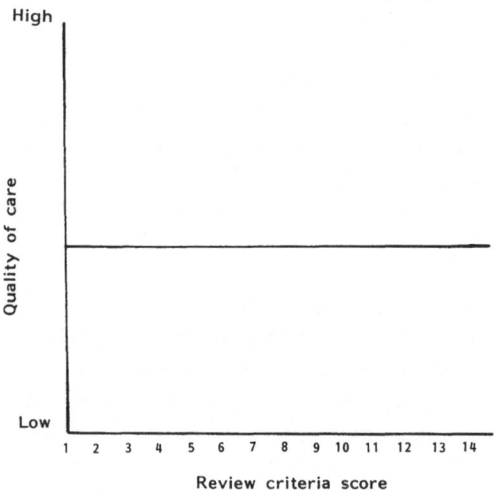

FIGURE 2. The relationship between initial review criteria and some measure of quality of treatment.

the basis of sharp cutting scores—and, in some cases, even relatively sharp cutting scores—the opportunity exists to make some assessment of the effects of that decision. The regression–discontinuity design can, in fact, be a fairly powerful design, often providing quite persuasive evidence for a causal effect of the intervention examined. The statistics for detecting regression–discontinuity effects are reasonably well worked out, although further improvements are likely, especially for the case where the cutting score may be somewhat "fuzzy." Such improvements will be especially welcome in attempts to evaluate peer review systems, as a great many of the cutting scores are likely to be fuzzy. For example, cases may be reviewed before 24 sessions for various reasons, and others may be allowed to go beyond 24 sessions without review. Nonetheless, cases are either assigned or not assigned to peer review and, once reviewed, are either continued or discontinued, paid or not paid, on the basis of some type of cutting score, however fuzzy that score may be. In these cases, the opportunity should be taken to use this design, especially when the cutting scores happen to be fairly sharp. Trochim's book (1984) is the standard for the field.

QUALITATIVE METHODS AND PERSONAL KNOWING

Although not really representing a methodology *per se*, it is important to call attention to the role of qualitative methods and personal knowing in understanding the developing program. As outlined in the evaluation succession model, personal knowledge ranges from intuition to systematic ethnography. As Tharp and Gallimore (1979) noted, personal knowing is not a substitute for other methods. Rather, it may serve the function of understanding and interpreting the results gleaned from other methods, of providing ideas translatable into variables, and of complementing more quantitative methods. Especially in the developing program, data

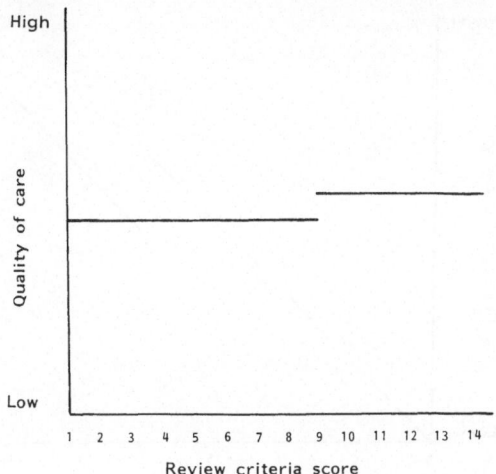

FIGURE 3. The relationship between initial review criteria and the measure of quality of care with intervention.

without interpretation and variables without ideas are bound to lead to dead ends and frustration.

OUTCOME EVALUATION

We have devoted most of this paper to the process evaluation of peer review, noting that a program must be in a reasonable state of effectiveness (e.g., approaching climax) before an overall evaluation of the program can take place. Of course, a great many outside pressures may make necessary early outcome evaluations (e.g., limited funding sources), and repeated outcome assessments are desirable as the program continues to change. It is important to note, however, that an outcome evaluation is not likely to be helpful in isolating how specific elements in the program need improvement or even which elements need improvement (Cronbach, 1975; Tharp & Gallimore, 1979). Remember that, when we are discussing the outcome evaluation of a program, the program itself becomes the independent variable. The issues of strength and integrity of treatment discussed earlier apply to the program as a whole. Assessing a weak program is likely only to confirm the null hypothesis.

The conceptual dependent variables for an outcome evaluation of a peer review system are fairly straightforward and boil down to (a) enhancing the quality of care and (b) increasing efficiency so as to decrease costs. Operationalizing these variables is more difficult. Judging the effects of peer review on treatment is especially difficult because assessing the effects of treatment at all is by no means easy. At the very least, a program should provide evidence that it is not harming patients. Assessing costs is similarly difficult. There may well be hidden costs in the system.

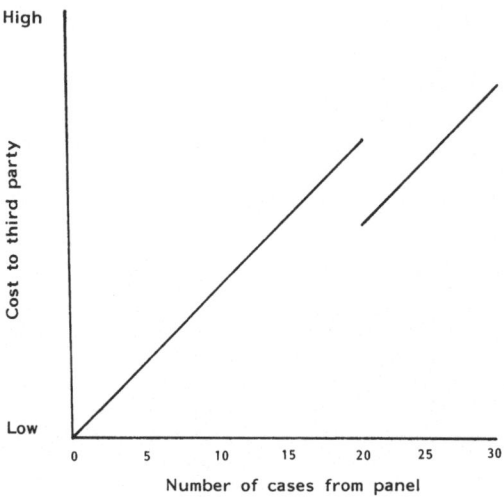

FIGURE 4. The relationship between the number of cases from a third party's panel and the cost to that third party with intervention.

For example, if it takes therapists excessive time to fill out treatment reports, they may increase their fees to compensate for that time. Other potential outcomes of peer review include effects on patterns of practice and educational functions (Sechrest & Hoffman, 1982).

Methodologically, the true experiment is strongly desired when evaluating the evolved program (see McCall & Rice, 1983, for an example of experimental methodology applied to a large-scale mental-health program). Multiple-time-series designs and nonequivalent comparison-group designs may also be used. The evaluator should keep in mind, however, that the audience for an outcome evaluation is likely to be different from the audience for a process evaluation. Outcome evaluations must often be plausible to third-party payers, funding sources, and a variety of policymakers as well as the public. Process evaluations are likely to be "in-house" affairs designed to improve the program, but outcome evaluations must usually prove the worth of the program.

CONCLUSION

Peer review as a means of quality assurance in psychology is in its infant stages. As with any program, especially new programs, the initial challenge is to create a program that has enough strength and integrity to have the potential for showing some successful results. Failure to do so will lead to a confirmation of the null hypothesis, a "no-results" conclusion about the efficacy of the program as a whole. If individual peer-review programs fail to prove their efficacy, then peer review in general may be deemed ineffective and may be prematurely discarded as a means of quality assurance. If this becomes the case, the costs to a profession needing to establish and/or to maintain credibility with third-party payers, politicians, and the public may be great.

To maximize the chances of avoiding such an unfortunate end, peer review programs need to conduct ongoing process evaluations. This chapter has provided a rudimentary introduction to methods and issues pertinent to the process evaluation of peer review programs. However, the reader should be aware that conducting a process evaluation requires considerable flexibility and inventiveness. The evaluator may need to use multiple methodologies or even create new ones. The difficulties involved are substantial.

Yet, the task is necessary and, as past instances of successful evaluations demonstrate, feasible. The alternative is to virtually doom future outcome evaluations of peer review to failure. Finally, there are considerable side benefits to such evaluations. Each evaluation, successful or unsuccessful, paves the way for future evaluations of the same or other programs. Each variable created, each outcome measure devised, and each research method invented contributes to the science of peer review evaluation, and ultimately to the science of evaluating quality assurance.

REFERENCES

Aronson, S. H., & Sherwood, C. C. (1967). Researcher versus practitioner: Problems in social action research. *Social Work, 12,* 89–96.

Babbie, E. (1983). *The practice of social research.* Belmont, CA: Wadsworth.

Baum, M. L., Amish, D.S., Chalmers, T. C., Sacks, H. S., Smith, H., & Fagerstrom, R. M. (1981). A survey of clinical trials of antibiotic prophylaxis in colon surgery: Evidence against further use of no-treatment controls. *New England Journal of Medicine, 305,* 795–799.

Bejar, I. I. (1980). Biased assessment of program impact due to psychometric artifacts. *Psychological Bulletin, 87,* 513–524.

Borgatta, E. (1955). Research: Pure and applied. *Group Psychotherapy, 8,* 236–277.

Campbell, D. T., & Stanley, J. C. (1966). *Experimental and quasi-experimental designs for research.* Chicago: Rand McNally.

Clairborn, W. L., Biskin, B. H., & Friedman, L. S. (1982). CHAMPUS and quality assurance. *Professional Psychology, 13,* 40–49.

Cohen, L., & Holstein, C. (1982a). Characteristics and attitudes of peer reviewers and providers in psychology. *Professional Psychology, 13,* 66–73.

Cohen, L., & Holstein, C. (1982b). Year of degree and psychologists' attitudes toward peer review. *Professional Psychology, 13,* 175–180.

Cohen, L., & Oyster-Nelson, C. (1981). Clinicians, evaluations of psychodynamic psychotherapy: Experimental data on psychological peer review. *Consulting and Clinical Psychology, 49,* 583–589.

Cook, T. D., & Campbell, D. T. (1979). *Quasi-experiments: Design and analysis issues for field settings.* Chicago: Rand McNally.

Cronbach, L. J. (1964). Evaluation for course improvement. In R. W. Heath (Ed.), *New curricula.* New York: Harper & Row.

Cronbach, L. J. (1975). Beyond the two disciplines of scientific psychology. *American Psychologist, 30,* 116–134.

Fisher, K. (1985). Possible deficits threaten future of APA peer review. *APA Monitor, 16,* 1.

Gilbert, J. P., Light, R. J., & Mosteller, F. (1977). Progress in surgery and anesthesia: Benefits and risks of innovative therapy. In J. P. Bunker, B. A. Barnes, & F. Mosteller (Eds.), *Costs, risks and benefits of surgery.* New York: Oxford University Press.

Gottman, J. M., & Glass, G. V. (1978). Analysis of interrupted time series experiments. In T. R.

Kratochwill (Ed.), *Single-subject research: Strategies for evaluating change.* New York: Academic Press.

Hirschi, T., & Selvin, H. (1973). *Principles of survey analysis.* New York: Free Press.

Holland, R. S. (1984). *Perceived truthfulness and perceived usefulness of program evaluations by direct services staff.* Unpublished doctoral dissertation, University of Michigan.

Judd, C. M., & Kenny, D. A. (1981). *Estimating the effects of social interventions.* New York: Cambridge University Press.

Kazdin, A. E. (1974). Methodological and interpretive problems of single-case experimental designs. *Journal of Consulting and Clinical Psychology, 4,* 629–642.

Konecni, V. J., & Ebbesen, E. B. (1979). External validity of research in legal psychology. *Law and Human Behavior, 3,* 39–70.

Luft, L. L., Sampson, L. M., & Newman, D. E. (1976). Effects of peer review on outpatient psychotherapy: Therapist and patient follow-up survey. *American Journal of Psychiatry, 13,* 891–895.

McCall, N., & Rice, T. (1983). A summary of the Colorado clinical psychology/expanded mental health benefits experiment. *American Psychologist, 38,* 1279–1291.

Neale, J. M., & Libert, R. M. (1973). *Science and behavior: An introduction to methods of research.* Englewood Cliffs, NJ: Prentice-Hall.

Peterson, O. L., Andrews, L. T., Spain, R. S., & Greenberg, B. G. (1956). An analytical study of North Carolina general practice, 1953–54. *Journal of Medical Education, 31,* 1–165.

Pizzirusso, C., & Cohen, L. (1983). Psychologist's evaluations of psychodynamic psychotherapy: A peer review analogue experiment. *Professional Psychology, 14,* 57–66.

Reichardt, C. S. (1979). The statistical analysis of data from nonequivalent group designs. In T. D. Cook & D. T. Campbell (Eds.), *Quasi-experimentation: Design and analysis issues for field settings.* Skokie, IL: Rand McNally.

Scriven, M. (1967). The methodology of evaluation. In R. W. Tyler, R. M. Gagne, & M. Scriven (Eds.), *Perspectives of curriculum evaluation,* (AERA Monograph Series on Curriculum Evaluation, No. 1.). Chicago: Rand McNally.

Sechrest, L. (1984). *Evaluating health care.* Unpublished manuscript, University of Arizona.

Sechrest, L., & Hoffman, P. E. (1982). The philosophical underpinnings of peer review. *Professional Psychology, 13,* 14–18.

Sechrest, L., & Redner, R. (1979). *Strength and integrity of treatments in evaluation studies.* Washington, DC: National Criminal Justice Reference Service, National Institute of Law Enforcement and Criminal Justice, Law Enforcement Assistance Administration, U.S. Department of Justice.

Sechrest, L., & Yeaton, W. H. (1981a). Assessing the effectiveness of social programs: Methodological and conceptual issues. In S. Ball (Ed.), *New directions for program evaluation.* Beverly Hills, CA: Sage.

Sechrest, L., & Yeaton, W. H. (1981b). Empirical bases for estimating effect size. In R. F. Boruch, P. M. Wortman, & D. S. Cordray (Eds.), *Reanalyzing program evaluations.* San Francisco: Jossey-Bass.

Sechrest, L., West, S. G., Phillips, M. A., Redner, R., & Yeaton, W. (1977). Some neglected problems in evaluation research: Strength and integrity of treatments. In L. Sechrest, S. G. West, M. A. Phillips, R. Redner, & W. Yeaton (Eds.), *Evaluation studies review annual* (Vol. 4). Beverly Hills, CA: Sage.

Sechrest, L., Ametrano, I. M., & Ametrano, D. A. (1982). Program evaluation. In J. R. McNamara & A. G. Barclay (Eds.), *Critical issues, developments and trends in professional psychology.* New York: Praeger.

Stricker, G., & Sechrest, L. (1982). The role of research in criteria construction. *Professional Psychology, 13,* 19–22.

Sutcliffe, J. P. (1980). On the relationship of reliability to statistical power. *Psychological Bulletin, 88,* 509–515.

Tharp, R. G., & Gallimore, R. (1979). The ecology of program research and evaluation: A model of evaluation succession. In L. Sechrest, S. G. West, M. A. Phillips, R. Redner, & W. Yeaton (Eds.), *Evaluation studies review annual* (Vol. 4). Beverly Hills, CA: Sage.

Trochim, W. M. K. (1984). *Research design for program evaluation: The regression-discontinuity approach.* Beverly Hills, CA: Sage.

Weiss, C. H. (1971). *Organizational constraints on evaluation research.* New York: Bureau of Applied Social Research.

Weiss, C. H. (1972). *Evaluation research: Methods of assessing program effectiveness.* Englewood Cliffs, NJ: Prentice-Hall.

Weiss, C. H., & Bucuvales, M. J. (1980). Truth tests and utility tests: Decision makers, frames of reference for social science research. *American Sociological Review, 45,* 302–313.

Yeaton, W. H., & Sechrest, L. (1981). Critical dimensions in the choice and maintenance of successful treatments: Strength, integrity and effectiveness. *Journal of Consulting and Clinical Psychology, 49,* 156–167.

5

Education for Quality Assurance

Russell J. Bent

The author's involvement with the quality assurance of psychological services in a number of settings since the mid-1960s has led to his conviction that psychology students must be introduced to the concepts and practice of quality assurance throughout their training. Without such an introduction, only the most watered-down or self-serving service-quality assurance system is likely to emerge because of the political and self-centered interests of many practicing psychologists. The author's 7 years of experience with the American Psychological Association (APA) CHAMPUS peer-review system have reflected the confusion and the lack of knowledge and experience that practicing psychologists have of peer review and quality assurance concepts. The APA still has no coherent policy on quality assurance, nor on peer review. Similarly, reflecting their APA/CHAMPUS experience, Shueman and Troy (1982), in addressing quality assurance, concluded that

> to be successful, psychologists must be introduced to such procedures early in their graduate training programs. Accountability procedures must become an integral part of the professional preparation of psychological service providers so that they will have the experience necessary to practice accountability and they will accept the necessity for such procedures. (p. 64)

The author, with faculty support and contributions, has directed an effort to introduce the concepts related to quality assurance and quality assurance activities as part of a doctoral training program in clinical psychology. This chapter summarizes the major areas of the curriculum that are part of this effort.

QUALITY: A GENERAL CONCEPT, A SPECIFIC ACTIVITY

Quality is a definable characteristic of an object, a product, or an activity. The characteristics attributable to quality may be evaluated as positive or negative, and as highly specific or very general. When the term *quality* is applied to mental health

---

RUSSELL J. BENT • School of Professional Psychology, Wright State University, Dayton, Ohio 45435.

services or psychological services, it is implied that quality is controlled, monitored, or ensured, a dimension of management and accountability emerges.

To prevent confusion, *quality* must be thought of as a generic term, which, in turn, must be specific if it is to have any value in practice. After all, there is no one indicator of a good psychological service provider or of what is an acceptable psychological service. A number of qualities must be present to indicate a general quality. It is helpful, then, to think of the term *quality* as always requiring specific definitions and as having multiple meanings as determined by the definer. In something as imprecise as psychological practice, this definition is critical if one is to avoid confusion and to enhance the usefulness of quality assurance activities.

Quality assurance activities are those activities designed to manage, monitor, evaluate, and enhance a quality or qualities. In any approach to quality assurance, the definers select qualities that should be present and organize activities that lead to evaluative conclusions about whether such qualities are present in some degree or are not present to a sufficient degree.

It is also helpful to think of some qualities as associated with the training, or designation, of the psychologist and other qualities as associated with the services rendered by the designated psychologist. Confusion often results when this distinction is not recognized. For example, if *quality* is defined solely as the designated characteristics of the psychologist and the psychologist meets those definitions (e.g., has graduated from an APA-accredited professional program, is licensed, is experienced, and has advanced certification) then the psychological services provided by that psychologist would always be judged as being of good "quality." However, with the growing complexity of practice and the reality of individual instances of service by psychologists being questioned by consumers, professionals, and third parties (Morton, 1982), there is a need for quality assurance about services rendered by fully designated psychologists. Both quality assurance of the designation of the provider and quality assurance of the service activities of the provider are necessary.

Until recently, organized psychology has been most concerned with the quality of the psychologist as determined by education, training, statutory recognition, and advanced certification. Ethical standards and standards of practice have extended beyond provider designation. Only recently (Claiborn, Stricker, & Bent, 1982) has organized psychology focused on the quality assurance of individual instances of service provision or patterns of practice rendered by fully designated psychological service providers. Psychology training must cover the psychologist's designation, practice, and ethical standards in order to gain accreditation by the APA (APA, 1979). There are no requirements by the APA related to quality assurance in individual practice other than the traditional requirements of proper course work, the supervision of students' practice, and the adequate observation of student practice by the faculty. The widely used quality-assurance activity of formal review by peers is not a requirement, nor is it taught, much less practiced, in many curricula.

This chapter describes the curriculum related to quality assurance of the School of Professional Psychology at Wright State University. The teaching of quality assurance as it relates to the designation and the standards of psychologists is described in summary, as it is not atypical of most curricula. The teaching of quality assurance as it relates to individual practice or patterns of practice is described in more detail. In this latter regard, several features are covered, including a management information system, the psychological-services review process (documented and face-to-face), and the psychological-practice competency evaluation.

QUALITY ASSURANCE IN THE CURRICULUM OF THE CLINICAL PSYCHOLOGIST

In teaching the designation aspects of quality assurance as related to the psychologist, the curriculum covers most of these issues in the 1st and 2nd years of a four-quarters-per-year, 3-year curriculum. The issues covered in the 1st year of the curriculum start with the ethical principles of psychologists of the APA. When new students begin the first quarter at the school, a convocation is held for them. A central theme of the convocation is the public acceptance of a professional credo, part of which includes acceptance by them of the ethical principles. Students are expected to guide their values and behavior according to these professional standards while in the program and, of course, after graduation. Adherence to the ethical principles at this early juncture and during graduate work promotes the students' socialization of the values of the profession. The ethical principles on competence, moral and legal standards, and professional relationships are particularly relevant in guiding both students and faculty in the educational process itself. In this way, a committee on ethics with both faculty and student representation can address issues that might be handled by academic policy alone. The school's credo and adherence to the ethical principles are reaffirmed on graduation. Graduates of the program and students alike comment that this is one of the most meaningful affirmations they have in their careers.

Educational standards related to APA accreditation, standards for psychological tests and measures, and the history of various conferences on training, as well as the changing role of the psychologist, are covered in the first year through the school's seminar in professional development. Just about all of the designation and general-standards quality-assurance issues are taught in the professional development seminar. The professional development seminar was devised to be a seminar forum for professional issues and for the modeling of professional practice by the full-time clinical faculty of the school. The seminar is labor-intensive, each faculty member being assigned five or six students for the three quarters in each year that the seminar continues. Students are assigned to a different faculty member during each of their 3 years in the program.

The professional development (PD) seminar also includes clinical practice presentations by students, faculty supervision, student personal and professional support, school administration issues, and developmental, professional issues, such as using supervision when starting a field practicum or selecting an internship setting. Students rate PD groups as being very meaningful, educational, and personal.

In the 2nd year of the curriculum, the PD group is used to carry out both documented quality-assurance peer review and face-to-face peer review. Early in the program, separate review groups were organized, but it proved to be less cumbersome and more natural to teach peer review in the PD groups starting early in the second year. The format of these reviews is described in the chapter. During the 2nd year, the standards for providers of psychological services and the related specialty standards are reviewed, together with licensing standards and issues, the American Board of Professional Psychology certifying process, and other specialty examinations. Many of the key issues, or themes, are presented to the full class and are followed by a "personalization" of themes in the small PD seminar format. The PD leaders and students alike return to "themes" when presenting practice samples and supervision material.

Contemporary designation and standards are presented in a 3rd-year professional issues seminar, which is a mandatory course for all students. In this seminar, current APA committee agenda items are selected by the faculty. Small student groups are then organized as the boards or committees considering these issues, and the group is required to deal with the issues as if it were the committee. The class observes and later interacts with the "board" or "committee." The professional issues seminar is team-taught by senior faculty.

In the middle of the 3rd year, each student must complete the practice proficiency evaluation, a mandatory evaluation of knowledge, attitudes, and skills related to psychological practice. The practice review in the PD groups and in the practice proficiency evaluation are quality assurance activities related to the individualized review of the psychological services. These quality-assurance activities include both documented review and face-to-face review, which are described in detail later in this chapter.

There are two formal courses related to quality assurance of psychological services that are required in the 2nd year of the curriculum. These courses are "Individualized Service Planning and Quality Assurance" and the course entitled "Service Delivery: Management and Administration." The latter course presents topics on incorporating quality assurance methods into service delivery, particularly preferred-provider organizations and employee assistance plan organizations. The former course introduces the student to individualized service planning for clients using a broad, eclectic service model. The course presents a management information system, service by objectives (SBO), which leads to psychological service review. Students document cases using the SBO method and learn to review psychological services using the students' own field-practice clients and reviewing each other's clients in a peer review format. Each student receives feedback on his or her review style from the reviewers. An openness and a natural use of quality assurance that have consultative, educational feedback are developed. There is a genuine lack of defensiveness, which often dominates the practicing psychologist's relationship to quality-assurance service review. The next section presents a more detailed description of the SBO system, which is the system of documentation supporting much of the individualized service-review activity.

SERVICE BY OBJECTIVES: A GOAL-ORIENTED MANAGEMENT INFORMATION SYSTEM

GENERAL INTRODUCTION

The quality assurance of psychological services has two main aspects: the specification of the quality to be monitored or evaluated and the activity or process that is used to assess that quality.

Because of the costliness and resistance by many providers in directly observing practice, the use of information in documented form is a more realistic, practical way to proceed with quality-assurance practice review. Most quality-assurance activities by hospitals, mental health centers, and third-party intermediaries use quality assurance methods that rely on documented information and judgments by providers of a like discipline, or of like expertise, serving as "peer" reviewers. Psychologists are invariably a part of one or more of these settings and are subject to review. As Baker (1983) pointed out in his fine chapter on continuing professional development, "Now, as psychology acquires greater parity, the professional must take a responsibility to self-monitor not only for intrinsic reasons, but also with the realization that if we do not monitor ourselves, we will certainly be monitored by others" (p. 1367).

As a start in self-monitoring, the student is introduced to a psychological "record" system. In order to carry out documented review, there must be some documentation to review. There is no psychological "reporting" system generally accepted by, or recommended for, the practicing psychologist. Psychologists, usually working in agencies or in medical settings, use existing record systems with minor modifications or with no changes at all.

It is surprising, perhaps, that the psychological discipline, with emphasis on empiricism, operational definitions, and psychometric and scientific values related to practice, has not developed record systems incorporating this emphasis. Kiesler (1977) pointed out the great potential that agency, hospital, and other record systems would have for research and individual case evaluation if those record systems incorporated more aspects from the discipline of psychology into their format. The author further contends that better formulated documentation systems can also aid client services by contributing to the assessment, planning, and quality-assurance review process, thereby promoting service management and quality. Quality enhancement procedures can be reasonably carried out if the documentation system provides the elements of information and organization required by quality assurance procedures. In order for documented information to be used, the required information elements must be present and readily available in a systematized format that promotes easy, efficient access to that information. Further, the information required by the system should be the minimum required to make the system work. Information items should be repeated only if absolutely necessary. The system should be flexible and capable of very wide application, including review for various factors and utilization concerns.

Service by objectives (SBO) is a comprehensive management information (record) system designed to provide a documentation method that promotes quality psychological practice. As defined earlier, quality is what the provider system cares to (operationally) define as quality. The SBO system lends itself to incorporating quality assurance considerations into the documentation procedures. For example, SBO requires a basic client-information base, enough information analysis to result in service goals, a service strategy, and a systematic review of the service plan, followed by service replanning, if indicated (Bent, 1982).

As developed, SBO has a history rooted both in scientific values and in the realities of service delivery ranging from independent office practice to various forms of group practice in either independent or public-supported services. The SBO system to be described has been incorporated into mental hospital units, community-mental-health centers, and independent practice settings.

In developing SBO, a number of customary record-system approaches were rejected, such as using such diagnostic categories as those that appear in the third edition of *The Diagnostic and Statistical Manual of Mental Disorders* (DSM-III; American Psychiatric Association, 1980) as being central to the system or what is referred to as source (discipline-oriented) information formats. A departure from source records, the problem-oriented medical record (POMR) was promoted by Lawrence Weed (1964). The POMR centers the record on the patient's problem(s), not on the patient's diagnosis, followed by a series of endless, disparate, unrelated information entries. In some respects similar to the POMR, SBO centers on the planned, expected objectives, or outcomes, of psychological service. The attainment of service goals or objectives, rather than problem resolution, serves as a table of contents serves in a book.

In many respects, SBO is a variant of a problem-solving model. The problem-solving model is broad (eclectic) in its format, essentially consisting of:

- Designation of a problem
- Analysis of the problem
- Proposed and implemented activities to solve the problem
- Evaluation of the problem-solving effort (often feeding back to more refined problem designation, analysis, implementation, and resolution)

Such a general paradigm was proposed in 1971 by Urban and Ford as an integrating model for the many psychotherapy orientations in our field and as a step toward the development of a more coherent psychology of human change resulting from constructive interventions. The problem-solving paradigm has been proposed as an intervention "orientation" (Dixon & Glover, 1984).

The SBO modifications of the problem-solving model are several. First, the emphasis is more on the attainment of resolution of problems than on reasonable goals related to problems. In psychological intervention work, some clients presenting problems are not appropriate for psychological work at all, and others' problems can be only partially resolved, not solved. The reasonable attainment of objectives designed to resolve problems is the approach rather than a dichotomous notion of problem and problem solution. The psychologist often overemphasizes problems (pathology), virtually excluding or seriously minimizing the client's strengths. A focus on goals and objectives provides a better balance by bringing to bear the client's assets (Kiresuk & Lund, 1977). Problems must be balanced with a consideration of assets, and both, in turn, must be subjected to an analysis resulting in attainable service objectives.

Second, new human skills or competencies may be the focus of the intervention activity in a strictly growth mode, that is, not related to problems. Simply, developing new competencies is a proper activity of the service system, which, in turn, may be related to the prevention of problems or the development of new human effectiveness. Whether the focus is on attaining objectives or developing new competencies, diagnosis and therapy are reframed as activities used to describe and analyze client problems and then to set reasonable objectives or new competencies through the application of intervention activities clearly related to the service objectives.

Service by objectives is decidedly prospective in thrust. Record systems are thought of as retrospective documents, emphasizing services that have been done and diagnoses that have been made. The major thrust of SBO is to indicate what is planned and what is being done, as well as what has been done.

The SBO system modifies the problem-solving model, adhering to a goal-attainment model, by emphasizing the goals and objectives aspect of the problem-solving model. The goal-attainment model has the following requirements:

- Description of the problem or the new competency
- Analysis of the problem or the new competency
- Establishment of reasonable goals and objectives related to problem resolution or new competencies
- Designation of the intervention methods in a planned strategy to attain goals and objectives, or new competencies
- Systematic review and evaluation of the degree of goal and objective attainment, and possibly the development of a modified service plan

Each of the above requirements is incorporated into the three main sections of the SBO documentation system. Each main section has a number of documentation formats designed to satisfy the section's information requirements. Each format has guidelines for completion. The sections are separated to promote easier use of the system and to increase the reliability of information in the various sections.

Guidelines aiding in the design and use of SBO have their bases in the technologies and the research results of several idiographic measurement methods.

Idiographic methods have been relatively neglected by psychologists in favor of nomothetic measures. The essence of providing service for a client resides in the individuality of that client. It would appear natural for measures of individuality to be clearly related to a documentation system whose primary purpose is to communicate the client's individual characteristics and individualized service plan.

Essentially, idiographic methods require effort and care in the descriptive aspects of client assessment, service planning, and subsequent review and evaluation. Most generally considered, idiographic methods adhere to the demands of an operational definition. Unlike nomothetic measures, idiographic methods do not rely on statistical, normative assumptions and inferences for their value.

The many strengths of idiographic methods include:

- The ability to be tailored to unique intervention situations with greater relevance and outcome validity
- Recording formats that are "standardized," yet open-ended and requiring individualized information
- Measures that include clients and other system respondents, allowing flexibility in recording broad system variables
- Methods that may be applied repeatedly for evaluation and replanning activity
- Modification and amplification that can occur as new data appear throughout the intervention process
- Although open-ended, descriptions that require operational definition insofar as possible, allowing professional and lay observers (including clients) to reliably agree about the designation of problems, competencies, and service objectives

The standardized, relatively open-ended formats of SBO require that broad guidelines be followed by the service provider and be reflected in his or her documentation using the system. As examples, the guidelines related to problem designation and problem analysis have drawn heavily from the problem-oriented medical-record system and related research (Grant, 1972; Weed, 1964) as well as behavioral assessment methods (Haynes & Wilson, 1979; Hersen & Bellack, 1981); guidelines for goal and objective setting from the target complaints method (Battle, Imber, Hoehn-Saric, Stone, Nash, & Frank, 1966); goal-attainment scaling methods (Kiresuk & Lund, 1977); and the work of Carkhuff and Anthony (1979). Service procedure and service strategy designations have relied on the initial American Psychological Association's CHAMPUS project, expert opinion by various therapists, and the analysis of service procedures of the major psychotherapies by Beutler (1983). Guidelines do not have a single narrow orientation but attempt to be honestly generic. In effect, the orientation of SBO, if there is one at all, is eclectic or systems, allowing the use of any single or combination of specific orientations, or procedures. Often, a global or a very general description of the service process is suggested by a particular orientation (some psychodynamic approaches, for example), and such a description is appropriate and acceptable. Behavioral descriptive methods are not required, but a clarity of description is required. The position taken by some service providers that certain therapeutic orientations or service plans defy any description is not acceptable.

Space does not allow the inclusion of full guideline descriptions, but a guideline summary is presented at the end of the next section in this chapter.

THE SBO DOCUMENTATION FORMAT

The reader is requested to refer to Appendix A of this chapter for copies of the documentation formats to be described.The SBO documentation format is organized under the following three sections:

The Client Information Base

The client information base is a section in which specific client information is filed. The section has a file listing indicating those information elements and reports that contribute to the formulation, development, and evaluation of the clients' service episode. Two forms are required under this section: the initial-contact-documentation form and the client initial-evaluation-information form.

The initial-contact-documentation form gathers only very basic identifying information related to the client, which is often gathered by a receptionist, and the basic reason for referral, gathered by a short telephone contact with the client. The purpose of this transaction is to connect the client to an approprioate psychologist (or other resource, if indicated) as soon as possible.

During and following the client's initial consultation with the psychologist, the client initial-evaluation-information form is completed. This form very carefully defines the problem situation and a quick estimate of client suicidal risk, or lethality, and need for immediate (crisis) services. An analysis of client functional systems is included to better imbed the problem situation in the proper context. Substance use is also included here. A professional analysis of the problem in context is formulated, together with an estimate of intervention, service actions, and client acceptance or compliance with the recommendation made. Generally, just before the initial evaluation, the client, with the receptionist's help, completes the basic client-information form. This form has been carefully devised to identify the client, to gather information for third-party billings, to authorize an information release for billing purposes, to authorize an emergency contact, and to provide brief information about related previous services or current health, legal, or other human services. This required information, together with any other information related to the client (a past or to-be-completed psychological-test-battery report, for example), is systematically recorded and filed in the client-information-base section. This section is the last section in the folder collating the SBO forms.

The Individualized Psychological-Service Plan (ISP)

This service plan form is filed first in the SBO folder. It is, after all, the major current and future plan guiding service for the client, and it is based on all the other information in the system. The single form is the only one in this section. The form contains a carefully formulated problem-analysis summary related to the initial evaluation interview(s) or subsequent service contacts that were necessary to formulate the ISP. A final goal, the best reasonably estimated outcome of service, is formulated with a checkoff of the final outcome goal category and an estimate of the time variable in weeks and units of service.

Next, the intermediate goals are listed (if possible) and are related to each goal (if appropriate) and the intervention procedure(s) related to that intermediate goal. An estimate of the client's concurrence with the plan and the client's signature (if this will not have a negative influence on the client), with any comments, complete this format.

Students receive several hours' training in each aspect of the ISP. A summary of the guidelines related to properly completing the ISP is in Appendix B of this chapter. These summary guidelines are also used from time to time for the evaluation of providers in the peer review process.

The Individualized Psychological-Service Review (ISR)

This form provides for periodic review of the service plan. The service provider anticipates the next date when the service plan is to be reviewed. The date should be related to the plan and the anticipated goal attainment (e.g., perhaps several days at most for goals related to crisis intervention services and months for long-term supportive therapy). The actual date of review, the units of service, and the reviewer are indicated. The reviewer may be the student's supervisor (usually) or a peer review team (periodically). Both the client and the provider estimate global progress toward the service goals.

When a review is completed, another date is set for the next review, and so on, until service is terminated or continuity of service by another provider is accomplished. Termination information and disposition should be entered and appear as the last review entry. The ISP and the last entry in the ISR constitute a relatively complete summary of the service episode of a given client and can be used for information summary purposes.

Progress and observations are a subsection of the ISR. In the current information system, billing, accounting, and service-transaction monitoring are handled by an automated (computer) program. A summary of these transactions appears on the left side of the SBO folder. Various supervisors and providers use this format in different ways. Some write "newsy" notes for each service session. The author discourages such notes, preferring that they be part of the provider's personal notes, not part of the record. Significant progress, not merely the process of each service transaction, is helpful as it relates to specific service goals. Lethality information or the rationale for significant changes in the service plan should have a referent in this subsection.

It is important to point out that documentation in the SBO system should occur during the service hours with the client's involvement, if possible, or directly after the service session. Subsequential ISP's are encouraged so that, as knowledge of the client, the client's progress, and a changing context occurs, evaluation and replanning can take place.

THE QUALITY-ASSURANCE SERVICE-REVIEW PROCESS

The review process is of two types. Students periodically review documented service plans in the PD seminar groups and in the quality assurance course. The form used for this review can be seen in Appendix B of this chapter. Just written information is exchanged.

Each student presents one or more client service plans and progress reports in the PD seminar group. Three PD seminar members (one member is sometimes the faculty member) serve as peer reviewers while other members observe. The student personally presents the ISP and the ISR (if appropriate) and other information to satisfy his or her need to present sufficient information about the client service. The information presented is augmented by reviewer comments and answers to questions brought up by reviewees in a personal, but formal, interaction. Documented information is forwarded to the reviewers before the meeting of the PD seminar's peer review panel.

Review is carried out in a cordial, facilitative atmosphere, with an emphasis on allowing the reviewee to demonstrate his or her knowledge, skills, values, and use of supervision.

The review panel allocates a little less than an hour for a quality-assurance service review, with the following format:

1. *Review focus and strategy plan.* The panel reviews the service information documentation (the ISP, perhaps the ISR, and other material—sometimes a recorded client–provider tape segment) and defines the focus of the review process. In effect, those aspects of quality that are to receive special focus during the review are specified. The reviewee does not personally take part in this process. The time allowed for this focused review definition is 10 minutes.

2. *Brief presentation by reviewee.* This part of the review format allows the reviewee to put the client work sample into context, for example, where the service was rendered, the nature of the supervision, broader background material, and recent progress or issues. The reviewee presents for about 5 minutes.

3. *Panel-member–reviewee interaction.* A collegial discussion and elaboration of the work sample according to the focus and strategy are determined by the panel. Questions, suggestions, elaborations, clarifications, and reactions between the reviewee and the reviewers are discussed. The reviewee is assured of the "last word." The time allotted is about 15 minutes.

4. *Summary individual feedback.* Feedback is given by each reviewer regarding that individual's evaluation and consultative comments relate to the review focus. Student interaction follows the reviewer feedback. The total time allotted is 15 minutes.

5. *Summary consensus feedback.* The panel develops a short written consensus, an evaluative summary of the work sample and the reviewee's presentation. The reviewee is an observer only. The time allotted is 10 to 15 minutes.

The summary evaluation comments are communicated on a form similar to the ISP guidelines in Appendix B. Each work sample may be rated in the five designated quality areas or in any other designated quality area. The six "standard" quality areas are problem description, problem analysis and intervention rationale, service goals and objectives, service strategy, professional role, and other focuses. A rating scale may or may not be used in the process.

The learning experience for students exposes them to a more formal, external review process than supervision provides. The experience also lets them experience the work of a reviewer with reviewee and faculty feedback. During the 3rd (last) year in the academic program, the student has a client in therapy reviewed by a senior psychologist practitioner who is a member of our voluntary clinical faculty.

A particularly exciting development in the program has been the "consider guidelines." These guidelines represent suggestions derived from evaluative re-search and expert clinical, consensual opinion related to various practice activities. The student is encouraged to review these guidelines when involved in the activity or the general problem or goal addressed by the guidelines. The student is to care-fully individualize his or her service, considering strongly acceptable general analysis suggestions as a guide. The guidelines relate to such activities as:

- Suggestions for the analyis of neuropsychological problems
- Suggestions on how to write an individualized treatment plan
- Approaches to intervention with treatment goals and objectives related to
  — "depression"
  — overeating
  — severe head pain
  — feelings of worthlessness and poor college productivity
  — increased ability to relate and be at ease with marital partner
  — development of "survival skills" in a repeatedly hospitalized individual

The program has a plan to computerize these guidelines and continuously up-date them through our word-processing system.

It should be noted that a service activity, billing, and accounting system (in-cluding case scheduling) has been computerized and will not be covered in this pre-sentation. There are no immediate plans to computerize other parts of SBO, with the exception of the basic-client-information form.

THE CLINICAL-PRACTICE PROFICIENCY EVALUATION

The clinical-practice proficiency evaluation is the most central, verifying quality-assurance activity in the school's curriculum. Rather than having only the rather elaborate set of procedures, committees, and quality indicators connected with the dissertation, the program affords equal attention to an "orals" type of prac-tice evaluation.

Because the primary aim of the school is to graduate practitioners in keeping with the empirically verified careers of most clinical psychologists, it appears appro-priate to emphasize a clinical proficiency evaluation recognized as a culmination and verification of several years of clinical training.

The format of the evaluation is similar to the American Board of Professional Psychology's specialty certification examination in clinical or counseling psychol-ogy. The student must select two work samples; a third sample is optional. One sample must emphasize assessment and treatment planning. Another sample must emphasize intervention knowledge and skills. Both samples represent recent or concurrent clients of the student.

The school has experimented with criteria, procedures, and requirements for this evaluation since 1980. The current methods work well for us and have been very well received by students and faculty. The carefully worked-out procedures are presented in Appendix C of this chapter.

Rather than present a detailed accounting of the proficiency evaluation here, we advise the reader to study Appendix C. As a guide to that detailed presentation, a summary of the process is described here.

The assessment work sample must be arranged in a clinical setting and is observed and rated by a member of the full faculty or of the clinical faculty. The student submits a report in 48 hours. An SBO-documented case with a video- (preferred) or audiotape of a therapy hour is also submitted. Copies of this material are made and forwarded to the proficiency-panel members.

The proficiency panel consists of at least two faculty, a full faculty member as chair and a clinical faculty member, with an optional third member. Current supervisors cannot serve on the panel. A student member (at the student's class level) is also appointed to the panel.

The student is rated on aspects of the work samples. The quality designations are relatively global and are indicated in the committee-member report and worksheet guide in Appendix C. The general quality areas include assessment, intervention, psychological knowledge and values, and professional role development.

Students are rated against what the faculty believes a 3rd-year student should be like in the quality areas addressed. Further, the student should have the basic qualities that would prepare him or her for an internship. A very global judgment of whether the student is prepared for an internship is made at the end of the panel's evaluation. Immediate feedback is given to the student following the evaluation. In a minority of instances, remediation requirements are demanded, which the student must complete before being approved for the internship year.

The proficiency quality review provides valuable feedback to faculty as well as students, importantly involving senior full-time psychologist practitioners in the process. The evaluation represents a critical verification of the student as ready to cope with internship-level work, pointing out the current strengths and limitations of the student in the judgment of the faculty. The student is also familiarized with what it is like to have the responsibility of making peer judgments on important aspects of practitioner quality requirements.

MONITORING STUDENT PRACTICE

A final feature of our program related to the quality assurance of psychological services is the training activities report. The detailed information gathered by this report can be found in Appendix D.

In summary, this report gathers information on the student's pattern of practice (training). Pattern-of-practice analysis is, in many ways, more meaningful than single-client reviews. The monitoring of student practice provides the student, the supervisor, and the faculty members with detailed summary information on the practice (training) activities of each student.

The activities report is submitted each month for the previous month's activities by each student. The form currently used is not the one presented in the format, as it is in optical scan form. All information is scanned and entered into a computer. A report writer produces a statistical, cumulative summary with a number of graphs. The report is by month and includes a cumulative summary for up to 1 year of a student's practice activities. These reports represent excellent self-correcting stimuli for students and faculty to better balance practice and to meet general training guidelines. Many internship program applications request detailed information about prior practice experience, and students use these summaries to provide such detailed summaries.

This type of monitoring and reporting allows the school to better manage supervisory resources and to keep an accurate record of what students and supervisors actually do on the dimensions used in the report.

SUMMARY

Our curricula in training practicing psychologists should include the concepts and practices involved in quality assurance. *Quality assurance* is a broad term that must be specified through clear definitions and through clearly described activities that determine the degree to which various qualities are present.

This chapter describes a broad approach that includes quality assurance concepts and activities in a curriculum designed to train professional (clinical) psychologists. Quality assurance related to the designation of the psychologist and related to the day-to-day practice of psychology has been developed. Several examples of this latter type of quality assurance were described. The information system used by the practitioner is particularly important in supporting and enhancing documented review, face-to-face review, and individual proficiency evaluation. Such an information system, service by objectives, is presented, together with a peer-review quality-assurance process, a practice-proficiency evaluation process, and a pattern-of-practice (training) reporting system.

There is no particular difficulty in incorporating quality assurance into a training curriculum. Thus far, such incorporation has been very well accepted by and very rewarding for students, clients, and faculty.

APPENDIX A

WRIGHT STATE

Wright State University
Dayton, Ohio

Individualized Psychological Service Plan

Psychological Services Center
School of Professional Psychology

Client ID

Client _____ Age ____ Sex ____ ☐ Initial/Service plan

Provider _____ Date of plan _____ ☐ Revised service plan

_____ Service units to date

Problem analysis summary

Final goal

☐ Crisis stabilization
☐ Growth: community/role competency
☐ Growth: personal/health enhancement
☐ Other: _____
Estimate of goal attainment within a time period of _____
weeks with _____ units of service.

Intermediate goals *(please number)*	Intervention strategy/procedure
1.	1.

Client concurrence ☐ Full ☐ Partial   Client's signature _____

Comments _____

WRIGHT STATE

Wright State University
Dayton, Ohio 45435

 **Individualized Psychological Service Review**

School of Professional Psychology
Psychological Services Center

Client ID

Client   Provider

Review

Anticipated review date _____/_____/_____   Actual review date _____/_____/_____
                                                          Units of service _____ Reviewer _____

Global progress/Client's estimate   Global progress/Provider's estimate
minimal   some   moderate   extensive   minimal   some   moderate   extensive

Review

Anticipated review date _____/_____/_____   Actual review date _____/_____/_____
                                                          Units of service _____ Reviewer _____

Global progress/Client's estimate   Global progress/Provider's estimate
minimal   some   moderate   extensive   minimal   some   moderate   extensive

Review

Anticipated review date _____/_____/_____   Actual review date _____/_____/_____
                                                          Units of service _____ Reviewer _____

Global progress/Client's estimate   Global progress/Provider's estimate
minimal   some   moderate   extensive   minimal   some   moderate   extensive

WRIGHT STATE

Wright State University
Dayton, Ohio 45435

Progress and Observations Notations

School of Professional Psychology
Psychological Services Center

Client ID

Client name	Provider

Directions

In this section of the record, *significant* progress toward service goals/objectives occurring after a *particular* service transaction may be entered. Periodic progress summaries are to be entered in the Individual Service Review Section of the record.

Observational notes, particularly related to lethality, crisis, required behavioral recording, or other specific timely recording indicated in the Individualized Service Plan should be entered under this section.

Do not enter supervisory notes, process information, rationale, or theory in this section.

Progress and Observations

Enter chronologically the date, goal/objective number, notation, or observation.

WRIGHT STATE

Wright State University
Dayton, Ohio 45435

Client Information Base Section

School of Professional Psychology
Psychological Services Center

Client ID

Client	Provider

Included under this section of the client record/management file are information elements and reports which contribute to the formulation, development, and evaluation of the client's service episode.

Examples of such information are
- Client complaint/problem inventory
- Role functioning scale
- Consultation reports and psychological assessment reports based upon psychological tests, test data, relevant interview summaries, information from other agencies, and so on.

Such information should be filed following this face sheet with a code letter and short content description in the space below.

Initial Contact Documentation

A _____

Client Initial Evaluation Information

B _____

C _____

D _____

E _____

F _____

G _____

H _____

I _____

J _____

K _____

L _____

WRIGHT STATE

Wright State University
Dayton, Ohio 45435

Initial Contact Documentation

School of Professional Psychology
Psychological Center
Community Branch

Date _____

Prospective client:

_____
Last name   First name   Middle initial

Male   Female   Age _____
(Circle)

Inquiry made by:

_____
Last name   First name   Middle initial

Relation to client

Referral source:

_____
Name   Agency or affiliation

Telephone number:

Residence _____   Work _____

If it is necessary to contact you at work, may we call there? The caller will not identify reason for calling.

To be completed by Psychological Center personnel

Information to caller

1. We would like to make you aware that the center is staffed by advanced professional psychology students under the close supervison of members of the professional psychology faculty.

2. This is basically a low-cost center, serving clients who cannot afford to pay private practice fees.

Referral information:

Action taken:

Resource person _____   Supervisor _____

WRIGHT STATE

Wright State University
Dayton, Ohio

 **Client Initial Evaluation Information**

School of Professional Psychology
Psychological Services Center

Client ID

Client	Provider	Date

Presenting Problem/Situation   Include duration, intensity, frequency, contextual factors.

Estimate of Immediacy of Service Need

☐ Immediate   ☐ Within a few hours to a day   ☐ Within a few days to a week   ☐ Within a month

Estimate of Lethality to

Self   ☐ High   ☐ Moderate ☐ Low   ☐ None

Others   ☐ High   ☐ Moderate ☐ Low   ☐ None

Comments Include ideation/threats

_____

_____

_____

Functional Areas Summary information of functional life areas relevant to presenting problem/situation

1 Physical health

2 Vocational/Occupational (School if currently attending)

3 Current living arrangement and relationship to immediate family

4 Significant others (intimate relationships)

5 Friends/Community involvement

6 Leisure time

7 Self-image/system, life "philosophy"

8 Substance use: Drug, alcohol, and prescribed drugs

Type	Present Use (Amount)	Past Use	Duration of Use
a.			
b.			
c.			
d.			
e.			

Comments, particularly if misuse is indicated

Summary/Professional Formulation

Estimate of Client Suitability for Psychotherapy (Or other type of intervention if indicated)

Service Implications/Actions

Estimate of client's compliance with planned actions

Provider's signature: _____ Supervisor's _____
(If applicable)

124 RUSSELL J. BENT

WRIGHT STATE

Wright State University
Dayton, Ohio 45435

Basic Client Information

School of Professional Psychology
Psychological Services Center

Client ID

Client Identification

Last name	First name	Middle initial

Street address		

City	State	Zip code

Date of birth	Sex ☐ F ☐ M	Age	Telephone

Employer	Work telephone number	Social Security number

Party Responsible for Reimbursement

Last name	First name	Middle initial

Street address		

City	State	Zip code	Telephone number	Date of birth

Employer	Work telephone number	Social Security number

Third Party

Subscriber name	Subscriber Social Security number

Subscriber/Group number	Employer name	Insurance company name

Subscriber name	Subscriber Social Security number

Subscriber/Group number	Employer name	Insurance company name

Medicare I.D. number	Medicaid/Welfare I.D. number

Industrial Claim	Claim number	Employer name	Date of injury

Number of children	Number of dependents	Number in household	Yearly income

Fee established at $_____ per unit of service	Comments

Authorization for Release of Information for Billing Purposes

I hereby authorize the release of any information necessary for third-party claim submission and/or payment for services. I authorize payment of third-party benefits to

_____ for services described herein.

_____ 19____

Date   Signature of client (Parent or guardian if minor)

Marital status		Spouse's name	
School currently attended	Grade	Highest level of educational attainment	
If client is a child, indicate legal guardian		Guardian's relationship to client	

Emergency Contact

Name	Telephone	Relationship
Address	City/State/Zip code	

Service-Related Information

Previous mental health services. Briefly describe what, when, intensity, outcome

Significant current legal problems

Current medical/health problems	Physician(s) name:
	Address:

Current involvement with other human service providers/agencies

Other relevant basic information

APPENDIX B

PEER REVIEW GUIDELINES - Initial Client Services

PROFICIENCY LEVELS

I: None or minimal knowledge and skill.
II: Some knowledge and some skill necessary to accomplish the T/R in
         well defined circumstances.
III: Adequate knowledge and basic skill necessary to carry out T/R in
         usual circumstances.
IV: Adequate to considerable knowledge and skill necessary to apply
         T/R in usual, new and complex circumstances.

- -

| REVIEW FOCUS | PROFICIENCY LEVEL |

Problem Description   I   II   III   IV   N/A

- reasonable information gathering procedure.
- evidence functional impairment/personal distress.
- description includes circumstances, frequency,
     consideration or degree of disruption, and
     point of onset.
- broader context within which problem exists.
- clear, descriptive communication of problem.

Problem Analysis/Intervention Rationale   I   II   III   IV   N/A

- reasonable analysis/interpretation of problem.
- consistent, coherent rationale moving toward
     development of intervention plan.

Service Goals and Objectives   I   II   III   IV   N/A

- goal related to problem.
- goals specific, concrete, expressed in terms of
     expected change to an estimated time point.
- goals reasonably attainable through psychological
     services.
- awareness of client strengths, potential
     limitations and context.

Service Strategy   I   II   III   IV   N/A

- service strategy is related reasonably to problem
     and problem/analysis.
- service procedures are specific and
     "usual/customary".
- trainee shows knowledge and skill in the service
     procedures indicated.
- there is an awareness of other service procedures
     which may be appropriate.
- limitations and strengths of selected service
     procedures are recognized.

Professional Role Behavior   I   II   III   IV   N/A

- trainee relates to professional role (Ethics,
     scientific and professional values and
     standards, identification as a psychologist,
     integration of diverse knowledge and skills,
     dedication to human welfare.

Other

<u>PEER REVIEW REVIEWER REPORT FORM</u>

<u>Questions to be addressed:</u>

Necessity of service
Appropriateness of service
Effectiveness of service
Specific:

Client/Reviewer I.D.

Client # _____

Reviewer: _____

<u>Peer Reviewer Report:</u>

WSU - SCHOOL OF PROFESSIONAL PSYCHOLOGY
Clinical Proficiency Evaluation - Committee Member Report/Worksheet

Student's Name:_____Committee Member:_____

		RATING KEY		
0	1	2 3 4	5	6
Unacceptable Performance	Below Average, Marginal	Low - Mid - High Acceptable	Higher Than Average	Exceptional Performance

RATINGS   **NOTES**

ASSESSMENT WORK SAMPLE

1. Your global rating of the detailed Assessment Observer's Report:

   0   1   2 - 3 - 4   5   6

2. Adequacy of assessment purpose and procedures:

   0   1   2 - 3 - 4   5   6

3. Communication through written report:

   0   1   2 - 3 - 4   5   6

4. Quality of conceptual skills and inferences in formulating judgment used in the assessment process:

   0   1   2 - 3 - 4   5   6

INTERVENTION WORK SAMPLE (#1): REQUIRED

1. Intervention plan, work sample #1

   0   1   2 - 3 - 4   5   6

2. Proficiency of intervention: skills/procedures

   0   1   2 - 3 - 4   5   6

INTERVENTION WORK SAMPLE (#2): OPTIONAL

1. Comments:

PSYCHOLOGICAL KNOWLEDGE AND VALUES

1. Intergration of the scientific, scholarly, theoretical knowledge base to practice:

   0   1   2 - 3 - 4   5   6

2. Ethical and legal implications of practice:

   0   1   2 - 3 - 4   5   6

PROFESSIONAL ROLE DEVELOPMENT-GLOBAL IMPRESSION   0   1   2 - 3 - 4   5   6

1. Summary Comments:

APPENDIX C

WSU - SCHOOL OF PROFESSIONAL PSYCHOLOGY

PROFICIENCY EVALUATION COMMITTEE

SUMMARY EVALUATION

Student's Name: _____   Date:_____

Evaluation Committee Members

Chairer: _____
                                            signature

Member: _____
                                            signature

Member: _____
                                            signature

Level of Performance

Would you accept this student as prepared to enter   YES   NO
a general, pre-doctoral internship program.

Quality of Performance

If yes is checked above, this student would be rated at a quality level of:

Just Acceptable   Acceptable   Strongly Acceptable   Exceptional

Comments (regarding level, quality or both):

TRAINING ACTIVITIES REPORT
OFFICE OF PROFESSIONAL AFFAIRS AND SERVICES - SCHOOL OF PROFESSIONAL PSYCHOLOGY
Wright State University, Dayton, Ohio

Trainee: _____ Period from ___/___/___ to ___/___/___.

Facility/Program: _____.

Primary Supervisor: _____.

DIRECT SERVICE CLIENTS   Total Hours at Facility/Program [   ]

	Adult	Older Adult	Child	Adolescent
Continued				
Discontinued				
New				

(Separately identifiable
clients excluding clients
in program activities)

NUMBER OF TRANSACTIONS

_____ individual   _____ (number of individual in long-term therapy)   _____ couples

_____ family   _____ group   _____ individual assessment   _____ other

DIRECT SERVICE TIME GOALS   (Percentage direct service total hours in:)

_____ assessment   _____ crisis stabilization   _____ restoration

_____ habilitation   _____ development   _____ % Other:_____

ASSESSMENT ACTIVITIES   SUPERVISION
                                                          (Name of Supervisor)

No.	TESTS
	Rorschach
	WAIS-R
/	Bender-Gestalt, MFD
/	MMPI, CPI
/	TAT, CAT
/	WISC-R, McCarthy Scales
/	Drawings: DAP, HTP
	Sentence Completion
/	WRAT, PIAT
	Strong Campbell II
	Stanford-Binet
	WPPSI
	Behavioral Indices
	"Battery" General
	"Battery" Neuropsych.
	Kaufman

_____ hrs. Individual w/_____

_____ hrs. Individual w/_____

_____ hrs. Group w/_____

_____ hrs. Group w/_____

_____ hrs. In-service training, describe:

_____ hrs. Workshops/Continuing Education,
           describe:

INDIRECT SERVICE ACTIVITIES

_____ hrs. consultation   _____ hrs. evaluation   _____ hrs. training/workshops presented

_____ hrs. administration   _____ hrs. program development   _____ hrs. needs assessment

_____ hrs. other/Explain: (Use back of this form if needed)

COMMENTS

APPENDIX D

PROCEDURES

For The

CLINICAL PRACTICE PROFICIENCY EVALUATION

School of Professional Psychology-WSU

A. ASSESSMENT CONSULTATION: PRELIMINARY TASKS

   1. Locate facility/site to carry-out assessment consultation.

   2. Identify a client new to you and establish approval (by facility and client)
       to assess the client. The client may be an individual, a couple, or a
       family -- no groups.

   3. Select an Assessment Observer.

       The Observer is generally available to carry-out one or two
       assessment observations and should be contacted as soon as
       possible upon approval by the Office of Professional Affairs
       & Services. Most clinical faculty may serve as Assessment
       Observers. You are free to select a faculty member to
       serve as your Observer, or if you so desire, an Observer
       will be assigned to you. When there is agreement, you may
       forward to the Observer a form titled Assessment Observer's
       Report.

   4. Set-up time and place for the assessment which is mutually agreeable
       among client/trainee/observer.

B. ASSESSMENT CONSULTATION: CLINICAL FORMAT/PROCEDURES

   1. The trainee is to establish an optimal assessment consultation situation.

   2. The assessment consultation procedures must:

      a) involve a (preliminary) review of available "records" and relevant
         information;

      b) formulate the focus and intended purpose of the assessment
         consultation;

      c) complete an assessment interview with the client (and others if the
         trainee so decides);

      d) select and administer at least one or more cognitive tests and
         one or more personality tests from the list to follow. Any other
         optional tests may be used as determined by the trainee.

COGNITIVE TESTS

Intelligence and Achievement Tests

Goodenough Draw-a-Man Test
McCarthy Scales of Children's Abilities
Peabody Picture Vocabulary Test
Stanford-Binet Intelligency Scale, Form L-M
Vineland Social Maturity Scale
Wechsler Adult Intelligence Scale (WAIS-R)
Wechsler Intelligency Scale for Children (WISC-R)
Wechsler Preschool and Primary Scale of Intelligence (WPPSI)
Wide Range Achievement Test (WRAT)

Neuropsychological and Screening Tests

Bender Visual-Motor Gestalt Test
Halstead-Reitan Neuropsychological Battery for Adults (15 and older)
Halstead-Reitan Neuropsychological Battery for Older Children (9-14)
Luria-Nebraska Neuropsychological Battery
Reitan-Indiana Neuropsychological Battery for Children (5-8)
Illinois Test of Psycholinguistic Abilities (ITPA)
Memory for Designs Test (Graham-Kendall)
Berry-Buktenica Developmental Test of Visual-Motor Integration (VMI)

PERSONALITY TESTS

Projective Personality Tests

Children's Apperception Test (CAT)
Figure Drawings, Draw-a-Person, House-Tree-Person, Projective Drawings
Holtzman Inkblot Technique
Rorschach Test
Sentence Completion Tests
Thematic Apperception Test (TAT)

Self-Administered Tests

California Psychological Inventory (CPI)
Edwards Personal Preference Schedule (EPPS)
Minnesota Multiphasic Personality Inventory (MMPI)
16 Personality Factor Questionnaire (Cattell)
Strong-Campbell Interest Inventory

e) submission of the formally written (typed) assessment consultation report to the Office of PA&S must be submitted within 48-hours of the assessment. Absolutely, no report shall be accepted after the 48-hours. The process shall have to be repeated if this time limit is exceeded.

a (clear/clean) photocopy of tests, protocols and test data must be submitted with the assessment report. All client personally identifying information must be deleted from the report and test data. A copy of the report and test protocols/data should be retained by the trainee. The original report with client personal identifiers included shall be submitted to the facility through the Observer at the earliest possible date.

3. The assessment consultation format may vary as appropriate and is to be determined by the trainee.

4. The client should sign an Informed Consent Agreement indicating he/she understands the evaluation will be shared with a faculty evaluation committee (names deleted) for a one-time evaluation (confidential), then all client material shall be destroyed.

C. INTERVENTION PRACTICE SAMPLE(S)

1. Select a practice sample from among your field practice clients, making sure you obtain informed consent from the client/facility. (If you do not have an intervention sample at the site and need a client, let this Office know so that one may be assigned.) The practice sample should fit the definition of general practice.

The work sample should be one unedited audio-cassette (or video) intervention "hour" between the trainee and the client. Tapes that are of poor quality are not acceptable.

2. Submit the practice sample and:

a) the Psychological Service Center's Client Initial Evaluation Information form;

b) an Individualized Service Plan for that client;

c) a summary of at what point (session) the practice sample has occurred in your treatment plan and how the sample relates to the treatment plan. This summary is prepared on the Individualized Psychological Service Review form.

D. OTHER CONSIDERATIONS

The trainee should review possible theoretical/scientific, professional,
ethical, and legal aspects of the work sample(s) in preparation for
discussion during the oral evaluation of the practice sample.

E. THE PROFICIENCY REVIEW PANEL

The Panel consists of at least three members. These members must include
one fully-affiliated faculty member, one clinical faculty member, and
one third-year student member. The student may elect to add a second
clinical faculty member to the panel, thereby increasing the panel size
to a total of four members. This may be done to increase the "depth"
of the committee, or to add expertise in a particular proficiency area.
Including the testing/evaluation Observer as the extra clinical faculty
member is encouraged. However, if the Observer is a full-faculty member,
or a current supervisor, they are not eligible to sit on the Panel, but,
of course, submit the Assessment Observer's Report of their evaluation
to the Panel.

The Office of PA&S must make the final approval and assignments
of all panels (members). The student may elect to allow the
Office to simply assign a panel, or the student may suggest names
of panel members. The Student Suggestion List for Clinical
Proficiency Panel Members must be completed to organize the
submission of such names. This "suggestion list" must be
submitted, on/or before, the close of business Wednesday,
January 9, 1987. Student/peer panel members will be assigned
by the Office of PA&S. The student shall be informed of the panel
membership in as timely a manner as possible.

Following panel membership assignment, the student must arrange the
tentative time and place of the evaluation acceptable to all panel
members. Help in room assignment is available from the Office. All
practice sample materials will be sent to panel members by the
Office of PA&S. To Assure timely delivery of evaluation materials to
panel members, do not confirm the panel meeting date until coordinating
with the Secretary of PA&S (in consideration of Office documentation, etc.).

F. THE PROFICIENCY EVALUATION OUTCOME

The purpose of the proficiency evaluation is to allow the trainee to
demonstrate an appropriate pre-internship (residency) level of general,
psychological practice consistent with the standards and expectancies
of the School. The evaluation allows the degree candidate much latitude
and a facilitative atmosphere in which to demonstrate his/her level of
competence.

3. Optionally and additionally, the trainee may submit a second practice
    sample to demonstrate his/her range of practice skills. The sample may
    not involve taping, but should involve a written summary exposition of
    the practice. The practice may involve a special proficiency, group,
    programmatic, community, consultative, staff development, preventive or
    other methods. The additional sample is primarily for the purpose of
    demonstrating either a special proficiency, or an additional general
    proficiency work sample.

    The evaluation shall result in an inventory of strengths and limitations
    as reflected in the evaluation, and the trainee shall receive formal
    feedback of the results. There is no pass/fail, or letter grade, for
    the evaluation. If necessary, and in the interest of the trainee,
    remedial or continuing educational requirements may be prescribed if
    significant limitations are present.

G. TIME LINES

    To assure less overload on faculty and students, all assessment
    consultation materials and intervention work samples are due in the
    Office of Professional Affairs & Services on the dates established
    each year by that office. These dates usually fall into the early
    part of the Winter quarter in Year III. Panel evaluations should
    be completed by the latter part of the Winter quarter in Year III.

REFERENCES

American Psychological Association. (1979). *Criteria for accreditations of doctoral training programs and internships in professional psychology.* Washington, DC: Author.

Baker, J. W. (1983). Continuing professional development. In C. E. Walker (Eds.), *The handbook of clinical psychology.* Homewood, IL: Dow Jones-Irwin.

Battle, C. C., Imber, S. D., Hoehn-Saric, R., Stone, A. R., Nash, E. R., & Frank, J. D. (1966). Target complaints as a criteria of improvement. *American Journal of Psychotherapy, 20,* 184–192.

Bent, R. J. (1982). The quality assurance process as a management method for psychology training programs. *Professional Psychology, 13,* 98–104.

Beutler, L. E. (1983). *Eclectic psychotherapy: a systematic approach.* New York: Pergamon Press.

Carkhuff, R. R., & Anthony, W. A. (1979). *The skills of helping.* Amherst, MA: Human Resource Development Press.

Claiborn, W. L., Stricker, G., & Bent, R. J. (1982). Special Issue: Peer review and quality assurance. *Professional Psychology, 13,* 166.

Dixon, D. N., & Glover, J. A. (1984). *Counseling: A problem solving approach.* New York: Wiley.

Grant, R. L. (1972). The problem-oriented record for psychiatry. *Journal of the National Association of Private Psychiatric Hospitals, 3,* 27–32.

Haynes, S. N., & Wilson, C. C. (1979). *Behavioral assessment.* San Francisco: Jossey-Bass.

Hersen, M., & Bellack, A. S. (Eds.). (1981). *Behavioral assessment: A practical handbook* (2nd ed.). New York: Pergamon Press.

Kiesler, D. J. (1977). *Use of individualized measures in psychotherapy and mental health program evaluation research: A review of target complaints, problem oriented record, and goal attainment scaling* (Report PLD-04317-77CK). Rockville, MD: Clinical Research Branch, National Institute of Mental Health, ADAMHA, U. S. Department of Health, Education and Welfare.

Kiresuk, T. J., & Lund, S. H. (1977). Goal attainment scaling: Research, evaluation and utilization. In H. C. Schulberg & F. Baker (Eds.), *Program evaluation in the health fields* (Vol. 2). New York: Behavioral Publications.

Morton, S. I. (1982). Peer review: A view from within. *Professional Psychology, 13,* 141–144.

Shueman, S. A., & Troy, W. G. (1982). Education and peer review. *Professional Psychology, 13,* 58–65.

Urban, H. B., & Ford, D. H. (1971). Some historical and conceptual perspectives on psychotherapy and behavior change. In A. E. Bergin & S. L. Garfield (Eds.), *Handbook of psychotherapy and behavior change: An empirical analysis.* New York: Wiley.

Weed, L. L. (1964). Medical records, patient care and medical education. *Irish Journal of Medical Science, 6,* 271–282.

6

The Effects of Contemporary Economic Conditions on Availability and Quality of Mental Health Services

ALEX R. RODRIGUEZ

It will be as profound and turbulent as any economic or social
upheaval in U.S. history.
Joseph Califano, 1986

INTRODUCTION

A revolution is surely occurring in the American way of health (Califano, 1986). It is neither silent, subtle, nor sudden. Like any true revolution, it has been looming for years, waiting for the historical moment. Following a definite developmental process, this one has been evolving in the breeding ground of American culture, political systems, and socioeconomic conditions for many years. Future historians will find this revolution both intriguing and historically logical. It will be analyzed as a course of multiple conflicting and converging forces that ultimately could not all be individually or collectively sustained. Fortunately, it will not be a story of cataclysm, but it will be one of dramatic change. This chapter reflects on that revolution as it is now unfolding and focuses on its expression in, and its impact on, the availability and quality of mental health services.

THE ECONOMIC DIALECTIC IN AMERICAN HEALTH CARE

Revolutions follow the poor abilities of social systems to anticipate and adequately manage the multiple economic, political, geophysical, and other conditions to which they are subjected. Although disease and illness have historically been a significant party to the revolutionary process, never has the attainment of broad health accomplishments led to disruptions in social systems functioning—until recent years in Western nations. In the midst of failing economic fortunes, Great Brit-

---

ALEX R. RODRIGUEZ • Preferred Health Care, Ltd., 15 River Road, Wilton, Connecticut 06897.

ain, over the past few years, has become a national case study for the difficult choices that a country that has valued and accommodated the financial requirements of quality health services must make when it comes on hard times. The British revolution in reordering health priorities has been a generally uneventful one, despite the protestations of physicians.

Only time will tell how much *Sturm und Drang* will be generated by the American version of this economically driven rethinking of the popular Western nation that health—through access to quality health service—is a basic human right. It is by the collective national experience and consensus about the right to health care that the nature and direction of this "revolution" will ultimately be determined. For now, the economic mandate—delivered by the payers for health care services—is that a crisis in the financing of care requires significant, if not revolutionary, changes in the American health-care system. Thus, the fortunes of health status will continue to follow the fortunes of the economy.

THE COST CRISIS: REAL AND IMAGINARY

By any estimation, the costs of health care have been escalating at unprecedented rates since 1965 (Figure 1 and 2, Tables 1 and 2), paralleling the national and world trends toward recession and inflation during this period (R. M. Gibson, Levit, Lazenby, & Waldo, 1984). Forecasters of health care costs, especially hospital costs, have been signaling alarm for several years (Congressional Budget Office, 1977, 1979a; Council on Wage and Price Stability, 1977), even though the problem has seemingly been a more recent one (Goldfarb, Mintz, & Yeager, 1982; "Hospital

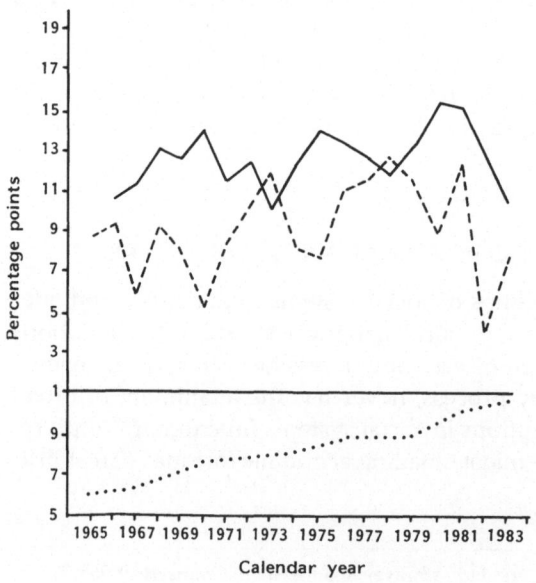

FIGURE 1. National health expenditures and gross national product and national health expenditures as a percentage of the gross national product, 1965–1983. Source: Health Care Financing Administration. ⎯ = percentage change in national health expenditures; --- = percentage change in gross national product; · · · = national health expenditure as a percentage of gross national product.

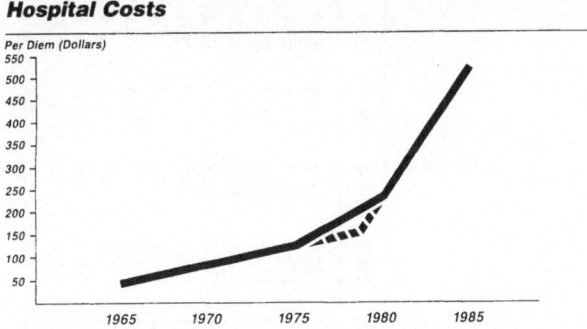

Hospital Costs

FIGURE 2. Hospital costs. Source: American Hospital Association and Equitable Life Assurance Society.

Costs," 1983; "Medical Care," 1983). Some would even contend that there isn't a real "crisis," arguing that 10% or 11% of the gross national product is not a particularly high level of social or economic commitment, given the high level of health in the United States afforded by our $390 billion in expenditures and the greater relative commitment—as high as 20% in Scandinavian countries—to health services (Lindsay, 1982).

In addition to strains within the national economy, a number of factors have contributed to this extraordinary growth in health costs. Notable among the causes is the growing availability and amount of health insurance coverage in recent years (Enthoven, 1984; Feldstein, 1981; Sloan, 1982). The growth of health benefits has resulted from a combination of employers' and society's enlightenment about the ties between health, productivity, and benefits; the use of competitive benefits to secure desired employees during the rapid expansion of the postwar American economy; and the influences of employees through unions, consumer organizations, and legislative bodies. These factors led to a pattern of progressive benefit escalations during the 1960s and 1970s, both in comprehensiveness of benefits and in dollar coverage. Not surprisingly, in an era of prosperity, provider-induced demand for services (National Center for Health Services Research, 1983; Wilensky & Rossiter, 1983) and beneficiary influences (Ware, Johnston, Davies-Avery, & Brook, 1979) were mediated through the relative luxury of insurance coverage. In the zeal to provide comprehensive and competitive insurance coverage, and in setting up the components of reimbursement policies (Roe, 1981, 1985), payers did not anticipate the combined power of the provider, the beneficiary, inflation, and other factors to raise health costs beyond most projections.

Among the other causes of the rise in health care costs are the following:

- Excessive numbers of hospital beds (Congressional Budget Office, 1979a; Goldfarb et al., 1982), associated with the availability of Hill-Burton and other loans, investor-owned and other market-driven factors, and the uneven effectiveness of health systems agencies
- "Defensive medicine," spurred by the increasing threat of malpractice claims (Bovbjerg & Havighurst, 1985), coupled with "compulsive medicine" and

TABLE 1. National Health Expenditures, by Type of Expenditure: Selected Years, 1929–1983[a]

Type of expenditure	1983	1982	1981	1980	1979	1978	1977	1976	1975	1974	1973	1972
Total[b]	355.4	322.3	285.8	248.0	215.1	190.0	170.2	150.8	132.7	116.3	103.4	93.9
Health services and supplies	340.1	308.1	272.7	236.1	204.6	180.2	161.0	141.8	124.3	108.9	96.5	87.4
Personal health care	313.3	284.7	253.4	219.1	189.6	167.4	149.1	132.8	117.1	101.5	89.0	80.5
Hospital care	147.2	134.9	117.9	101.3	87.0	76.2	68.1	60.9	52.4	45.0	38.9	35.2
Physicians' services	69.0	61.8	54.8	46.8	40.2	35.8	31.9	27.6	24.9	21.2	19.1	17.2
Dentists' services	21.8	19.5	17.3	15.4	13.3	11.8	10.5	9.4	8.2	7.4	6.5	5.6
Other professional services	8.0	7.1	6.4	5.6	4.7	4.1	3.6	3.2	2.6	2.2	2.0	1.8
Drugs and medical sundries	23.7	21.8	20.5	18.5	17.1	15.4	14.1	13.0	11.9	11.0	10.1	9.3
Eyeglasses and appliances	6.2	5.5	5.6	5.1	4.7	4.2	3.7	3.4	3.2	2.8	2.5	2.3
Nursing-home care	28.8	26.5	23.9	20.4	17.4	15.1	13.0	11.3	10.1	8.5	7.2	6.5
Other health services	8.5	7.6	7.0	5.9	5.1	4.8	4.2	4.0	3.8	3.3	2.7	22.6
Expenses for prepayment and administration	15.6	13.4	10.6	9.2	8.6	7.5	7.5	5.1	4.0	4.7	5.3	4.8
Government public health activities	11.2	10.0	8.6	7.7	6.4	5.3	4.3	3.8	3.2	2.7	2.2	2.0
Research and construction of medical facilities	15.3	14.2	13.2	11.9	10.4	9.8	9.2	9.0	8.4	7.5	6.8	6.6
Research[c]	6.2	5.9	5.6	5.4	4.7	4.4	3.9	3.7	3.3	2.8	2.5	2.4
Construction	9.1	8.3	7.6	6.5	5.7	5.3	5.3	5.3	5.1	4.7	4.3	4.2

	1971	1970	1969	1968	1967	1966	1965	1960	1955	1950	1940	1929
Total[b]	83.5	75.0	65.6	58.2	51.5	46.3	41.9	26.9	17.7	12.7	4.0	3.6
Health services and supplies	77.4	69.6	60.8	54.1	47.6	42.6	38.4	25.2	16.9	11.7	3.9	3.4
Personal health care	72.2	65.4	57.1	50.3	44.5	39.7	35.9	23.7	15.7	10.9	3.5	3.2
Hospital care	31.0	28.0	24.2	21.1	18.4	15.8	14.0	9.1	5.9	3.9	1.0	.7
Physicians' services	15.9	14.3	12.6	11.1	10.1	9.2	8.5	5.7	3.7	2.7	1.0	1.0
Dentists' services	5.1	4.7	4.2	3.7	3.4	3.0	2.8	2.0	1.5	1.0	.4	.5
Other professional services	1.6	1.6	1.5	1.4	1.3	1.2	1.0	.9	.6	.4	.2	.3
Drugs and medical sundries	8.6	8.0	7.1	6.4	5.8	5.5	5.2	3.7	2.4	1.7	.6	.6
Eyeglasses and appliances	2.0	1.9	1.7	1.5	1.3	1.3	1.2	.8	.6	.5	.2	.2
Nursing-home care	5.6	4.7	3.8	3.4	2.8	2.4	2.1	.5	.3	.2	.0	.0
Other health services	2.3	2.1	1.9	1.7	1.5	1.5	1.1	1.1	.7	.5	.1	.1
Expenses for prepayment and administration	3.5	2.8	2.5	2.7	2.2	2.1	1.7	1.1	.8	.5	.2	.1
Government public health activities	1.8	1.4	1.2	1.0	.9	.8	.8	.4	.4	.4	.2	.1
Research and construction of medical facilities	6.1	5.4	4.8	4.1	3.8	3.7	3.5	1.7	.9	1.0	.1	.2
Research[c]	2.1	2.0	1.9	1.9	1.8	1.6	1.5	.7	.2	.1	.0	.0
Construction	4.0	3.4	2.9	2.3	2.1	2.1	2.0	1.0	.7	.8	.1	.2

[a] Source: Office of Financial and Actuarial Analysis, Bureau of Data Management and Strategy, Health Care Financing Administration.
[b] Amounts expressed in billions of dollars.
[c] Research and development expenditures of drug companies and other manufacturers and providers of medical equipment and supplies are excluded from "research expenditures," but they are included in the expenditure class in which the product falls.

TABLE 2. Blue Cross and Blue Shield, Family
Coverage per Month (Includes Surgical and Major
Medical)[a]

Year	Cost per month ($)
1958	11.16
1968	34.10
1978	111.68
1984	172.44
2000	1,250.00 (est.)

[a]Source: Blue Cross and Blue Shield (Washington, D.C.).

"lazy medicine," which cumulatively results in excessive ancillary services
(General Accounting Office, 1983), admissions, and lengths of stay
• Administrative waste, now estimated to be 22% of all spending for U.S.
  health care, compared to 8.2% in Canada and 10.1% in Great Britain (Himmelstein & Woolhandler, 1986)
• Health fraud, now estimated at $1 million daily and involving 2% of all providers (McIlrath, 1985)
• Excessive costs of training excessive numbers of physicians, particularly
  specialists (Sloan & Schwartz, 1983), and other health care providers
• Regional practice and billing variations, influenced by training, reimbursement, and social factors (Apsler & Bassuk, 1983; Chassin, 1983; Davidson,
  1986; "Dramatic Variations," 1984; Mezzich & Coffman, 1985; Roe, 1981;
  Showstack, Stone, & Schroeder, 1985; Sloan, 1982; Wennberg, 1986)
• The increasing sophistication, availability, and use of expensive medical
  technologies in evaluation and treatment (S. H. Altman & Blendon, 1979;
  Showstack et al., 1985)

All of this is fueled by the increasing complexity and specialization of health
care markets and systems, and has led to what Relman (1980) termed the "medical-industrial complex," a labyrinthine complex of services that collectively comprise
the second largest business in the United States. Despite its size, this system is
poorly integrated at a personal level and not uncommonly leads to excessive and
duplicated services with uneven quality (Chassin, 1983). This system's size is formidable at a financial (Ginzberg, 1984) and political (Relman, 1983) level, often
thwarting any major governmental or other attempts to reorganize or reprioritize it.
Understandably, there is great pressure on individual and institutional providers,
which are torn between financial incentives, professional commitments to patients,
and habitual practice patterns. Although providers are not the only cause or cure for
rising costs in mental health care (Biegel & Sharfstein, 1984) or other health services, nevertheless they are frequently viewed as "the problem." Yet, they have contributed to dramatic and unprecedented gains in the health status of Americans,
who are generally satisfied with their health care (Englehart, 1984), even if they
want even better quality health services, especially care that is compassionate as
well as competent (Cousins, 1985).

A manifestation of the industrialization of health care has been the increasing investment attraction of the financial and business sectors to health services. Although it has no doubt had beneficial results (e.g., capital can provide care quality in the way of technology development, attractive salaries that lure qualified people into health careers, and improved health education and services), it has also considerably increased the costs of care. Investor-owned hospitals, both in general medical (Relman, 1983b) and psychiatric (Biegel & Sharfstein, 1984; Cousins, 1985; Englehart, 1984; Gaylin, 1984; Levenson, 1982, 1983; Relman, 1983b; Schlesinger & Dorwalt, 1984) care, have clearly raised the ante for health care costs, while not demonstrating either enhanced operational efficiencies (Watt, Derzon, Renn, Schramm, Hahn, & Pillari, 1986) or better health outcomes (Pattison & Katz, 1983) when compared with not-for-profit hospitals. In the overall scheme of things, hospitals, responsible for a major component of rising health costs, demonstrate the economic dialect in health care. Not only has it been increasingly necessary to acknowledge this economically driven system as health costs have risen, but there is a clear need of some external checks on its cost-accumulative nature. Thus, institutional and professional providers and other contributors to the health cost crisis have become both a solution to a problem—the need for improved quality in health care—and a problem requiring solutions.

THE COST-CONTAINMENT IMPERATIVE: REGULATION AND COMPETITION

Based on the assumption that health care costs are indeed excessively and unacceptably high, a broad consensus has developed that costs should be "contained" or at least restrained. The method has lacked consensus, however. On the one hand, many view cost containment and health services allocation as a process that shouldn't be left open to the market-driven forces that have helped to create the cost crisis. Concerns about the "right and proper" role of government in protecting the rights of individuals to quality health care, while regulating excessive drains on the economy, have led to support for a governmental regulatory course of cost containment (Gephardt, 1984; Waxman, 1983).

THE GOVERNMENT AS PAYER AND REGULATOR

Proponents of regulatory approaches cite the traditional role of federal and state governments in steering health costs through regulation (Comptroller General of the United States, 1980; National Center for Health Services Research, 1981) in ensuring that equal access to quality health services will be protected (Congressional Budget Office, 1979b), and in establishing successful applications of regulatory approaches (Greenfield, 1983; Kinzer, 1983). They would also note the many federal and state programs that emanated from the largesse of the Great Society initiatives, which, although very expensive and aggravating the trend toward inflation, resulted in undeniable gains in health status for all Americans, especially the underprivileged. The apogee of many years of federal support of mental health programming was the report of the President's Commission on Mental

Health (1978). This landmark report acknowledged the accomplishments of the myriad programs in mental health research, training, consultation, direct services, and public education that had improved mental health in the United States, but it also recognized the many unserved and underserved—notably ethnic minorities, children, and the elderly—who required better services. This report resulted in passage of the Mental Health Systems Act of 1980, which would have initiated a substantial transfer of federal support to state and local authorities and would have provided for a higher level of coordination of services. However, after a troubling five years of inflation and recession, American voters in 1980 indicated a need for a change in leadership. With this change came a "New Federalist" shift from a primarily regulatory approach to health services and cost control to one emphasizing "free market" competition. This shift signaled both a promising and troubling period of transition.

Regulatory approaches to ensuring quality and economic efficiencies have been known for some time to have limitations, notably the need for further and more complex regulatory definitions over time, the corollary need for bureaucratic evaluation and monitoring systems, and the inherent problems in promoting initiative and self-regulation within the regulated (Gaumer, 1984; McClure, 1981). Nevertheless, some continuing regulation has been needed in health care, just because it follows somewhat different economic influences in competitive environments (Pollard, 1983). Thus, rate regulation as a mechanism of cost control, especially for hospitals, has become a hallmark of contemporary thinking by payers (Sloan, 1983; Young & Saltman, 1983).

In 1981, the federal government, facing very troubling projections of insolvency in the funding for Medicare and Medicaid, initiated an ambitious system of reimbursement that was geared to reducing initially, Medicare hospital costs and then those associated with professional fees. The impact of this initiative of reimbursement, according to rates tied to a case mix approach (Schumacher, Clopton, & Bertram, 1982) under a system of diagnosis-related groups (DRGs), has been substantial to date (Ziegenfuss, 1985). Overall, costs have been reduced under this system, and a defined program of utilization and quality review has been installed through designated peer review organizations (PROs) (Grimaldi & Micheletti, 1984). However, there have been outcomes that underscore the limitations of such an approach. Hospitals have learned to "game" the system—"cream-skimming" the diagnoses that result in the greatest revenue, readmitting patients over time for continued care, and limiting patient care services that might enhance comfort or quality but that are "not necessary" for reimbursement (Ziegenfuss, 1985). Further, "cost shifting" has occurred: hospitals charge insurance carriers incrementally greater amounts to offset Medicare revenue losses (Ginsburg & Sloan, 1984). This phenomenon has further increased the need for other payers to check their losses through their own cost-containment initiatives, which will be discussed later in this chapter.

The imposition of the DRG system, with its attendant consequences for quality and effective management of illness over time, has raised numerous troubling questions about the social, professional, and ethical implications of systems that are aimed only at containing immediate health-care costs (Kapp, 1984; Mariner, 1984). Concerns about the rationing of care, equal access under the laws to quality, and the establishment of different standards and loci of care based on ability to pay are among the many vexing issues currently being debated in the wake of this initiative's being claimed to be "successful" by some analysts (Ziegenfuss, 1985). Critics of this particular system who acknowledge the appropriateness of a prospective payment approach to reimbursement point to other methods that better account for homogeneity and severity-of-illness factors, and that reduce the risk of arbitrary determinations of medical necessity of care (Brewster, Jacobs, & Bradbury, 1984; Gonnella, Hornbrook, & Louis, 1984; Horn, Sharkey, & Bertram, 1983).

The evaluation of psychiatric services for reimbursement under a prospective payment system presents special challenges because of diagnostic, symptomatic, severity-of-illness, and functional-level variables (Jencks, Goldman, & McGuire, 1985; Mezzich & Sharfstein, 1985). Although psychiatric services under Medicare are currently exempt from the DRG system, the development of a psychiatric prospective payment system has been mandated by the federal government. Various alternatives to psychiatric DRGs have been suggested (R. G. Frank & Lave, 1985; National Association of Private Psychiatric Hospitals, 1985), but currently, no clear path to psychiatric service reimbursement has been established (Jencks *et al.*, 1985), largely because of the significant evidence that no scheme yet developed can equitably assess care for reimbursement without adversely affecting the outcome. A great deal of interest in various psychiatric rating scales is being generated by payers (R. E. Gordon, Jardiolin, & Gordon, 1985a; R. E. Gordon, Vijay, Sloate, Burket, & Gordon, 1985b; Spitzer & Endicott, 1975; Spitzer, Endicott, Fleis, & Cohen, 1970; Spitzer, Gibbon, & Endicott, 1973), since the scales might offer a means of trying severity of illness to the appropriate level, intensity, and duration of psychiatric service. However, no consensus or tangible systems have yet been developed that would allow an empirically justified reimbursement system to be established.

In this circumstance, then, payers—including federal and state governments—are looking for alternative ways of controlling costs by passing their financial risk on to the other parties in the health care relationship. They are finding that regulatory approaches are inherently limited in controlling their risk and are increasingly attracted to the power of a more open-risk economic environment where the health care provider and the beneficiary will assume financial risks. By supporting the loosening of regulatory influences, these payers believe that the fuller influences of the economic dialectic will be expressed through financial risk assumption and will result in reduced "unnecessary" health-services use. Although this presumption is risky, it has clearly reached its day, as political forces are now supporting reduced regulatory intervention by payers and increased provision of health care systems regulated by market influences.

COMPETITION AND COST CONTAINMENT

One of the most debated economic issues in recent years has been the need to turn health services toward a "free market," where payers would promote the determination of negotiated market-clearing prices for health services. In such an approach, health services would be only another market commodity to be provided by health services vendors, who would engage "prudent buyers" in an economically balanced supply-and-demand negotiation. In such an idealized market, these parties would optimally use competitive circumstances to establish the requirements for the costs and quality of care. The government role would be to use judicious law and regulation to foster the competitive market environment's most efficient performance.

Federal policy changes have promoted more market competition as a means of containing health costs, both by active government initiatives (Costilo, 1981; Frech, 1981) and by severely limiting traditional federal roles (Bedrosian, 1983), thus allowing market forces to predominate over regulatory forces. The "new frontier" in health care created by these actions is one that is both refreshingly innovative ("Mental Health Market," 1986; Meyer, 1983; Scheier, 1986) and disquieting (Brown, 1981; Council on Medical Service, 1983; Fein, 1985; Tresnowski, 1983; Winsten, 1981; Wyszewianski, 1982). Procompetitive programs and proposals have been initiated with limited experience, serious thought about the longer-term effects on the cost and benefits of care, or demonstrations of their impact on quality or health outcomes. The promises of reduced costs and quality care through enlightened free-market approaches is considered a "grand illusion" by Ginzberg (1983b). He and others (Brown, 1981; Council on Medical Service, 1983; Fein, 1985; Wyzsewianski, 1982) have grave concerns about the capacity of an environment driven by competitive self-interest to be attentive to the greater social requirements of training, research, and quality in the health care system.

In any competitive enterprise, there are "winners" and "losers." Reasonable questions arise about the protection of those who are not able competitors lest they be afforded care that is limited in comprehensiveness, accessibility, and quality. Thus, the promise of the competition in health care is that it may result in more innovative, independent, and resourceful care. The possibility lingers, however, that without protections, it could become a vehicle for social Darwinism, by which the unable, the unwary, and the unprepared would be taken advantage of or neglected.

THE RACE TOWARD COST CONTAINMENT

In a race with no formal rules and no established starting point or end, the scramble called cost containment in health care is well under way. Whether the initiatives that will be subsequently discussed are popular or unpopular, are based on sound assumptions or not, or are reasonable or ill conceived now is irrelevant. Payers have rapidly taken off on a number of paths that they hope will reduce their financial liability for health care costs, signaled by a building imperative to act, which has finally exploded (Blendon & Rogers, 1983; Stuart & Stockton, 1973). This surge of activity has most dramatically affected hospitals (Abramowitz, 1986; Blendon, Schramm, Moloney, & Rogers, 1981; Hart & Hart, 1980; Schorr, 1983), although it is also aimed at health professionals, notably physicians (Eisenberg & Williams, 1981; "Name of New Game," 1983), who both support the concept of judicious cost containment ("More M.D.'s," 1984) and resist the free-for-all nature of the cost-containment programs currently underway (Platt, 1983). This reservation about an uncoordinated race toward short-term cost-cutting goals is supported by some economists, who view cost savings as contrived and short-lived if they will result in other costs to society (Ginzberg, 1983a). A constant theme in these counters to ambitious cost-containment schemes is the need for assurance of quality with changes in the health-care reimbursement and delivery systems. In fact, ensuring quality care "can reduce the cost of care by eliminating useless or potentially harmful care" (Donabedian, 1984) and is thus compatible with what is likely to be a primary goal of most future medical cost-containment programs: payment only for effectively evaluated and rendered care ("Care Quality," 1981; Shaughnessy & Kurowski, 1982).

The cost-containment imperative, in all of its rapidly proliferating forms, is creating the expected reaction from the provider community, which has adopted the various strategies of active resistance, alternative proposals, and accommodation to cost-driven changes in the health care environment (Fuchs, 1982; Ricardo-Campbell, 1982; Sorkin, 1986; Thurow, 1985). Organized professional counterbalancing to public and private cost-containment initiatives has been expressed through political, public-education, and other means (A. R. Nelson, 1984; Sammons, 1983) and is either self-serving or genuinely concerned about the adverse impact on quality of care and accessibility that is possible with cost-orientated planning and programming. Such changes pose threats to the time-honored professional and social values of quality and commitment in health care services. Further, to deny anyone less than the best possible health care has serious ethical meaning and potential legal ramifications.

The moral issues that surround the inevitable rationing decisions in a cost-containment, competitive market environment go beyond current levels of consensus about one's "rights" to quality health (Evans, 1983; Kohlmeier, 1983; Leaf, 1984; Lundberg, 1983; Mechanic, 1985; Woodward & Warren-Boulton, 1984). Such issues remain the most vexing and unresolved in the current race for cost containment and will ultimately require a higher social-political resolution. What the citizen might agree to in intellectually conceptualizing a justification for cost containment will not be acceptable at the moment of personal pain. In this world of pain dwells the practitioner, who must increasingly weigh the seeming contradictions of commitment to patient need, professional ethics, and economic self-interest. The conflicts generated by these often-incompatible goals have all of the makings of a collective professional neurosis. And in fact, the health professions have not faced such a ponderous moral dilemma in modern history. Such profound reservations about the cost-driven changes in health care will require profound resolutions—by both the health professions and the greater society that they serve. Unilateral initiatives by payers to command changes in providers' behavior will clash with the providers' ethics in such a way that both quality and cost benefits will be threatened.

EMPLOYER, INSURER, AND UNION INITIATIVES TO CONTAIN COSTS AND ASSURE QUALITY

American businesses have been paying for approximately one third of all health care in recent years (R. M. Gibson et al., 1984). Although many companies in more prosperous times prided themselves on generous health-benefits allowances, in more recent years, they have watched with alarm as increasing health costs have cut into profits and health-care and related-benefits funds. They have been particularly hard hit by the cost shifting reverberating from Medicare under DRGs and other cost-containment programs (Grimaldi & Micheletti, 1984). Broad benefits, reimbursement policies that have allowed increased charges, and limited capacity to monitor the true medical necessity of services have allowed their costs to increase disproportionately compared with such public programs as Medicare, which have tended to have a higher level of oversight. Thus, it would seem inevitable that, when the sleeping giant awoke, it would make great noises (Bernstein, 1986; Califano, 1986; Egdahl, 1984; Inglehart, 1982; Sapolsky, 1981). The consequences of this awakening awareness have been actions based on inherent characteristics of American business: aggressiveness, resourcefulness, and innovation.

They run the gamut from restructuring benefits to redefining the conditions under which care will be reimbursed (Caulfield & Haynes, 1981; Fox, 1984; Sullivan & Ehrenhaft, 1984; Tell, Falik, & Fox, 1984). Corporations and unions are increasingly showing an initiative in communicating with one another about cost control developments through health coalitions, a spate of local and national meetings on benefits management, and new publications such as *Business and Health*.

American businesses are not only becoming knowledgeable through reading about the initiatives of other payers of health care programs but are also showing an unprecedented resolve and creativity in starting up their own demonstration projects and implementing more empirical benefit analyses. All of this activity makes the private sector the current hotbed of new developments in health care. In a 1986 survey of American businesses, 87% of employers indicated that controlling health care costs was important, and 62% had set cost-containment goals for 1985; the majority of these achieved or exceeded those goals (McMartin, 1986). Benefit management strategies have included the following:

1. *Benefit redesign and structuring.* Although the rising costs of care have tempted many payers to severely cut or cap benefit coverage as their only strategy, others have taken a more thoughtful approach. Evidence is clear that a reasonable employee copayment is an effective means of limiting unnecessary use of medical and psychiatric services while not limiting access (Jensen, Feldman, & Dowd, 1984; Manning, Wells, Duan, Newhouse, & Ware, 1984), so 43% of companies have increased deductibles and 33% have increased employee contributions to health insurance premiums. Other redesigning options have included the elimination of first-dollar coverage; a reduction in the level of pay-related stop-loss coverage for catastrophic costs; and implementation of flexible spending accounts of a fixed amount of dollars for employee-health-service use, credit, or payback (Fielding, 1984a; Fox, Goldbeck, Spies, 1984; Nazemetz, 1985; Tavernier, 1983; Wolfson & Levin, 1985).

2. *Utilization review and management.* Although insurers have been conducting some limited levels of retrospective claims-based review for a number of years, in recent years a number of more focused preadmission, concurrent, and retrospective review programs have been instituted (Anderson, 1985; Fielding, 1985; "Industries Use," 1982; B. Thompson, 1985), emphasizing more detailed chart review, discussions with the provider, and direct evaluation of the patient. Review activities have tended toward using more qualified professional reviewers, and toward focusing review on specific treatment-plan goals and other details shown to be rele-

vant to length of stay, such as early-discharge planning (Cunningham, 1981). The mental health professions have been leaders in developing and providing peer review standards and services to third-party payers (Hamilton, 1985; Rodriguez, 1983b, 1984) and have spawned a trend toward private review services (Sandrick, 1982), both not-for-profit and proprietary. These review experiences have shown that care can be even more effectively provided and assessed for reimbursement purposes through intensive professional review and assistance. Such programs are generally referred to as *case management* and are increasingly being shown to be effective both in reaching accurate benefit decisions and in potentiating quality outcomes, both for medical-surgical services (Berenson, 1985; Hembree, 1985; Merrill, 1985; Spitz, 1985) and for mental health services (Perlman, Melnick, & Kentera, 1985; Rodriguez & Maher, 1986; Schultz, Greenley, & Peterson, 1983).

3. *Provision of work-site programs for employees and families.* A major shift in benefits management has been toward educating and counseling employees about how to effectively use health services and benefits (Fielding, 1984a; Fox, 1984). Increases in the provision of on-site employee health services (Rodriguez & Maher, 1986), employee counseling and referral services (Hembree, 1985; Lewin, 1986), and work-site risk assessment, prevention education, and related health promotion activities (Fielding, 1984a; Fox, 1984; Lewin, 1986) are among the many directions being taken by employers to more directly account for, manage, and reduce health care costs. In all of this, they are becoming much more sensitive to the emotional needs of their employees and families, as well as to the requirements for quality, timely, and adequate mental-health benefits and services.

4. *Orientation of care to alternative levels of treatment.* Employers as payers are increasingly showing an interest in reducing more expensive hospital services by providing incentives for the use of alternative levels of treatment, where care can be provided as safely and efficaciously as in a hospital (Berk, 1980; Fox & Spies, 1984; Gerber, 1983; Lister, 1983; McNeil, Pauker, Sox, & Tversky, 1982; Moxley & Roeder, 1984; "New Trends," 1983). Examples of alternative medical-surgical settings would be hospices, ambulatory surgery centers, home-health-care services, and nurse-midwife birthing centers. Likewise, benefits managers are becoming both more knowledgeable about and more interested in alternatives to psychiatric hospitalization (Ferber, Oswald, Rubin, Ungemack, & Schane, 1985; Gudeman & Shore, 1984; Gudeman, Shore, & Dickey, 1983; Kiesler, 1982; Mosher, 1983; Weisman, 1985). They are being influenced by evidence that long-term care does not necessarily result in a more favorable therapeutic outcome than shorter term care (T. Gordon & Breakey, 1983; Kiesler, 1982), and that psychosocial rehabilitation programs and partial psychiatric-hospital programs oriented to the community treatment model can reliably reduce both numbers of admissions and lengths of stay in inpatient settings (Bond, 1984; Bond, Witheridge, Setze, & Dincin, 1985; Ferber *et al.*, 1985; Jordan, 1985; Law, 1981; Weisman, 1985). Given the rate of growth of both inpatient and outpatient services in recent years (Table 3) and, in particular, the increase in longer term hospital admissions for children and adolescents (Figure 3), these less expensive alternative-care settings are enticing to strategic benefit planners. Although some might question the consistency in clinical training, competence, and quality among alternative psychiatric-treatment settings, outcome

TABLE 3. Trends in Psychiatric Treatment[a]

Inpatient admissions[b]

1955 : 1.7   1981 : 6.4

  Percentage increase = 376[c]

Outpatient care[b]

1955 : 0.5   1981 : 16.4

  Percentage increase = 3,280[c]

[a]Source: Alcohol, Drug Abuse, and Mental Health Administration.
[b]Expressed in millions of persons.
[c]40% U.S. population increase, 1955–1981.

evidence indicates equivalent rates of quality outcome across populations of patients who otherwise are not so ill that they cannot be treated in "step-down" community-based programs.

5. *Reimbursement of care through alternative delivery and financing systems.* Employers as payers are increasingly acting, with financial self-interest and concern about quality of care, to define the systems that will ensure the highest likelihood of providing cost-beneficial care, that is, a "product with value." As they become more knowledgeable about the merits of organization, competitive purchasing, and management in health care, employers are more and more specifying that health benefits be provided through organized and managed care systems such as health maintenance organizations (HMOs) and preferred-provider organizations (PPOs), which they see as more accountable for the costs of care through the passing on of financial risks. The decisive trend of employers' and unions' offering employees financial incentives to enroll in HMOs and PPOs has caused significant shifts of care from fee-for-service systems to these systems and has sent major reverberations throughout the insurance, provider, and financial communities (Barkholz, 1986; Etheredge, 1986; Graham, 1986; Memel, 1986; Staver, 1986; Wegmiller, 1983). Although PPOs offer some attractions in cost reduction and quality assurance (Acquilina & Stowe, 1985; Billet & Cantor, 1985; Cowan, 1984; Gabel & Ermann, 1985; Kralewski, 1984), most employers are not promoting them (McMartin, 1986) and see them as a relatively short-lived phenomenon in their current form. HMOs pre-

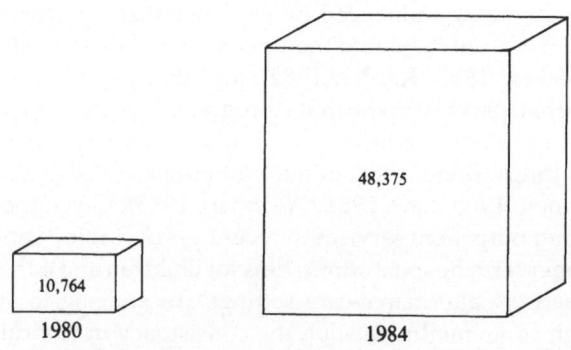

FIGURE 3. Psychiatric admissions of children under 18 in the United States.

sent a very different picture to employers, who are more comfortable with the capitated, at-risk nature of an HMO contract. Although there are a number of both concerns and affirmations about the quality of care, the effectiveness of management by "gatekeeper" functions, and the effect of some HMOs on the provider–patient relationship ("Gatekeeper Payment," 1984; Luft, 1985; Manning, 1984; Moore, Martin, & Richardson, 1983; Schroeder, Clarke, & Webster, 1985; Simenstad, 1985; D. Smith & Metzner, 1970), especially for psychiatric services within HMOs (Bennett, 1987; Bittker, 1985; Wells & Manning, 1985), the current and projected growth of HMOs indicates that they will become the predominant system of health care delivery during the next 20 years (Memel, 1986).

If anything can generally be said about the scope and magnitude of these alternative delivery and financing initiatives, it is that they are so profound that they will thoroughly affect the way in which health care is defined, reimbursed, and delivered for the foreseeable future (Califano, 1986; "The Future," 1985). The "third-party" designation of insurers is quickly losing its identity, as employers and unions are increasingly assuming self-insurance and benefits-administration functions and amplifying their once-distant "fourth-party" role. Their independence and power is manifested by their increasing intelligence about the costs and benefits of care—that is, care that is quality- and cost-effective—as well as by their boldness and creativity in defining the conditions under which they will pay. Their influence is further extended by the collective purchasing and other initiatives of local business coalitions (Davis, 1983) and their significant influence over local, state, and federal health policies through their political and economic might. The speed and scope of the changes instituted by these payers of care are paralleled only by the impact of the federal government on health care in the 1960s and 1970s. What should not be minimized is the reality that these changes are already occurring, and that they are just beginning.

THE EFFECT OF CHANGES ON MENTAL HEALTH SERVICES

Mental health benefits, never really stable—even with favorable signs such as increased dollar and length of treatment coverages (American Psychiatric Association, 1983) and increased earnings for practitioners (White, 1983) in recent years—are now clearly in trouble again ("Mental Health Benefits," 1981; Sharfstein & Biegel, 1984; Sharfstein & Taube, 1982; Stein, 1983). In spite of the relatively successful efforts of mental health associations to increase public understanding and acceptance of mental illness and its treatments, payers continue to have reservations about the focus, length, and effectiveness of cognitive psychotherapies. Moreover, traditional cultural biases against the mentally ill continue, related to fears of, and philosophical connections of moral inferiority to, those with mental and substance-abuse disorders. These biases and reservations are manifested by diminished absolute and relative budgetary priorities for mental health services and by the scandalous consequences: homeless mentally ill too numerous to count, prisons overflowing with possessed and dispossessed mentally disordered persons, and emotional problems experienced by children and youth that can only be termed a crisis, if not an epidemic. The great momentum in this nation in the overcoming of the social burdens of mental illness now seems to be going in reverse. These evident reversals are tied to a troubled economy and a now troubled health-care benefit.

This gear-crunching shift in momentum to cost containment in mental health services is having significant effects on professionally valued and traditional treatment concepts, which have been based predominantly on psychoanalytic-psychodynamic principles and the corollary "paying for time." The reductions in mental health coverage are being felt especially in inpatient settings (Gould, 1981) and are resulting in a crisis in the access of middle-class families to quality mental-health services (Sharfstein & Patterson, 1983). Given the well-known effects of troubled economic conditions on the mental health of those affected (McGuire & Weisbrod, 1981), it would seem that more mental health benefits, rather than fewer, would be both needed and provided. Although mental health associations struggle to educate payers, to inform their members of the crisis in reimbursement, and to support legislatively mandated mental-health benefits (Committee for Financing Mental Health Care, 1981; McGuire & Montgomery, 1982; Muszynski & Brady, 1983; Sharfstein, Muszynski, & Gattozzi, 1983; Sharfstein, Muszynski & Myers, 1984b), these efforts are clearly not enough to allow mental health services to be provided by a traditional reimbursement or in a traditional setting.

The future changes in the organization, financing, and management of mental health care are beginning to be seen now (Flynn, 1985; Luckey, 1983; Rodriguez, 1985; Sharfstein & Beigel, 1985) and are being created largely in the private sector (Carone, Yolles, Krinsky, & Keiffer, 1985; Fielding, 1984b; General Mills American Family Forum, 1980; Herrington, 1986a; Katz, 1986; Lee Schwartz, 1984; Patterson, 1985; Rosen & Orr, 1986). Employers and unions are understanding better their need to provide effective mental-health benefits, out of both moral commitment to employees and financial self-interest. They know better how to define and demand that that care be focused, timely, efficient, and of good quality. They continue to be troubled by the relatively general standards and criteria by which care can be assessed (American Psychiatric Association, 1985; American Psychological Association, 1980) but are encouraged by more objective, research-based approaches to defining mental disorders (American Psychiatric Association, 1980) and evaluating them (R. E. Gordon et al., 1985a,b; Spitzer & Endicott, 1975; Spitzer et al., 1970; 1973). Both public and private payers continue to question the efficacies of many psychotherapies (J. D. Frank, 1975; London & Klerman, 1982; Parloff, 1982; Schacht & Strupp, 1985; M. L. Smith, 1982; Stein, 1980), despite the acknowledged technical problems in conducting efficacy research (Hine, Werman, & Simpson, 1982; Siegel & Kunzel, 1982). Their concerns raise questions about the missed agenda in recent years of pressing for well-controlled efficacy research, which now leaves advocates of equitable mental health benefits looking both for researchers (Weissman, Rounsaville, & Chevron, 1982) and for convincing clinical and epidemiological data (Rubin, 1982).

Of course, a significant body of efficacy data does already exist (Borus, Olendzki, Kessler, Burns, Brandt, Boverman, & Henderson, 1985; Collins, Ellsworth, Casey, Hickey, & Hyer, 1984; Houpt, Orleans, George, & Brodie, 1979; Mumford,

Schlesinger, Glass, Patrick, & Cuerdon, 1984; Piersma & White, 1985; Sharfstein, Muszynski, & Arnett, 1984a; Smith, Glass, & Miller, 1981; Taintor, Widem, & Barrett, 1984). It needs to be collated into a multiprofessional consensus, and then publicaly endorsed. It needs much complementary research, and the challenge to the professions working with the public and private sectors is to develop a common research agenda and budgets that will reflect the shared responsibility of each to determine what should be provided to whom, where, and for how long. Only with such data will mental health benefits become equitable, the necessity of care justified, and the mandate for quality care socially established. In the gaps of such collaboration among the mental health professions and payers and in the absence of more precise efficacy data and care criteria, the momentum of change in health care will carry the mental health professions to an uncertain future, determined by people who are tired of waiting for clear professional guidance.

THE IMPACT OF CHANGES ON QUALITY ASSURANCE SYSTEMS

With the rapid changes in the organization and financing structure of health care and the implicit demand for quality and accountability in a cost-driven competitive environment, special demands will be placed on quality assurance systems. They will need to evolve in concert with the complexities of the systems that they are both monitoring and serving. Moreover, they will need to be as cost- and data-oriented as the other management systems within the care systems, to both substantiate the cost–benefit nature of the care and their operating costs.

Quality assurance activities are already reflecting the management and prospective orientations that will increasingly be seen as providers link up with managed and organized care programs and networks (Lang, 1984). Quality assurance systems are being used to track utilization trends and the completion of treatment-planning goals, and to reduce liabilities through structured risk-management programs. They are being integrated into providers' formal utilization forecasting (MacStravic, 1984) and marketing activities through assessments of patient therapeutic and attitudinal responses to quality of care (Morrison, Rehr, Rosenberg, & Davis, 1982; R. E. Thompson & Rodrick, 1982). Effective quality-assurance systems will more and more become barometers of staff productivity, by assessing problems related to their providing focused, timely, and quality care through monitoring activities (Forguer & Anderson, 1982; Luke & Boss, 1981). Thus, these systems will be integral to the smooth transition to newer delivery arrangements, which may be quite variable depending on provider contract or other reimbursement requirements. These systems will then serve as modulators of organizational change, while maintaining some predictable levels of quality control over services (Hetherington, 1982; Jessee, 1981; Luke, Kreuger, & Modrow, 1983; Restuccia & Holloway, 1982; Scott, 1982; Shortell & LoGerfo, 1981). Quality assurance systems will thus become a vehicle of organizational effectiveness and, concomitantly, of economic survival in a competitive environment.

The realization of these future needs is occurring in the current circumstances of change. Accountability mechanisms are being more widely instituted for staff credentialing, continuing professional education (Atchinson & Brooks, 1981; R. E. Thompson, 1981; Youel, 1983), cost-containment education (General Accounting Office, 1982; Williamson & Associates, 1982), and performance evaluation of quality and utilization efficiency. Care systems in competitive circumstances will need to ensure that cost-beneficial care will be provided through careful documentation of practitioner decisions. Hence, in such situations, consistently ineffectual practitioners will become a liability and will increasingly either be dismissed or have their privileges limited to their areas of competence. Because medical records as well as defined therapeutic outcomes will be vehicles of quality assessment, more emphasis will be placed on detailed, organized, objective, and relevant clinical information that can be assessed against standards of care. However, patient-care management systems tied to review activities will continue to be challenged by psychiatric care practices, which are not yet readily correlated with highly objective review indices or patient characteristics, and which struggle with confidentiality issues in the documentation process (Henisz, Levine, & Etkin, 1981; Hiller & Seidel, 1982; Perlman, Schwartz, Paris, Thornton, Smith, & Weber, 1982).

Quality assurance systems will have to make both conceptual and operational leaps to accommodate to managed and organized care arrangements, especially networks. These systems will need to be integrated with those of the multiple levels and providers of care through which patients will pass over time, in effect to account for and direct their efficient care. Data and communications systems will necessarily become more complex and operations more standardized, to meet overall goals for resource management and therapeutic effectiveness. This need coordination and integration will drive care systems to expand quality assurance activities beyond the walls of hospitals (Reeder, 1981) and will ultimately require an expansion of the scope of the current Joint Commission on Accreditation of Hospitals quality-assurance standards.

The environment of market change is already affecting the quality assurance field. The rapidly evolving requirements of payers are spawning the growth of numerous management vendors that provide a range of specialty services. For instance, local hospitals and professional groups that are forming a preferred-provider organization are finding the need to contract with the firms specializing in organizing and managing the PPO, including marketing, quality–utilization management, and data services.

As in any new market, the capabilities of these vendors is variable, so providers organizing in competitive times are finding a need for consultants who will guide them to an effective management entity. Of course, the capabilities of consultants are likewise variable, leaving providers very vulnerable. Besides the vulnerability arising from developing services that are not economically competitive, some groups are experiencing the strains of organizational change that affect their professional commitments to quality and that result in quality assurance programs that are primarily geared to cost containment. For instance, persons who need long-term medical or psychiatric care (Kane, 1981; Rzaha, 1983) are very vulnerable in systems that are cost-driven, especially where there is a profit goal.

Practitioners can feel caught in a terrible crosscurrent in such circumstances, which are becoming more common in the modern era of competition. This crosscurrent will increasingly result in professional conflicts with business interests and could serve to undermine the inherent professional ethical function of conventional quality-assurance activities. Thus, the business side of quality assurance poses both creative challenges for improving quality and standards and a threat to health itself. It will be imperative for the health professions to be involved in the setting of standards for the newer quality-assurance systems and in educating payers about the need for quality standards for these evolving care systems.

PROFESSIONAL ISSUES IN CURRENT AND FUTURE QUALITY ASSURANCE

Over the evolution of modern medicine and health care, quality assurance has gradually shifted from the decentralized individual practitioner, guided by his or her own personal and professional values and knowledge, to more centralized administrative sources. Hospital-care review committees and local professional associations assuming responsibility for local professional ethics and competence have been the traditional local institutions for ensuring quality. Over time, state licensing, national accreditation, certification, and other systems have been established to review and recognize the structural components related to quality of care. The process aspects of quality—related to specific treatment episodes—have increasingly become the province of payers for care. Although the payers have often sought input from established professional leaders and their associations in setting reimbursement policies in the past, they are now more frequently developing their own in-house "professional" resources for assessing "medical necessity" and quality.

Unfortunately, these internal activities are too often alienated from professional association liaisons and are too focused on a reimbursement decision as the end point of responsibility. In situations where either patterns or significant single instances of aberrant care indicate a threat to quality outcome or competence, few payers have tended to assume responsibility for the practitioner's actions or the patient's health. They have assumed that their legal responsibility is only to pay or not to pay and have been unsettled by the threat of suit by the provider. This attitude and general historical circumstance that payers are in the business of benefits and not quality assurance have led to a disquieting alienation from the professional practitioners, who view their colleagues who work for payers as "hired guns," and have resulted in professional consultants to payers feeling denigrated within the business environment. This historical trend should become of greater concern to professional associations and practitioners as commercial interests predominate even more within health care. Quality assurance, rarely understood or accepted by payers as a business function, must be defined as a required commodity in the new environment of health services. The definitions of quality and quality assurance, then, are ultimately professional issues. Further, the demand for quality must come from the professions—as part of both patient and professional advocacy—working within the institutions of government and public opinion.

Professional issues arising from the reorganization of health care are many and include the following:

- The ability of providers to serve concurrently as advocates for the patient and his or her family and as gatekeepers of their health-care resources (Korcok, 1986)
- The uncertain future of teaching centers and programs, which, because of their greater operating costs and frequent treatment of indigent or less financially able patients, will experience serious economic liabilities in a competitive environment
- The threat to quality that can occur if practitioners are so unfulfilled by their work—because of benefit decisions that question and interfere with their autonomous function with patients—that less able persons are the only ones attracted to the health professions
- The ethical problems inherent in professional practitioners' becoming so caught up in the business "ethic" that they risk abrogation of their professional ethics (Relman, 1985a,b; Shore & Levinson, 1985; Wendorf, 1982)
- The need for professionally developed criteria and professional roles in the quality assurance of confidentiality of medical information in current and future health-benefits-management initiatives (Borenstein, 1985; Schuchman, Nye, Foster, & Lanman, 1980), especially in an automated era (Hiller & Beyda, 1981)
- The necessity of peer review as a component of benefit determinations, and the essential roles of professional associations in establishing peer review standards and criteria and liaison activities with proprietary review programs (Hamilton, 1985; Luke & Modrow, 1982; Rodriguez, 1983b, 1984)
- The absolute requirement that professional associations cooperatively develop and promote national professional standards of practice and review, which would provide guidance to training programs, practitioners, and payers and would establish a quality-of-care base on which arbitrary payer denials of care could be legally challenged (L. K. Altman, 1986; "National Utilization," 1983; S. H. Nelson, 1979; Richman & Barry, 1985; Rodriguez, 1983a; "Test Program," 1981; Ver Berkmoes, 1986)

POLICY ISSUES AFFECTING QUALITY IN MENTAL HEALTH

The longer term future of health care internationally and nationally is more uncertain (Doremus, 1984; Goldschmidt, 1982) than the immediate future, which is both certain and already occurring (Flagg, 1985; Furrow, 1983; R. W. Gibson, 1985; Relman, 1983a; Talbott, 1984; Tarlov, 1983). Despite the ambiguities of longer term economic, sociopolitical, and related trends, it is clear that health care in general and mental health services in particular will not be provided in the ways in which most current practitioners have been trained and practiced to date. Hopefully, the human interactional aspects of psychotherapy that promote healing and change—empathy, support, hope, and practical assistance—will not disappear with the changes in delivery systems. Likewise, essential professional roles in defining and ensuring efficacy and quality must continue. Quality assurance systems must be ingrained at all levels of health care through the active involvement of qualified professionals. This must occur through business insight, legislation, or any other means that can be mobilized by the professions.

In addition to the professional roles in developing newer quality-assurance systems (Brook & Lohr, 1981; Brook, Williams, & Davies-Avery, 1976; Duncan, 1980; Rodriguez, 1987), a number of public policy issues must be worked through. Through democratic institutions, the health professions must cooperatively shape this nations' commitment to the entitlements of quality, equality, and accessibility in health care. No area requires more effort than that of ensuring such rights in mental health care. Very likely, this will be achieved through a combination of professional associations' funding public education, research, and lobbying; policy development through crises and insight; and a collective societal affirmation of mental health as an essential part of existence.

Political and other financing systems will continue to struggle with budget priorities and to look for creative ways to meet the dual demands for cost restraints and quality. Thus, the swelling influence of the private sector in effecting changes in health systems will not escape the greater societal roles in defining—and paying for—the right to health. Such changes in the public and private sectors will result in a continuing multilevel development of policy options (Etheredge, Reinhardt, Marmor, Dunham, Davis, & Blumenthal, 1985; Jessee, 1983; Merwin & Ochberg, 1983; Rappleye, 1983; Relman, 1982; Rubin, 1980; Snoke, 1982). It will be only through the substantial *collective* commitment of professional organizations that the assurance of quality in care will become part of any societal resolution of its real and imagined economic limitations.

Thus, the ultimate policy issue affecting future quality in mental health is that now confronting society and its health professionals: the collective commitment to quality. A society that tolerates undertreated or untreated mentally ill individuals, who are either incarcerated or set adrift in the streets, is a society lacking insight, compassion, commitment, or competence. The true measure of our time, then, is not how we will deal with the economic throes affecting our trade and prosperity but the level of mental health we will require or tolerate in our society (President's Commission on Mental Health, 1978). A look at our jails, nursing homes, runaway centers, and morgues should tell us how much more successful the professions and society will need to be in the future than we are now. In this ongoing assessment of policy commitments to quality, we will discover the ultimate bottom line.

> The dogmas of the quiet past are inadequate to the stormy present . . . . As our case is new, so we must think anew and act anew. We must disenthrall ourselves.
>
> Abraham Lincoln
> Address to Congress, 1862

REFERENCES

Abramowitz, K. (1986, January–February). Hospital role on decline. *Physician Executive*, pp. 2–6.

Acquilina, D., & Stowe, J. B. (1985, December 6). Increasingly sophisticated customers forcing PPOs to address market needs. *Modern Healthcare*, pp. 105–106.

Altman, L. K. (1986, March 25). Yardstick of medical performance is sought. *New York Times*, p. C4.

Altman, S. H., & Blendon, R. (Eds.). (1979). *Medical technology: The culprit behind health care costs?* (DHHS Publication No. PHS 79-3216). Washington, DC: U.S. Government Printing Office.

American Psychiatric Association. (1980). *Diagnostic and statistical manual of mental disorders* (3rd ed.). Washington, DC: Author.

American Psychiatric Association. (1983). *Economic fact book for psychiatry.* Washington, DC: American Psychiatric Press.

American Psychiatric Association. (1985). *Manual of psychiatric peer review* (3rd ed.). Washington, DC: Author.

American Psychological Association. (1980). *APA/CHAMPUS outpatient psychological peer review manual.* Washington, DC: Author.

Anderson, J. (1985). Industry looks at utilization review. *Alabama Journal of Medical Science, 22*, 403–407.

Apsler, R., & Bassuk, E. (1983). Differences among clinicians in the decision to admit. *Archives of General Psychiatry, 40*, 1133–1137.

Atchinson T. A., & Brooks, E. T. (1981). Centralized continuing education programs. *Quality Review Bulletin, 7*, 13–16.

Barkholz, D. (1986, January 31). Investor-owned chains in race to introduce insurance products. *Modern Healthcare*, p. 50.

Bedrosian, J. C. (1983, December). Market forces creating change, absent a national policy. *Financier*, pp. 51–54.

Bennett, M. J. (1987). Quality assurance activities for mental health services in health maintenance organizations. In G. Stricker & A. Rodriguez (Eds.), *Handbook of quality assurance in mental health.* New York: Plenum Press.

Berenson, R. A. (1985). A physician's perspective on case management. *Business and Health, 2* (8), 22–25.

Berk, A. (1980). Cost and efficacy of the substitution of ambulatory for inpatient care. *New England Journal of Medicine, 303*, 393–397.

Bernstein, A. (1986, March 31). Chopping health care costs: Labor picks up an axe. *Business Week*, pp. 78, 80.

Biegel, A., & Sharfstein, S. S. (1984). Mental health care providers: Not the only cause or only cure for rising costs. *American Journal of Psychiatry, 141*, 668–672.

Billet, T. C., & Cantor, J. A. (1985, Autumn). Employers' experience with preferred provider organizations. *Compensation and Benefits Management*, pp. 21–26.

Bittker, T. E. (1985, December). Psychiatry's future in HMOs. *HMO Mental Health Newsletter*, pp. 1–5.

Blendon, R. J., & Rogers, D. E. (1983). Cutting medical care costs. *Journal of the American Medical Association, 250*, 1880–1885.

Blendon, R. J., Schramm, C. J., Moloney, T. W., & Rogers, D. E. (1981). An era of stress for health institutions. *Journal of the American Medical Association, 245*, 1843–1845.

Bond, G. R. (1984). An economic analysis of psychosocial rehabilitation. *Hospital and Community Psychiatry, 35*, 356–362.

Bond, G. R., Witheridge, T. F., Setze, P. J., & Dincin, J. (1985). Preventing rehospitalization of clients in a psychosocial rehabilitation program. *Hospital and Community Psychiatry, 36*, 993–995.

Borenstein, D. B. (1985). Confidentiality. In J. Hamilton (Ed.), *Psychiatric peer review.* Washington, DC: American Psychiatric Press.

Borus, J. F., Olendzki, M. C., Kessler, L., Burns, B. J., Brandt, V. C., Broverman, C. A., & Henderson, P. R. (1985). The "offset effect" of mental health treatment on ambulatory medical care utilization and charges. *Archives of General Psychiatry, 42*, 573–580.

Bovbjerg, R. R., & Havighurst, C. C. (1985). Medical malpractice: Update for noncombatants. *Business and Health,2* (9), 38–42.

Brewster, A. C., Jacobs, C. M., & Bradbury, R. C. (1984). Classifying severity of illness by using clinical findings. *Health Care Financing Review, 6,* 107–108.

Brook, R. H., & Lohr, K. N. (1981). Quality of care assessment: Its role in the 1980s. *American Journal of Public Health, 71,* 681–682.

Brook, R. H., Williams, K. N., & Davies-Avery, A. (1976). Quality assurance today and tomorrow: Forecast for the future. *Annals of Internal Medicine, 85,* 809–817.

Brown, L. (1981). Competition and health care cost containment: Cautions and conjectures. *Milbank Memorial Fund Quarterly—Health and Society, 59,* 145–189.

Califano, J. A. (1986, March 25). A revolution looms in American health. *New York Times,* p. A31.

Care quality linked to lower cost. (1981, October 23). *American Medical News,* p. 14.

Carone, P. A., Yolles, S. F., Krinsky, L. W., Keiffer, S. N. (1985). *Mental health problems of workers and their families.* New York: Human Sciences Press.

Caulfield, S. C., & Haynes, P. L. (1981). *Health care costs: Private initiatives for containment.* Washington, DC: Government Research Corporation.

Chassin, M. R. (1983). *Variations in hospital length of stay: Their relationships to health outcomes* (Office of Technology Assessment, Technology Case Study No. 24). Washington, DC: U.S. Government Printing Office.

Collins, J. F., Ellsworth, R. B., Casey, N. A., Hickey, R. B., & Hyer, L. (1984). Treatment characteristics of effective psychiatric programs. *Hospital and Community Psychiatry, 35,* 601–605.

Committee for Financing Mental Health Care. (1981). *Mental illness and third party payment.* Washington, DC: American Psychiatric Association.

Comptroller General of the United States. (1980). *Report to the Congress: Rising hospital costs can be restrained by regulating payment and improving management.* Washington, DC: U.S. Government Printing Office.

Congressional Budget Office. (1977). *Expenditures for health care: Federal programs and their effects.* Washington, DC: U.S. Government Printing Office.

Congressional Budget Office. (1979a). *Controlling rising hospital costs.* Washington, DC: U.S. Government Printing Office.

Congressional Budget Office. (1979b). *Profile of health care coverage: The haves and have-nots.* Washington, DC: U.S. Government Printing Office.

Costilo, L. B. (1981). Competition policy and the medical profession. *New England Journal of Medicine, 304,* 1099–1102.

Council on Medical Service. (1983). Effects of competition in medicine. *Journal of the American Medical Association, 249,* 1864–1868.

Council on Wage and Price Stability. (1977). *The rapid rise of hospital costs.* Washington, DC: U.S. Government Printing Office.

Cousins, N. (1985). How patients appraise physicians. *New England Journal of Medicine, 313,* 1422–1424.

Cowan, D. H. (1984). *Preferred provider organizations.* Rockville, MD: Aspen.

Cunningham, L. S. (1981). Early assessment for discharge planning. *Quality Review Bulletin, 7*(10), 11–13.

Davidson, J. (1986, March 5). Research mystery: Use of surgery, hospitals vary greatly by region. *Wall Street Journal,* p. 35.

Davis, E. C. (1983, December). Local coalitions best hope to control delivery costs. *Financier,* pp. 27–30.

Donabedian, A. (1984). Quality, costs and cost containment. *Nursing Outlook, 32,* 142–145.

Doremus, H. (1984, December). Health care 2000. *Health Care Strategic Management,* pp. 4–10.

Dramatic variations in medical practice blamed for misspent health dollars. (1984, August 13). *Medical World News,* pp. 18–19.

Duncan, A. (1980). Quality assurance: What now and where next? *British Medical Journal, 280,* 300–302.

Egdahl R. (1984). Should we shrink the health care system? *Harvard Business Review, 62,* 125–132.

Eisenberg, J. M., & Williams, S. V. (1981). Cost containment and changing physician's practice behavior. *Journal of the American Medical Association, 246,* 2195–2201.

Englehart, J. K. (1984). Opinion polls on health care. *New England Journal of Medicine, 310,* 1616–1620.

Enthoven, A. C. (1984). The Rand experiment and economical health care. *New England Journal of Medicine, 310,* 1528–1530.

Etheredge, L. (1986). The world of insurance: What will the future bring? *Business and Health, 3(3),* 5–9.

Etheredge, L., Reinhardt, U., Marmor, T. R., Dunham, A., Davis, K., & Blumenthal, D. (1985). *Health care: How to improve it and pay for it.* Washington, DC: Center for National Policy.

Evans, R. W. (1983). Health care technology and the inevitability of resource allocation and rationing decisions. *Journal of the American Medical Association, 249,* 2208–2219.

Fein, R. (1985). Choosing the arbiter: The market or the government. *New England Journal of Medicine, 313,* 113–115.

Feldstein, M. (1981). *Hospital costs and health insurance.* Cambridge: Harvard University Press.

Ferber, J. S., Oswald, M., Rubin, M., Ungemack, J., & Schane, M. (1985). The day hospital as entry point to a network of long-term services. *Hospital and Community Psychiatry, 36,* 1297–1301.

Fielding, J. E. (1984a). *Corporate health management.* Menlo Park, CA: Addison-Wesley.

Fielding, J. E. (1984b). Managing mental health benefits. In J. E. Fielding (Ed.), *Corporate health management.* Menlo Park, CA: Addison-Wesley.

Fielding, J. E. (1985). A utilization review program in the making. *Business and Health, 2(7),* 25–28.

Flagg, D. C. (1985, October). A business perspective on the future of health care. *American College of Utilization Review Physicians Newsletter,* pp. 1–2, 7–8.

Flynn, T. J. (1985). *Issues and trends in the organization and financing of mental health benefits.* Washington, DC: American Enterprise Institute.

Forguer, S. L., & Anderson, T. B. (1982). A concerns-based approach to the implementation of quality assurance programs. *Quality Review Bulletin, 8(4),* 14–19.

Fox, P. (1984). *Synthesis of private sector health care initiatives* (DHHS Contract No. 100-82-0031). Washington, DC: U.S. Government Printing Office.

Fox, P., Goldbeck, W., & Spies, F. (1984). *Health care cost management.* Ann Arbor, MI: Health Administration Press.

Fox, P., & Spies, F. (1984). Weighing alternative systems. *Business and Health, 1(3),* 5–10.

Frank, J. D. (1975). Evaluation of psychiatric treatment. In A. M. Freedman, H. I. Kaplan, & B. J. Sadock (Eds.), *Comprehensive textbook of psychiatry/II* (2nd ed., Vol. 2). Baltimore, MD: Williams & Wilkins.

Frank, R. G., & Lave, J. R. (1985). A plan for prospective payment for inpatient psychiatric care. *Hospital and Community Psychiatry, 36,* 775–776.

Frech, H. E. (1981). *The long-lost market in health care: Government and professional regulation.* Washington, DC: American Enterprise Institute.

Fuchs, V. (1982). The battle for control of health care. *Medical Director, 8(6),* 5–9.

Furrow, B. R. (1983, Fall). Will psychotherapy be transformed in the 1980's? *Law Medicine and Health Care,* pp. 96.

The future of health care delivery in America. (1985). New York: Sanford C. Bernstein.

Gabel, J., & Ermann, D. (1985). Preferred provider organizations: Performance, problems and promise. *Health Affairs, 3(1),* 24–40.

Gatekeeper payment system could result in lower quality of care. (1984, May). *Hospital Peer Review,* p. 61.

Guarner, G. L. (1984). Regulating health professionals: A review of the empirical literature. *Milbank Memorial Fund Quarterly—Health and Society, 62,* 380–416.

Gaylin, S. (1984). The coming of the corporation and the marketing of psychiatry. *Hospital and Community Psychiatry, 36,* 154–159.

General Accounting Office. (1982). *Physician cost containment training can reduce medical costs* (GAO/HRD Publication No. 82-36). Washington, DC: U.S. Government Printing Office.

General Accounting Office. (1983). *GAO finds many ancillary services unnecessary* (GAO/HRD Publication No. 83-74). Washington, DC: U.S. Government Printing Office.

General Mills American Family Forum. (1980). *Proceedings: Private sector initiatives to promote mental health.* Washington, DC: Author.

Gephardt, R. (1984, December). The need for legislation. *Financier,* pp. 6–11.

Gerber, D. (1983). Community programs for affordable health care. *Inquiry, 20,* 127–133.

Gibson, R. M., Levit, K. R., Lazenby, H., & Waldo, D. R. (1984). National health expenditures 1983. *Health Care Financing Review, 6,(2),* 1–29.

Gibson, R. W. (1985). The future of the practice of psychotherapy. *The Psychiatric Hospital, 16*, 155–159.

Ginsburg, P. B., & Sloan, F. A. (1984). Hospital cost shifting. *New England Journal of Medicine, 310*, 893–898.

Ginzberg, E. (1983a). Cost containment—Imaginary and real. *New England Journal of Medicine, 308*, 1220–1224.

Ginzberg, E. (1983b). The grand illusion of competition in health care. *Journal of the American Medical Association, 249*, 1857–1859.

Ginzberg, E. (1984). The monetarization of medical care. *New England Journal of Medicine, 310*, 1162–1165.

Goldfarb, D. L., Mintz, R., & Yeager, M. S. (1982). Why did hospital costs increase in 1981? *Hospitals, 16*, 109–114.

Goldschmidt, P. G. (1982). *Health 2000.* Baltimore, MD: Policy Research.

Gonnella, J. S., Hornbrook, M. C., & Louis, D. Z. (1984). Staging of disease: A case mix measurement. *Journal of the American Medical Association, 251*, 637–644.

Gordon, R. E., Jardiolin, P., & Gordon, K. K. (1985a). Predicting length of hospital stay of psychiatric patients. *American Journal of Psychiatry, 142*, 235–237.

Gordon, R. E., Vijay, J., Sloate, S. G., Burket, R., & Gordon, K. K. (1985b). Aggravating stress and functional level as predictors of length of psychiatric hospitalization. *Hospital and Community Psychiatry, 36*, 773–774.

Gordon, T., & Breakey, W. B. (1983). A comparison of the outcomes of short and standard-stay patients at one year follow-up. *Hospital and Community Psychiatry, 34*, 1054–1056.

Gould, M. A. (1981). The effects of societal and legislative pressures on inpatient treatment. *Journal of the National Association of Private Psychiatric Hospitals, 12*(1), 6–15.

Graham, J. (1986, January 31). Insurers develop new products to protect their market share. *Modern Healthcare,* p. 51.

Greenfield, W. M. (1983). New approaches to using the determination-of-need process to contain hospital costs. *New England Journal of Medicine, 309*, 372–734.

Grimaldi, P. L., & Micheletti, J. A. (1984). Utilization and quality review under the prospective rate system. *Quality Review Bulletin, 10*(2), 30–37.

Gudeman, J. E., & Shore, M. F. (1984). Beyond deinstitutionalization: A new class of facilities for the mentally ill. *New England Journal of Medicine, 311*, 832–836.

Gudeman, J. E., Shore, M. F., & Dickey, B. (1983). Day hospitalization and an inn instead of inpatient care for psychiatric patient. *New England Journal of Medicine, 308*, 749–753.

Hamilton J. (Ed.). (1985). *Psychiatric peer review: Prelude and promise.* Washington, DC: American Psychiatric Press.

Hart, D. K., & Hart, J. D. (1980). Hospitals and economic turbulence in the 1980's. *Journal of Contemporary Business, 9*, 97–110.

Hembree, W. E. (1985). Getting involved: Employers as case managers. *Business and Health, 2*(8), 11–14.

Henisz, J. E., Levine, M. S., & Etkin, K. (1981). The psychiatric record and quality review. In C. Siegel & S. K. Fischer (Eds.), *Psychiatric records in mental health care.* New York: Brunner/Mazel.

Herrington, B. S. (1986a, March 7). Payment schemes for care predicted. *Psychiatric News,* pp. 25, 31.

Herrington, B. S. (1986b, March 7). Psychiatrists face uncertainty of changing practice systems. *Psychiatric News,* pp. 1, 24.

Hetherington, R. (1982). Quality assurance and organizational effectiveness in hospitals. *Health Services Research, 17*, 185–201.

Hiller, M. D., & Beyda, V. (1981). Computers, medical records and the right to privacy. *Journal of Health Politics, Policy and the Law, 6*, 463–487.

Hiller, M. D., & Seidel, L. F. (1982). Patient care management systems, medical records and privacy: A balancing act. *Balancing Health Reports, 97*, 332–345.

Himmelstein, D. U., & Woolhandler, S. (1986). Cost without benefit: Administrative waste in U.S. health care. *New England Journal of Medicine, 314*, 441–445.

Hine, F. R., Werman, D.S., & Simpson, D. M. (1982). Effectiveness of psychotherapy: Problems of research on complex phenomena. *American Journal of Psychiatry, 139*, 204–208.

Horn, S. D., Sharkey, P., & Bertram, D. A. (1983). Measuring severity of illness: Homogenous case mix group. *Medical Care, 21*, 14–25.

Hospital costs for psychiatric care up sharply. (1983, June 6). *American Medical News*, p. 7.

Houpt, J. L., Orleans, C. S., George, L. K., & Brodie, H. K. (1979). *The importance of mental health services to general health care*. Cambridge, MA: Ballinger.

Industries use preadmission screening to cut hospitalization. (1982, January). *Hospital Peer Review*, pp. 1–2.

Inglehart, J. (1982). Health care and American business. *New England Journal of Medicine, 306*, 120–124.

Jencks, S. F., Goldman, H. H., & McGuire, T. G. (1985). Challenges in bringing exempt psychiatric services under a prospective payment system. *Hospital and Community Psychiatry, 36*, 764–769.

Jensen, G., Feldman, R., & Dowd, B. (1964). Corporate benefit policies and health insurance costs. *Journal of Health Economics, 3*, 275–296.

Jessee, W. F. (1981). Approaches to improving the quality of health care: Organizational change. *Quality Review Bulletin, 7*(7), 13–18.

Jessee, W. F. (1983). Assuring the quality of health care: Policy and perspectives. In C. Sager & L. C. Jain (Eds.), *Policy issues in personal health services*. Rockville, MD: Aspen.

Jordan, D. D. (1985). Impact of an alternatives program on client utilization of services in a county mental health system. *Hospital and Community Psychiatry, 36*, 1291–1296.

Kane, R. A. (1981). Assuring quality of care and quality of life in long term care. *Quality Review Bulletin, 7*(10), 3–10.

Kapp, M. B. (1984, December). Legal and ethical implications of health care reimbursement by diagnosis related groups. *Law Medicine and Health Care*, pp. 245–253.

Katz, D. M. (1986, January 18). Outpatient benefit hikes key to mental health plans new look. *National Underwriter*, p. 4.

Kiesler, C. A. (1982). Mental hospitals and alternative care. *American Psychologist, 37*, 349–360.

Kinzer, D. M. (1983). Massachusetts and California—Two kinds of hospital cost control. *New England Journal of Medicine, 308*, 838–841.

Kohlmeier, L. M. (1983, December). Question of morality, economics now also confront U.S. medicine. *Financier*, pp. 9–13.

Korcok, M. (1986, April 4). From patient advocate to gatekeeper. *American Medical News*, pp. 21–23.

Kralewski, J. (1984, July–August). Employers' perspectives on the PPO concept. *Hospital and Health Services Administration*, pp. 123–139.

Lang, D. A. (1984). Prospective quality assurance. *Quality Review Bulletin, 10*(5), 143–145.

Law, W. (1981, October). *Host home: Providing a family-community alternative to traditional residential placement for the emotionally disturbed child*. Paper presented at the meeting of the American Academy of Child Psychiatry, Dallas.

Leaf, A. (1984). The doctor's dilemma—and society's too. *New England Journal of Medicine, 310*, 718–720.

Lee, F. C., & Schwartz, G. (1984). Paying for mental health care in the private sector. *Business and Health, 1*(10), 12–16.

Levenson, A. I. (1982). The growth of investor-owned psychiatric hospitals. *American Journal of Psychiatry, 139*, 902–907.

Levenson, A. I. (1983). Issues surrounding the ownership of private psychiatric hospitals by investor-owned hospital chains. *Hospital and Community Psychiatry, 34*, 1127–1131.

Lewin, M. E. (1986). *New strategies in the payment and organization of mental health benefits*. Washington, DC: American Enterprise Institute.

Lindsay, C. M. (1982, May). Is there really a crisis in the cost of health care? *Colloquium, 3*, pp. 1, 2, 8.

Lister, J. (1983). Current controversy on alternative medicine. *New England Journal of Medicine, 309*, 1524–1526.

London, P., & Klerman, G. L. (1982). Evaluating psychotherapy. *American Journal of Psychiatry, 139*, 709–717.

Luckey, J. W. (1983). The changing mental health scene. In C. Sager & L. C. Jain (Eds.), *Policy issues in personal health services*. Rockville, MD: Aspen.

Luft, H. S. (1985). HMOs: Friends or foes? *Business and Health, 3*(2), 5–9.

Luke, R. D., & Boss, R. (1981). Barriers limiting the implementation of quality assurance programs. *Health Services Research, 16,* 305–314.

Luke, R. D., & Modrow, R. (1982). Professionalism, accountability and peer review. *Health Services Research, 17,* 113–123.

Luke, R. D., Kreuger, J. C., & Modrow, R. E. (Eds.). (1983). *Organization and change in health care quality assurance.* Rockville, MD: Aspen.

Lundberg, G. D. (1983). Rationing human life. *Journal of the American Medical Association, 249,* 2208–2219.

MacStravic, R. S. (1984). *Forecasting use of health services.* Rockville, MD: Aspen.

Manning, W. G. (1984). A controlled trial of the effect of a prepaid group practice on use of services. *New England Journal of Medicine, 310,* 1505–1510.

Manning, W. G., Wells, K. B., Duan, N., Newhouse, J. P., & Ware, J. E. (1984). Cost-sharing and the use of ambulatory mental health services. *American Psychologist, 39,* 1077–1089.

Mariner, W. K. (1984, December). Diagnosis related groups: Evading social responsibility? *Law Medicine and Health Care,* pp. 243–244.

McClure, W. (1981). Structure and incentive problems in economic regulation of medical care. *Milbank Memorial Fund Quarterly—Health and Society, 59,* 107–144.

McGuire, T. G., & Montgomery, J. T. (1982). Mandated mental health benefits in private health insurance. *Journal of Health Politics, Policy and the Law, 7,* 380–406.

McGuire, T. G., & Weisbrod, B. A. (1981). *Economics and mental health* (DHHS Publication No. PHS 79-55071). Washington, DC: U.S. Government Printing Office.

McIlrath, S. (1985, October 18). Health fraud's cost: $1 million a day. *American Medical News,* pp. 2, 25, 26.

McMartin, E. J. (1986, February 15). A. S. Hansen Inc. health care survey. *Medical Benefits,* pp. 1–2.

McNeil, B. J., Pauker, S. G., Sox, H. C., & Tversky, A. (1982). On the elicitation of preferences for alternative therapies. *New England Journal of Medicine, 306,* 1259–1262.

Mechanic, D. (1985). Cost containment and the quality of medical care: Rationing strategies in an era of constrained resources. *Milbank Memorial Funds Quarterly—Health and Society, 63,* 453–475.

Medical care costs rose twice as fast as other prices. (1983, April 18). *American Medical News,* p. 4.

Memel, S. L. (1986, February). *Competitive strategies: HMO's, PPO's, IPO's.* Paper presented at the meeting of the American College of Health Care Executives, Chicago.

Mental health benefits: Shrinking reimbursement. (1981, November 9). *Washington Report on Medicine and Health,* pp. 1–4.

Mental health market undergoing change. (1986, February 21). *American Medical News,* pp. 17–18.

Merrill, J. C. (1985). Defining care management. *Business and Health, 2*(8), 5–9.

Merwin, M. R., & Ochberg, F. M. (1983). The long voyage: Policies for progress in mental health. *Health Affairs, 2,* 96–127.

Meyer, J. (Eds.). (1983). *Market reforms in health care.* Washington, DC: American Enterprise Institute.

Mezzich, J. E., & Coffman, G. A. (1985). Factors influencing length of hospital stay. *Hospital and Community Psychiatry, 36,* 1262–1264, 1270.

Mezzich, J. E., & Sharfstein, S. S. (1985). Severity of illness and diagnostic formulation: Classifying patients for prospective payment systems. *Hospital and Community Psychiatry, 36,* 770–772.

Moore, S. H., Martin, D. P., & Richardson, W. C. (1983). Does the primary care gatekeeper control the costs of health care? *New England Journal of Medicine, 309,* 1400–1404.

More M.D.'s support proposals to control costs. (1984, January 20). *American Medical News,* p. 13.

Morrison, B. J., Rehr, H., Rosenberg, G., & Davis, S. (1982). Consumer opinion surveys. *Quality Review Bulletin, 8*(2), 19–24.

Mosher, L. R. (1983). Alternatives to psychiatric hospitalization., *New England Journal of Medicine, 309,* 1579–1580.

Moxley, J. H., & Roeder, P. C. (1984). New opportunities for out-of-hospital health services. *New England Journal of Medicine, 310,* 193–197.

Mumford, E., Schlesinger, H. J., Glass, G. V., Patrick, C., & Cuerdon, T. (1984). A new look at evidence about reduced cost of medical utilization following mental health treatment. *American Journal of Psychiatry, 141,* 1145–1158.

Muszynski, S., & Brady, J. (1983). *Coverage for mental and nervous disorders: Summaries of 300 private sector health insurance plans.* Washington, DC: American Psychiatric Press.

Name of new game: Allocation of resources. (1983, January 7). *American Medical News,* pp. 1, 7, 8.

National Association of Private Psychiatric Hospitals. (1985). *A proposal on prospective reimbursement for psychiatric hospitals.* Washington, DC: Author.

National Center for Health Services Research. (1981). *Options, incentives and employment-related health insurance coverage.* Washington, DC: U.S. Government Printing Office.

National Center for Health Services Research. (1983). *Economic incentive and physician practice: An examination of Medicare participation decisions and physician-induced demand* (Publication No. PB-83-208132). Washington, DC: Author.

National utilization standard urged. (1983, July). *Hospital Peer Review,* p. 96.

Nazemetz, P. M. (1985). Health benefits redesigns to stay atop the competition. *Business and Health,* 2(8), 40–42.

Nelson, A. R. (1984, April). *Health care and its effect on the economy.* Statement of the American Medical Association to the Joint Economic Committee, United States Congress, Washington, DC.

Nelson, S. H. (1979). Standards affecting mental health care: A review and commentary. *American Journal of Psychiatry, 136,* 303–307.

New trends in care seen as reimbursements cut. (1983, January 7). *American Medical News,* pp. 1, 24.

Parloff, M. B. (1982). Psychotherapy research evidence and reimbursement decisions. *American Journal of Psychiatry, 139,* 718–727.

Patterson, D. Y. (1985, May). *Patients and psychiatrists in HMOs and PPOs.* Paper presented at the meeting of the American Psychiatric Association, Dallas.

Pattison, R. V., & Katz, H. M. (1983). Investor-owned and not-for-profit hospitals. *New England Journal of Medicine, 309,* 347–353.

Perlman, B. B., Schwartz, A. H., Paris, M., Thornton, J. C., Smith, H., & Weber, R. (1982). Psychiatric records: Variations based on discipline and patient characteristics, with implications for quality of care. *American Journal of Psychiatry, 139,* 1154–1157.

Perlman, B. B., Melnick, G., & Kentera, A. (1985). Assessing the effectiveness of a case management program. *Hospital and Community Psychiatry, 36,* 405–407.

Piersma, H. L., & White, M. (1985). An evaluation of treatment outcome for psychiatrically hospitalized adults. *The Psychiatric Hospital, 16,* 121–126.

Platt, R. (1983). Cost containment—Another view. *New England Journal of Medicine, 309,* 726–730.

Pollard, M. R. (1983). Competition or regulation. *Journal of the American Medical Association, 249,* 1860–1863.

President's Commission on Mental Health. (1978). *Report to the President.* Washington, DC: U.S. Government Printing Office.

Rappleye, W. C. (1983, December). System for rational choices. *Financier,* pp. 19–29.

Reeder, M. P. (1981). The benefits of areawide quality review. *Quality Review Bulletin,* 7(12), 2–3.

Relman, A. S. (1980). The new medical-industrial complex. *New England Journal of Medicine, 303,* 963–970.

Relman, A. S. (1982). An institute for health care evaluation. *New England Journal of Medicine, 306,* 669–670.

Relman, A. S. (1983a, July). The future of medical practice. *Health Affairs,* pp. 5–19.

Relman, A. S. (1983b). Investor-owned hospitals and health care costs. *New England Journal of Medicine, 309,* 370–372.

Relman, A. S. (1983c, December). Powerful new policy force: The medical industrial complex. *Financier,* pp. 40–44.

Relman, A. S. (1985a). Antitrust law and the physician entrepreneur. *New England Journal of Medicine, 313,* 884–885.

Relman, A. S. (1985b). Dealing with conflicts of interest. *New England Journal of Medicine, 313,* 749–751.

Restuccia, J. D., & Holloway, D. C. (1982). Methods of control for hospital quality assurance systems. *Health Services Research, 17,* 182–197.

Ricardo-Campbell, R. (1982). *The economics and politics of health.* Chapel Hill: University of North Carolina Press.

Richman, A., & Barry, A. (1985). More and more is less and less. *British Journal of Psychiatry, 146,* 164–168.

Rodriguez, A. R. (1983a, October). *National standards of review.* Paper presented at the meeting of the American College of Utilization Review Physicians, Denver.

Rodriguez, A. R. (1983b). Psychological and psychiatric peer review at CHAMPUS. *American Psychologist, 38,* 941–947.

Rodriguez, A. R. (1984). Peer review program sets trends in claims processing. *Business and Health, 1*(10), 21–25.

Rodriguez, A. R. (1985). Current and future directions in reimbursement for psychiatric services. *General Hospital Psychiatry, 7,* 341–348.

Rodriguez, A. R., & Maher, J. J. (1986). Psychiatric case management at CIBA-Geigy. *Business and Health, 3*(3), 14–17.

Roe, B. B. (1981). The UCR boondoggle: A death knell for private practice? *New England Journal of Medicine, 305,* 41–45.

Roe, B. B. (1985). Rational remuneration. *New England Journal of Medicine, 313,* 1286–1289.

Rosen, R. H., & Orr, F. C. (1986). Occupational mental health - A continuum of care. In A. Riley (Ed.), *Occupational stress and organizational effectiveness.* New York: Praeger.

Rubin, J. (1980). A survey of mental health policy options. *Journal of Health Politics, Policy and the Law, 5,* 210–232.

Rubin, J. (1982). Cost measurement and cost data in mental health setting. *Hospital and Community Psychiatry, 33,* 750–754.

Rzaha, C. B. (1983). Quality assurance in long term care. *Quality Review Bulletin, 9*(8), 229–232.

Sammons, J. H. (1983, Dec.). Doctor's active role in shaping tough policy choices. *Financier,* pp. 46–50.

Sandrick, K. M. (1982). Private review—Its impact on health care and physicians. *Quality Review Bulletin, 8*(4), 5–6.

Sapolsky, H. M. (1981). Corporate attitudes towards health care costs. *Milbank Memorial Fund Quarterly— Health and Society, 59,* 561–585.

Schacht, T. E., & Strupp, H. H. (1985). Evaluation of psychotherapy, In H. I. Kaplan & B. J. Sadock (Eds.), *Comprehensive Textbook of Psychiatry* (4th ed., Vol. 2). Baltimore: Williams & Wilkins.

Scheier, R. L. (1986, February 21). Marketplace is the new frontier. *American Medical News,* p. 7.

Schlesinger, M., & Dorwalt, R. (1984). Ownership and mental health services: A reappraisal of the shift toward privately owned facilities. *New England Journal of Medicine, 311,* 959–965.

Schorr, B. (1983, December 2). Hospitals scramble to track costs as insurers limit reimbursements. *Wall Street Journal,* p. 2.

Schroeder, J. L., Clarke, J. T., & Webster, J. R. (1985). Prepaid entitlements: A new challenge for physician-patient relationships. *Journal of the American Medical Association, 254,* 3080–3082.

Schuchman, H., Nye, S. G., Foster, L. M., & Lanman, R. B. (1980). Confidentiality of health records: A symposium on current issues, opportunities and unfinished business. *American Journal of Orthopsychiatry, 50,* 639–677.

Schulz, R. I., Greenley, J. R., & Peterson, R. W. (1983). Management, cost and quality of acute inpatient psychiatric services. *Medical Care, 21,* 911–928.

Schumacher, D. N., Clopton, C. J., & Bertram, D. A. (1982). Measuring hospital case mix. *Quality Review Bulletin, 8*(4), 20–26.

Scott, W. R. (1982). Managing professional work: Three models of control for health organizations. *Health Services Research, 17,* 198–211.

Sharfstein, S. S., & Beigel, A. (1984). Less is more? Today's economics and its challenge to psychiatry. *American Journal of Psychiatry, 141,* 1403–1408.

Sharfstein, S. S., & Beigel, A. (Eds.). (1985). *The new economics and psychiatric care.* Washington, DC: American Psychiatric Press.

Sharfstein, S. S., Muszynski, S., & Gattozzi, A. (1983). *Maintaining and improving psychiatric insurance coverage: An annotated bibliography.* Washington, DC: American Psychiatric Press.

Sharfstein, S. S., Muszynski, S., & Arnett, G. M. (1984a). Dispelling myths about mental health benefits. *Business and Health, 1*(10), 7–11.

Sharfstein, S. S., Muszynski, S., & Myers, E. (1984b). *Health insurance and psychiatric care: Update and appraisal.* Washington, DC: American Psychiatric Press.

Sharfstein, S. S., & Patterson, D. Y. (1983). The growing crisis in access to mental health services for middle-class families. *Hospital and Community Psychiatry, 34,* 1009–1014.

Sharfstein, S. S., & Taube, C. A. (1982). Reductions in insurance for mental disorders: Adverse situation, moral hazard and consumer demand. *American Journal of Psychiatry, 139,* 1425–1430.

Shaughnessy, P. W., & Kurowski, B. (1982). Quality assurance through reimbursement. *Health Services Research, 17,* 157–182.

Shore, M. F., & Levinson, H. (1985). On business and medicine. *New England Journal of Medicine, 313,* 319–321.

Shortell, S. M., & LoGerfo, J. P. (1981) Hospital medical staff organization and quality of care. *Medical Care, 19,* 1041–1055.

Showstack, J. A., Stone, M. H., & Schroeder, S. A. (1985). The toll of changing clinical practices in the rising costs of health care. *New England Journal of Medicine, 313,* 1201–1207.

Siegel, K., & Kunzel, C. (1982). Barriers to the integration of psychotherapy research and practice. *The Psychiatric Hospital, 13,* 149–155.

Simenstad, P. O. (1985). Health care delivery in a multispecialty group. *Business and Health, 2*(8), 27–30.

Sloan, F. A. (1982). Effects of health insurance on physicians' fees. *Journal of Human Resources, 17,* 533–557.

Sloan, F. A. (1983). Rate regulation as a strategy for hospital cost control. *Milbank Memorial Fund Quarterly—Health and Society, 61,* 195–221.

Sloan, F. A., & Schwartz, W. B. (1983). More doctors: what will they cost? *Journal of the American Medical Association, 249,* 766–769.

Smith, D., & Metzner, C. (1970). Differential perceptions of health care quality in a prepaid group practice. *Medical Care, 8,* 264–275.

Smith, M. L. (1982). What research says about the effectiveness of psychotherapy. *Hospital and Community Psychiatry, 33,* 457–463.

Smith, M. L., Glass, G. V., & Miller, T. (1981). *The benefits of psychotherapy.* Baltimore: Johns Hopkins University Press.

Snoke, A. W. (1982). The need for a comprehensive approach to health and welfare. *American Journal of Public Health, 72,* 1028–1034.

Sorkin, A. L. (1986). *Health care and the changing economic environment.* Lexington, MA: Lexington Books, D. C. Heath.

Spitz, B. (1985). Medicaid case management: Lessons for business. *Business and Health, 2*(8), 16–20.

Spitzer, R. L., & Endicott, J. (1975). Psychiatric rating scales. In A. M. Freedman, H. I. Kaplan, & B. J. Sadock (Eds.), *Comprehensive textbook of psychiatry/II* (2nd ed., Vol 2). Baltimore: Williams & Wilkins.

Spitzer, R. L., Endicott, J., Fleis, J., & Cohen, J. (1970). The psychiatric status schedule: A technique for evaluating psychopathology and impairment in role functioning. *Archives of General Psychiatry, 23,* 41–55.

Spitzer, R. L., Gibbon, M., & Endicott, J. (1973). *Global assessment scale.* New York: New York State Department of Mental Hygiene.

Staver, S. (1986, February 7). Companies market variety of prepaid plans. *American Medical News,* pp. 1, 19–20.

Stein, J. (1980, December 13). Does psychotherapy really work? The government tries to find out. *National Journal,* pp. 2113–2115.

Stein, J. (1983). Stagnation threatens psychiatric care. *Business and Health, 1*(1), 34.

Stuart, B., & Stockton, R. (1973). Control over utilization of medical services. *Milbank Memorial Fund Quarterly—Health and Society, 51,* 341–394.

Sullivan, S., & Ehrenhaft, P. M. (1984). *Managing health care costs: Private sector innovations.* Washington, DC: American Enterprise Institute.

Taintor, Z., Widem, P., & Barrett, S. A. (Eds.). (1984). *Cost considerations in mental health treatment: Settings, modalities and providers* (DHHS Publication No. ADM 84-1295). Washington, DC: U. S. Government Printing Office.

Talbott, J. A. (1984). Psychiatry's agenda for the 80's. *Journal of the American Medical Association, 251,* 2250.

Tarlov, A. R. (1983). The increasing supply of physicians, the changing structure of the health services system, and the future practice of medicine. *New England Journal of Medicine, 308,* 1235–1244.

Tavernier, G. (1983). Companies prescribe major revisions to medical benefits programs to cut soaring health care costs. *Management Review, 72*(8), 9–19.

Tell, E. J., Falik, M., & Fox, P. (1984). Private-sector health care initiatives: A comparative perspective from four communities. *Milbank Memorial Fund Quarterly—Health and Society, 62,* 357–379.

Test program to assess physicians in Maryland, Virginia, Washington, D.C. (1981, May). *Hospital Week,* p. 3.

Thompson, B. (1985). The merits of tracking provider performance. *Business and Health, 2*(7), 30–31.

Thompson, R. E. (1981). Relating continuing education and quality assurance activities. *Quality Review Bulletin, 7*(1), 3–6.

Thompson, R. E., & Rodrick, A. G. (1982). Integrating patient concerns into quality assurance activities. *Quality Review Bulletin, 8*(2), 16–18.

Thurow, L. C. (1985). Medicine versus economics. *New England Journal of Medicine, 313,* 611–614.

Tresnowski, B. (1983). Competition: Good news and bad news. *Inquiry, 20,* 199–200.

Ver Berkmoes, R. (1986, April 4). Variation studies gain support. *American Medical News,* pp. 3, 19.

Ware, J. E., Johnston, S. A., Davies-Avery, A., & Brook, R. H. (1979). *Conceptualization and measurement of health for adults in the health insurance study: Vol III. Mental health* (R-1987/3-HEW). Santa Monica, CA: Rand Corporation.

Watt, J. M., Derzon, R. A., Renn, S. C., Schramm, C. J., Hahn, J. S., & Pillari, G. D. (1986). The comparative economic performance of investor-owned chain and not-for-profit hospitals. *New England Journal of Medicine, 314,* 89–96.

Waxman, H. (1983). In defense of government and health. *Business and Health, 1*(2), pp. 15–18.

Wegmiller, D. (1983). Financing strategies for non-profit hospital systems. *Health Affairs, 2*(2), 48–54.

Weisman, G. K. (1985). Crisis-oriented residential treatment as an alternative to hospitalization. *Hospital and Community Psychiatry, 36,* 1302–1305.

Weissman, M. N., Rounsaville, B. J., & Chevron, E. (1982). Training psychotherapists to participate in psychotherapy outcome studies. *American Journal of Psychiatry, 139,* 1442–1446.

Wells, K. B., & Manning, W. G. (1985, May). *Mental health care in HMO and fee-for-service.* Paper presented at the meeting of the American Psychiatric Association, Dallas.

Wendorf, B. (1982). Ethical decision-making in quality assurance. *Quality Review Bulletin, 8*(1), 4–6.

Wennberg, J. (1986). Which rate is right? *New England Journal of Medicine, 314,* 310–311.

White, J. S. (1983, February 7). Psychiatrists' earnings: Will the boom last? *Medical Economics,* pp. 89–95.

Wilensky, G. R., & Rossiter, L. F. (1983). The relative importance of physician-induced demand in the demand for medical care. *Milbank Memorial Fund Quarterly—Health and Society, 61,* 252–277.

Williamson, J. W., & Associates. (1982). *Teaching quality assurance and cost containment in health care.* San Francisco: Jossey-Bass.

Winsten, J. (1981). Competition in health care. *New England Journal of Medicine, 305,* 1280–1282.

Wolfson, J., & Levin, P. J. (1985). *Managing employee health benefits.* Homewood, IL: Dow Jones-Irwin.

Woodward, R. S., & Warren-Boulton, F. (1984). Considering the effects of financial incentives and professional ethics on "appropriate" medical care. *Journal of Health Economics, 3,* 223–237.

Wyszewianski, L. (1982). Market-oriented cost containment strategies and quality of care. *Milbank Memorial Fund Quarterly—Health and Society, 60,* 518–550.

Youel, D. B. (1983, February). Patient care, continuing health professions education, and quality assurance. *American College of Utilization Review Physicians Newsletter,* pp. 1–2.

Young, D., & Saltman, R. (1983). Preventive medicine for hospital costs. *Harvard Business Review, 61,* 126–133.

Ziegenfuss, J. T. (1985). *DRGs and hospital impact.* New York: McGraw-Hill.

7

The Role of the Consumer in Quality Assurance

AGNES B. HATFIELD AND H. BERNARD SMITH

The need for accountability in the provision of mental health services is receiving considerable attention. It is generally agreed that, at a minimum, services should be safe and effective, accessible to those who need them, and reasonably priced. The term *quality assurance* is usually given to the efforts that mental health providers make to ensure these minimum standards. Peer review, boards of examiners and licensure, professional organizations, and educational institutions all play a part in ensuring that these standards will be met. Considerable effort goes into the process of identifying objectives, creating appropriate evaluation devices, and recommending and implementing change. This effort is to be expected from responsible professionals, and the public would be truly ill served if they did not meet these expectations. However, mental health professionals alone cannot ensure that all people will be well served; they are providers of service and necessarily reflect the biases of those who have personal financial gain at stake.

It is a matter of considerable concern that those who use and pay for mental health services, and whose personal lives may be critically effected by the outcome, have so little to say about the quality and appropriateness of these services. In the marketplace, generally, customers have influence over how products and services are supplied; however, in the field of mental health services, the power of the consumer has been very limited, With partial costs paid by third parties, the actual cost to the consumer is obscured. The consumer may settle for less adequate treatment, erroneously believing that he or she is not paying much for it. It is a matter of further concern that mental health professionals, even though they have considerable self-interest in the process, serve as advisors to major insurance carriers. CHAMPUS, which is the Defense Department's Civilian Health and Medical Program of the Uniformed Services, contracts with the American Psychiatric Association to ensure quality of service and cost containment. Consumers are not thought to be an authority on these matters, which will affect them so vitally.

The purpose of this chapter is to explore possible reasons for the consumers

---

AGNES B. HATFIELD • Department of Education, University of Maryland, College Park, Maryland 20742.   H. BERNARD SMITH • Private practice, Washington, D.C. 20005.

passive-dependent role and to suggest ways in which consumers may be more fully empowered to work in behalf of their own interests.

THE GROWTH OF THE CONSUMER MOVEMENT

The Western tradition of the helping professions has maintained a hierarchical arrangement of power by which consumers have been excluded from significant participation in determining service provision. Ours has been a paternalistic system in which the professional defines what is good for the consumer and dispenses what the professional determines is needed (Steinman & Traunstein, 1976). Recipients of services have tended to rely on the wisdom of experts who are presumed to know better what is good for them than the recipients can discern for themselves. Doctors and other professionals have typically taught families that the professional is in charge and is the expert, and that his or her prescriptions are to be followed. The consumer role is relegated to accurately describing symptoms and complying with the ministrations recommended by the omniscient professional (Hollander, 1980).

• Lenrow and Burch (1981) summed up the reasons for the consumer's apparent powerlessness as follows: (a) professionals are presumed to have possession of scarce resources; (b) only other professionals similarly trained can understand these resources; (c) special knowledge entitles them to much independence in decision making; (d) professionals are entitled to set the ground rules for relating to clients; and (e) unilateral exercise of power is justified for efficient, effective service.

• The growing resentment on the part of consumers due to this unilateral use of power goes unrecognized. Consumers perceive themselves as dependent on the power of the professional and therefore as vulnerable to the professional's abuse of power. The mistrust by some consumers is based on their beliefs that professional are essentially exploitative and more concerned with their economic interests and privileges than they are with the well-being of the persons they are supposed to serve. Others perceive professionals as academically cloistered, bureaucratically oriented, and socially privileged.

What is especially frustrating is that, when professionals are confronted they tend to mask their self-interest and declare that they are speaking for the welfare of their clients. Arbuckle (1977) is critical of those who take it upon themselves to speak on behalf of the consumers. He likens this attitude to the situations in which white people speak for black people and the rich speak for the poor. Torrey (1978) sees a similar thing happening to mental health lobbies consisting mainly of providers, who constantly try to rationalize their self-interests as the interests of the consumer. The net result, he insists, is no small degree of hypocrisy.

There is a growing opposition to the virtual monopoly of service providers in our society. Miller, Brodsky, and Bleechmore (1976) made some of the strongest statements. They said that "those whose business is service to the public must be prepared to answer to the public. Any operation or even entire profession, acting for long without external scutiny tends to develop expedient idosyncratic procedures of possible detriment to the public" (p. 277). Another sharp critic, Robinson (1973), focused his concern specifically on the issue of therapy: "Without the guidance and protection of law each therapist is judge, jury, and ethicist. To make matters worse, it is in the nature of his formal education that issues of law and ethics have been avoided with nearly brazen assiduousness" (p. 133).

According to the analysis of Gartner (1976), it was in the 1960s that the nature and practice of the human services began to be questioned—most vocally by those considering themselves ill served. Gartner quoted Margaret Mead as saying that there is "a revolt of all people who are being done good to" and that we are at the end of an era when great numbers of people were willing to believe that professionals always "knew best and did good." Gartner, however, believes that it may be premature to declare the era of professional control to be at an end, though he does feel that a heightened sense of accountability and a greater effort to make services responsive are in evidence. The development of advisory and governance boards and the demand for competency-based performance, together with consumer efforts to bring budgets under scrutiny, have weakened the professional's monopoly over control of services.

Hollander (1980) believes that we are already in a "third revolution in mental health" (p. 561). It is marked by a new culture-wide service ideology that assumes that consumers are capable of being the primary directors of their own growth and development. The new professional, in Hollander's view, is now a valued member of a community of common concerns but no longer dominates the helping process. This change in service ideology is deeply rooted in the major intellectual and social concerns of the present historical period. "Many people now believe," Hollander noted, "that uncritical acceptance of expert direction is an important factor in the intransigence of many problems and that expert intervention often creates its own set of iatrogenic difficulties, often leading to cycles of 'disabling help'" (p. 563). Expert opinion, it is important to recognize, can no longer be assumed to be part of an apolitical system; the professional provider represents a particular interest which is not always that of the consumer. The new ideology, Hollander wrote, does not question the desirability of professional assistance, but it does reject professional dominance and supports more egalitarian relationships between consumers and professionals.

It was in such a climate of change, and in keeping with the new ideology, that a new consumer-advocacy organization, the National Alliance for the Mentally Ill (NAMI), was formed in 1979. The NAMI brought together, under one umbrella organization, a grass-roots movement of families and patients seeking more responsive service for those with severe long-term mental illnesses. Since 1979, it has fostered the growth and affiliation of nearly 700 local self-help groups involving over 40,000 persons. The NAMI has become a well-recognized voice for those 2–3 million families that are advocates for persons with serious disorders. If it continues its rapid rate of growth of approximately 125 new affiliates each year with its present intensity of effort, its clout should become considerable.

QUALITY ASSURANCE: A CONSUMER PERSPECTIVE

The consumer and the provider are probably in agreement that the goals of quality assurance are (a) safe and effective service; (b) fair availability and access; and (c) cost-effectiveness. However, they are quite likely not to agree about what each of these terms means. The purpose of this section is to examine each of the three goals of quality assurance from a consumer perspective.

SAFE AND EFFECTIVE SERVICE

The client is typically persuaded to enter treatment for needs that he or she has identified, but just as typically, he or she will learn to see his or her problems from the view of the helper. This is what Dewar (1978) called the "professionalized client." This kind of client—probably most of us—learns about himself or herself and his or her circumstances in terms of the counselor's theory of behavior, in which the counselor selects, from the myriad of the knowable things about a person, the ones that are most important to that theory. In learning the helper's system, the client learns the helper's definition of the situation; this then becomes the reality in which both operate. "The fact that today's science is tomorrow's superstition," wrote Dewar somewhat cynically, "does not diminish the power of present beliefs" (p. 250). Each provider socializes each client into his or her own belief system. The "good client," Dewar noted, accepts his or her status as subordinate to the helper and does not challenge the helper.

The power of professionals lies in the fact that *they* define the problem, *they* determine the solutions, and *they* evaluate safety and effectiveness. Clearly, to redress this imbalance, clients must insist on their own problem definitions and claim the right to speak for themselves about treatment effectiveness. As a matter of fact, many clients do make choices by using the only option available: they drop out of treatment. It is generally conceded that the therapy dropout rate is very high. The reason, many writers say, is the lack of congruence between the goals and expectations of the patient and of the therapists (Boghi, 1968; Heine & Trosman, 1960; Lazare, Cohen, Jacobson, Williams, Mignone, & Zisook, 1972). Their concepts of reality and problem definitions do not match. Although dropping out of treatment is one way of exercising choice, it is a costly one. An individual can spend important insurance dollars going from one provider to another and, even more important, can be significantly delayed in getting effective help.

In a study of the consumer behavior of family caregivers for persons with severe mental illnesses, Hatfield, Fierstein, and Johnson (1982) and Hatfield (1983) were unable to find any significant relationships between what families perceived as their needs in therapy and what they had got from therapy. Families rated therapists high on their ability to listen, their appreciation of the families' difficulties, and their respect for the families' expertise and skills. They rated therapists rather low on helping families to set goals, or using time in therapy sessions effectively, and on keeping families informed of what was going on in these sessions. Most families indicated that they had had little or no information about the service before they engaged it. They commented that "it was our only choice," "we were told it would help our daughter," or "a friend recommended him or her."

Too little attention has been given to issues of safety in individual and family treatment. Graziano and Fink (1973) warned that we cannot assume that the patient is never harmed. Therapists and clients are two highly complex systems in which both positive and negative effects can occur. The assumption of the benign nature of therapy has been challenged, and consumers must ask about potential negative effect in treatment and evaluate what is happening to them as treatment proceeds. Hare-Mustin (1980) summarized some of the potential risks in family therapy that all potential users ought to weigh carefully: the therapist may assume that he or she is nonaligned with members even though there has been a tacit acceptance of the priority of needs of one member over those of others; laws of privileged communication may not be protected; and family members may suffer from embarrassment, anxiety, and loss of respect in the eyes of other family members when they are pressured to make disclosures in family session.

Billions of dollars are spent each year in this country for mental health treatment. How much of that amount is purely wasted because consumer and provider were poorly matched and no value came to the client? We suspect that the loss in dollars is staggering. We hear routinely from families that it may take, literally, dozens of trials with different therapists and a loss of 10 to 12 years to find the appropriate help for patient and family. What is wrong, we believe, is the consumer's reluctance or inability to ask enough of the right kinds of questions and the providers' inability or unwillingness to access their own competence and to provide candid answers about what they can and cannot do.

Hare-Mustin, Maracek, Kaplan, and Liss-Levinson (1979) believe that it is the ethical responsibility of practitioners to ensure that clients will make informed choices about entering and remaining in therapy. Although we support the view that the practitioner should be helpful in this manner, we would urge consumers to assume their own responsibility for securing the information they need in order to make informed choices. Although it is true that many people seek service in times of crisis, when it is difficult for them to sort things out, they can school themselves not to make long-term commitments during a time of crisis. Too many providers have too much self-interest at risk for consumers to be able to depend on them for objective guidance. The ultimate hope for consumers is that they will be well informed, assertive, and able to take responsibility for themselves. Clients will learn to think of themselves as customers capable of shopping until they find the service that fits their needs (Bymer, 1977).

Nevertheless, we do agree with Hare-Mustin et al. (1979) about the kinds of information that consumers need. First, they need knowledge of the procedures, goals, and possible side effects of therapy. Second, they need information about the qualifications of the therapist. Hare-Mustin et al. pointed out that "providing a description of skills and experience and responding to clients' queries safeguard both therapist and client from unrealistic expectations" (p. 6).

A third point that Hare-Mustin et al. make is that clients should be told alternatives to therapy. The Standards for Providers of Psychological Services of the American Psychological Association state that therapists have a responsibility to inform themselves about the network of human services in their community and to refer clients to them. Some of these alternatives may be self-help groups, crisis centers, religious activities, hospitalization, and therapeutic residences. Persons in need of help find themselves invariably shunted in the direction of therapy, but as Graziano and Fink (1973) reminded us, "psychotherapy is only one of several possible avenues for solving personal problems—and it is not even certain it is one of the most effective. By pulling persons into the treatment system, we may preclude their use of other potentially useful resources" (p. 359). Many other disciplines—education, nursing, recreation, and religion—are prepared to offer useful help to people. But consumers themselves must become educated to the available resources and must use them to their advantage. It may be difficult for therapists, whose livelihood depends on their clientele, to enthusiastically recommend alternative resources.

As one means of protecting the interests of both consumer and provider, Hare-Mustin et al. (1979) recommended the use of contracts to define the therapeutic relationship. A contract is a negotiation that takes place between partners of equal power. Usually, the following should be specified: the method of therapy, its goals, the length and frequency of sessions, the duration of the treatment, the cost and the method of payment, provisions for cancellation and renegotiation of the contract, and the degree of confidentiality. Techniques to be used at odds with the client's values should be specified.

When the client has a highly disabling mental illness is dependent on his or her family for care, the provider–consumer relationship becomes more complex. The primary consumer is, of course, the patient; however, the caregiver may be instrumental in selecting the services to be used, in negotiating with the provider, and in paying for the service. In that way, the caregiver is also playing a consumer role. However, the family caregiver also plays a provider role in seeing that the patient complies with recommended treatment, in providing a protected living environment, and in involving himself or herself in the case management and social rehabilitation. A successful treatment outcome may depend on how well both roles can be worked out with the service provider. As providers, family caregivers need access to knowledge about mental illness, and they need to learn skills in managing disturbing behavior. This does not mean that most families want to assume a third role, that of patient, as well. Because mental health services have been unprepared to teach families how to cope with mental illness, they have offered patienthood and therapy instead, and thus, they have alienated the families with whom they needed to collaborate.

It is clear from this discussion that providers of services cannot speak for consumers, as the provider perspective may be substantially different from that of the consumer. Mental health professionals have traditionally minimized client satisfaction as an outcome measure; they have, instead, tended to view client opinions as laden with psychodynamic distortion (Schulberg, 1981). We, however, agree with Schulberg that clients should be the focal point for outcome studies. Program quality, which is usually assessed by the providers of the service, is not the same thing as efficacy of treatment. If treatment outcome is to be fully assessed, consumers must play a part in formulating the questions to be asked. Clients must be able to convey fully how they have benefited and how they may have been hurt by the treatment. Most important, there must be follow-up of those who dropped out of treatment after initiating it. From these groups, much can be learned where the service is failing to meet people's needs (Buck & Hirschman, 1980; Langsley, 1975; Reeder, 1972). Once evaluation data are collected, their effect on policy should be maximal.

FAIR ACCESS TO SERVICE

Much has been written about the lack of access to services in rural areas and in economically depressed sections of the city. This is a serious problem that the community mental-health movement is attempting to address. It is discussed fully in other parts of this book. For persons with severe psychiatric disorders, there are additional issues of adequate access to relevant services. For the most part, professionals have been trained to work in hospitals and in one-to-one therapy, and few have been trained in social-rehabilitation and social-skills development. Too many dollars are going into costly psychotherapy because that is what providers are trained to do. Yet, there is considerable agreement that psychotherapy has limited value for those with psychotic disorders. Patients do not have fair access to what they need because present reimbursement policies make it impossible to provide what is most effective for them. Consumers need to involve themselves in the complex issue of third-party payment and to encourage carriers to develop a system of benefits that would result in more appropriate and more cost-effective treatment.

Further problems of access to care arise because the mentally ill are seen as a homogeneous population. Mental health planners and providers of service become committed to one model of rehabilitation service or supportive housing, and because this model serves one kind of clientele, the thought is that it should be good for all. Those clients who do not find the service appropriate to their needs are seen as refusers of service. Providers feel that they are "off the hook" because they have offered something, and they are quick to remind us that they cannot compel clients to avail themselves of the service offered. More adaptability is expected from the psychiatrically disabled population than from the population generally. More understanding is needed of the great diversity in this population, and more creativity is necessary to devise new and different models before we can begin to meet the standard of equal access. Our impression from relationship with consumers in NAMI is that only a very small fraction of patients have an optimum program for their growth and rehabilitation.

Many people, some even in the field of mental health, have a distorted notion of just how well the full range of the psychiatrically disabled is being served. They are misled by the occasional well-staffed, well-financed demonstration model into thinking that this is how most patients are, or could be, cared for. A consumer movement must necessarily concern itself with the needs of all its members. Members do not harbor the illusion that society is about to make a radical change and divert additional billions to care for the mentally ill. The reality is that there will never be enough money. Although consumers continue to advocate for whatever they can get, they recognize that costs must be contained and that money must be used much more efficiently if all are to have access to good care. A system of service delivery that produces very high quality for some and nothing for others is a poor system (Sechrest & Hoffman, 1983). Even though billions are being spent on the care of the mentally ill, too few benefit.

Cost-Effectiveness

The nation's medical bill continues to rise. In 1980 it rose the most in 15 years: a total of 15.2% over 1979, which exceeded the rate of general inflation (Sharfstein & Taube, 1982). Expenditures for mental health care (outside of nursing homes) were estimated at $14 billion in 1980. More than half, or $8.3 billion, was spent for active care in inpatient facilities, predominantly in state and private mental facilities and nonfederal general hospitals. More than $3 billion was spent on therapy at mental health facilities. The expenditure for service of psychiatrists, other physicians, and psychologists in private practice totaled $2 billion. One way or another, the citizen or consumer had to pay these costs (McGuire, 1981).

Consumers need to develop new strategies for cost containment, but they will surely encounter many problems, one of the most difficult of which is to define just what kinds of needs it is appropriate to cover. The problem, McGuire(1981) asserted, is that emotional distress is ubiquitous, anyone can qualify as needy, and treatment, once begun, could be continued indefinitely. The President's Commission on Mental Health (1978) noted that there were problems. "The boundaries of psychological therapies . . . seem to be expanding without limits. Psychotherapists are expected to do more than treat mental disorders; positive mental health is the new cry, presumably marked by 'self-actualization,' growth, and even spiritual oneness with the universe" (Volume IV, p. 1747).

Mechanic (1978) explained the problems further: Knowledge about the etiologies and treatments of mental health disorders remains uncertain; therefore, it is difficult for policymakers to develop appropriate boundaries for differentiating mental illness from abundant frustrations, problems, and disillusionments of everyday life. The typical approach, as in the medical arena, has been to define as legitimate illnesses whatever conditions are treated by designated experts—psychiatrists, psychologists, and social workers. Insurance carriers avoid the quagmire of defining insurable illnesses; rather they protect themselves by high co-insurance, deductables, limited visits, and limited total benefits covered. These restrictions on benefits result in a real hardship for severely disabled persons who have a desperate need for service and cannot manage without it. Furthermore, as Mechanic made very clear,

> we have developed insurance programs for psychiatric benefits that effectively limit total expenditures for psychiatric services, but also reinforce traditional, ineffective, and inefficient patterns of mental health care, inhibit innovation and use of less expensive mental health personnel, and reinforce medical as compared with social or educational approaches to patients' psychological problems. (p. 483)

There is no coverage for long-term programs in the community. Present policies serve to encourage hospital use and psychotherapy.

"Fiscal accountability is expected to move from the shadows to the forefront in the next few years," Shulberg (1982, p. 526) predicted. The tendency will be to err on the side of caution rather than to continue to spend large amounts of money for treatments of limited or unknown value. Mental health professionals have resisted facing up to economic issues; they have tended to think of money issues as irrelavent and dehumanizing. Members of the consumer movement, however, cannot afford to ignore the problem; they will insist on fiscal accountability.

In his recent book on financing psychotherapy, McGuire (1981) pointed out that coverage for psychotherapy has been controversial since it was first offered as a part of major medical policies of commercial insurers in the 1950s. Until recently, psychotherapy was well insured by the Blue Cross and Blue Shield High Option Plan for federal employees, but commercial carriers' early experiences caused them to draw back quickly. Large benefits went to those less disabled, who were able to carry on their usual functions without psychotherapy. There were many complaints about improper utilization.

The fact that insurance coverage reduces the price to the user stimulates demand; therefore, the user is more likely to seek care and to stay in treatment longer, and the provider is more likely to recommend longer treatment. The term *moral hazard* has been used to describe the temptation of the policyholder to be careless in controlling insured loss. It is not the insurance carrier that is exploited, for it charges premiums according the benefits that it expects to pay out. The consumer ultimately pays, either through premiums, deductibles, or coinsurance requirements. If these become too high, the marketability of insurance is destroyed, and the mental health industry loses clients.

In the immediate future, the consumer movement must devote considerable study to determining what can be done to protect its members from disastrous financial costs due to mental illness and yet get care that is safe and effective for loved ones. Experts in the field have identified some possible approaches.

1. *Deductibles and coinsurance.* These are the most common approaches now being used to deal with the "moral hazard." They make the consumer pay more of the costs; therefore, he or she becomes more cautious about use. These approaches do not solve the problem of what is an appropriate need for service; minor tensions are reimbursed on the same basis as major psychoses. Those with long-term illnesses and those who cannot function without help have little choice; they must have service however financially disastrous.

2. *Regulation of form and length of treatment.* It is conceivably possible that careful differential diagnoses could be developed that would allow for different predetermined lengths of care. McGuire (1981) used as a hypothetical case a diagnosis of bipolar illness of moderate impairment that would be entitled to 20 visits per year. This would work, of course, only if diagnoses can be, and will be, made objectively and sensibly.

3. *Peer review.* Attempts at cost containment are currently made through the process of peer review. This brings mental health professionals into the business of reviewing claims and making recommendations about standards of care, presumably to ensure quality and cost-effectiveness. Sechrest and Hoffman (1982) posed doubts about the effectiveness of peer review because peers may not be in agreement about what is necessary, peers are unlikely to report unnecessary care,

and providers may not cooperate with peer review. Costs due to unnecessary care, whether caused by inefficiency, lack of knowledge, over cautiousness, or a desire for financial gain, are not likely to be reduced. McGuire (1981) agreed. Unless peer review works in the interest of individual providers, it will encounter difficulties; providers will be able to manipulate the peer review formula to obtain the treatment that they and their patients need. Peer review would work only if providers had an interest in being restrained; therefore, McGuire wrote, "It would be reckless to rely on peer review as the exclusive or even general instrument for control of moral hazard" (p. 165).

4. *Provider organizations.* It may be in the interest of consumers to have a system of prepaid mental-health services that would be provided in an organized care setting following a pattern set by health maintenance organizations (HMOs). In a prepaid service, mental health staff would be on a salary; there would be little financial incentive in their extending treatment unnecessarily (Gorman, 1978). More nonmedical staff at less cost could be used, and treatment goals would be oriented toward return to function. McGuire (1981) feels that the idea of a prepaid plan is promising but not without hazards. In his opinion, in some community mental-health centers where staff works on salary, a high fraction of time is spent in conferences; in these systems, there was no great incentive to treat patients.

5. *Benefits for alternate forms of care.* As we have already noted, present patterns of third-party payment tend to encourage over use of such costly services as hospitals and individual psychotherapy because less costly alternative services are not ordinarily covered by insurance. For those with severe and chronic disorders, this is unfortunate, for social rehabilitative services may be of the most therapeutic help, and yet, as is so often the case, this kind of help may be unavailable. Among the reasons given for this state of affairs are the lack of a nationally developed accrediting body and the lack of quality assurance procedures (Young, 1982). Young believes that these limitations could be overcome by a process of certification of facilities and program approval. Programs could undergo professional review in which objectives are clearly delineated, strategies for meeting the objectives are described, and progress toward goals is assessed. We would add that there should be considerable consumer involvement in the whole process. The organized consumer movement would surely support such an effort.

THE EMPOWERMENT OF CONSUMERS

Our efforts up to this point have focused on reasons that consumers should be involved in the total planning for the services they use and in evaluating their outcome. The rest of this chapter is concerned with how consumers can develop the necessary skills and the power to go with them to actually affect the way services are given. Of first importance is consumer education, to be discussed next.

CONSUMER EDUCATION

Buyers and sellers are better matched when all come to the marketplace with accurate information. Traditionally, in the marketplace of mental health services, this has not been the case. Consumers have not been equipped to ask the right kinds of questions, and socialization in our culture has discouraged an assertive posture with professionals. Consumer education can change all that. Willett (1977) made a very convincing argument:

> Essentially a preventive tool, consumer education can make the difference between ig-
> norance and effective use of personal resources, between marketplace exploitation and fair
> competition, between lost opportunity, wasted dollars, inhibited social programs and
> healthy development, conservation, equity, and efficient social progress and healthy self-
> development. . . . It teaches the art of questioning. It encourages responsible consumer
> behavior. . . . It equips the individual to understand and use, and thereby enforce, the laws
> of the land. (p. 2)

Consumer education, Willett further asserted, can be a change agent on the model of the civil rights and women's movements, a truly revolutionary tool. Consumers of mental health services will learn to identify and define their own problems and to know the full array of treatment possibilities (Hornstra, Lubin, Lewis, & Willis, 1972). They will learn to ask the right questions and to persist until they have enough information to make wise choices. They will want to know both the possible benefits and the possible negative effects of any treatment modality and how to negotiate meaningful verbal or written contracts.

Educated consumers will understand that mental health services are expensive and that the consumer pays, one way or another, either through taxes, insurance, or out-of-pocket dollars. They will need to understand insurance and how benefits and premiums are determined. Above all, consumers must understand that only they can play the role of consumer; no one else can look out for consumer interests or speak in behalf of the consumer. Similarly, consumers must understand that it is the nature of providers to look out for their self-interests. Although consumers and providers may have many interests in common, may cooperate in many endeavors, and may like each other as persons, in their specific roles of consumers and providers there is conflict of interest. Consumers need to become comfortable with that fact.

BOARDS AND COMMISSIONS

In the human service sector, there has been little questioning of the value of services; for the most part, they have been considered a self-evident good (Gartner, 1979). Typically, consumers have a tolerance of public sector agencies and nonprofit components of the private sector that they would not have of business in general. There is a myth that the nonprofit organization must be involved in service of a higher order and therefore should not be judged by the market model. In the public sector, the inefficient are not driven out as they are in the private sector (Bymers, 1977). Often, there are no alternatives, and there is therefore an effective monopoly. Services have no competition, so consumers must "take it or leave it," and providers can act independently of their consumers (Meeneghan & Mascari, 1971).

Community mental health centers have been required to have citizen participation since the 1975 Amendments to the Community Mental Health Centers Act. This requirement has not always been adequately implemented for many reasons, one of which is that consumers do not know this is law, do not feel they can make a difference, or are not assertive enough to insist on their rights. The development of consumer-advocacy organizations across the country should give members the kind of backing they need to insist on meaningful participation. They need to know some of the politics of service provision so they can deal with them. Wilder (1975) examined some of the problems. There is resistance on the part of staff to consumer boards. The staff feels that citizens do not know enough about mental health or their own needs to be effective. Every staff member is sure that he or she can speak for the community, especially when it is for the staff member's own best interest. Wilder found, however, that there are unsettled questions: Who is the community? Who are its representatives? Are they consumers or citizens? Most important, what should be their authority? In a model described by Wilder, a citizen board is given power through veto power on certain issues. The board is involved in an ongoing assessment of clients for their perceptions of quality and helpfulness and feeds the results back to the staff.

Advisory boards of rehabilitation centers, halfway houses, hospitals, and any other services provided by the community should have consumer involvement in effective numbers. The dangers of being merely "token members" are very real, and consumers should be alert to that possibility. Consumers' power comes from their strong commitment and the expertise with which they can speak from the vantage point of those who are the recipients of service. Consumers can serve a vigorous oversight function that involves policing and ensuring that laws will be obeyed and policies enforced (Willett, 1977). Although it is expected that having relevant knowledge and being persistent and assertive should be enough clout in most situations, consumers should be willing, if necessary, to use the media, litigation, or the legislature to ensure the rights of clients and quality service. Vattano (1972) saw a real "power-to-the-people" social movement in the rapid growth of self-help organizations. Their actions have been changing many basic institutions and posing a new challenge to the helping professions. In them, Vattano saw a new egalitarianism—an effective and meaningful extension of democracy.

MUTUAL HELP AS AN ALTERNATIVE SERVICE

Central to the concept of the self-help movement is the idea and practice of mutual help. One who is faced with an especially painful and difficult situation in life needs to share this dilemma with someone else who is in a similar situation. There is a kind of empathic understanding unavailable elsewhere. Persons with mental illness or their families can now choose to join a mutual help group as a way to cope with problems in addition to, or in place of, professionally provided services. Mutual help groups have affected service because they demonstrate a workable model, whose principles a provider can adopt, and which can be of low cost. Mutual help groups combat the idea that families are disorganized and dysfunctional, and that only mental health professionals know how to help them.

Traditionally, mental health professionals have assumed that, if a family has a mentally ill relative, other members of the family are pathological. What is required, if one believes that, is long, costly treatment of the family as well as the patient.

Although there is no clear evidence that families have benefited from whichever of the various treatment modalities they were assigned, practitioners have profited. When families get together in mutual help groups and come to know each other, things look very different to them. They discover that families with mental illness are very much like other families except for the heavy burden that they carry in coping with mental illness. They also discover that, by assigning a sick role to families, professionals have created an inordinate amount of anxiety and guilt, which has an iatrogenic effect that actually interferes with their family's functioning.

Families have actively opposed the negative image that professionals have had of them over the years and have begun making their own definitions of the dilemmas they are facing. Many find that they can get the knowledge and management techniques that they need from their self-help groups; others search out education and support from practitioners who are skilled in this kind of service. Fewer seek out the long-term, costly types of individual and family therapy. Thus, NAMI and its various affiliates are having, and will increasingly have, an effect on ensuring that treatments will be effective and economical.

The Lay Referral System

Consumer groups have considerable influences over which professionals and which services are used. Gottlieb (1976) called this a "lay referral network." It means that members of mutual support groups consult each other in order to locate a resource that may work for them. With time, consumers may gain considerable influence over the direction that professional practice takes.

SUMMARY AND CONCLUSION

The need for accountability in the provision of mental health services is unquestionably a matter of deep concern to consumers. How deeply perturbed the users of service have been has gone unnoticed because they had no voice with which to express their disenchantment. Providers of services have maintained almost sole power over service provisions, and our system has been such that they needed to be accountable only to each other. Such exclusive rights to defining service, setting the costs, and evaluating the outcome are now being challenged by such consumer groups as the National Alliance for the Mentally Ill.

REFERENCES

Arbuckle, D. (1977). Consumers make mistakes too: An invited response. *Personnel and Guidance Journal,* 56, 226–228.

Borghi, J. (1968). Premature termination of psychotherapy and patient–therapist relationship. *American Journal of Psychotherapy, 22,* 460–473.

Buck, J., & Hirschinan, R. (1980). Economics and mental health services: Enhancing the power of the consumer. *American Psychologist, 35,* 653–660.

Bymers, G. J. (1977). The public section monopoly. *Section Policy, 8,* 106–109.

Dewar, T. R. (1978). The professionalization of the client. *Social Policy, 8,* 4–9.

Gartner, A. (1976). *The preparation of human service professionals.* New York: Human Service Press.

Gartner, A. (1979). Consumers in the service society. In A. Gartner, C. Greer, & F. Riessman (Eds.), *Consumer education in the human services.* New York: Pergamon Press.

Gorman, M. (1978). Mental health insurance: Problems and prognosis. *American Journal Public Health, 68,* 444–446.

Gottlieb, B. (1976). Lay influence on the utilization and provision of health services: A review. *Canadian Psychological Review, 17,* 126–136.

Graziano, A. M., & Fink, R. S. (1973). Second-order effects in mental health treatment. *Journal of Consulting and Clinical Psychology, 40,* 356–364.

Hare-Mustin, R. (1980). Family therapy may be dangerous to your health. *Professional Psychology, 1,* 935–938.

Hare-Mustin, R., Maracek, J., Kaplan, O. A., & Liss-Levinson, (1979). Rights of clients, responsibilities of therapists. *American Psychologist, 34,* 3–16.

Hatfield, A. B. (1983). What families want of family therapy. In W. McFarlane (Eds.). *Family therapy in schizophrenia.* New York: Guilford.

Hatfield, A. B., Fierstein, R., & Johnson, D. M. (1982). Meeting the needs of families of the psychiatrically disabled. *Psychosocial Rehabilitation Journal, 6,* 27–40.

Heine, R., & Trosman, H. (1960). Initial expectations of the doctor–patient interaction as a factor in continuance in psychotherapy. *Psychiatry, 23,* 275–278.

Hirschowitz, R. G., & Levy, B. (Eds.). (1976). *Introduction: The changing mental health scene.* New York: Spectrum.

Hollander, R. (1980). A new service ideology: The third mental health revolution. *Professional Psychology, 11,* 561–566.

Hornstra, R., Lubin, B., Lewis, R., & Willis, B. (1972). Worlds apart: Patients and professionals. *Archives of General Psychiatry, 27,* 872–883.

Langsley, D. G. (1975). Community health: A review of the literature. In W. E. Barton & C. J. Sanborn (Eds.), *An assessment of the community mental health movement.* Lexington, MA: Lexington Books.

Lazare, A., Cohen, F., Jacobson, A., Williams, M., Mignone, R., & Zisook, S. (1972). The walk-in patient as a customer: A key dimension in evaluation and treatment. *American Journal of Orthopsychiatry, 42,* 872–883.

Lenrow, P. B., & Burch, R. W. (1981). Mutual aid and professional services—Opposing or complementary? In B. H. Gottlieb (Ed.), *Social networks and social support.* Beverly Hills, CA: Sage.

McGuire, T. G. (1981). *Financing psychotherapy—costs, effects, and public policy.* Cambridge, MA: Ballinger.

Mechanic, D. (1978). Considerations in the design of mental health benefits under national health insurance. *American Journal of Public Health, 68,* 482–488.

Meenaghan, T., & Mascari, M. (1971). Consumer choice, consumer control in service delivery. *Social Work, 16*(4), 50–57.

Miller, H. L., Brodsky, S. L., & Bleechmore, J. F. (1976). Patient's rights: Who's wrong? The changing roles of mental health professionals. *Professional Psychology, 7,* 274–276.

Raymond, M., Slaby, A., & Lieb, J. (1975). *The healing alliance.* New York: W. W. Norton.

Reeder, L. G. (1972). The patient client as a consumer: Some observations on the changing professional–client relationship. *Journal of Health and Social Behavior, 13,* 406–412.

Robinson, D. (1973). Therapies: A clear and present danger. *American Psychologist, 28,* 129–133.

Schulberg, H. C. (1981). Outcome evaluations in the mental health field. *Community Mental Health Journal, 17,* 132–142.

Schulberg, H. C. (1982). Evaluating community and mental health programs. In H. C. Schulberg & M. Killilea (Eds.), *The modern practice of community mental health.* San Francisco: Jossey-Bass.

Sechrest, L., & Hoffman, P. E. (1982). The philosophical underpinning of peer review. *Professional Psychology, 13,* 14–18.

Sharfstein, S. S., & Taube, C. A. (1982). Reductions in insurance for mental disorders: Adverse selection, moral hazard, and consumer demand. *American Journal of Psychiatry, 139,* 1425–1430.

Steinman, R., & Traunstein, D. (1976). Redefining deviance: The self-help challenge to human services. *Journal of Applied Behavioral Science, 16,* 347–361.

Torrey, E. F. (1978 September). The mental health lobby: Providers vs. consumers. *Psychology Today, 12,* 17–18.

Vattano, A. J. (1972). Power to the people: Self-help group. *Social Work, 17,* 7–15.

Wilder, J. F. (1975). Strengths of the community mental health movement in urban areas. In W. Barton & J. Sanborn (Eds.), *An assessment of the community mental health movement.* Lexington, MA: Lexington Books.

Willett, S. (1977). Consumer education or advocacy—or both? *Social Policy, 8,* 2–8.

Young, H. H. (1982). Alternatives in mental health care and criteria for quality assurance. *Professional Psychology, 13,* 91–97.

8

Alternative Futures for Assuring the Quality of Mental Health Services

PETER G. GOLDSCHMIDT

INTRODUCTION

This chapter explores quality assurance in mental health under four alternative futures. What do we mean by *mental health quality assurance?* How can we think about the future? What is the alternative-futures approach? How can it help us to explore the future? How might the quality of mental health services be assured in various alternative futures?

MENTAL HEALTH SERVICES

In this chapter, mental health is viewed as a component of health, which may be quantified as health status. Health status is a measure of a person's health throughout life which includes the number of years lived and the health quality of life, considering morbidity, institutionalization, and functional ability (Goldschmidt, 1978). Persons suffering from mental health problems may not be able to do usual activities, may exhibit other dysfunctions, or may experience affect changes that detract from the health quality of their life. The objective of mental health services is to maintain and improve an individual's mental health, thereby contributing toward a person's health status. Such services are interventions designed to achieve this objective. They involve professionals, people who are trained explicitly for the purpose and who derive their living from their practice. The determinants of health and the relationship between health outcomes and health services are complex (Goldschmidt, 1982b). Nevertheless, we assume implicitly that health services improve health outcomes.

PETER G. GOLDSCHMIDT • World Development Group, 5101 River Road, Bethesda, Maryland 20816.

QUALITY ASSURANCE

For the purposes of this chapter, quality encompasses explicit mechanisms for assuring that mental health services will achieve the objective of maintaining and improving mental health. Importantly, such mechanisms do not exist independent of services, objectives, and their social context. One obvious example of the social context is the value placed on quality assurance. To what extent are patients, providers, and society concerned about quality? Health services may be seen as serving many functions—providing employment, for example—as well as their intended function of maintaining and improving health. Our values define the outcomes we want from health services (their objectives); our expectations are an acceptable standard of achievement.

Both consumers and providers may support quality assurance mechanisms, but for different reasons. Consumers may want reassurance about the quality of a service. Providers may want the satisfaction of knowing that they are performing to standards. Quality assurance mechanisms may act as barriers to entry, protecting the providers' economic interests. Professions may use such mechanisms to control their members; governments may use them to control expenditures. Quality assurance systems may alter consumers' expectations and providers' behavior. For example, knowing that one's work is subject to scrutiny may result in higher quality. This sentinel effect may be more important than the information derived from the quality assurance mechanism.

Essentially, quality assurance mechanisms must (a) measure a criterion attribute (e.g., the performance of a health system, service, or provider); (b) compare it to some normative standard of performance; and (c) indicate whether corrective action should be taken. By extension, quality assurance mechanisms may (d) encompass the means of identifying the appropriate corrective actions to be taken and to implement them. Further, an ideal system would incorporate feedback mechanisms to monitor the results of such corrective actions. Moreover, in the larger context, an ideal system would contain elements designed to ensure the quality of the quality assurance mechanism itself. The application of corrective actions assumes knowledge not only about the determinants of failure to achieve standards but also about the interventions that would correct them, to say nothing of the will and means to effect the interventions and their ethical and socioeconomic consequences. To what extent can provider behavior be changed? Aside from provider ability and willingness to change, incentives and sanctions are important aspects of quality assurance. Not only may corrective interventions not be coupled closely with other aspects of quality assurance mechanisms, but such interventions may be only one of many competing incentives and sanctions impinging on providers.

One approach to measuring quality examines structures, processes, and outcomes, and their relationships. Measuring structures is the simplest and least expensive. However, in using such measures for quality assurance purposes, one is forced to make many assumptions about the relationship between structural characteristics and intended outcomes, unless they have been validly established empirically. For example, licensure of professionals who graduate from accredited schools and pass examinations is a common quality-assurance mechanism. Although intuitively appealing, such mechanisms may be necessary but insufficient to ensure quality of care, which inherently involves processes. Measuring processes is both more difficult and more expensive and, if one does not know how they relate to outcomes, of limited value—again, because the relationship between process and outcome must be assumed. Measuring outcomes is the most difficult and expensive. Further, one must determine what produced them: To what extent did the service produce the outcome? How do the structures (e.g., the characteristics of the providers and the facilities) relate to the processes (the technology and the nature of intervention) in these regards?

Quality assurance costs money. Empirically, the extent to which it is cost-effective depends on measuring its impact on maintaining and improving provider performance and on rating this improvement. Within this larger context, the value of quality assurance must depend on the value of the service whose quality is being assured. Insofar as outcomes are concerned, money spent on making sure that an effective service is performed according to guidelines would seem wasted. However, providers may draw satisfaction—and the public, reassurance—from knowing that services are being performed as intended. Further, even if we know that care is substandard, the cost of the actions needed to correct the problem may outweigh the value of the benefit to be gained by the correction. Because an empirical determination of cost-effectiveness is often not feasible or is economically imprudent, we are left with our assumptions in these matters, and our assumptions may be no more than our hopes.

Nevertheless, decisions must be made in practice, whether or not solid information exists. Such information may result from intuition, experience, or research. If research information is to contribute to decision making, the apparatus must exist to identify critical research questions, to conduct the research, to assure its quality, to disseminate the results to the decision maker, and to facilitate the use of the result in practice. Producing information for quality assurance raises questions and presents policy choices, too, for example: Who should provide the requisite information? To whom should information about the provider's performance be available? Should individual patients, the profession, or the society act on the information? Should actions be effected through marketplace mechanisms, or through professional or governmental regulation? Who should determine the value of quality assurance? Who should assess the impacts of alternative mechanisms? Who should decide between trade-offs? How is efficiency to be balanced against equity? The answers to these questions will shape the future mechanisms for assuring the quality of mental health services.

THINKING ABOUT THE FUTURE

The present is ephemeral; the past, permanent; the future, malleable. We create our future. We no longer believe that it is preordained, with each of us living his or her own life according to some master script. What we choose to do and not to do, the choices others make, and random occurrences interact to shape what comes to pass. One difficulty is separating what we *want* to occur from what *will* occur. Preferences and predictions are intertwined. To the extent that we can, it is useful to consider future possibilities divorced from preferences.

Since ancient times, people have been interested in *predicting* the future as a way of reducing uncertainty—and of maximizing gains, if others are not privy to the prophecy. The difficulty, of course, lies in making predictions. Except in the cases of self-fulfilling prophecy or coincidence, people have not had much luck in predicting events (with the exception of certain natural phenomena). In recent years, we have begun to develop techniques for *forecasting* the future. Their objective is to provide some insight into what is *likely* to occur in specified situations, given certain constraints, and under fixed assumptions. Forecasts do not imply the certainty of predictions. In health planning, the most common forecasting technique is the extrapolation of past trends. Trend extrapolation assumes that present forces will continue to act in the same direction and with the same intensity and will give rise to the same interactive effects. Trends may interact; some may be diminished; others may be enhanced. Even the most complicated computer-based extrapolations may assume that the future is an extension of the present; it is not. Discontinuities, such as political revolutions, economic shocks, or technological breakthroughs, may be important future-shaping events. Thus, forecasting models are more useful for the short run than for the long run.

MANAGING THE FUTURE

Speculating about the future can improve decision making. Every decision involves assumptions about the future. By revealing our implicit assumptions, we can examine their credibility. In this way, we can gain insight into what parameters are important in shaping the future, and into how we might respond to them. The result may permit us to better order ends and means to achieve objectives in a way conducive to our worldview. Decisions involve trade-offs among future benefits and risks under conditions of uncertainty. Moreover, some benefits realizable in a favorable future may be traded off for the flexibility to respond to possible unfavorable futures. We need to identify robust decisions—those expected to yield good outcomes under a range of alternatives.

The future is uncertain, but it is not unmanageable. The alternative futures approach represents a potent technique for managing uncertainty and for improving decision making. To begin with, we must either imagine alternative futures for society and the health system or identify those trends and potential discontinuities that may differentiate one future from another. Using this information, we can create any number of alternative futures, including the plausible sequences of events that could lead up to each one. Collectively, these scenarios delineate the range of futures that we expect for planning purposes. What eventually comes to pass may not have been represented by any of the scenarios, or it may contain elements from all of them. The fact that scenarios are not predictions does not diminish the value of the alternative futures approach for planning purposes. Two principle approaches exist for creating scenarios of alternative futures:

1. Structure the various futures envisioned by pundits and other forecasters.
2. Select dimensions with driving forces and possible discontinuities that may differentiate the future, identify focal points along these dimensions, and construct coherent scenarios (using different combinations of focal points along the dimensions) to capture the desired range of plausible alternative futures.

The greater the number of dimensions considered, the richer the scenarios; the more numerous the scenarios, the finer the differentiation of alternative futures. One must also decide on the scope of the scenarios: alternative worlds, societies, health systems, components of the health system, or elements within a component. One should begin at least one level above the one of interest, to provide the proper context. In the case of alternative health systems, one must at least begin with alternative societies.

ALTERNATIVE FUTURES FOR THE UNITED STATES

James Dator, a political scientist and futurist at the University of Hawaii, delineated four alternative societal futures based on published images of the future created by futurists and pundits (personal communication, 1979):

1. Continued growth and prosperity
2. Collapse, resulting in a "lifeboat" economy
3. A conservor society, with enforced lowered expectations
4. A transformation, resulting from the emergence of values stressing cooperation, frugality, and inner growth

Starting with these four images, I have constructed scenarios of these alternative U.S. societies in the year 2025, including their health-care systems (Goldschmidt, 1982a). The four scenarios of the health care system describe, among other things, the personal health services, public health programs, research and technology, resource development, and national health policy that might exist, if these societies prevailed in 2025. What mental health services might exist? How would their quality be assured? For this chapter, I expanded the scenarios to include possible answers to these questions. The results follow.

2025: CONTINUED GROWTH

The economic growth scenario is an extension of the successful past; it represents progress toward fulfillment of the traditional American dream. Success has been achieved through competent management, an expanding private economy, and increased world trade, facilitated by transnational corporations, often involving private-sector–government partnerships. A national social consensus and international tranquillity have permitted the application of sophisticated technology and have facilitated the development of needed resources, including energy. Militarily, the United States remains strong but has little need to exercise its power—a tribute to the deterrent effect of that power.

The material standard of living has reached unprecedented heights. People have remained achievement-oriented, despite some transitory fluctuations in values. The education system has expanded and adapted to the economy's needs; education pays. The transition from traditional to technological industries has been smooth. Unemployment, although higher than traditional definitions of full employment, is low and, although not distributed equally, does not overburden one region or social class; however, underemployment is a growing phenomenon. Increased productivity permits substantial leisure and recreation activities. Many recreation areas are overtrafficked and despoiled, despite an increase in their number. Roads, beaches, national parks, and amusement areas are always crowded. Broadcast, cable, and in-home television—particularly video games and activities—consume many leisure hours. Cities have been largely refurbished, and population densities are lower; more people live in suburbs and exurbs; small towns have continued to grow.

The nation and the world are characterized by global communications, high mobility, high consumption, and political and ecological stability, despite continued violence, terrorism, regional conflicts, and minor ecological disasters. Confidence in institutions runs high; political and investment decisions have generally been wise. Impressive technological gains have been made in all sectors: agriculture, communications, energy, and medicine. Increased wealth has been distributed equitably through transfer programs that the public considers fair and socially responsible corporate decision-making. People exude satisfaction with progress in the material culture and confidence in the future. Even the disadvantaged feel that their children will fare better than they have. Some sectors of society, particularly among the upper and middle classes, express concern about the narrow focus of the material culture and wonder if America's success can be sustained in the long run, especially in the face of changing values, international competition, and global resentment.

HEALTH SYSTEM

Medical megaplexes, bastions of high technology, dominate the nation's health-care system. In the fight against disease, more doctors, nurses, and other personnel are employed per patient than ever before. Most doctors conduct their practices from the hospital, to take advantage of the latest technology. Transplants, gene therapy, and telemedicine are used widely. A plethora of drugs and devices exists. The hospital-on-the-wrist is the apex of medical technology. This device not only monitors body functions but also adjusts them to maintain homeostasis. Prevention is secondary to cure. More and more social problems and personal disorders are seen as being within the purview of the health care system. Drug abuse, alcoholism, and other behavioral disorders as well as mental illness, are serious problems; stress-related disorders have proliferated. The population suffers from chronic diseases. Health care expenditures are very high, financed by a variety of nonintegrated mechanisms. There is no national health policy. Government regulations constitute a maze of conflicting incentives and constraints.

MENTAL HEALTH

Society is achievement-oriented and pluralistic and values individualism. The material culture is emphasized, and change is rapid and is accepted as inevitable. There is high tolerance of diversity, and there are many opportunities. Freedom of choice is moderate. Employment is high and, for some, involves high technology and technochange; others toil at menial tasks. Management styles vary tremendously; the workplace is stressful for many. Leisure is extensive and recreation is varied. Families are small and are viewed primarily as economic units. Family ties are weak, and abuse of family members is not uncommon. Individuals are as likely to draw support from distant friends via telecommunications as they are from nearby family members. Mental health problems are seen as resulting from various stresses and the socioeconomic order, as well as from biopathology. The problems are many: substance abuse, behavioral disorders, "angst," disorders of introspection, suicide, and violence, including homicide. Substance use is accepted and abuse is common.

MENTAL HEALTH SERVICES

All types of behavioral, personality, and affective problems are seen as being within the domain of mental health services. There is a common perception of the need for mental health services, and they are provided extensively. They share the common goal of helping the individual to cope and to be a productive member of society.

Interventions are mostly diagnostic and therapeutic. Preventive services are limited. Screening is confined mostly to the military, where emphasis is on selecting out people who should not be given certain positions, and to combating substance abuse. School systems and employers undertake limited screening, using mostly computer-mediated batteries of questions to identify persons under excessive stress or at high risk for mental disorder. Often, such questions are part of a broader health assessment or health risk appraisal. Screening for genetic markers of mental disorder is beginning to be used, especially in the military. Elsewhere, their use is inhibited by concern for civil liberties.

Once a person presents with a possible mental disorder, a vast array of physical and psychological tests are brought to bear to rule out so-called physical problems or to identify specific lesions, the result of which is mental illness. PEARS (particle emission, absorption, and resonance) imaging techniques are used extensively, as are analyses of genetic markers—either directly or through their biochemical manifestations. Practitioners in medical centers place ever-increasing reliance in such techniques, at the expense of traditional history-taking. Drugs are the mainstay of therapy. Their number and potency is far greater than ever before, leading to problems of iatrogenesis. However, extensive computerization of medical records has brought such problems under control. Infusion pumps, diffusion patches, and time-release implants and ingestants have reduced medication frequency and have increased compliance. Ablation therapies are used in selected cases and have been made easier by new laser technologies and stereoscopic imaging devices.

Psychotherapy, in its various manifestations, is also an accepted way of coping with the stress created by an achievement-oriented, rapidly changing society, with ephemeral support systems. However, the use of talking therapies is declining in the wake of the newer biotechnologies, which many regard as the first real therapies for mental illness. Rehabilitation services continue to be adjuncts to therapy. Many are offered by independent practitioners, of various traditions, each plying their trade. Their use is limited because of reimbursement constraints. Many hold the view that they are a useful part of treatment; others, that they are nonessential, and that consumers should bear the cost, as they do for health clubs and various stress-management programs. Prevention advocates say that these should be covered services and, if need be, that less money should be spent on high-technology therapies.

Services are provider-mediated for the most part. Cadres of professionals from various traditions use drugs, individual and group therapy, counseling, and a plethora of other modalities determined by a provider's particular training and perceptions of patient's needs. Services are delivered under various auspices, mostly private, with separate public services, especially for criminals and the poor. Practitioners function in private offices, clinics, hospitals, and other facilities. Many of them work for or under contract with, organizations such as health care corporations, health maintenance organizations, and companies that administer preferred private organizations and other managed-care plans. Often, to remain competitive, a practitioner may belong to one or more plans, as well as to an independent practice association. However, independent practitioners still abound, especially for nonmedical therapies. There is little integration of services—separate drug and alcohol dependence clinics exist, for example—and they are mostly unrelated to other health or social services.

Extensive rationing of services occurs implicitly through one's ability to pay or, more usually, to have someone pay for him or her. Some employers provide on-site services with extensive coverage of mental health services; most do not. Public third-party programs provide very limited coverage. Fees are high. The majority of costs are met through private third-party payments, but because of limited benefits and coinsurance and deductibles, out-of-pocket expenses are significant. The limited services available to the poor are free of charge, if one can get in. The courts have ruled recently that providing a narrowly defined array of services is sufficient for a community to discharge its responsibilities for the mental health of the population. Greater emphasis has to be placed on services to criminals, because they are a manifest danger and society has to be protected from them.

QUALITY ASSURANCE

The public has high expectations of the quality of care but also great concerns about its cost. Quality assurance programs are operated for their public-relations and economic value. They account for less than 1% of health-care expenditures. Quality assurance programs are mandated by central government regulation, but their form and operation are influenced largely by provider associations, third-party payers, and other interest groups. Third-party payers', including governments', primary concern is cost containment. Yet, the number of malpractice claims continues to rise; awards continue to break records.

There is an extensive, expensive patchwork of quality assurance systems. Third-party payers are the driving force behind these systems, but their administration is left largely to large institutional providers or to professional associations. All types of mental health services are encompassed by some form of quality assurance program. The principal mechanism is peer review that concentrates on structure and process, with the ostensible goal of improving care. Requisite bodies certify facilities to deliver various types of services. Practitioners must pass licensing or certifying examinations set by professional associations. In addition, they must periodically obtain recertification, its frequency varying by profession and locality. However, third-party payers have their own set of rules and may impose additional qualifying requirements, as well as conditions for reimbursement. Health care organizations that provide services through prepaid plans follow their own procedures.

In some plans, after a first visit or admission, the patient's need for care must be certified by a practitioner panel selected by the third-party payer. Because of excellent telecommunication, these panels may never meet yet may be constituted of practitioners from around the country. Other plans have negotiated admission or service criteria with institutional providers. They generally place great emphasis on the practitioner's use of services and therapeutic practices. Most often, such standards are prescribed by the institution, the result of negotiations between its practitioners and third-party payers. For independent practitioners, criteria are selected and standards set by various professional associations, always mindful of third-party payers' requirements. Quality assurance practices are usually focused on services involving large or increasing expenditures.

Practitioners must submit detailed reports of patients' problems and therapy to justify reimbursement. The automation of record keeping and billing makes these reports easy to do. However, concern abounds about the potential for invading privacy as third-party payers demand more and more information. Functional assessments are used to determine a patient's severity of illness, which partially determines reimbursement rates. A potpourri of instruments is used for this purpose, depending on the institutional provider or third-party payer. This multiplicity is not a problem, however, as the instruments are all computerized. Their reliability and validity is a separate question, especially as there are few checks on how they are used or, in fact, on whether they are used at all. Increasingly, third-party payers are insisting that functional assessments be used to monitor progress in treatment and are making payment contingent to their use, in any attempt to realize value for money. Practitioners are paid only if progress is satifactory, but once the patient maintains a certain functional level, coverage ceases. Considerable controversy exists about what are appropriate treatment goals, what constitutes satisfactory progress, and at what functional levels people should be covered for services.

Institutions or practitioners collects the necessary data, and groups of peers make whatever judgments are needed about covered services, about whether a practitioner has performed according to standards, and if not, whether the explanations are sufficient to warrant an exception for reimbursement. If a deficiency is identified, improvement suggestions are given to the practitioner, but there is little monitoring to see if these have been adopted or tried. Practice associations and managed-care plans are clamping down increasingly: practitioners who consistently do not follow quality assurance mandates are dropped from membership. Third-party payers may rule a practitioner or an institution ineligible for reimbursement if the number of deficiency reports exceeds a preset limit. However, only rarely are restrictive measures, such as decertification, imposed.

There is little research on quality assurance mechanisms. Their assessment and improvement is informal and heuristic. Large health-care systems and other institutional providers undertake periodic reviews of their systems. These reviews are mostly concerned with improving the reliability of data for administrative decision-making and in support of negotiations with third-party payers. These, in turn, impose substantial reporting requirements on institutions and practitioners to monitor the care rendered and to attempt to control expenditures. Such data are viewed as essential to keeping practitioners in line.

2025: ECONOMIC COLLAPSE

During the last part of the twentieth century, both external and internal forces adversely affected the country. The world economy was depressed, but with inflationary prices for natural resources. Global stagflation set in; the U.S. economy stagnated, eventually to depression levels. Managerial competence, resources, and technology all fell short of requirements. Ultimately, world trade dwindled in the face of regional conflict and the pursuit of narrow national interests. Fueled by international chaos, national dissension, violence, and ecological problems, the U.S. economy collapsed.

The material standard of living plummeted as the nation's sophisticated economic and sociopolitical infrastructures unraveled. Violence, hunger, and death were rampant, until locally self-sustained communities emerged. Survival became the essence of daily life. Communities resorted to hunting and gathering, establishing farms and a few locally viable industries. Every able-bodied person is pressed into service; there is no unemployment. Those who do not work survive at best only on the charity of others. Education is largely by seeing and doing. The prevailing technologies are those of a bygone era that could be adapted successfully within the prevailing local circumstances. The concept of leisure has largely been erased. However, once people have attended to the survival tasks, they spend their remaining time in talking, sleeping, and sex.

Institutions collapsed along with the economy. Strong, strict, survival-oriented family and social values prevail locally. Few decisions remain. When they are made, all decisions reside in a local strongman, who usually consults a council of elders. Local strongmen, their henchmen, and their families constitute the only semblance of privilege remaining. They extort tribute from those for whom they see themselves responsible, and they occasionally plunder neighboring communities.

Despair reigns. For most people, the only future with which they are concerned is surviving today and, if they are lucky, perhaps tomorrow. The privileged few are concerned with clinging to power or with preserving their relative advantage. No one seems to have the energy to rebuild the economy. Some think the golden age has gone forever; others hope that, one day, America will be rebuilt.

HEALTH SYSTEM

Village societies exist throughout America. The fortunate few have a makeshift hospital run by a self-proclaimed physician assisted by a few paramedics, all of whom live in the building or close by. In some villages, the doctor has established a series of health posts, each staffed by a paramedic. The physician trains the paramedics. One day, one of them will take the doctor's place, when he or she can practice no longer. The patients are usually children or young adults, as most people do not live long. Epidemics are common, despite an emphasis on hygiene. Trauma accounts for most hospital practice; stays are very short. The competence of medical practitioners varies from village to village and relies on inheriting the technology of a previous age, supplemented by trial and error and local resources. Old textbooks are revered as bibles of medical practice when available; access to them is guarded jealously. Practitioners concentrate on the practical. Sometimes, they try a new intervention, but meaningful results are hard to come by. Most care takes place in peoples' homes. Payments for services are made in kind. Where prepayment plans exist, paramedics periodically visit each house, collecting food and other contributions and dispensing health education and medical advice. The doctor and the paramedics supplement their income by operating a small collective farm. There is no federal regulation or interference of any kind in the practice of medicine, and patients do not expect much.

MENTAL HEALTH

Society is survival-oriented. Self-reliance and individualism are valued. The material culture is emphasized, but there is little of anything to go around. Diversity is tolerated, but there are few opportunities and freedom of choice is limited. Change is slow. Most work is routine, but workplace stress is perceived to be quite low. Leisure is undifferentiated from work, and few recreational diversions exist. Families vary in size, but individuals look first to family members for support. Ties are strong, but abuse of family members is not uncommon. Few mental-health problems are perceived to exist. Those that do—personality disorders, substance abuse, and homicide—are believed to be the result of interactions between biophysiology and various stresses. Work is universal; management, autocratic. Substance use and abuse are accepted, as is suicide.

MENTAL HEALTH SERVICES

Only personal behavior is seen as being within the domain of mental health. The goal of what few services exist is to minimize the impact of an individual's aberrant behavior. Disordered people viewed as harmless are often left to wander around. All-purpose health centers provide diagnostic and therapeutic interventions, although many services are informal and are provided by family members.

Community leaders sometimes identify individuals whom they consider mentally ill or otherwise destabilizing. Such individuals may be recommended for treatment, confined to their homes, or even excluded from society. This constitutes the only preventive service, other than practitioners' dispensing of advice as they go from house to house to collect contributions. Practitioners use drugs and counseling. The drugs are dispensed, and usually made, by the practitioner. They are limited both in number and potency, but patients nonetheless take the potions as directed. In selected, refractory cases, the doctor may use electroconvulsive or ablative therapy, but the risks are high and the outcomes are viewed as unrewarding. Because all health services are provided from a makeshift center, usually by a lone practitioner, they are highly integrated. Lack of or only minimal record keeping impairs continuity of care.

QUALITY ASSURANCE

People are skeptical of the need for mental health services and have low expectations of those that are provided. There are no funds for quality assurance. Whatever quality assurance exists is initiated by the provider himself or herself for reasons of self-interest. Consumers focus on outcomes: they may be able to avoid bad practitioners, but they cannot pick out good care. There is no meaningful review or research. Data collection is informal, and deductions are experiential.

2025: CONSERVOR SOCIETY

Faced with a distressed economy, resource constraints, and increasing social unrest, the federal government has acted decisively to restore order and to revive the economy. The United States has withdrawn gradually from an increasingly turbulent and hostile world; military preparedness has grown concomitantly. This U.S. power is used to openly maintain narrowly drawn national interests, principally to buttress close allies and to preserve peace among the nation's important trading partners.

America has abandoned traditional individual freedoms to preserve its material culture. The federal government has entered into an alliance with business to control the economy and to regulate sociopolitical activities. Modern management, computer, information-processing, and behavior-control technologies have all been used to maximal effect. Despite increased resource constraints and ecological deterioration, material living standards have been preserved through enforced conservation and environmentalism, and through improved technology, management, and social discipline; expectations have been brought into line with socioeconomic realities. The education system has been adapted both to enhance social control and to provide the economy with the needed workers. Entrance into the system is controlled to conform strictly with federally determined socioeconomic policy. Personnel planning permits smooth adjustments to new technologies and changing economic circumstances. Unemployment remains a problem, especially among workers with few or unneeded skills. The chronically unemployed are placed in community work programs that provide training and jobs for everyone, no matter how unemployable they may appear to be. Work incentives are extremely strong.

Workers are rewarded strictly according to their contributions, defined by a federal government–business council. Managers and bureaucrats are selected and promoted by their respective corporations or agencies. These technocrats constitute the elite, enjoying a variety of perquisites. Small entrepreneurial businesses flourish in niches untouched by corporations, but their proprietors never gain the status or wealth afforded to the system's technocrats, managers, and administrators.

The population is increasingly aggregated into metropolitan areas. Recreation is spent in a variety of activities, mostly in-home entertainment and local activities. Intercommunity travel is restricted. Although professional sports teams are generally recognized by and affiliated with particular cities, matches actually take place in television studios where computers are used for simulation.

The United States is administered by a technocratic managerial-bureaucratic elite that is diffuse and faceless. Although the President is elected popularly, the office is mostly ceremonial, as is the Congress; both institutions are seen as largely irrelevant to the proper functioning of society in the technology–communications– information era. The judiciary decides cases according to a federally mandated code and computer-mediated precedence, revised periodically to fit socioeconomic realities. The prime central directive emphasizes local community self-sufficiency— recycling, preservation of the environment, and maintenance of the economy.

Despite an increasing awareness of the price paid in the reduction of individual freedoms to maintain the material culture, and of disparities in distribution of income and economic opportunities, people are generally content. The value of social order is much appreciated because of the economic benefits. People are resigned to the present order and, at the same time, are guardedly optimistic about the future. Perhaps some traditional freedoms will return if economic management improves or if ecological conditions take a favorable turn.

HEALTH SYSTEM

Medical megaplexes offering tertiary or highly specialized care are an integral part of a larger health care system delivering a complete spectrum of services. The nation has a patchwork quilt of such government-run systems, each with its assigned service area. Everyone is enrolled in the federal prepaid national health plan. Satellite clinics, located at work sites and in neighborhoods, emphasize health education and preventive care, referring patients to other system units only when necessary. They are staffed by paramedics supervised by a physician who makes weekly visits. Many procedures are done here or in other system units such as surgicenters, outside the hospital's confines.

Computers are at work everywhere: counting, monitoring, and controlling. They link all health system units; telemedicine is routine. Computers maintain everyone's medical record, mediate mandatory checkups, track high-risk people, monitor patients, and alert them when treatment is needed. Genetic engineering is in vogue as a preventive measure, producing healthy people, not merely repairing defects. Chronic conditions are still common despite prevention, and costly care is withheld at the extremes of life. People suffer from stress-related diseases; suicide rates are rising. The entire health-care system is managed centrally so that health can be achieved at least cost. Health care expenditures are kept in line; bureaucrats run the system. Unhealthy products are banned; healthy life-styles are encouraged. Only cost-effective interventions are permitted; treatments are denied to people who fall below the threshold preset by the health system.

MENTAL HEALTH

Social stability and order are the hallmarks of the conservor society. Although individuals are achievement-oriented, collectivism is valued. The material culture is emphasized, and living standards are high. There is little tolerance of diversity, and civil liberties are subordinated to social needs. There is little crime or terrorism. Opportunities for individuals are controlled, as is freedom of choice. The rate of technochange is moderate, employment is very high, and technological change affects the workplace more than any other sphere of life. Management is bureaucratic. The workplace is moderately stressful. Leisure is available to all, but recreational activities are somewhat limited. Families are small and supportive, family ties are moderately close, and there is little abuse of family members. Social support networks are mostly personal and local. Mental disorders are believed to result from various stresses, biophysiology, and maladaptation of the individual. They encompass behavioral problems, substance abuse, depression, and suicide. Substance use is common, although discouraged.

MENTAL HEALTH SERVICES

Mental health problems are seen as widespread, and the need to treat them is accepted universally. Services are extensive. All behaviors are viewed within the domain of mental health services. Most services are provider mediated, and even self-help efforts tend to be encouraged by or organized around one or more practitioners. Their goal is to help individuals function in society and to prevent social disorder. Although diagnostic and therapeutic interventions are designed for individuals, prevention interventions are often oriented toward communities.

Screening programs for identifying individuals at high risk for mental problems are universal. A multitude of technologies is used to monitor risk, to identify cases in need of treatment, and to track progress in treatment. They include extensive data collection about each individual's background, genetic makeup, physiology and biochemistry, and medical history, as well as current status. Moreover, the results of extensive research on the determinants of mental health, risk indicators, and intervention outcomes are used immediately; they have been validated as improving the effectiveness of prevention. Recent advances in artificial intelligence, especially new ways of structuring knowledge made possible by progress in computers and other fields, have facilitated the development of such universal screening systems. Now artificial intelligence systems are being used to identify and design the research needed to perfect these systems. People identified as being at high risk for mental disorders are provided with instructions on how to reduce the risk. If work stress is identified as a factor, employers cooperate by assigning workers to other jobs. Employers like the screening systems, as they help to place workers in the role where they are most productive.

People identified as being in need of treatment are provided with immediate therapy. Artificial intelligence, expert systems, and a variety of other decision-support technologies search out such people, provide a definitive diagnosis, prescribe treatment, and facilitate its continuous follow-up. Diagnostic interventions include PEARS (particle emmission, absorption, and resonance) devices, more sophisticated genetic marker and biochemical studies than can be used in screening systems, as well as history taking. Increasing reliance is being placed on technology because it is perceived to be more cost-effective, especially with the advent of usable artificial intelligence systems. Now under consideration is entering genetic information into a person's lifelong medical record that is created at birth, as well as its being updated through routine screening systems, which will be expanded to include sophisticated diagnostic interventions. What interventions are used among each population will depend on the results of cost-effectiveness analysis using information generated by the existing systems.

All types of treatments are used, although drugs have long been the principal agent. Genetic engineering is being used increasingly to correct defects in a person's makeup, including those defects that result in mental illness. More primitive treatments, such as ablation and psychotherapy, are used where indicated. Counseling is an acceptable way of coping with stress, However, some people do not avail themselves of these services because they fear that these services will affect their prospects for individual advancement, little realizing that such decisions are based on much harder data already in their records. Long-term follow-up is routine and is facilitated by the universal computerization of records. Criminals subjected to therapy are put under surveillance for the rest of their lives. Such systems are well developed and reliable, using sophisticated computer-pattern-recognition programming for detection and implanted devices and satellite triangulation tracking for surveillance.

Services are delivered as part of a comprehensive health-care system under public control, providing a high degree of service integration. They are provided by specially trained personnel at worksites and other clinics, and in institutions. Advances in therapy and social policy have virtually eliminated the need for long-term institutionalization. All costs are met from patients' monthly prepayments for health care. Everyone in need receives some therapy, the modalities depending on socioeconomic as well as technical factors. The rationing of care depends on cost effectiveness consideration. The fraction of health care dollars expended for mental health services is now much higher than it used to be.

QUALITY ASSURANCE

Mental health services, like all others, are expected to be adequate, not excellent. Nevertheless, quality assurance mechanisms are pervasive to ensure both the adequacy of services and their cost effectiveness. Over 3% of health care expenditures is devoted to quality assurance activities and another 4% to health services research, in order to provide needed information. Legal malpractice claims are rare, although the possibility of pursuing them still exists.

A separate quality-assurance subsystem monitors all health care. It is mandated by government regulations; everyone sees the value. The subsystem's operational form depends on local circumstances, within prescribed limits. The administration is left largely to regional health care managers. The goal is to improve cost-effectiveness and to remedy deficiencies that affect it negatively. The focus is on outcomes in relation to processes. Facilities are maintained at certified levels, and all practitioners must attain and maintain the minimal competency requirements that are a product of the quality assurance subsystem. Criteria and standards are set by a central government board with input from the quality assurance subsystems. They encompass indications for treatment, as well as appropriate modalities, and whether or not short-or long-term institutionalization is indicated. Extensive use is made of standardized functional assessments, both to determine treatment and to monitor progress.

The subsystem collects and verifies data from computerized files and compares practices with standards; its judgments are final. Great effort is given to ensuring the reliability of such data. Severe sanctions can be imposed for failure to report accurately. Practitioners not performing to the standards are provided with remedial training or are subject to other corrective interventions. Progress is monitored closely. Nonperformers are ultimately reassigned or retrained for other work. Practitioners performing exceptionally well are rewarded with bonuses and merits that are good toward career advancement. Increasingly, decision support technology is being used to assure quality prospectively, for example, to prevent practitioner's ordering unnecessary tests or prescribing treatments not considered cost-effective. Health system managers view the quality assurance subsystem as essential to the cost effective delivery of care. Not only does it provide incentives and sanctions for performing to standards, but it also permits the assessment of practices.

Extensive research is conducted on the quality system itself to improve its design and function. Priorities are based on the subsystem's assessment of achievable health benefits or financial economies not currently being achieved. Every facet of the subsystem is subject to research, including the development of improved functional-assessment instruments to chart patient progress and provider performance. Additionally, formal mechanisms exist to monitor quality-assurance-subsystem performance and to provide feedback to ensure the accomplishment of its purpose.

2025: TRANSFORMATION

Confronted by repeated failures to restore economic growth—and growing doubts about its utility—social values have changed gradually. Past a certain point, the presumed benefits of economic growth are not commensurate with the costs of its pursuit, even if it were attainable. Personal and social development have become ideals. By 2025, most people have subscribed to a life-style exemplified by simplicity, cooperation, and inner growth. Internationally, the United States has become less concerned about global events, acting only when narrowly drawn national interests are apparently threatened. Global communications provide instant access to remote events but reduce everything to insignificance. Increasing self-sufficiency permits the United States to become an observer rather than a participant in events, leaving regional and local conflicts and power struggles to the nations involved. Any U.S. involvement is remote, either through sanctions or client states; direct involvement seems unthinkable.

According to most traditional measures, the material standard of living has remained largely unchanged from levels achieved in the late twentieth century. However, life has changed in many significant ways. Basic needs are satisfied as economically as possible, consistent with sound ecological practices. Necessary, but unpleasant, low-skill jobs are often rotated and assigned by lot. Technological advances in agriculture, computers, information processing, and robotics have contributed significantly to achievement of this goal; only appropriate technology is fostered. Not only is the economy sustainable in a finite environment, but much time is available for leisure and personal development, viewed in spiritual rather than material terms. For many people, the separation between work and leisure has blurred; for some, there is no distinction. Many people work at home, using extensive computer and telecommunications networks. Although they may never physically leave their community, they interact instantly with colleagues or friends anywhere in the world. The movement of information has largely replaced the movement of people. Education, too, has been changed by information technology. Adaptability is emphasized, not the retention of knowledge or the learning of era-limited skills. Unemployment is hard to define, given the economic, educational, and social changes that have occurred. However, not everyone is plugged fully into an economic network, and despite community-run welfare programs, some people find the benefits of the new order elusive, and a few are essentially excluded from society.

People, production, and power have all been decentralized. Individuals and local communities have become increasingly self-sufficient and autonomous. Government services have diminished because of greater personal and community self-reliance. The main functions of the federal government are coordination and national defense. Confidence and trust in institutions run very high, fostered by frequent and open communications and the fact that institutions, especially government, perform a limited set of activities deemed appropriate by popular consensus.

Future prospects look excellent. The distributed nature of society, the use of appropriate and autonomous technologies, living within local environmental limits, and the apparent lack of limits to personal development augur well for a viable self-sustaining society. Life is pleasant, if simple and frugal. The detractors from the prevailing values include die-hard neomaterialists, fundamental traditionalists, anti-hedonists (who see society becoming narcissistic, purposeless, and a-ideological), and those for whom frugality springs less from choice than from economic necessity. Some political leaders and intellectual dissonants advocate halting the trend to abdication of moral and ideological leadership, and reasserting America's position in the world. However, most people fear the assumed consequences: social alienation, increased inequity, environmental deterioration, and other ill effects that the transformation was designed to avoid. They maintain that simplicity is the only sustainable life-style.

HEALTH SYSTEM

The hospital has become the community's workshop, not the doctor's. This medical megaplex, offering only tertiary care services, is a core element in an integrated health-care system, but not its central focus. The nation's health-care system is decentralized; each community has several competing systems. Almost everyone is enrolled in one or more of these systems, and a rich variety of providers practice both within and outside their confines.

The hospital has become a communications center, linking computers in people's homes to the latest medical knowledge, monitoring devices worn by patients, and integrating system elements. The information gap between providers and consumers has narrowed; everyone is his or her own physician. Most primary care takes place in the home, supported by community-based services and resources. People pursue healthy life-styles and focus on personal development. Acute conditions are few; behavioral and stress disorders have been reduced. Aging is accepted as inevitable, and fewer heroic measures are used to stave off death. Health care expenditures are high because of all the personnel employed and the high technology involved. However, funds are expended more efficiently now than they were in the past. Both government and people afford a high priority to health. Much effort is devoted to wellness. The federal government issues guidelines and provides information but does little else, leaving policy decision-making to communities and individuals.

MENTAL HEALTH

Personal development and social harmony are the goals of the transformational society. Although living standards are high, the material culture is not emphasized. Diversity is encouraged and opportunities for personal development abound. Freedom to choose is an ideal. Civil liberties are promoted; crime is controlled. Change is viewed as unending and proceeds at a variable rate through different segments of society. Employment is high. Although much economic activity takes place in the home, work hours are limited, and some people's participation is marginal. Most decisions are decentralized and management is largely democratic. Technological changes are incorporated at a moderate rate. Leisure time is extensive and recreational pursuits are varied. The integration or lack of differentiation of work, home, and leisure is a source of stress, as are the complexity and anonymity of society. Families are small and ideally nurture personal development. Ties are generally weak, but variable; abuse of family members is uncommon. Interpersonal relationships are many and highly developed. Most are remote, although personal relationships with those who are nearby are also encouraged. Mental health problems are believed to result primarily from biophysiological causes interacting with life stresses. Prevalent problems include disorders of introspection, substance abuse, personality disorders, and alienation.

MENTAL HEALTH SERVICES

The domain of mental health is confined largely to personal and social problems. The goal is to provide for personal growth, to facilitate development, to enhance the quality of life, and to encourage autonomy.

A vast mosaic of services designed to maintain and improve mental health has been created by the rich variety of practitioners and the multitude of technologies they use. Meditation, substance use, and computer-aided digital-video counseling and psychotherapy are all well-accepted ways of coping with stress. Exercise, life-style programs, and a plethora of preventive services abound. Screening is routine in schools, in prepaid health plans, and among some employers, often as part of a general wellness orientation. Questionnaire, demographic, and genetic-physiological measurements are captured on an individual's life record. These records act as keys accessing computer programs that examine an individual's profile in relation to his or her goals, and that dispense helpful advice on how these goals can be achieved or modified to allow greater harmony. The diagnostic and therapeutic interventions used also depend on the person's perceptions of appropriate treatments. The vast proliferation of decision-support technology facilitates these choices. Sophisticated imaging and laboratory analysis systems are used to rule out organic lesions or those that can be repaired with genetic therapy or attenuated with drugs or devices. Expert systems and artificial intelligence are used extensively. Some people prefer to rely on drugs or devices for therapy; others prefer alternative means of therapy.

A variety of services are delivered by different providers under mostly private, decentralized auspices. Some practitioners work for large, integrated health systems; some are members of professional associations or managed-care plans; and others remain as solo practitioners, providing personalized services, although they may use computers and video support systems in their practice. The development and marketing of such systems are largely in the hands of relatively few suppliers, although innovations are the work of small-scale entrepreneurs. Self-help networks facilitate coping with stress and the diagnosis of health problems. Therapy takes place in the home by computer, or in practitioners' offices or in clinics. Because of the variety and the segmentation of the services available, they are only moderately integrated, through portable electronic records, for example. Further, the extent of individual self-help efforts militates against provider-mediated integration of mental health and other services.

Because of the plethora of providers, fees are generally low; few people seek intensive and expensive psychotherapy. Many expenses are paid out-of-pocket because the available prepaid plans mostly offer coverage that is limited in various ways. The wide range of services available usually ensures that everyone will receive some form of mental health service, even if not practitioner-mediated. The poor are particularly hard-hit, having limited access to both practitioners and computer-mediated self-helps. Some mental disorders go untreated because the person is unwilling to submit to treatment or cannot afford it. Third-party coverage for service is sparse, and government programs are limited, consisting mainly of reimbursement for certain covered services. Counting all forms of services, a substantial number of resources are allocated to mental health.

QUALITY ASSURANCE

Expectations of mental health services are high; some people expect more than others. The driving force of most quality assurance activities is the desire to improve, although the possible loss of clients through adverse reports is also a factor.

Quality assurance accounts for somewhat less than 3% of health expenditures. Generally people appreciate the need for quality assurance and often play an important part through either through the marketplace or various other regulatory mechanisms. The government guides quality assurance activities but leaves implementation entirely to local health-care systems, professional associations, or other requisite bodies. Health plans, third-party payers, and professional and consumer associations all play a role in shaping quality assurance mechanisms. Their relative influence varies according to local issues and circumstances. For the most part, quality assurance activities fall within the health care system; there is little malpractice litigation.

The type of quality assurance system in place varies tremendously, from the extensive systems used by large, integrated health-care systems to virtually unregulated practitioners. Because of various reporting requirements, practitioners must often keep extensive records. However, the effort involved is small, as most information is computerized from patients' lifelong records or some other already-automated service and is processed electronically. Data are transmitted daily from the provider to the information compilers' computers. A number of private organizations exist to test and approve information services, computer programs, videos, and self-help technologies. However, they have no regulatory authority.

Quality assurance of services focuses on outcomes in relation to processes and attempts to facilitate achieving the achievable. Information on health care systems, plans, facilities, and practitioners is freely available, partly as the result of disclosure laws and partly through the efforts of private services seeking to satisfy providers' legal requirements and consumers' thirst for information. However, there are no licensure requirements, and anyone is free to practice, to operate a facility, or to provide a service, as long as they disclose the nature of their training, experience, results, and so on. Some third-party payers impose their own rules about whom they will reimburse, for what health problem or interventions, and under what circumstances.

Locally established mechanisms, which vary considerably in complexity and composition, select criteria and set standards. Often the criteria include a person's functional ability, and the standards are expressed in terms of the improvement that has resulted from intervention. For the most part, special quality-assurance units collect data and judge performance. They also suggest training or other interventions to remedy deficiencies, to adjust practice, or to alter provider education. Local quality-assurance units can prohibit practices in select cases but rely mostly on corrective actions by practitioners or by patients, who have access to practitioner performance assessments.

Some research on quality assurance is done locally; some centrally. Using research results, the central quality-assurance institute provides guidelines to local systems to help them monitor their own quality-assurance efforts. One of the institute's key roles is aiding health service management: assessing what information is needed and what is available and assigning priorities for generating missing information; validating existing and needed information; and disseminating it to users. Like other decision-making organizations, however, the institute is largely decentralized, and the extent of monitoring varies according to local wishes and circumstances.

CONCLUDING REMARKS

The magnitude and the nature of mental health services demanded and supplied depend on many factors. Chief among them are views on what is a mental health problem, what is its cause, and how it might be prevented, treated, or ameliorated; prevailing assumptions and knowledge about health care services and who should deliver them; and economic realities and competing priorities. Similarly, what we demand from mental health services—and the quality assurance mechanisms we institute to try to fulfill our expectations—also depends on various factors. The future is uncertain but not unmanageable. We should not assume that the future will be an extension of the present: it is not. Further, we can create alternative futures, as was done in this chapter. Because every decision involves assumptions about the future, we can improve present decision-making by making our assumptions explicit and examining their implications. Thus, the alternative-futures approach is a useful tool for examining and designing quality assurance systems for mental services.

ACKNOWLEDGMENTS

The author thanks Dr. John Williamson and Dr. Alex Rodriguez for their helpful comments on the draft manuscript, and Dr. Jim Lundy for manuscript production.

REFERENCES

Goldschmidt, P. G. (1978). *A model for measuring health status: Application to the U.S. population.* Baltimore: Policy Research Incorporated.
Goldschmidt, P. G. (1982a). *Alternative futures for the U.S. health care system.* Baltimore: Policy Research Institute.
Goldschmidt, P. G. (1982b). *Health conditions of the year 2000.* Baltimore: Policy Research Institute.

9

Therapeutic Issues and Quality Assurance Efforts

Milton Theaman

Peer review, particularly its application to outpatient treatment, is a recent addition to quality assurance procedures. Its introduction has raised concern about the effect it will have on clinical practice. This discussion will address these concerns. The justification for these concerns will be evaluated by means of early experience with the peer review process. The issues that will be considered are (a) the relation of peer review to standards of practice; (b) confidentiality; (c) usual and customary practice versus necessary and appropriate treatment as a review criterion; (d) the theoretical orientation of provider and reviewer; (e) reimbursement of providers for completion of report forms; and (f) reactions to peer review.

STANDARDS

The purpose of peer review is to monitor the performance of providers, using as criteria the standards that the professions have set for themselves. It is not the responsibility of peer review to set these standards.

Standards are formulated in broad, generalized statements (American Psychological Association, 1974). They acquire meaning through the manner in which they are implemented by providers. Implementation is influenced by training, theoretical orientation, experience gained in practice, and research findings. In the mental health area, particularly because of the diversity of theoretical orientations, these influences result in a considerable variety of approaches to the treatment of similar conditions. Additional variety is introduced to accomodate local circumstances. A lack of resources in the area may prevent the use of the most efficacious or

Throughout this discussion reference is made to developments in the field of psychology and documents and procedures adopted by the American Psychological Association. These are meant to be illustrative of the situation in the mental health field, in general, and should be equally pertinent to the other mental-health professions.

MILTON THEAMAN • 565 West End Avenue, New York, New York 10024.

most efficient treatment procedure. In monitoring performance, peer review must accommodate the variety of treatment approaches that the profession accepts as valid. Its task is not to dictate which of the acceptable approaches the provider should undertake but to monitor the adequacy with which the provider is carrying out what he or she has undertaken to do.

Where the standards are clear and explicitly set forth, the review is uncomplicated. Thus, the *Standards for Providers of Psychological Services* (American Psychological Association, 1974) states that "there shall be a written service delivery plan for every consumer for whom psychological services are provided . . . which will analyze the problems, set priorities among established goals, and outline systematic procedures for implementation of the plan" (p. 17). The structure of the Mental Health Treatment Report (MHTR) currently in use in the peer review process derives from this standard. It calls for a statement of the conditions for which treatment is required, the goals of treatment, the interventions that will be undertaken to achieve these goals, and as treatment continues, the progress that has been made. If a practitioner has not already formulated a written treatment plan, he or she will, in effect, do so in completing the MHTR. As an alternative to completing the MHTR the provider has the option of submitting the information requested in narrative form. Such a narrative statement would contain the essential elements of a treatment plan. By structuring the review process in this manner, the provider is requested to adhere to the standards of good practice established by the profession.

To be consistent with the objectives of monitoring, not promulgating, standards, the report form must avoid bias for or against any of the diverse theoretical orientations adopted by different providers. The basic outline of the report form that calls for presenting problems, goals, proposed interventions, and progress appears to be a structure that is orientation free.

The structure of the report may also influence the structure of treatment records. Practitioners may choose to organize these records with a view to making the completion of the MHTR simpler.

The standard for the introduction of innovative practices is another that is monitored by peer review. A profession rooted in a science is continually exploring ways to discover new and improved ways of serving the public. The development of innovative theories and procedures is encouraged. However, before they are recognized as appropriate interventions, they must be supported on a theoretical and/or empirical basis ("Specialty Guidelines," 1981, p. 644). Peer review is a vehicle for monitoring the requirement that innovative procedures must be properly supported before they are introduced into general practice.

Because standards are expressed as broad statements of policy and objectives, they may be implemented by different providers in different ways, each provider asserting the validity of his or her procedures. How does peer review deal with this dilemma? Two examples from current experience speak to this issue.

When the review program of the Civilian Health and Medical Program of the Uniformed Services (CHAMPUS) was first established, it adopted procedures based on the belief that individual psychotherapeutic treatment of more than one member of the same family by the same therapist was rare and questionable practice. When such a treatment procedure was encountered, the case was routinely referred for peer review to determine whether there were any special circumstances that might justify this procedure. Objections arose from some psychologists who claimed that this procedure was neither rare nor contraindicated, and that it contained advantages that made it, in many instances, the procedure of choice. A survey of practitioners (Benedict & Stricker, 1983) disclosed that a significant minority did engage in this procedure, which they justified both on a theoretical basis and on the basis of circumstances that were not at all rare. As a result, this procedure was eliminated as a criterion for routine referral for review.

The issue of obtaining the client's signature on the treatment report also elicited divergent views. At the outset, this signature was required unless, in the therapist's judgment, it would be harmful to the client to sign the report. A MHTR that did not comply with this requirement was referred for peer review. Some providers observed that discussing the report with the client was a useful procedure that advanced the therapeutic process. Pearce and Newton (1969, pp. 429–432) believe that a periodic review with the client of the status and progress of treatment is a productive procedure. Others, in significant numbers, found this procedure to be intrusive and interruptive of the therapeutic process. In acknowledgment of this diversity, the client's signature was made an optional requirement, and the decision was left to the provider.

It would appear that, when standards are clear and explicit, peer review procedures will be determined by them. When they are not, customary practice will prevail. In sum, the profession's standards, whether determined by explicitly stated consensus or by customary practice, will dictate peer review procedures. Peer review will not dictate what those standards should be.

CONFIDENTIALITY

Because confidentiality is so central to the therapeutic relationship, it is crucial to determine the effect of peer review on the confidentiality of the information obtained during treatment. To make this determination, it is necessary to understand the extent to which confidential communications between a provider and a client are privileged under licensing laws.

The New York State licensing law for psychologists is typical. Under this law, a psychologist may not disclose a confidential communication between her- or himself and the client in any trial, hearing, proceeding, or administrative action, without the client's prior consent (Mariano & Feldman, 1983). There are, however, some exceptions. The privilege does not apply when the information relates to the commission of a crime or a civil wrong. In *Tarasoff v. The Regents of the University of California* (1976), this exception was discussed with respect to potential crimes. The case states that if the provider has reason to believe that the client presents a danger to another party, the provider must take "reasonable steps" to prevent the danger, which includes the duty to warn the person who may be harmed. There are other exceptions. A psychologist may be required to testify without the client's consent in guardianship, custody, child protective, or competency proceedings when ordered to do so by the court, or else be subject to contempt of court. A psychologist is also required, by law, to report cases of child abuse. In general, a court may suspend the privilege if it determines that the public interest would be better served by disclosure than by confidentiality.

Insurance coverage for health services is viewed as benefiting the public because it makes these services available to more people by spreading the costs among the members of a group. For this benefit, the insured is required to relinquish some degree of confidentiality. The third party, which is now involved in the service transaction, needs access to some information in order to administer the program. The choice between waiving the benefit or waiving the privilege remains with the insured.

The insurance contract requires the insured to submit sufficient information to enable the carrier to determine whether the condition is covered under the contract; whether the treatment is necessary and appropriate to remedy the condition; and whether the treatment is being given by an authorized provider. By entering into this contract, the insured consents to the release of this information, which is transmitted on a standard claims form. In the absence of peer review, if a carrier has questions about the necessity or appropriateness of treatment that are not resolved by the information on the claims form, it may require the insured to submit to a consultation with a professional chosen by the carrier to obtain an independent opinion. The consultant submits his or her report to the carrier, and all information obtained in this review is kept in the claims file.

If a carrier wishes to use the American Psychological Association's peer review procedures, it enters into a contract with the APA. This contract requires the carrier to

> institute mechanisms to ensure protection for private patient and provider information so that there is no significant increase in the risk of inadvertent disclosure of private information. Peer review information shall be maintained securely and separately from other claims information, will not be microfilmed or entered into a computerized data base and will not be used for purposes other than the peer review. (American Psychological Association, 1983, p. 1)

Thus, if a carrier wishes to confirm necessity and appropriateness through the APA's peer review procedures, it is required, by contract, to adhere to more secure procedures than would apply if it sought additional information on its own.

It is feared by some that if this information is in the files of the insurance carrier it is more accessible to subpoena. If a court wishes to obtain this information, it may subpoena it from the provider as readily as it may from the carrier. As the contract with the carrier forbids the use of the information for any purpose other than peer review, the carrier is constrained from releasing it unless compelled to do so by law or by court order.

These contracts also call for an additional authorization by the client for the release of information for peer review. This authorization echoes the language of the contract, stating that the information will be used solely for peer review purposes and will not be released under this authorization for any other purpose unless expressly permitted or required by law. This form is under constant review to ensure that it will clearly inform the client of her or his right to waive, or to refuse to waive, the privilege of confidentiality, and to ensure that it will not promise a greater privilege than the law provides. The objective is to be sure that the client's consent is truly informed.

These precautions serve as a restraint on the *carrier's* release of information. However, it has been suggested that, by authorizing the release of information to the insurance carrier, the client forfeits her or his right to confidentiality in any future action, even if it is unrelated to insurance reimbursement. Thus, if a client wished to invoke her or his claim to confidentiality in a subsequent divorce proceeding, she or he would have forfeited her or his right to do so. The question is whether a limited waiver applies everywhere, nowhere, or only in some states. An opinion provided by the attorneys for the American Psychological Association (Ennis, Friedman, Bersoff, & Ewing, 1983) states that there are significant arguments, supported by judicial precedent, that may sustain a limited waiver in all jurisdictions. Other legal opinion (Lowenstein, Sandler, Brochin, Kohl, Fisher, Boylan, & Meanor, 1983) questions whether a limited waiver is permitted in New Jersey. If there are states in which a limited waiver is not sustained, it will be necessary to seek legislative redress. It should be noted, however, that this question arises not as a consequence of the release of information for peer review, but as a consequence of the prior authorization for the release of information for insurance reimbursement. If this forfeiture is verified, it would place the confidentiality privilege in jeopardy on the acceptance of any health insurance coverage, not only mental health coverage, even in the absence of peer review.

It has also been suggested that a provider may be liable for damages to the client that result from use, in some other proceeding, of the information submitted for peer review. If the information is divulged pursuant to a valid authorization or court order, it would be difficult to find a legal basis for provider liability. It should be noted that, for a client's authorization to be valid, it requires truly informed consent.

The review procedure itself calls for the anonymity of both client and provider. Before the record is sent for review, any information that will identify the client or the provider is deleted. The provider is instructed to avoid including in her or his report any information that will identify the client. If the provider believes this is not possible, she or he may request an accommodation from the usual procedures to meet the special circumstances of that case.

Does peer review, then, increase the risk of exposure of private information? Despite all the contractual and procedural safeguards used, the fact that peer review requires the information to be handled by more people than would be the case in its absence increases the possibility of slip-ups, even if inadvertent. Vigilant application of these safeguards may minimize the consequences of human fallibility but will not dispel them. The question here is, as it is with so many things in life, such as buying packaged foods, driving a car, and using nuclear energy: Does the benefit derived make this a risk worth taking? The continuous expansion of health insurance coverage indicates that the public is content to cede some measure of confidentiality for the benefit derived from this coverage. Will the public be willing to accept the additional risk to confidentiality posed by peer review, in order to retain health insurance coverage? As peer review will, in time, apply to all health coverage, the question is not limited to mental health. The answer will always rest with the insured.

USUAL AND CUSTOMARY VERSUS NECESSARY AND APPROPRIATE

Should the reviewers' task be limited to determining whether the treatment procedures are usual and customary, or should they be asked to judge whether the procedures are necessary and appropriate? *Usual* is defined as a practice in keeping with the individual provider's general modus operandi. *Customary* is defined as that range of usual practices provided by psychologists of similar training and experience for the same service within a specified geographic or socioeconomic area. The early voluntary review programs developed by both psychologists and psychiatrists in cooperation with private insurance carriers invoked *usual* and *customary* as review criteria. The review programs sponsored by the federal government, as exemplified by the CHAMPUS program, use *necessary* and *appropriate* as the criteria.

It is alleged that these criteria lead to different kinds of reviews. However, an examination of the review process refutes this allegation. Essentially, reviewers must judge whether a problem exists and whether the interventions adopted and the manner in which they are being implemented are likely to attain the goal of ameliorating the problem. Thus, the reviewer must consider whether the treatment is necessary and appropriate. This judgment is based on the reviewer's training and experience, knowledge of the literature, and acquaintance with the usual and customary practices in the community. Reviewers are deliberately selected from the same community as the providers to ensure that they will be familiar with local custom.

No profession can accept the validity of usual and customary practices that are not necessary and appropriate. Conversely, a determination of necessary and appropriate treatment cannot ignore the usual and customary practices in the community, particularly as they may be dictated by local circumstances. No matter which of these two sets of criteria is declared to be the basis for review judgment, the review process cannot fail to take both into consideration.

THEORETICAL ORIENTATION

Can a reviewer of one theoretical orientation effectively and fairly review the practice of a provider who subscribes to a different theoretical orientation? Current procedures assume there will be consistency among reviewers of different theoretical orientations on such basic issues as necessity of treatment, goals, appropriateness of the intervention, and progress.

Matching provider and reviewer orientations raises several questions: 1. How narrowly should theoretical orientation be defined? Are such broad categories as psychodynamic, cognitive, behavioral, and psychopharmacological sufficient? Would a Freudian regard a Sullivanian as being of the same theoretical orientation, as both are psychodynamic? Would either accept a follower of Kohut? Would a Minuchin-oriented family therapist feel fairly reviewed by someone who adheres to Bowen's views? Is an eclectic orientation to be considered different from or equivalent to all others? 2. What are the practical problems of such a procedure? Would it create too great an administrative burden for the system to bear? 3. One feature of the review system is that the reviewers practice in the same geographic area as do the providers they are reviewing, so that they are familiar with the customary practices in that area. If it is not possible to find someone in that area with a similar theoretical orientation, which takes precedence, geography or orientation? Is it more important to know the customary practices in the area than to be of similar theoretical orientation?

The basic question is whether the reviews of those with different theoretical orientations differ in any significant ways. This question should be answered empirically. How it is studied is influenced by whether one is looking for the cost-containment or the quality-assurance consequences of peer review.

Studies that look for differences in recommendations for benefit payments focus on cost containment. A series of studies by Cohen and his associates (Cohen, 1981; Cohen & Oyster-Nelson, 1981; Cohen & Pizzirusso, 1982) and one by Biskin (1983) are in this category. The results are equivocal. Methodologically, the Cohen and Pizzirusso (1982) study is the best of that series. That study was in two parts, the first using 78 psychologist/CHAMPUS reviewers belonging to three orientation groups (psychodynamic, eclectic, and behavioral), and the second using 27 additional reviewers representing the same three orientations. Cohen and Pizzirusso found no differences among the orientation groups in their benefit recommendations for past care in either part of the study. There were differences in the recommendations for future care in the first part of the study. Psychodynamic reviewers recommended reimbursement for more future care than did eclectic or behavioral reviewers. In the second part of the study, there were no differences in the recommendations for future care.

Biskin (1983) used the CHAMPUS review format in obtaining reviews from 141 university-counseling-center staff members from 35 universities. Two treatment reports were used, one by a provider with a psychodynamic orientation, the other by a provider with a cognitive-behavioral orientation. The reviewers were categorized by these same two theoretical orientations. The study found no reliable evidence that either the provider or the reviewer orientations affected the reimbursement recommendations for either past or future care.

In an examination of the quality assurance consequences of peer review, different questions need to be asked. A beginning was made by Cohen and Pizzirusso (1982), who asked for recommendations about such significant aspects of practice as the use of alternative procedures, the need for consultation, referral to another therapist, and the need for future treatment. They found no differences among the orientation groups in their responses to these questions. Perhaps the question should be put directly: Is this good treatment or poor treatment? Would reviewers of different orientations agree in this judgment, regardless of the orientation of the provider?

A focus on quality assurance consequences would recommend asking providers whether the review was useful either in confirming the validity of the treatment procedures being used or in providing ideas for improving them. Would a provider find the review by someone with a similar orientation more insightful or more redundant? Would the review of someone with a different orientation be less meaningful, or would it offer suggestions for additional useful interventions? Of course, the quality of the review is critical to its usefulness, but is theoretical orientation a significant variable?

Although it would be prudent to await more data before settling this issue, decisions are often made on the basis of other considerations. The pressure for matching provider and reviewer orientations comes primarily from psychodynamically oriented providers who believe that behaviorally oriented reviewers are unsympathetic to long-term treatment. Current procedures require three reviewers for each case. Recognizing the practical difficulty of arranging for all reviewers to be of the same theoretical orientation as the provider, the recommendation has been made that at least one reviewer be of similar orientation. If this recommendation is adopted, it is not likely to result in any significant change in the composition of the review panel for psychodynamically oriented providers. Because a large percentage of the current panel of reviewers is psychodynamically oriented, the random assignment of reviewers adhered to in the past has met the recommended standard in almost all cases.

REIMBURSEMENT FOR COMPLETING TREATMENT REPORTS

As currently implemented, the peer review process is a document-based review. To be successful, it requires a conscientiously prepared treatment report. There is consensus that the provider should be reimbursed for the time spent preparing the report. Opinion differs about how to charge for this activity. Two procedures have been proposed: (a) direct billing, either as a flat fee or a fee based on the time spent on this task, and (b) billing for this task by including it with other overhead costs when the provider determines what his or her service fee should be.

Advocates of direct billing assert that including these costs in the service fee unfairly penalizes those without insurance coverage by requiring them to subsidize costs incurred by those with such coverage. Overcoming this unfairness by establishing different fee schedules for those with and without insurance coverage is said to be cumbersome and impractical. Moreover, it is advantageous to identify the costs attributed to peer review, wherever possible, so that the cost effectiveness of peer review can be determined.

Advocates of service fee billing cite the customary practice of a generalized allocation of overhead costs—rent, utilities, and secretarial and custodial services—to all clients without regard to differential use. If a different fee schedule for those with and those without insurance coverage is preferred, it would not be an unusual deviation from current practice. Most providers vary fees, generally in response to the clients' needs.

The official position of most insurance carriers is that the insured is responsible for submitting the information necessary to support a claim. Therefore carriers are not obligated to pay (direct billing) for obtaining this information. They have not raised objections to including these costs in the service fee as a component of overhead costs. The carriers, however, do not consistently adhere to this position in practice. They have been known to make exceptions to their official position and, at times, to pay a flat fee for the submission of reports.

Among the professions whose members qualify for insurance reimbursement (psychology, psychiatry, social work, and psychiatric nursing), only the American Psychological Association has adopted a position in support of direct billing. Thus far, this action has not persuaded insurance carriers to change their official position on this issue.

REACTION TO PEER REVIEW

The public at large, as represented by consumer groups and legislative bodies, views peer review as a legitimate responsibility of the health service professions. Consumer groups have long advocated greater accountability on the part of these professions. Legislators are expressing their views by mandating peer review in legislation establishing or extending health-service-delivery programs (California, 1981–1982; Ohio, 1977–1978). Legislators have added peer review to training, research, ethics, and licensing as elements that identify a responsible profession.

The reactions of the two professions that have established peer review programs, psychology and psychiatry (social work and nursing, until now reviewed by psychiatry, are just beginning to establish their own review programs), have been markedly different. Some psychologists have objected to the introduction of peer review as a quality assurance mechanism, claiming that licensing and the profession's code of ethics are sufficient guarantors of quality. Some have objected to the way the program is being implemented. There has been no significant protest among psychiatrists. Indeed, the American Psychiatric Association has been actively promoting its peer review program as a useful service to all three parties in the service transaction: the client, the provider, and the fiscal intermediary. As of January 1, 1984, the American Psychiatric Association has entered into contracts with 27 insurance companies for the provision of peer review services. The American Psychological Association had contracted with 10 insurance companies, generally in response to a request from the company (S. Shueman, personal communication, January 1984).

It is difficult to attribute the difference in the response of the two professions to differences in the way the programs are implemented. Both organizations developed their programs initially for CHAMPUS, which promoted collaboration between them to maximize uniformity and consistency in their procedures. The forms used in both programs are the same. What is different is the prior experience of these organizations with peer review. As a branch of medicine, which has had tissue committees and utilization review in hospitals for a long time, psychiatry has had previous involvement in professional review. The professional standards review organizations (PSROs) established by the government to review all health services supported by federal funds have been organized and administered by physicians. Although they were authorized to review both inpatient and outpatient services, their first effort was directed at short-term hospital stays, with which psychologists have had little involvement. Despite early objections to the PSRO program, medicine now appears to have accepted the reality of peer review even for outpatient services. The Institute of Medicine, a branch of the National Academy of

Science, has accepted a contract from the U.S. Department of Defense to develop proposals for the introduction of peer review into all branches of medicine.

Psychology has not had a comparable history of peer review. For psychologists, peer review represents a qualitative extension of the concept of accountability. In the past, quality assurance mechanisms focused on the qualifying criteria for admission to practice: the completion of training and the meeting of licensing requirements. Even continuing education, a postlicensing accountability measure, deals with training and preparation. Only in the area of ethics was there an examination of performance. However, the occasional review of performance for ethical transgressions touches many fewer practitioners than do the systematic procedures of peer review. For psychologists, particularly those in private practice, peer review represents a distinct change in practice.

For clients already in treatment, the introduction of peer review may raise such questions as: Is the adequacy of my treatment being questioned? Are my benefits in jeopardy? Who will get to see the information that is submitted? The client's reaction generally turns on the attitude of the provider. If the provider presents peer review as a systematic procedure used by professionals to ensure that services will be of good quality and describes the steps taken to protect confidentiality, the client doesn't feel singled out for special scrutiny. Indeed, the efforts of a profession to monitor the quality of the services delivered by its members can be reassuring to those in need of those services. If the provider is hostile to peer review, this attitude is frequently conveyed to the client, who gets caught up in the provider's conflict with the review program. If the conflict reaches the point of refusal by the provider to participate in it, the client's benefits are truly in jeopardy.

SUMMARY

Peer review monitors the performance of providers to determine whether it adheres to the standards of good practice promulgated by their professions. In professional training programs, accountability to a supervisor is a continuing presence. Peer review makes collegial accountability a continuing presence in professional practice. In anticipation of possible review, providers are likely to ask themselves how their colleagues would review the procedures they are using. More prevalent than review by colleagues will be self-review by providers. They will be monitoring their own practice. This self-review is likely to be peer review's most far-reaching effect on professional practice.

REFERENCES

American Psychological Association. (1974). *Standards for providers of psychological services.* Washington, DC: Author.

American Psychological Association. (1983). *Standard agreement with insurance carriers for provision of peer review for outpatient psychological services.* Washington, DC: Author.

Benedict, J. G., & Stricker, G. (1983). *Report on survey of clinical practice.* Unpublished manuscript.

Biskin, B. (1983). *The effects of theoretical orientation and experience on the quality of peer review of outpatient psy-*

chological services and reimbursement recommendations. Unpublished doctoral dissertation, University of Maryland, College Park.

California State Legislature. (1981–1982). Regular session. Assembly bill No. 3480.

Cohen, L. (1981). Peer review of psychodynamic psychotherapy: An experimental study of the APA/ CHAMPUS program. *Professional Psychology, 12,* 776–784.

Cohen, L., & Oyster-Nelson, C. (1981). Clinicians' evaluations of psychodynamic psychotherapy: Experimental data on psychological peer review. *Journal of Consulting and Clinical Psychology, 49,* 583–589.

Cohen, L., & Pizzirusso, D. (1982). Peer review of psychodynamic psychotherapy: Experimental studies of the APA/CHAMPUS program. *Evaluation and Health Professions, 5,* 415–436.

Ennis, B. J., Friedman, P. R., Bersoff, D. N., & Ewing, M. F. (1983, August 23). Memorandum of the effect of a patient's release of confidential treatment information for peer review purposes, submitted to the American Psychological Association.

Lowenstein, A. V., Sandler, R. M., Brochin, M. D., Kohl, B. M., Fisher, A., Boylan, M. P., & Meanor, M. C. (1983, July 1). Memorandum on confidentiality and peer review submitted to New Jersey Psychological Association.

Mariano, W., & Feldman, S. (1983). Privileged communications and confidentiality. In *New York State Psychological Association annual reference diary.* New York: New York State Psychological Association.

Ohio State Legislature. (1977–1978). Regular session, 112th General Assembly. LSC 112 0423-6.

Pearce, J., & Newton, S. (1969). *The conditions of human growth.* New York: Citadel Press.

Specialty guidelines for the delivery of services by clinical psychologists. (1981). *American Psychologist, 36*(6), 640–651.

Tarasoff v. The Regents of the University of California, 520 P. 2d 553 (1976).

III

LEVEL OF CARE

10

Quality Assurance in Acute Inpatient Services

JACK A. WOLFORD

INTRODUCTION

Improved quality of care for patients has been sought by physicians and has been their goal since the time of Hippocrates. Although certainly not binding, the Hippocratic oath has served as an ethical guide for the medical profession since its beginning. Perhaps the most quoted sentence and dictum from the oath is "I will prescribe regimen for the good of my patients according to my ability and my judgment and never do harm to anyone" (Dorland, 1981). Although the term *quality of care* is not mentioned, it is implied in the Hippocratic oath. Hippocrates and those who followed over the next several centuries during the height of the Greek and Roman Empires tried to understand mental illness and contributed positively to care of the mentally ill. It was not long, however, before ignorance, superstition, and stigma played important roles in the neglect and maltreatment of the "insane." From time to time, a voice was raised on the side of the mentally ill and their needs, but it was largely unheard or disregarded. The mentally ill or insane were beaten, accused of witchcraft, and left to starve and were housed with vagrants, criminals, and other of society's rejects.

In 1566, San Hipolito was founded in Mexico City. It was the first hospital in the New World devoted totally to the mentally ill. In 1547, the old monastery of St. Mary of Bethlehem, which was first mentioned as a hospital in 1330 and housed inmates in 1403, was handed over to the city of London as a hospital for the insane by Henry VIII. It later became known as "Bedlam." This hospital epitomized the inhumane care that many of the mentally ill received at that time (Slaby, Tancredi, & Lieb, 1981).

Hospital care of the mentally ill in the United States began in 1756, when the Pennsylvania Hospital accepted its first mental patients, and in 1773 the first public hospital was established in Williamsburg, Virginia.

The father of American psychiatry, Benjamin Rush, published his text *Medical*

JACK A. WOLFORD • Western Psychiatric Institute and Clinic, University of Pittsburgh, School of Medicine, Pittsburgh, Pennsylvania 15213.

Inquiries and Observations upon the Diseases of the Mind in 1812 (1812/1962). Although some of his treatment techniques using emetics, purgatives, and bloodletting and his special rotating chair make the quality of care suspect, he was active in the moral treatment movement. In 1792, Philippe Pinel released the mentally ill from their chains at Bicêtre hospital, where he was director, and ushered in the humanitarian tradition in psychiatry in Europe, and the movement quickly spread to the United States. A fervent optimism developed in relation to the curability of mental illness. Although the optimism would not be sustained, it was important in the organization of mental hospitals and the development of specialized personnel with an emphasis on the interpersonal aspect of patient care and the importance of the milieu of the hospital units. In the beginning of the hospital era, the hospitals were small, and the claims of cure were in the 80% to 90% range. Slowly, a grouping of forces began to change the small hospital. Population increased by normal growth and also by soaring immigration rates. There were many complications of change for the immigrants and a higher rate of illness for these new citizens. Dorothea Dix (1802–1877) began a crusade for public hospitals and their use for those improperly incarcerated in jails and almshouses and otherwise neglected in society. She convinced many legislators to start the trend toward the large public hospital which later proved to be so difficult to manage. A professional organization in medicine had evolved, and specialty groups developed within it. These groups began to "regulate" practice by developing codes of ethics. These codes, although not legally binding, became powerful forces in practice. The organizations were concerned about the quality of the practice of medicine. This concern became particularly evident with specialty board certification and, in some instances, recertification.

Medical education was conducted in a laissez-faire way until revolutionized by the Flexner Report in 1910 (1910/1972). Education, schools, and their curricula were thrust into the arena of quality concern. Accrediting bodies with standards began to shape the future. Just as in medicine, other professional organizations were setting standards for nursing, social work, and psychology education and practice.

PSYCHIATRIC HOSPITALS

By the end of World War II, there were a growing number of private hospitals, but almost no general hospitals with psychiatric units. Most patients were in state or county public hospitals or veterans' hospitals. Most psychiatric hospitalization was for extended periods of time.

Several events occurred to promote a change in the scene: the introduction of psychoactive drugs in 1955, the Joint Commission's report in 1961 entitled "Action for Mental Health," and President John F. Kennedy's "Message from the President of the United States" relative to mental illness and mental retardation in 1963. This message was followed by the Mental Retardation Facilities and Community Mental Health Centers Construction Act of 1963, and later an amendment that created staffing funds.

The community mental-health movement and the philosophy of shorter hospitalization closer to home, coupled with more effective treatment, were quite influential in the shift from the use of the large state hospital to the acute psychiatric units of general hospitals and private hospitals. In 1965, the passage of Medicare and Medicaid laws and their increased funding provisions provided further impetus for growth.

By 1981, there were 29,384 beds in psychiatric units of general hospitals, 23,000 in private hospitals, 33,796 in the Veterans Administration hospitals, and 184,079 in the state and county hospitals. In spite of the numerically greater number of beds, the inpatient care episodes were about the same: 572,000 for the general-hospital acute units and 574,000 for the state and county hospitals (Potlash, Gold, Bloodworth, & Extein, 1982; Visotsky & Plant, 1982).

The advent of the acute inpatient service, whether it was located in the general hospital, the private psychiatric hospital, or a community mental-health center, brought with it more public and private funding and stimulated an increase in regulations. These regulations arose from inside the profession and from external forces of government.

From within the hospital system, regulation through bylaws, which outline and define criteria for practice, took place both from internal pressures and external influences. Specialty board certification evolved within professional organizations. The need to keep abreast of new knowledge stimulated continuing medical education (CME), and some societies initiated the requirement of specified hours of CME in order for the individual member to retain membership. Accrediting bodies have had a major influence on hospital bylaws.

JOINT COMMISSION OF ACCREDITATION OF HOSPITALS

The Joint Commission of Accreditation of Hospitals (JCAH) grew out of the American College of Surgeons, and together, they have been setting standards since 1918. The JCAH, as an independent unit, was founded in 1952. As JCAH accreditation is now required by Medicaid, Medicare, CHAMPUS, and numerous state licensing bodies, it has really ceased to be a voluntary, self-regulatory body. Numerous national organizations have participated, and continue to participate, in the formulation of JCAH standards.

In 1970, the Accreditation Council for Psychiatric Facilities was formed as a Categorical Council of the Joint Commission of Accreditation of Hospitals. It is composed of seven member organizations:

American Academy of Child Psychiatry
American Association on Mental Deficiency
American Hospital Association
American Psychiatric Association
National Association of Private Psychiatric Hospitals
National Association of State Mental Health Program Directors
National Council of Community Mental Health Centers

Although the 1972 *Accreditation Manual* did not mention quality assurance, it was not long before a statement was issued. Writing in *Hospitals* in April 1974, John D. Porterfield III noted that the JCAH Board of Commissioners, on December 8, 1973, approved the following statement of quality of patient care:

> The Board considers the quality of patient care to be the central purpose of its entire accreditation process. It has generally agreed that the enhancement and protection of quality is a function which is primarily a responsibility of the Medical Staff with respect to medical care and supportive medical service. In the hospital accreditation standards, therefore, is the expressed policy of the Board that adequate performance of this function shall be a substantial factor in the accreditation status of the hospital. (Porterfield, 1974, p. 20)

The medical staff was charged with:

1. A retrospective medical audit
2. A continuing medical education program using findings of the audit
3. A utilization and review procedure geared to the findings of the audit
4. A credentials committee, which takes audit findings into consideration

In 1974, JCAH required evidence of initiation of the processes aimed at evaluating quality of care, and in 1975, it expected to find substantial performance.

In the JCAH consolidated standards for child, adolescent, and adult psychiatric, alcoholism, and drug abuse programs (1979, 1981), a major emphasis was placed on specific quality-assurance programs.

PEER REVIEW AND PROFESSIONAL STANDARDS REVIEW ORGANIZATIONS

The medical profession had started aspects of peer review in its hospitals as early as 1949–1951 with the tissue committee concept. The autopsy is another example of peer review. It was not until 1971, however, that the American Medical Association (AMA) and its Council on Medical Services issued a comprehensive, second-volume *Peer Review Manual*. The American Psychiatric Association (APA) stated its position on peer review in 1973.

Although part of the impetus for peer review has come from within the profession, much of the drive came from a physician Senator from Utah, Wallace F. Bennett, who was the author of the PSRO Amendment to Public Law 92-603 enacted in November, 1972. An unsuccessful attempt in 1970 probably signaled both the AMA and APA about the upcoming amendment. Although not all inpatient psychiatric units came under PSROs, many did, including the acute inpatient services, and a real impetus for self-scrutiny by peers began with this amendment. The PSRO review is based on:

1. Concurrent utilization and review based on admission and continued stay criteria.
2. Medical audit studies.
3. Norms of care based on regional problems of practice.
4. Ultimate responsibility rests with the Secretary of the Department of Health, Education, and Welfare in all areas of the PSRO operation, but operation may be delegated to regional PSROs and medical staffs.

Peer review has grown beyond the PSRO. Third-party payers, including CHAMPUS and many commercial insurance companies, have joined with the professional organizations in numerous projects of peer review, including both record review and on-site studies.

THE CONCEPT OF QUALITY AND QUALITY ASSURANCE

The concept of *quality* is an elusive one and a workable definition of *quality assurance* has been difficult to formulate. The word *quality* is a noun, and *Webster's* (1969) defines it as:

> 1. That which makes a being or thing such as it is; a distinguishing element or characteristic. 2. The characteristics of anything regarded as determining its value, place, worth, rank, position, etc., kind; when unqualified, peculiar excellence. 3. Degree of excellence; relative goodness; grade; as, high quality of fabric.

At times, in defining QA, adjectives have been used with the noun to denote superior quality, optimum quality, high quality, and the opposites. In general, however, the noun *quality* alone denotes a condition that indicates excellence. *Webster's* also defines *quality control* as

> an aggregate of functions designed to insure quality by initial critical study of designs, materials, processes, equipment and workmanship followed by periodic inspection and analysis of the results of inspection to determine causes for defects and by removal of such causes.

Extant quality-assurance programs have many of the features implied by this definition (Richman, 1975).

Donabedian set the stage for our present-day quality-assurance programs with his article on "Evaluating the Quality of Medical Care" in 1966. He pointed out that there are three broad areas for research: structure, outcome, and process. Programs today involve all three of these areas. The structure, or input, is primarily a listing of the essentials in staff requirement in both number and qualifications, space and safety requirements for buildings and facilities, and required essential services. These are outlined in the JCAH requirements for accreditation and are a significant part of the evaluation, even though ideal staff-to-patient ratios and professional staff distributions are unsettled issues.

The process of care and its analysis are addressed by utilization review, patient care audits, record audits, and peer review. Outcome studies, as pointed out by Liptzin (1974), "provide the most important and direct measures of quality. But, for several reasons, they are the most difficult to carry out" (p. 1375). He pointed out the many and major difficulties in doing outcome studies. These are essentially related to whether the outcome is the result of the treatment. Many factors other than treatment influence outcome. The variables in long-term treatment are legion, and although those in an acute inpatient unit are fewer, there are nevertheless many. There are, however, outcome measurements that are of value and that can be used in an inpatient setting. These pertain to obtaining objectives of treatment planning, monitoring changes in problem behavior (McLean & Craig , 1975), discharge readiness measurements (Hogarty & Ulrich, 1972), and other measurements of goal attainment (Tischler & Reidel, 1973). It should be emphasized that outcome evaluations also rely on how well the patient feels and the opinion of significant others, even though these may not be "scientific" measurements. Glick & Hargreaves (1979) used adjustment scales that formalize these subjective feelings with instruments to measure them.

Reinhardt (1983), an economist, pointed out that a source of confusion regarding quality care is the variety of contexts to which the concept can apply. He discussed three broad areas: (a) the impact of the entire health-care system on the well-being of those it is intended to serve; (b) the efficiency and effectiveness of the medical treatment (the management of particular medical conditions); and (c) the efficacy of particular medical procedures. The first he called "macro-quality of health care," the second and third, "micro-quality of care." These include the quality of technical care and the quality of the "art of care." The latter is defined as "the milieu, manner and behavior of the provider in delivering care to and communicating with the patient" (p. 6). The "art of care" (defined in 1976) was just beginning to be measured in 1983 and continues to be at present. (Brook, Williams, & Avery, 1976).

Schulberg (1976) emphasized that any definition and set of standards that are bureaucratically administered will conflict with the evidence that professional work suffers when detailed directives are passed down by administrators. He also pointed out the error of "the imposition of subjective values in the guise of pseudo-scientific knowledge" (p. 1048). Another stricture pointed out by Schulberg is that

> it must be emphasized that evaluating psychiatric effort (prospectively, concurrently or retrospectively) on the basis of standards rests upon the premise that prescribed levels of clinical activity produce beneficial outcomes. To the degree that this premise is untested or untenable, conformance to standards is meaningless and even misleading. (p. 1048)

The obvious implication of this article is to involve physicians and other professionals in standard setting at all levels: nationally, statewide, and in local hospital quality-assurance programs.

In a thought-provoking article in 1977, Menninger emphasized that "quality care is care which acknowledges individual differences as more important than similarities" and further that

> quality might be defined as the goodness of fit between:
>   1. The problem requiring therapeutic attention
>   2. The desired outcome (the goal or purpose of treatment)
>   3. The treatment used
>
> as sensed or experienced by the patient, as judged by the physician and his colleagues, as measured by outcome studies.(p. 480)

THE ACUTE INPATIENT SERVICE

The beginning of the acute inpatient service for psychiatric patients was discussed earlier. The concept of acute care is based more on function than on structure, and it should be emphasized that an acute inpatient service may exist in a general hospital, a private hospital, a veterans' hospital, a state mental hospital, or a community mental-health center. The philosophy of function is the important differentiating factor.

Use of an acute inpatient service should be conceptualized as a cross-sectional period in the life of a patient. In a few instances, it may cover the total period of a person's illness episode, but in the majority of cases, the inpatient care is only but a brief part of the overall continuous treatment of an individual. The reasons for hospitalization must meet criteria that establish the necessity for the admission and continued stay through utilization review evaluation. Thus the acute inpatient service indicates an extremely important part of the illness and a time for refining diagnosis and assessment, establishing a treatment plan primarily directed at the acute phase of the illness, and carrying out the plan in as effective a way and as short a time as is compatible with quality care.

To carry out this philosophy of care, every acute inpatient unit should have access to an emergency room, either a general medical emergency room with psychiatric consultation or its own psychiatric emergency room or service. This service should be evaluated in looking at quality assurance.

The key to successful therapy is usually assessment and correct diagnosis. This has become particularly true with the advent of new treatments that are more specific than some of the new therapy of the past. With this development in mind, the assessment process becomes paramount, and there is a need for professional staff and equipment that allow for a complete evaluation using the biopsychosocial model. They must either be present on the premises or easily and immediately available. Anything short of their presence cannot provide a quality assessment.

The concept of *acute treatment* implies that the program carried out on the unit could not be done as well elsewhere. It implies adequate numbers and training of staff to provide safety for the acutely ill, who on occasion show untoward behavior and aggression toward the self or others. The professional expertise must be available to develop, monitor, and carry out the treatment plan, whether it calls for biological, psychological, social, or, usually, a combination of these approaches.

Although length of stay (LOS) on an acute inpatient unit may vary a great deal from patient to patient, the implication is that, in general, the LOS is shorter. Short hospitalization is most often thought of as 21–28 days, although brief hospitalization and crisis hospitalization imply even shorter stays. A key element in the LOS requirement of patients is the accessibility and availability of aftercare. When a coordinated system of emergency care and triage, acute inpatient care, day hospital, rehabilitation programs, and outpatient programs exists, the acute inpatient service will function best.

Where cost containment leads to arbitrary mean LOS, as has been suggested for diagnostic-related groups (DRGs), difficulties will arise unless programs are in place. The result could be early discharge but an increase in readmission rates. Some of these DRGs seem to be arbitrary and too short, without any research data to back them.

THERAPEUTIC ENVIRONMENT

A therapeutic environment must be established to aid in patient interactions and emotional growth outside of designated specific therapeutic sessions. This milieu should be evaluated as part of a quality assurance program. The type of therapeutic milieu or organization of the unit need not be the same for each but will vary as determined by the type of patient in relation to diagnosis, age, and whether the acute episode is with or without a chronic illness with habit deterioration that interferes with daily living. Occupational therapy, vocational rehabilitation, psychiatric rehabilitation services, recreational therapy, activities therapy, dance therapy, music therapy, and other forms of activity serve an important purpose but should be used by the patient as outlined in his or her treatment plan and in conjunction with the acute treatment phase. Maxmen, Tucker, and LeBow (1974) wrote of the "reactive environment" of the hospital and defined it as "one in which the unit's overall structure maximally utilizes and coordinates the entire staff's efforts toward the rehabilitation of the patient's particular behavior problems so that ultimately he becomes a longstanding and productive member of society" (p. 33).

This type of environment provides many opportunities for conflict resolution and growth. Staff conferences around issues of conflict where there may be need to deal with anticipated or actual violence against the self or others provide educational experiences for the staff and therapy for the patient. Other conflicts around food service, recreational equipment, and therapeutic leave issues provide material for discussion. These incidents may also result in the need for a patient care audit or other measure if they are recurrent or the patient's complaints need further evaluation.

A therapeutic environment is a safe environment. Patients (and their families) must be secure in the knowledge that the patient's untoward behavior can be controlled. Policies for staff education in patient control issues and methods are mandatory. Policies and procedures for various levels of observation appropriate to various levels of disturbed behavior should be enunciated and written. Quality care on an acute inpatient service requires that all of these be in place. In addition, various safety measures, in relation to fire and physical health emergencies, must be addressed. In-service education will impart the knowledge to staff.

THE MEDICAL RECORD

This is not the place for a detailed description of the medical record. A few remarks are in order, however, because the medical record is so significant in helping to evaluate quality of care. Much of the literature of the early 1970s stressed the value of the patient care or medical audit but remarked that the medical records were so poor that audits were difficult, if not impossible, in many instances. Medical records have improved as regulations (JCAH) have emphasized them and have built in requirements regarding assessment and diagnosis, and treatment plan recording. Progress notes are required to reflect the treatment plan goals. Recording of treatment and changes, as they occur, must be described. Unusual incidents must be recorded. Discharge and aftercare plans are essential and need to be recorded. A discharge summary is very important in that it gives a total picture of the acute inpatient stay and is quite valuable for the next service, particularly if that service is out of the system.

Records provide the information for utilization review and are essential to finding problems in patient care audits and for carrying out studies. Charts are the basis of many peer review activities, both internal and external.

The record should also contain any outcome evaluations as obtained through psychiatric rating scales (Endicott & Spitzer, 1980) and observations of goal attainment as reflected in the progress notes and in the discharge summary.

It should be emphasized that records need not follow any rigid form. There is room for innovation. The problem-oriented record system (POS) has been touted by some as the answer to quality control, whereas others feel that a careful non-bandwagon approach should be taken. Some feel that certain aspects are antithetical to the practice of psychiatry (Grant, 1973; Lipp, 1973a,b; Spitzer, 1973; Taylor, 1973). A medical records committee takes responsibility for changes, updating, and monitoring the quality of the medical records.

FUNDAMENTALS OF A QUALITY ASSURANCE PROGRAM

The fundamental elements of a quality assurance program have already been alluded to, and there is a need to bring them together. A quality assurance program is designed to assist in the promotion of quality care through the analysis, review, and evaluation of the clinical practice that exists within the particular acute inservice program. The QA program not only evaluates the quality of care but, through established communication mechanisms, feeds back to those who carry out the treatment. Where deficiencies exist, remedial measures, including educational ones, are taken. At a later date, an analysis of the efficacy of the corrective measures is carried out. To accomplish the above, certain structures should be in place:

1. Any successful program, regardless of its structure and intensity of application, will be only as successful as the potential that exists for change in the system. Those in authority, such as the director of the unit, the president of the medical staff, and representatives of persons significant to the functioning of the committees and the various programs, must participate actively and, if the unit is part of a larger system (e.g., a general hospital), must represent the acute inpatient unit so that it receives its share of the available studies and resources.

2. There is also a need for the participation of the entire staff of the service in the quality assurance program. Without their involvement and an investment in the program, it will not be as successful as desired.

3. A program should be unique to match the uniqueness of the clinical program.

4. The organization should reflect a structure that encompasses all of the committees and persons involved who will have the authority and expertise to carry out the functions of the quality assurance program. A plan including an organizational chart should be available.

5. The program should be multidimensional and problem oriented. The sources of data are numerous, but data come most frequently from individual reports, utilization review, patient care audits, medical records evaluations, incidents that are reported involving patient care, psychological autopsies, newspaper reports concerning any patient, advisory committees, volunteers, community agency input, and patient input, either individually or through the patient rights coordinator.

6. Once a problem is identified and a study completed, either formally by audit or by discussion within the QA committee, there must be a way to carry out the necessary change. This need once again demonstrates the necessity of the involvement of people in authority who can initiate change. Those persons on the unit must be involved so they can carry out the change.

7. An annual evaluation of the quality assurance program is required. Not only does it give the committee a chance to review their work and assess it, but through the chairperson's report, which finally goes to the governing body, such an evaluation calls attention to the program. It highlights the significance of the program for the proper functioning of the acute inpatient service. Four committees are essential to any quality assurance program. First, there is the oversight committee which has responsibility to the governing authority. Second, there are the utilization review committee, the medical records committee, and the patient care audit committee. Other committees should certainly be involved when germane. The example of a quality assurance program that follows gives more details regarding the structure of one program.

PEER REVIEW

Formal programs of peer review are covered elsewhere (see, for instance, Chapter 20) in this book. The APA *Manual of Psychiatric Peer Review* (1985) is the peer review authority for psychiatry. Other articles contribute to an understanding of peer review as well as of other aspects of quality assurance (Conklin, 1983; Lalonde, 1982; Langsley & Lebaron, 1974; Mattson, 1984; Newman & Luft, 1974; Richman, 1975; Richman & Pensker, 1974). Some additional statements are warranted regarding an acute inpatient service.

Many formal and informal peer-review activities take place on the unit. These should be encouraged and recorded for the QA committee. Medical records committees frequently have quarterly chart reviews by peers. Utilization-review-committee activities are a form of peer review. A case conference involving an outside moderator is peer review. Asking a colleague to review a patient's treatment is asking for peer review. Having others sit in on team meetings and asking them for criticism involve peer review. All disciplines should be a part of informal peer review.

One further peer-review activity can be very helpful to the staff of an acute inpatient unit: inviting the staff of a similar unit in another service center to reciprocate in an informal evaluation of records, the milieu, conferences, and the structure and functioning of the staff of the unit. This activity requires a certain amount of courage but may pay reasonable dividends.

THE QUALITY ASSURANCE PROGRAM: AN EXAMPLE[1]

PURPOSE

The primary service goals of the Western Psychiatric Institute and Clinic (WPIC) of the University of Pittsburgh are the delivery of comprehensive and integrated mental-health services and the continual provision and improvement of quality patient care within the available resources. In order to carry out these goals, a Quality Assurance and Risk Management Program has evolved at WPIC. The inpatient services at WPIC are primarily devoted to acute short-stay programs.

In the beginning of the program, the risk management component was developed separately with the primary goal of reducing the potential for injury to people and property. It was developed in response to insurance requirements; however, it soon became a key element in improving hospital systems, educating employees, and enhancing the quality of patient care, and it was combined with our quality assurance program to form the Quality Assurance and Risk Management Program (Huber & Wolford, 1981). Elements of the program are designed to promote and maintain quality care through the analysis, review, and evaluation of the clinical practice within WPIC.

AUTHORITY

The Quality Assurance and Risk Management Committee (QARM) was established to implement the total QA and Risk Management Program, and it receives its authority from the board of trustees, which is the governing body and functions through the WPIC Executive Committee.

SCOPE OF RESPONSIBILITIES AND FUNCTIONS

The functions and responsibilities of the overall QA program include the following:

- Ascertain the quality of patient care provided
- Examine the mission and philosophy of WPIC system of care
- Observe the delivery of effective, efficient, safe, and clinically sound care in the least restrictive setting
- Identify and resolve problems that impact on the quality of care
- Assign high priority and resources to the investigation of problems that have the greatest impact on patient care and clinical performance
- Implement recommendations regarding changes in procedures or corrective actions such as the establishment of educational or training programs, the revision of WPIC policies and procedures, the revision of staffing patterns, and/or the improvement of the scheduling of resources
- Review and analyze all initial investigators' reports annually

[1]The author is indebted to the many persons of various disciplines who participated in the development and recording of this program at Western Psychiatric Institute and Clinic.

- Review and analyze all activities of the Medical Records, Utilization Review, Patient Care Audit, Patient Care Monitoring, and Quality Assurance Advisory Committees
- Assist in the implementation of recommendations resulting from licensure accreditation and certification reviews
- Seek input from sources outside the immediate WPIC system, such as community agencies, service consumers, and the WPIC Citizens Advisory Committee
- Ensure compliance with pertinent laws, regulations, and WPIC policies

ORGANIZATION

The organizational chart (Figure 1) shows the relationship of the Quality Assurance and Risk Management Committee to other components of WPIC. The committee is cochaired by the Chief of Adult Services and the Director of Child and Adolescent Services and contains key people who can both initiate and carry out change when necessary. In addition to the cochairpersons, the membership includes the following:

> Associate Administrator, Clinical Services
> Chairpersons of the following committees: Medical Records, Patient Care Audit, Utilization Review, Safety/Disaster
> Director of Base Service Unit (Community Mental Health Center)
> Director of Medical Records (Librarian)
> Director of Primary Medical Care
> Executive Committee members
> Information Systems Specialist
> Legal counsel
> Patient Rights Coordinator
> President of the medical staff
> Quality Assurance Coordinator
> Risk Manager
> A representative from the psychiatric residents
> A representative from each module or inpatient unit

MEETINGS

The QARM Committee meets monthly or more frequently if deemed necessary. Special meetings may be called in consultation with the cochairpersons or on the advice of the Quality Assurance Advisory Committee. The latter committee meets monthly and is chaired by the Coordinator of Quality Assurance. Subcommittees appointed by the cochairpersons meet on an on-call basis for immediate discussion of critical incidents.

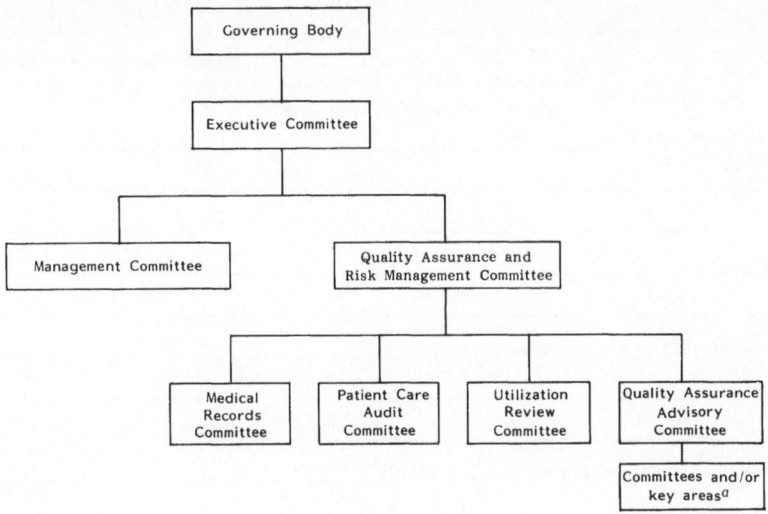

FIGURE 1. WPIC Quality Assurance and Risk Management Committee organizational chart. Committees and/or key areas include allied health professions, business offices, cost containment, infection control, medical emergency, Nursing and Social Work Discipline Committees, nutrition, Office of Education and Regional Planning, Patient Care Monitoring, pharmacy, research, Safety/Disaster, staff development, and volunteer and community services.

COMPONENT COMMITTEES

The QARM Committee is the decision-making, senior-leadership committee, which guides the quality assurance activities of the following four committees through the chairpersons and by participation of the members in many instances:

1. *Medical Records Committee*—responsible for ensuring the adequacy and accuracy of the medical record as the source document, which is indispensable to the evaluation of the patient's care.

2. *Patient Care Audit Committee*—responsible for the development of patient profiles and the conduct of patient care evaluation studies.

3. *Utilization Review Committee*—responsible for ensuring high-quality patient care through the effective and efficient use of the services and facilities of WPIC.

4. *QA Advisory Committee*—responsible for maintaining written records of centralized and decentralized quality-assurance activities and for functioning as an advisory body to the QARM Committee.

PROBLEM-SOLVING APPROACH

Each committee uses a problem-solving approach to patient care problems. Problems are identified in numerous ways, including, but not limited to, committee minutes and reports, patient comments, initial investigation reports, the Risk Managers or the Quality Assurance Coordinator's monitoring of the system, and individual or group concerns. The patient care problem report is used to seek input from any WPIC staff member (Figure 2).

(See Instructions on back)

QUALITY ASSURANCE PROGRAM
Patient Care Problem Report

Description of Problem: _____

_____

_____

_____

_____

How did you identify the problem? _____

_____

_____

What impact does problem have on patient care? _____

_____

_____

_____

What may be the cause(s) of the problem? _____

_____

_____

_____

_____

What were the attempts to solve the problem (if any)? _____

_____

_____

_____

How might this problem be corrected? _____

_____

_____

_____

_____

Submitted by:_____ Date _____

I request that my name be kept confidential. Yes_____   No_____
2/01/83

FIGURE 2. The Patient Care Problem Report.

The Patient Care Problem Report is designed to permit any WPIC staff member an opportunity to actively participate in the Quality Assurance Program. The primary goal of the program is to ensure that care and treatment provided for patients meet high standards of clinical practice yet remain efficient, affordable and safe. In order to reach this goal, the program must identify problems which in some way interfere with the service delivery system. Once a problem is identified, the Patient Care Audit Committee will determine its priority, impact on care and, if indicated, will design a study leading to recommendations, corrective action and follow up.

Completed Patient Care Problem Reports should be forwarded to the Secretary, Patient Care Audit Committee, WPIC, Room 30.

INSTRUCTIONS:

Description of the problem
A simple description of the problem is suggested. The types of problems which are relevant include, but are not limited to: (1) Problems directly or indirectly affecting patient care; (2) Problems that pose safety hazards to patients or staff; (3) Problems which affect either the underutilization or overutilization of the Institute's resources of time, money, staff, etc.

How did you identify the problem?
Please note if you discovered this through observation, report, experience, informal discussion. Was your source a meeting, incident report, etc.

What impact does problem have on patient care
In your opinion how many people are affected and how serious is the impact of this problem?

What may be the cause(s) of the problem?
Since you have taken time to think about this problem you may already have excellent ideas as to the potential causes. Please list if so.

What were the attempts to solve the problem (if any)?
So that we do not repeat corrective actions which did not work in the past, please list attempts made and the results.

How might this problem be corrected?
If you have any suggestions concerning the correction of this problem, please provide your recommendations.

FIGURE 2. (*Continued*)

PROBLEM REPORT

Problems are given a priority for investigation and resolution based on several aspects, namely, the scope of the problem: how many people are affected, the significance of impact on patient care, and the feasibility of gaining resolution. Critical incidents are handled immediately. Problems determined to need further assessment may require a tailor-made study design using clinically valid criteria, a review of current literature, and expert consultation in the field. This activity would be carried out by the Patient Care Audit Committee. Examples of Patient Care Audit Committee studies are as follows:

1. "Evaluation of Patient Seclusions." Problem and objective: To determine the use of seclusion of the acute inpatient services (January 1977, repeat August 1977).

2. "Patients Rights Compliance." Problem and objective: Are patients being given their rights and do they have ready access for complaints? (November 1977).
3. "Patients Leaving against Medical Advice (AMA)." Problem and objective: To determine the type of patient signing out AMA and is there a pattern? (March 1979, repeat November 1982, repeat January 1983).
4. "Adolescent/Young Adult Care Study." Problem and objective: Are adolescent and young adults receiving adequate treatment (quality care)? (May 1980, repeat May 1981).
5. "Minor Tranquilizer Study." Problem and objective: To determine the use of minor tranquilizers on the acute inpatient units (November 1981).
6. "Simultaneously Administered Neuroleptic Medication Study—Inpatient." Problem and objective: What is the use of more than one neuroleptic medication? (January 1982).
7. "Utilization of Adult and Children's Therapeutic Pass Study." Problem and objective: What is the pattern and appropriateness of the therapeutic pass? (August 1982).
8. "Continuity of Treatment between Inpatient/Outpatient Status Study." Problem and objective: To determine whether there is continuity of patient care within the WPIC modules, that is from the inpatient units to the ambulatory settings (August 1984).

When the cause and scope of the problem have been assessed, corrective action is implemented, if indicated. An educational program may be used. Problem monitoring occurs to determine if the issue has actually been resolved or reduced. The monitoring may be through a reassessment study, or it may be more informal, as in a report and through verbal feedback. Education through staff development is a primary method of accomplishing this mission. Figure 3 illustrates the ongoing staff development at WPIC. The efficacy of patient-care and medical audits has been questioned by some (Anderson & Shields, 1982). This author feels that audits are of value if they are carefully planned and carried out in relation to a definable, measurable problem. JCAH discontinued audit quotas in 1979, leaving problem-solving programs up to the local facilities.

The crisis control workshop grew out of a need to have staff become knowledgeable about techniques, both verbal and physical, of controlling disturbed patients. All members of the inpatient units are required to attend a workshop.

To anticipate problems in advance, a quality indicator form (Figure 4) that lists critical aspects of the unit and its atmosphere was designed by the nursing service personnel. These are filled out daily and discussed with supervisors.

The philosophy of primary nursing in psychiatry, as used at WPIC, is described in detail by Romoff and Kane (1982).

Peer review activities take place internally in relation to all disciplines through:

Utilization review activities
Medical Records Committee chart review

October

MONDAY	TUESDAY	WEDNESDAY	THURSDAY	FRIDAY
3	**4**	**5**	**6**	
*2:00-3:00 11th Fl. 503—AFFECTIVE DISORDER CASE CONFERENCE *2:00-3:00 7th Fl. 507—AYAM CASE CONFERENCE *2:15-3:15 8th Fl. 502—GERIATRIC CASE CONFERENCE	*10:00-12:00 Rm. 418 HEAD NURSE MEETING *1:30-3:00 Rm. 421 323—INTRODUCTION TO THE EXPRESSIVE ART THERAPIES AT WPIC J. Rubin; E. Irwin	*12:00-1:00 Rm. 421 ADVANCED NURSES MEETING P. Skorupka *2:00-3:00 Rm. 420 314A—CPR CLASS M. Payne *3:30-4:30 9th Fl. 506—SCHIZOPHRENIA CASE CONFERENCE	*9:00-10:00 DEC 506—DEC INSERVICE *1:30-2:30 Rm. Sem. A 342—CONTRAINDICATIONS OF PSYCHOTROPIC MEDICATIONS WITH OVER THE COUNTER DRUG: IMPLICATIONS FOR PATIENT EDUCATION W. Galarneau *2:00-5:00 Rm. 404A 314—CPR CERTIFICATION M. Payne; D. Hopp	*10:30-12:00 Auditorium GUEST LECTURE SERIES
10	**11**	**12**	**13**	**14**
*8:30-4:00 Rm. 404A & 404B CRISIS CONTROL WORKSHOP *2:00-3:00 11th Fl. 503—AFFECTIVE DISORDER CASE CONFERENCE *2:00-3:00 7th Fl. 507—AYAM CASE CONFERENCE *2:15-3:15 8th Fl. 502—GERIATRIC CASE CONFERENCE	*9:00-10:00 Auditorium GRAND ROUNDS *10:00-12:00 Rm. 418 HEAD NURSE MEETING *1:15-2:15 8th Fl. 319—NURSING ROUNDS Presentation by 8th Floor Nurses	*2:00-3:00 Rm. 413 PRECEPTORS' MEETING G. Cari *2:00-3:15 Rm. 1050 CRASH CART REVIEW L. DeJean *2:00-3:30 Rm. 420 314A—CPR CLASS Denise Russin *3:30-4:30 9th Fl. 506—SCHIZOPHRENIA CASE CONFERENCE	*9:00-10:00 DEC 506—DEC INSERVICE *10:00-11:00 Rm. 1050 CRASH CART REVIEW L. DeJean *2:00-5:00 Rm. 404A 314B—CPR RECERTIFICATION Denise Russin	*10:30-12:00 Auditorium GUEST LECTURE SERIES

FIGURE 3. Typical WPIC staff development program monthly calendar. *AYAM = Adolescent/Youth Adult Module.

MONDAY 17	TUESDAY 18	WEDNESDAY 19	THURSDAY 20	FRIDAY 21
*2:00–3:00 Rm. Sem. A 305—BURN-OUT SYNDROME AND METHODS OF COPING Joan Kyes	*9:00–10:00 Auditorium GRAND ROUNDS	*12:00–1:00 Rm. 420 324—PATHWAY TO HEALTH SERIES: WEIGHT CONTROL Kathy West	*9:00–10:00 DEC 506—DEC INSERVICE	*10:30–12:00 Auditorium GUEST LECTURE SERIES
*2:00–3:00 11th Fl. 503—AFFECTIVE DISORDER CASE CONFERENCE	*10:00–12:00 Rm. 418 HEAD NURSE MEETING	*2:00–3:00 Rm. 420 STUDENT COORDINATORS' MEETING G. Cari	*2:00–3:00 Rm. 420 342—WHAT DIRECTION ARE WE GOING IN NURSING AS A PROFESSION? Sue Reitz	
*2:00–3:00 7th Fl. 507—AYAM CONFERENCE	*2:00–4:00 a.m. 8th Fl. 314B—CPR CERTIFICATION L. Lenkner	*3:30–4:30 9th Fl. 505—SCHIZOPHRENIA CASE CONFERENCE		
*2:00–5:00 a.m. 8th Fl. 314—CPR CLASS & PRACTICE SESSION				
*2:15–3:15 8th Fl. 502—GERIATRIC CASE CONFERENCE				

24	25	26	27	28
*2:00–3:00 11th Fl. 503—AFFECTIVE DISORDER CASE CONFERENCE	*9:00–10:00 Auditorium GRAND ROUNDS	*2:00–3:00 Rm. 420 STAFF DEVELOPMENT LIAISON NURSES' MEETING G. Cari	*9:00–10:00 DEC 506—DEC INSERVICE	*10:30–12:00 Auditorium GUEST LECTURE SERIES
*2:00–3:00 7th Fl. 507—AYAM CASE CONF.	*10:00–12:00 Rm. 418 HEAD NURSE MEETING	*3:30–4:30 9th Fl. 505—SCHIZOPHRENIA CASE CONF.	*3:00–4:00 Rm. 404A 304—RELAXATION THERAPY AND TECHNIQUES Barbara Szekely	
*2:15–3:15 8th Fl. 502—GERIATRIC CASE CONF.	*2:00–3:30 Rm. 420 407—CONFLICT: WHAT IT IS AND APPROACHES TO UTILIZE FOR ACHIEVING RESOLUTION Karen Evanczuk			
*2:30–3:30 Rm. 420 341—DEVELOPMENTAL STAGES AND TASKS FOR PRE-SCHOOL THROUGH EARLY ADOLESCENTS J. Moser				

31
*2:00–3:00 11th Fl. 503—AFFECTIVE DISORDER CASE CONFERENCE
*2:00–3:00 7th Fl. 507—AYAM CONFERENCE
*2:15–3:15 8th Fl. 502—GERIATRIC CASE CONFERENCE

FIGURE 3. (Continued)

QUALITY INDICATOR FORM

DATE: _____   UNIT: _____   CENSUS: _____

I.   Diagnostic Ration

     a. _____Conduct Disorder   d. _____Anxiety Disorder

     b. _____Psychosis   e. _____Oppositional Disorder

     c. _____Affective Disorder   f. _____Other (Specify)

                                                     _____

                                                     _____

                                                     _____

II. Observation Status

     a. _____SOO   b. _____OO

III. Incident Reports

     a. _____Assaultive Acts (Patient to Patient)

     b. _____Assaultive Acts (Patient to Staff)

     c. _____Property Damage (Patient, Staff and/or Hospital Property)

     d. _____Custodial

     e. _____Critical

IV. Levels of Assistance

     a. _____Level I   c. _____Level III

     b. _____Level II   d. _____All Available Staff

V.   Seclusion

     a. _____Per Program

     b. _____Violence/Assaultive Behavior

VI. Call Offs   R.N.   N.A.   C.A.   P.A.   OTHER

     a. 7-3   ____   ____   ____   ____

     b. 3-11   ____   ____   ____   ____

     c. 11-7   ____   ____   ____   ____   ____

VII. Staffing Pattern R.N.   N.A.   C.A.   P.A.   OTHER

     a. ____7-3   ____   ____   ____   ____

     b. ____3-11   ____   ____   ____   ____

     c. ____11-7   ____   ____   ____   ____   ____

VIII. Census

     a. ____Admissions   c. ____Discharges

     b. ____Readmission   d. ____Length of Stay per Discharge

                                      ____

                                      ____

FIGURE 4. The Quality Indicator Form.

Grand rounds
Case conferences
Individual case review when problems of care exist
Patient care audits
Advanced nurses' meetings to discuss and review unit functioning

Faculty medical peer review has also recently been initiated on a formal basis. Each faculty or medical-staff physician is reviewed three times a year following the instructions in Figure 5; the review is recorded on the form in Figure 6.

External peer review is also used. Some peer reviews are solicted; others are imposed. For example:

Accreditation agencies through on-site review:
1. Medicaid
2. Medicare
3. Drug and Alcohol Review Board
4. JCAH
Third-party payers through chart review
Concurrent utilization review by Medicaid (state)
Program review by invited committee of peers as consultants to a program

QUALITY ASSURANCE GOALS

In addition to the condition of an active program as described above, additional goals for 1987 are:

1. Identify quality assurance activities institutewide.
2. Increase staff awareness of quality assurance programs. A quality assurance newsletter has been initiated.
3. Ensure "feedback" to WPIC staff regarding QARM Committee recommendations.

EVALUATION

The overall Quality Assurance and Risk Management plan is reviewed and/or revised on a formal basis at least annually. The goals of the QA program are compared to the actual activities of the previous year. Modification in the plan is made by the QARM Committee to improve the effectiveness of the program.

CONCLUSION

This chapter has looked at quality assurance from the view of the acute inpatient service. It appears from the literature that, with some exceptions, short-term hospitalization may be as effective as long-term. It is also apparent that the success of the acute inpatient service depends on its being part of a system for continuous

INSTRUCTIONS

This form is to be utilized in the process of Faculty Medical Peer Review. A faculty physician must conduct the actual review and complete this form accordingly.

The following questions are designed to assist in the review process. They are not intended to be all-inclusive and serve only as guidelines in assessing each of the stated categories.

A. Patient Assessment and Diagnostic Workup
   1. Is a faculty assessment of the patient present?
   2. Has information lacking at the time of initial assessment been obtained?
   3. Is there evidence of ongoing assessment of newly diagnosed problems?
   4. Has a final patient diagnosis been reached?

B. Treatment Planning
   1. Are problem statements written in clear and concise language?
   2. Do goal statements relate to the identified problems?
   3. Is there a stated time frame for achievement of goals?
   4. Is responsibility for treatment clearly specified?
   5. Is there evidence that the patient knows and consents to treatment plan (to extent possible)?

C. Medical Treatment Conducted in Accordance with the Treatment Plan
   1. Do the medical progress notes refer to the treatment plan?
   2. Is the treatment plan regularly updated?
   3. Is there evidence that the problems stated in the treatment plan are being addressed?

D. Appropriateness of Treatment
   1. Is there evidence that the proper medications are utilized, given the stated diagnosis?
   2. Is there evidence that medications are or are not properly titrated (e.g. patient falls, side effects, etc.)?
   3. Is there evidence that the faculty physician is engaging in providing active treatment to the patient (e.g. documentation in terms of content, length and frequency)?
   4. Are side effects of medication being monitored?

E. Follow-up of Physical Findings
   1. Is there documentation of follow-up concerning physical findings?
   2. Are results of tests which have been ordered present?

F. Supervision of Medical Care Provided
   1. Are patient medical problems identified?
   2. Are appropriate strategies stated for dealing with identified medical problems?
   3. Are medical student, physician assistant and resident progress notes appropriate in terms of content, length and frequency?
   4. Are notes dated and signed?

G. Discharge Planning
   1. Has the discharge planning process begun in a timely manner?
   2. Is the discharge planning process and final disposition appropriate for meeting the needs of the patient?
   3. Is there evidence that the patient and/or the patient's family (guardian, etc.) was involved in the discharge planning process?

H. Family Contact
   1. If family physician contact was requested, did it occur?
   2. Is the frequency of family meetings appropriate?

I. Quality of Record Keeping
   1. Are all clinical notes legible?
   2. Do clinical notes reflect the treatment provided?

J. Plan of Action to Correct Deficiencies

Any noted deficiencies should be discussed with the faculty physician who will then assure the implementation of any actions necessary to correct these deficiencies. This plan of action should specifically state the identified deficiency, a plan of corrective action, and the individual(s) or group who will implement the necessary measures.

FIGURE 5. Instructions for the Faculty Medical Peer Review Form.

FACULTY MEDICAL PEER REVIEW

Name of Patient:_____ Medical Record #_____

Date of Admission:_____Discharge:_____ Date Chart Review:_____

Attending Physician:_____ Primary Resident:_____

Name of Physician Conducting Review:_____

Name of Person Doing Physical Examination:_____

	Satisfactory	Unsatisfactory
A. Patient Assessment and Diagnostic Workup Comments & Suggestions:_____ _____ _____		
B. Treatment Planning Comments & Suggestions:_____ _____ _____		
C. Medical Treatment Conducted in Accordance with Treatment Plan Comments & Suggestions:_____ _____ _____		
D. Appropriateness of Treatment Comments & Suggestions:_____ _____ _____		
E. Follow-up of Physical Findings Comments & Suggestions:_____ _____ _____		
F. Supervision of Medical Care Provided Comments & Suggestions:_____ _____ _____		
G. Discharge Planning Comments & Suggestions:_____ _____ _____		
H. Family Contact Comments & Suggestions:_____ _____ _____		
I. Quality of Record Keeping Comments & Suggestions:_____ _____ _____		
J. Plan of Action to Correct Deficiencies Comments & Suggestions:_____ _____ _____		

cc: Q.A. Coordinator
     Director, Adult Services
     Chief, Child & Adolescent Services
     Attending Physician Reviewed

FIGURE 6. The Faculty Medical Peer Review Form.

care with emergency, inpatient services, partial hospitalization, and outpatient care.

The use of acute inpatient service is viewed as a brief, but very important, period in the continuous care of a patient. It is paramount for many reasons, as pointed out in the text, that the quality of care be maintained at a high level. The mental health disciplines have been devoted to quality assurance; however, the impetus of outside groups in requiring structure through organization and the development of formal programs of quality assurance has contributed greatly. Some aspects of the requirements are onerous and, in fact, questionable, but overall, they have had a positive impact.

The APA CHAMPUS project, which established for CHAMPUS beneficiaries a nationwide program of peer review, has been a major contribution and has led to other fiscal intermediaries' joining the program. This is an example, through both chart review and on-site peer review evaluations with follow-ups, of how lengths of stay can legitimately be decreased and how quality of care can be improved. Without the cooperation and participation of the mental health disciplines and their national organizations, CHAMPUS Peer Review could not function.

Quality assurance is the responsibility of each professional, and it is imperative that all professionals participate in their local program as well as through their organization at a state and national level.

REFERENCES

American Medical Association. (1971). *Peer review manual* (2 vols.). Chicago: AMA Division of Medical Practice.

American Psychiatric Association. (1983). Position statement on peer review. *Journal of the American Psychiatric Association, 130,* 381–385.

American Psychiatric Association. (1985). *Manual of psychiatric peer review.* (3rd ed.). Washington, DC: APA Committee on Peer Review (John M. Hamilton, Chairperson).

Brook, R. H., Williams, K. W., & Avery, A. D. (1976). Quality assurance in the 20th century: Will it lead to improved health in the 21st. In R. H. Egdahl & P. M. Gertman (Eds.), *Quality assurance in health care.* Germantown, MD: Aspen Systems Corporation.

Conklin, J. (1983). Quality assurance. In J. A. Talbott & S. R. Kaplan (Eds.), *Quality assurance in psychiatric administration.* New York: Grune & Stratton.

Donabedian, A. (1966). Evaluating the quality of medical care. *Milbank Memorial Fund Quarterly, 44,* 166–206.

Dorland's illustrated medical dictionary. (26th ed.). (1981). Philadelphia: W. B. Saunders.

Endicott, J. & Spitzer, L. (1980). *Psychiatric rating scales.* In H. I. Kaplan, A. M. Freedman, & B. J. Sadock. (Eds.), *Comprehensive textbook of psychiatry* (3rd ed., Vol. 1). Baltimore: Williams & Wilkins.

Flexner, A. (1972). *Medical education in the United States and Canada: A report to the Carnegie Foundation for the Advancement of Teaching.* New York: Arno Press. (Original work published 1910).

Glick, D., & Hargreaves, A. (1979). *Psychiatric hospital treatment for the 1980's—A controlled study of short versus long hospitalization.* Toronto, Lexington, MA: Lexington Books, D. C. Heath.

Grant, L. (1973). Enthusiasm no substitute for hard work. *International Journal of Psychiatry, 11,* 366–373.

Hogarty, G. E., & Ulrich, R. (1972). The discharge readiness inventory. *Archives of General Psychiatry, 26,* 419–426.

Huber, A., & Wolford, A. (1981). Investigative reporting cuts risk at psychiatric facility. *Hospitals, 55,* 3–76.

Joint Commission of Accredition of Hospitals. (1979, 1981). *Consolidated standards for child, adolescent and adult psychiatric, alcoholism and drug abuse programs.* Chicago: Author.

Joint Commission of Accreditation of Hospitals. (1979). Evolution of quality assurance reflected in new standard. *Quality Review Bulletin, 5,* 2–3.

Kennedy, J. F. (1963). *Message from the President of the United States Relative to Mental Illness and Mental Retardation.* (An address delivered to the Congress and the people of the United States.) Washington, DC: Government Printing Office.

Lalonde, B. (1982). Quality assurance. In M. J. Austin & W. E. Hershey (Eds.), *Handbook on mental health administration*. San Francisco: Jossey-Bass.

Langsley, G., & Lebaron, I., Jr. (1974). Peer review guidelines: A survey of local standards of treatment. *American Journal of Psychiatry, 131*(12), 1358–1362.

Lipp, M. (1973a). Author's reply. *International Journal of Psychiatry, 11*, 380–381.

Lipp, M. (1973b). Quality control of psychiatry and the problem oriented system. *International Journal of Psychiatry, 11*, 355–365.

Liptzin, B. (1974). Quality assurance and psychiatric practice: A review. *American Journal of Psychiatry, 131*(12), 1374–1377.

Mattson, M. R. (1984). Quality assurance: A literature review of a changing field. *Hospital and Community Psychiatry, 35*(6), 605–616.

Maxmen, J. S., Tucker, G. J., & LeBow, M. D. (1974). *Rational hospital psychiatry: The reactive environment.* New York: Brunner/Mazel.

McLean, P. D., & Craig, K. D. (1975). Evaluating treatment effectiveness by monitoring changes in problematic behaviors. *Journal of Consulting and Clinical Psychology, 43*(1), 105.

Menninger, R. (1977). What is quality care? *American Journal of Orthopsychiatry, 47*(3), 476–488.

Mental Retardation Facilities and Community Mental Health Center Construction Act (1963, October 13). Public Law 88-164, 88th Congress.

Newman, D. E., & Luft, L. L. (1974). The peer review process: Education versus control. *American Journal of Psychiatry, 131*(12), 1363–1366.

Porterfield, J. D., III. (1974). The Joint Commission on Accreditation of Hospitals and Quality Assessment. *Hospitals, 48*, 20.

Potlash, A. L. C., Gold, M. S., Bloodworth, R., & Extein, I. (1982). The future of private psychiatric hospitals, *Journal of Hospital and Community Psychiatry, 33*(9), 735–739.

Reinhardt, A. E. (1983). Quality assessment of medical care: An economist's perspective, *Quality Review Bulletin, 9*, 252.

Richman, A. (1975). *Methodology for quality review in psychiatry.* Paper presented at the Multiversity Conference—Quality of Treatment and Care, Greensboro, NC.

Richman, A., & Pensker, H. (1974). Medical audit by clinical records. *American Journal of Psychiatry, 131*(12), 1370–1373.

Romoff, V., & Kane, I. (1982). Primary nursing in psychiatry: An effective and functional model. *Perspectives in Psychiatric Care , 20*(2), 7–78.

Rush, B. (1962). *Medical inquiries and observations upon the diseases of the mind.* New York: Hafner Publishing. (Original work published 1812)

Schulberg, H. C. (1976). Quality of care standards and professional norms. *American Journal of Psychiatry, 133*(9), 1047–1051.

Slaby, A. E., Tancredi, L. R., & Lieb, J. (1981). *Clinical psychiatric medicine.* Philadelphia: Harper & Row.

Spitzer, R. L. (1973). Problem oriented medical records: Some reservations. *International Journal of Psychiatry, 11*, 376–379.

Taylor, R. L. (1973). Quality control of what? *International Journal of Psychiatry, 11*, 374–375.

Tischler, G. L., & Reidel, D. C. (1973). A criterion approach to patient care evaluation. *American Journal of Psychiatry, 130*(8), 913–915.

Visotsky, H. M., & Plant, E. A. (1982). Psychiatry in the general hospital: Visualizing the changes. *Journal of Hospital and Community Psychiatry, 33*(9), 739–741.

Webster's seventh new collegiate dictionary. (1969). Springfield, MA: G. & C. Merriam.

11

The Influence of External Forces on the Quality Assurance Process

Robert W. Gibson

In the past, mental health professionals have claimed total ownership of quality assurance and have asked only whether treatment met the usual standards of care. They have resisted intrusions by government regulatory agencies, third-party payers, the courts, and even professional and hospital associations. Professionals have perceived cost containment, safety, accessibility of services, patient satisfaction, and efficacy as irrelevant. Examination of these external elements of quality assurance has, at best, been deemed a necessary evil—more evil than necessary. In more recent years, these external forces have assumed preeminence because of the escalation of costs, consumer advocacy, civil rights litigation, proactive peer review by professional associations, and the acceptance of quality assurance standards set by the Joint Commission on Accreditation of Hospitals (JCAH). In this chapter, the dynamics, impact, and significance of these external forces are examined and discussed.

Quality assurance has been described as "the shared responsibility of health professionals and government to provide a reasonable basis for confidence that action will be taken, both to assess whether services meet professionally recognized standards and to correct any deficiencies that may be found" (Department of Health, Education, and Welfare, 1975, p. 143). Government has declared itself a partner. In an effort to be more specific about what elements are to be considered, *quality care* has been defined as offering patients "the greatest achievable health benefit, with minimal unnecessary risk and use of resources, in a manner satisfactory to the patient" (Department of Health, Education, and Welfare, 1975, p. 143). Therapeutic effectiveness, safety, cost, and patient satisfaction are the key elements identified in this definition of *quality care*. Standards of practice, so dear to the professional, are not even mentioned.

Cost containment has been the most persuasive and disturbing external force. Because of their orientation toward the treatment of an individual patient, health professionals have not tried to reduce the cost of care. As long as unlimited re-

ROBERT W. GIBSON • Sheppard and Enoch Pratt Hospital, 6501 North Charles Street, Baltimore, Maryland 21285-6815.

sources are available, cost poses no problem. Health planners and administrators seeking to maximize the benefits of a defined health care system or program having finite resources take a different view. If resources can be used more efficiently and if greater productivity can be achieved, more patients within the system can be treated, achieving in the aggregate a greater health benefit. The cost issue is especially prominent in long-term hospital-inpatient psychiatric treament.

QUALITY ASSURANCE UNDER THE JOINT COMMISSION ON ACCREDITATION OF HOSPITALS

Until the early 1970s, a single set of standards was applied by JCAH to both psychiatric and general hospitals. Because this approach did not do justice to the specialized needs of the psychiatric patient, an Accreditation Council for Psychiatric Facilities (ACPF) was formed by the JCAH in 1970 to develop standards that would be specifically for psychiatric facilities.

The very first standards developed by the ACPF stressed quality assurance even though that term had not come into common usage. The 1976 *Accreditation Manual for Hospitals* explicitly stipulated that "hospitals will demonstrate that the quality of care provided to all patients is consistently optimal" (Joint Commission on Accreditation of Hospitals, 1976, p. 27). The manual went on to require that members of the medical and other professional staffs must evaluate patient care by measurement of actual care against specific criteria. Various methodologies such as medical audit and focused reviews involving peers were developed "to ensure clinical practice of the highest quality."

The 1983 *Consolidated Standards Manual for Child, Adolescent, and Adult Psychiatric, Alcoholism, and Drug Abuse Facilities* requires hospitals to have in place "a well-defined, organized program designed to enhance patient care through the ongoing objective assessment of important aspects of patient care and the correction of identified problems" (Joint Commission on Accreditation of Hospitals, 1983, p. 33). This JCAH manual emphasizes that quality assurance must be a total institutional activity involving the governing bodies, the administration, and the professional staff. The five essential components are:

1. *Problem identification* through multiple data sources, chiefly care monitoring, safety, infections control, prescriptive practices, and profile analysis
2. *Problem assessment* by determining the cause and scope of problems, concurrent, or retrospective review
3. *Problem correction* through implementation of decisions designed to eliminate identified problems
4. *Monitoring of problem resolution* to ensure that an identified problem will be eliminated or at least satisfactorily reduced
5. *Program monitoring* to document the effectiveness of the overall program in enhancing patient care

The JCAH emphasizes quality assurance because the patient is unfamiliar with treatment procedures; has a need to believe in the treatment; and is uncertain about the outcome. The patient cannot judge the quality of medical treatment by the outcome of his or her single experience. Some patients recover even though their treatment has been poor; some patients do not improve even though they have received the best treatment available. Experienced professionals have difficulty agreeing on appropriate standards for the quality of treatment, the length of treatment, and the services needed. The inability of the patient to judge treatment and the need for protection of the patient have led JCAH to make quality assurance an essential part of the accreditation process. In general, professional organizations led by responsible professionals support the JCAH efforts to protect the rights of patients.

Although the main approach of the JCAH is the achievement of quality care through the establishment of standards of practice, inclusion of cost as an element of quality has for many years been reflected in the standards. The guiding principle of JCAH quality assurance in the *AMH/84 Accreditation Manual for Hospitals* indicates that "the hospital shall demonstrate a consistent endeavor to deliver patient care that is optimal *within available resources* and consistent with achievable goals" (Joint Commission on Accreditation of Hospitals, 1984, p. 147; italics added). The issue of cost is clearly identified.

UTILIZATION REVIEW UNDER THE JCAH

Utilization review (UR) was first introduced as a requirement for participation under Medicare and Medicaid and then as a JCAH standard to ensure that admission was medically necessary and that hospital services were efficiently used. Despite the emphasis on standards identifying the appropriateness of treatment in accordance with professional standards, the thrust of UR has consistently been cost containment.

Utilization review, a theme throughout the five components of quality assurance, bridges the gap between quality and cost. As required by the JCAH, UR is defined as "the process of using predefined criteria to evaluate the necessity and appropriateness of allocated services and resources to assure that the facilities' services are necessary, cost efficient, and effectively utilized" (Joint Commission on Accreditation of Hospitals, 1983, p. 211). The UR program must include analysis of the appropriateness and clinical necessity of admission, continued stays, and supportive services. Utilization review relies on techniques such as profile analysis, patient care studies, medication reviews, and infection control. Particular reliance is placed on the establishment of length-of-stay norms so that review dates can be established to prevent unnecessary and prolonged use of the facilities. Short-term stays of a few days are seldom at issue; longer-term stays of weeks or months are the concern. UR impacts primarily on long-term hospital care.

Unfortunately, this systematic scrutiny of the treatment process by UR has been perceived by mental health practitioners as an infringement of their professional prerogatives. This attitude reflects the belief of professionals that their judgment should be accepted without question. It must be acknowledged that quality assurance activities, particularly UR, have been designed to reduce costs by changing patterns of practice.

THIRD-PARTY PAYERS

In the 1960s, when increases in the cost of health care accelerated, third-party payers tried various measures to contain costs: limitations on the facilities to be paid, on the number of hospital days, on specific services; deductibles, higher coinsurance for psychiatric than for medical and surgical treatment; the exclusion of specific disorders such as alcoholism; and dollar limits per calendar year and for lifetime benefits. These restrictions are easy to administer; professional review is not needed; a clerk can deny a claim.

Another approach has been to establish limits based on the medical necessity for hospitalization, the appropriateness of the treatments provided, and the expectation that the patient will improve within a reasonable length of time. A clerk cannot make such judgments. Professional review based on standards of practice is necessary. Denials of payment based on such criteria, if challenged, can create adverse publicity. Defending such denials in court may be difficult.

By weaving cost containment into the very fabric of quality assurance, claims review by third-party payers has altered patterns of practice. Claims review forces the professional reviewer to decide if the hospital admission is medically necessary, if each specific service is medically necessary, and if the service is appropriate for the specific case and has been provided in accordance with accepted standards of practice. Not only must the overall treatment plan and services be medically necessary, but the level of intensity and the setting in which the services are provided must be necessary. In other words, there must be documentation that inpatient treatment is truly necessary, and that day treatment or outpatient care would not be adequate for the particular problem.

In general, professional care providers have objected to arbitrary limits based on a fixed number of days, deductibles, and coinsurance. These devices are not responsive to a given patient's need. The American Psychiatric Association (APA)[1] has fought vigorously for the evaluation of claims by peer review, using professionals recognized in their field.

BLUE CROSS FEDERAL EMPLOYEES PROGRAM

The influence of insurance decisions on the nature and quality of treatment is illustrated by events of 1972 with the Federal Employees Program (FEP) Blue Cross plan. Before that time, the Blue Cross FEP was viewed by psychiatrists as a model program because it provided coverage for mental illness on the same basis as for medical and surgical conditions (inpatient psychiatric hospitalization, when medically necessary, was covered 365 days a year, and there was a correspondingly liberal outpatient benefit).

In 1972, when newspaper accounts suggested that the Blue Cross FEP might, in a single year, lose some $60 million, denials of psychiatric claims increased. The

[1]Throughout this chapter the acronym APA will refer to the American Psychiatric Association alone. References to the American Psychological Association will be spelled out.

most common reason given for denials was that the treatment was primarily milieu therapy. But milieu therapy was never defined. Blue Cross had accepted recommendations made by a consultant (Lahar, 1973) that treatment modalities, such as psychosurgery, insulin coma, electroshock treatment, and high-dose drug therapy should be favored. The consultant had recommended denial of treatment approaches that stressed individual and group psychotherapy and drug therapy when used in low dosage. This latter therapeutic approach was equated with milieu therapy.

Paradoxically, the very programs the Blue Cross consultant advised against were advocated by the JCAH standards for psychiatric hospitals. Cost overruns, not quality, were the driving force for the Blue Cross denials. Cost and quality were confounded in such a way as to give the impression that the quality issue was being brought in as a device to contain costs. As might be expected, there was an intense response by both providers and patients. Denials of claims were appealed and many patients filed suit.

Political pressure, consumer protest, and professional criticism eventually reestablished a more reasonable balance, but with little clarification of the quality-versus-cost issue. Unfortunately, thousands of patients terminated treatment prematurely. To what degree providers were forced to modify their treatment practices is impossible to determine. It is undeniable, however, that the changes that occurred were the result of the denial of payment rather than of a consideration of the quality of the treatment provided.

Again, in the mid-1970s, Blue Cross FEP was confronted by accelerating costs. This time, however, the claims review technique was not used. Instead, a sharp reduction was made in benefits from the virtually unlimited inpatient benefit in any calendar year down to 60 days for the high-option and only 30 days for the low option plan. Again, there was a storm of protest, which led to discussions by top officials of the Blue Cross FEP plan and officers of the American Psychiatric Association. The fundamental distinction between the arbitrary-limit approach (exclusively cost containment) and the medical-necessity-and-appropriateness-of-care approach (utilization review to achieve cost containment and ensure quality) was articulated.

Dr. Robert Laur, Vice-President of the Blue Cross FEP plan, chose to go with the utilization review approach. This decision reflected his commitment as a physician to undertaking the difficult task of professional review rather than the simplistic administrative solution (Westlake, 1976). It was an effort to preserve quality while containing cost. The 60-day limit was withdrawn with a return to the prior benefit tempered by a $50,000 lifetime limit. An agreement was reached by Blue Cross and the APA to develop guidelines that could be used for peer review. This agreement gave added impetus to the development by the APA of a system for peer review with standards by which reviewers could judge practice.

Unfortunately, the inexorable escalation in costs of health care and its ever-increasing percentage of the gross national product continued. Again, in 1981, Blue Cross FEP reverted to the arbitrary limit approach, reducing inpatient benefits to 30 days in the low-option plan and 60 days in the high-option plan. This time, providers and consumer efforts failed to achieve a reversal. A compromise was reached that provides for catastrophic coverage, subject to peer review, in the 1984 benefit package. This catastrophic benefit is available after the beneficiary has paid $4,000. Still uncertain are the criteria for hospitalization under catastrophic benefit. Will the criteria be medical necessity, treatability, dangerousness, or what? Who will do the reviews? Most important, will the judgments of competent professional reviewers be considered good and sufficient reason to continue treatment?

CHAMPUS

Experiences over the past decade with the Civilian Health and Medical Program of the Uniformed Services (CHAMPUS) are a further illustration of the important influence of third-party payers on quality assurance. For years, the CHAMPUS program, just like the Blue Cross FEP, was viewed as a model with virtually unlimited mental health benefits, essentially the same as for medical and surgical conditions. In 1974, appalling abuses by providers treating CHAMPUS beneficiaries were discovered. These findings, coupled with serious cost overruns, led to an administrative decision to limit the CHAMPUS benefit for inpatient psychiatric treatment to a maximum of 120 inpatient days and 40 outpatient visits a year.

In response, many professional groups (the APA in particular) as well as representatives of military personnel and their families appealed for a reversal of this decision. This campaign was successful and led to the appointment by the National Institute of Mental Health (NIMH) of the Select Committee on Psychiatric Care and Evaluation (SCOPCE). Under this program, multidisciplinary peer reviews were conducted that, in short order, identified substandard facilities. Payments to substandard providers were eliminated. The SCOPCE program was a genuine quality-assurance effort that led to substantial improvement in the quality of care received by thousands of the CHAMPUS beneficiaries. Furthermore, the $100,000 spent on the SCOPCE project by eliminating substandard treatment led to savings estimated at $5 million.

The success of the SCOPCE program in fusing quality assurance and cost containment led to a contract between the U.S. Department of Defense and both the American Psychiatric Association and the American Psychological Association to develop a peer review system to deal with both quality assurance and cost containment. In developing the criteria and the system of review, the APA National Advisory Committee balanced the goals of quality assurance and cost containment. Elimination of unnecessary or prolonged hospitalization undeniably will improve quality. At the same time, cost containment can be achieved by the identification of treatment that is medically unnecessary or could be provided at a lower level of intensity and expense. The elimination of medically unnecessary hospitalization not only saves money; it also ensures that treatment will be provided in the least restrictive environment.

In the APA CHAMPUS Peer Review Program, quality issues have been dealt with directly. Reviewers do not simply say yes or no to claims. They examine the specific treatment plan, look at progress notes, and review consultants' notes—sometimes the whole medical record. Undeniably, the review of medical records is not the same as a direct examination of the patient. Nevertheless, experience has shown that quality can be assessed by a review of the medical record. Initially, the APA review system was challenged by psychiatrists. Before its implementation, the APA review system was extensively discussed with psychiatrists at the association's district branches, at the assembly of district branches, and with the board of trustees. The educational and collegial approach used by the review system has now been reasonably well accepted by those reviewed. This process highlighted the necessity for involvement and input from those who are to be reviewed.

Despite the success of the APA CHAMPUS Peer Review System, arbitrary limitations again were imposed by the Congress in 1983 (Authorization Act, 1984) as a reaction to cost overruns. The legislation first mandated a 60-day cap on psychiatric inpatient treatment. Vigorous lobbying by the APA resulted in a provision for an extension beyond 60 days subject to peer review. Extension beyond 60 days was limited, however, to severe dangerousness to self or others or to a serious medical problem. Under the revised standards, peer reviewers can do little more than review records to determine if the patient is dangerous or has a serious medical complication. Quality assurance has been severely curtailed in the CHAMPUS Peer Review Process by the imposition of this arbitrary limit.

CLAIMS REVIEW

In general, when arbitrary lengths of stay are imposed and treatment is confined within narrow parameters, quality assurance is limited. Conversely, when a third-party payer permits longer stays monitored by peer review for medical necessity and appropriateness of treatment, the opportunity for quality assurance as a part of claims review is greatly enhanced. Whether it is considered quality assurance, utilization review, or claims review, the review process has a decisive influence on patterns of practice. The usual process for the review of a claim is as follows: an individual seeks help; a provider renders a service; the provider submits a request for payment to a third-party payer; an administrative review is made to determine if the patient is eligible and the service covered; a review is conducted to determine if the treatment is medically necessary and appropriate; and the payment is either made or denied. The critical step in this review is the judgment of appropriateness of treatment. Over a period of time, pay–no-pay decisions about specific patients shape policy. Reimbursements encourage—and denials discourage—the use of specific treatment modalities.

In treating a specific patient, psychiatrists can be relied on to provide the treatment that, in their clinical judgment, is best for the patient despite an adverse decision by a third-party payer. Over a period of time, however, a continuing stream of denials dictated by the cost will inevitably modify treatment practices. Claims review, driven by cost cutting, will not improve the quality of treatment; more likely, quality will be sacrificed. For example, denials are often based on a bias unrelated to the merits of the treatment. Adolescents are seen as troublesome delinquents, alcoholics as morally weak, and drug abusers as criminals. Claims review dictated by cost alone does not reflect or reinforce contemporary state-of-the-art opinions about the quality of psychiatric care; Claims review is not intended to promote quality. It is designed to reduce costs in response to the economic needs of the payers.

If judgments are made by the administrators concerned primarily with cost containment or by practitioners who do not represent the mainstream of professional thinking, the quality of psychiatric practice will be eroded. The APA Peer Review system for CHAMPUS was developed to maintain and improve quality by using reviewers judged by their peers to be outstanding practitioners. The APA Peer Review System not only ensures that clinical judgments about the treatment of individual patients will be made by leaders of the psychiatric profession but validates their judgments by using more than one reviewer. Using outstanding psychiatric practitioners creates a dynamic system in which judgments are responsive to changing patterns of practice. These safeguards are essential to ensure that psychiatric practice will be shaped by day-to-day decisions made by competent professionals.

The APA Peer Review Program, as developed for CHAMPUS, does not rely on arbitrary limits to contain costs; rather, it relies on review emphasizing appropriateness of care. Psychiatrists are at the heart of the review system: (a) clinical judgments reflect current psychiatric concepts; (b) standards and criteria are established, refined, and modified by the psychiatric profession; (c) the standards and criteria are open to public scrutiny; and (d) the findings are used to improve psychiatric care as well as to control abuse and unnecessary expenditures.

IMPACT OF ARBITRARY LIMITS

It is impossible to judge the full impact of the elimination of benefits for long-term (beyond 60 days) psychiatric hospitalization. Before the 1983 Blue Cross FEP and the 1983 CHAMPUS cutbacks, knowledgeable professionals guessed that 20–25 million individuals in this country had financial resources through insurance and personal funds to support psychiatric hospital treatment beyond 60–90 days. In 1983, the Blue Cross and CHAMPUS cuts reduced this number by several million. Cuts by other insurers have probably reduced to less than 20% of the population those individuals who can afford private psychiatric hospital care beyond 60 days.

What will happen to those patients who need but cannot afford long-term hospital treatment? Presumably, severely ill patients who fail to respond to short-term hospitalization will be referred to public mental hospitals for treatment. But the deinstitutionalization programs of most states reduce costs by preventing long-term treatment. Perhaps the length of treatment of less severely ill patients can be shortened by more rapid diagnostic workup and more intensive treatment. Psychiatric units in general hospitals are treating more patients on a short-term basis; their programs are usually limited to symptom relief and stabilization. Perhaps some patients can be discharged from the hospital earlier if partial hospital services supported by transitional living and psychosocial programs are available. Unfortunately, transitional living and psychosocial programs have received only modest support to date from third-party payers.

In 1983, Blue Cross of Maryland funded a pilot study to determine the effect of a benefit offering their beneficiaries 75% coverage for 200 days in a halfway house. The pilot study will compare a control group and a treatment group of patients from a private psychiatric hospital (the Sheppard & Enoch Pratt Hospital, Baltimore, Maryland) and a public mental hospital (the Springfield State Hospital Center, Sykesville, Maryland). The study is designed to determine whether a benefit supporting halfway-house care will reduce recidivism. Of particular interest is whether insurance coverage for halfway houses will reduce hospital readmissions to a degree that will offset the costs of halfway-house care. If so, adding halfway coverage as a benefit might not require an increase in the premium paid by subscribers. In addition to the cost benefit, the pilot study will measure improvement to determine whether the treatment goals and improvement achieved during hospitalization are sustained.

Although the results of this study are not yet available, it will attempt to highlight two issues. First, the cutback in coverage for psychiatric hospitalization affects both private and public hospitals. Virtually all states have given a high priority to deinstitutionalization programs in their public facilities. By attempting to reduce costs, state governments as well as private insurers put presure on patients and hospitals for shorter lengths of stay. Second, it is not enough to look at just the bottom-line reduction in costs (i.e., the reduction of recidivism offsetting the cost of halfway houses or deinstitutionalization paying for community services). The therapeutic effect on patients must be assessed. Can treatment goals be achieved with shorter hospital stays coupled with a more effective use of partial hospitalization, outpatient services, psychosocial programs, and transitional living in the community, or will we reduce costs but pay a price in human suffering and disability?

The existing techniques of quality assurance have seldom dealt with the assessment of a total system. Quality assurance programs developed within the hospital setting usually address a discrete aspect of hospital care. Quality assurance studies rarely examine the interaction of the full spectrum of hospital treatment, transitional living, psychosocial programs, and community mental-health centers. Through the integration and cooperation of many diverse agencies assisted by sophisticated information systems, a state system has the potential to collect the data to conduct quality assurance studies assessing deinstitutionalization. The effect of shortened stays driven by cost reduction is difficult for the private psychiatric hospital to assess. Patients are often forced to leave the private hospital, so that it is impossible to evaluate the outcome and the long-range costs that are caused by inadequate treatment. In response to the reduced benefits that limit inpatient treatment, many private psychiatric hospitals are developing more comprehensive systems of care. At the very least, private psychiatric hospitals are strengthening discharge planning and establishing links with community resources than can provide the services needed by patients forced by cutbacks in private and public insurance programs to leave the hospital prematurely. In time, these changes may permit outcome studies that will identify the effect of arbitrary insurance limits on the quality of treatment.

COST REGULATION OF HEALTH SERVICES

In the early 1970s, encouraged by federal legislation, several states attempted to contain the escalation of health care costs by regulation of hospital charges.

Although the primary objective was the control of costs, it was recognized at the outset that hospital charges determine revenues, revenues determine resources, and resources are critical to quality of treatment. Maryland was on the cutting edge of cost regulation. The quality issues were addressed at least implicitly by legislation that created the Maryland Health Services Cost Review Commission (HSCRC).

Established in 1971, the commission was given 3 years to develop techniques for cost control that were expected

> to assure all purchasers of health care institutional services that the total *costs* of the institution *are reasonably related to* the total *services offered* by the institution; and that the institution's aggregate rates are set in reasonable relationship to the institution's aggregate costs; that rates are set equitably among all purchasers of services without undue discrimination. (1971 Md. Laws, p. 1312; italics added)

Determining the reasonable relationship between costs and services is clearly a quality assurance issue.

In actual practice, the commission has used a market basket approach to hospitals that offer comparable services. It is assumed that the market basket approach will reveal what resources are necessary to provide adequate services. Because hospitals are labor-intensive, with about two thirds of their expenditures for personnel, the determination of rates has its greatest impact on staffing patterns. Other facets of treatment are affected, however. For example, formulas have been developed to determine how many square feet per patient are reasonable for living accommodations, activities, and recreation.

Obviously, the market basket approach to staffing and other resources is, at best, a crude approach to achieving quality. In actuality, the goal of the HSCRC is to contain costs by eliminating the unnecessary use of resources, promoting greater cost-effectiveness, and increasing productivity. Particular emphasis is placed on forcing hospitals to examine whether additional resources yield benefits that are really worth the added cost. The need to retain quality is recognized but is left to each institution.

Early criticisms leveled by psychiatric hospitals were that many of the implicit standards derived from the market basket approach were applicable only to patients hospitalized for general and surgical conditions.As more psychiatric units in general were developed, these were sometimes used as the market basket but could not be directly compared to freestanding psychiatric hospitals because of differences in the patient population and the goals of treatment. In fairness, the staff of the HSCRC has been sensitive to these differences and responsive to arguments that identified the differences in patient needs. When needed, they have called on top professionals within the state system to evaluate questions of professional practice and quality.

The market basket approach is, at best, a rough benchmark for the resources needed to provide quality treatment. The individual hospital, in seeking increases above the market basket standard, has the best chance of success if findings generated from quality assurance studies can be presented. For example, a compelling argument would be a finding that more patients are being admitted who have made suicidal attempts, requiring more special observation and more staff. Even better would be a study that demonstrated how increased staffing permitting more intensive observation and more individualized intervention has led to a reduction over a period of time in suicidal attempts within the hospital.

Experience with cost review commissions has varied from state to state, partly because of differences in legislation, varied methodologies, the relationship between health associations and regulatory commissions, and the caliber of the commission's staff. There have been criticism, changing techniques, political pressure, and many lawsuits. For better or worse, this regulatory approach has been given considerable support. For example, in those states having cost review commissions, hospitals have been given waivers exempting them from prospective pricing regulations as long as the commissions keep the increase in hospital costs below the national average.

COURT DECISIONS IMPACTING ON QUALITY

Litigation, notably malpractice and negligence suits, as well as patients' rights suits, has been an important force affecting the quality of care, sometime for the better and sometimes for the worse. Malpractice suits, in many instances, have identified an individual practitioner and sometimes broad patterns of practice that constituted substandard quality. Several years ago, when such suits were infrequent, judgments relatively modest, and insurance inexpensive, malpractice suits did not have much effect on practice. As such suits increased in numbers, were more often decided in favor of the plaintiff, and had astronomical dollar awards, and as insurance premiums soared, both institutions and practitioners were forced to respond. On the positive side, institutions took remedial action to eliminate safety hazards, set up incident-reporting systems, and required continuing programs for professionals.

On the negative side, malpractice suits have caused many physicians to play it safe. Defensive medicine takes many forms: consultations, laboratory, and X-ray services may be overused; psychiatric patients may be handled in an unduly restrictive manner when there is only a remote possibility that some problem might occur; and psychiatrists may refuse to treat particularly difficult patients especially if there is potential for litigation. The end result may be that care will be given only by facilities required by the courts to provide it.

Still another serious problem is that the ultimate judgments are usually made by a jury of laypersons ill equipped to determine whether the treatment provided was appropriate and of adequate quality. The ability of attorneys to elicit a sympathetic response becomes the determining factor, rather than an objective assessment of the treatment provided. With the escalation of legal fees and the high costs of defending cases, many cases are settled out of court. These settlements may have little relationship to the quality of the treatment provided. Rather, they are based on factors such as the potential for adverse publicity, a balancing of a possible judgment of an unknown amount against a negotiated settlement, and the avoidance of the heavy legal expenditure needed even if the case is successfully defended.

One approach with some promise is the use of panels of experts to arbitrate such cases before they go to court. In general, arbitration panels give more reasoned decisions and offer some relief from the heavy legal costs. The threat of adverse publicity becomes less an issue. But malpractice litigation is hardly the optimal approach for quality assurance.

Within hospitals, corrective programs that lead to improved and safer treatment are the best solution and should become an integral part of quality assurance. In my judgment, it is somewhat unfortunate that such programs are frequently referred to as *risk management*. The implication of that phrase is that the major concern is the protection of the facility. To be sure, protection of the hospital or practitioner is an important issue. On the other hand, it should be even more important to the professional that the patient be protected from unnecessary exposure to hazardous or nontherapeutic interventions. There is no room in medicine for a decision that says in effect, we won't concern ourselves about a particular risk because it is cheaper to hire good lawyers and buy insurance.

PATIENTS' RIGHTS AND QUALITY ASSURANCE

Patients' rights litigation has been an important influence on quality of care. In the early 1970s, the *Wyatt-Stickney* case was a landmark decision. The Alabama Department of Mental Health was successfully sued by a patient who had been confined against his will within a state facility for treatment that the court deemed to be inadequate. The reasoning was that depriving a patient of his or her freedom by incarceration in a state mental hospital was a violation of the patient's constitutional rights unless he or she received an adequate treatment program. It was conceded at the time that the staffing and the physical facilities of the Alabama's state hospitals fell below what were considered reasonable standards. As a consequence, the court ordered the state to take remedial action. On the face of it, this seemed a simple and direct solution. Indeed, this judgment was hailed by mental health professionals as a forward step.

The *Wyatt-Stickney* decision, however, suddenly placed the court in the position of determining what constituted reasonable treatment. The court was obligated by its own decision to establish what amounted to a quality assurance program. Standards had to be developed not only for staffing, but for a variety of other organizational and procedural issues. A court monitor was appointed and a staff gradually assembled to monitor the hospitals. Unquestionably, the court was instrumental in causing significant improvement within the Alabama Department of Mental Health. Over a period of time, however, an adversarial climate developed, and all of the difficulties inherent in a court running the programs of hospitals by proxy became apparent. In more recent hearings on the case, the commissioner of the Department of Mental Health proposed that the department seek JCAH accreditation for state hospitals; and that such accreditation, when achieved, replace the court monitor system.

The legal arguments, the testimony of witnesses, and the extensive depositions of various experts highlighted two approaches to quality assurance. The court monitor system is essentially a regulatory approach that not only establishes standards for staffing but, in many instances, is applied to specific organizational plans as well as the details of the various treatment procedures. Monitors are stationed in the hospitals with free access to all areas so they can determine if there is compliance with their various rulings. The staff of a given hospital and even the central office staff of the state system have relatively little input into the standards or the process for enforcement.

The JCAH approach to quality assurance, by contrast, relies heavily on the development of a staff process for ongoing problem identification, analysis, correction, and monitoring of staff. It does not function as an external enforcer; it does stipulate that JCAH accreditation is contingent on evidence that a quality assurance program has been established and is being carried out. The guiding principle of JCAH quality-assurance programs is that the staff must be involved and held accountable.

Another type of litigation has occurred in those cases in which patients have insisted on the right to decide what treatments they will receive. In medicine, a patient generally has the right to refuse a treatment whether his or her decision seems reasonable or not. The problem arises when the patient's mental condition is deemed to be such that he or she cannot make a reasonable judgment. Before the many right-to-refuse-treatment cases, there was an implicit assumption that most psychiatric inpatients had limited ability to make reasonable judgments about what was in their best interest. For example, a patient who refuses to take medication by mouth might have the drug forcibly administered intramuscularly. Seldom was the question of mental competency carefully examined.

Now there have been numerous cases, some of which appear to be in conflict, that lead physicians to feel that they are taking a serious risk if they do not respect a patient's wish to refuse treatment. In an extreme emergency, when the administration of medication is necessary to calm a violently assaultive patient, there is relatively little legal risk. Much less certain is the situation in which the medication, over a period of time, can reasonably be expected to improve a chronically depressed patient who is not grossly psychotic or imminently dangerous. It is not the purpose of this discussion to deal with the many problems that arise from these decisions. The point is that these court decisions have a critical effect on the character and quality of patient care.

CURRENT REGULATORY APPROACHES

In addition to the cost controls initiated by third-party payers, federally sponsored programs, such as Medicaid and Medicare, have for 3 decades applied regulatory approaches to control costs. During the post–World War II era and the initiation of Medicare in the mid-1960s, increased access to services was emphasized. The financial incentives rewarded increases in the number of facilities, beds, and services along with improvements in quality through larger numbers and more highly trained practitioners as well as new technology (American Hospital Association, 1983).

Various efforts were made to contain costs. An early technique was to require hospitals to establish a utilization review program to be eligible for reimbursement under Medicare and Medicaid. Utilization review as applied under Medicare and Medicaid was marginally effective at best. In an attempt to strengthen this system, professional standards review organizations (PSROs) were developed to provide timely review of services to identify medically unnecessary admissions and services. On balance, the PSROs failed to meet expectations (*Medical World News*, 1977).

The utilization review approach was given new life in the Peer Review Improvement Act of 1982, which established professional review organizations (PROs). The duties and functions of these organizations include implementation and operation of a review system to eliminate unreasonable, unnecessary, or inappropriate care to Medicare beneficiaries, and to promote the quality of services.

Unfortunately, relatively few of the criteria relate to psychiatric hospitalizations; they are designed for general medical and surgical treatment. Indeed, the psychiatric hospital criteria proposed by the Maryland Foundation for Health Care (the statewide PRO) are designed only to establish standards for admission, the intensity of service required, and the need for continued stay. As yet, they do not contain standards to identify inadequate or poor treatment. Thus, the maintenance of quality will rest on the judgment of the psychiatrists treating the patient. These same psychiatrists must simultaneously resist pressures to limit the scope and duration of treatment.

Many states have passed legislation requiring a similar program for all patients (Maryland State Department of Health and Mental Hygiene, 1985). Large insurance carriers—Blue Cross and Blue Shield (Blue Cross and Blue Shield of Maryland, 1985), for example—have also established utilization review programs for their own beneficiaries. Again, the review process focuses on the medical necessity of admissions and continued stay. In Maryland, the private psychiatric hospitals and Blue Cross have established a closer collaborative relationship through this program. Psychiatrists for the hospitals and the Blue Cross staff jointly developed standards and criteria for review. At regular meetings, there is mutual feedback on the findings. This kind of dialogue provides an opportunity to balance quality considerations against cost cutting.

Amendments to the Social Security Law in 1983 established a prospective pricing system for treatment provided under the Medicare program. Prospective pricing is designed to reward hospitals that keep their costs below Medicare's present prices. At the heart of this system is a methodology known as the *Diagnosis-Related Groups* (DRGs) that will establish norms for the total costs of the treatment of a specific group of diagnostically related conditions. There was such limited experience in psychiatry with this approach that psychiatric hospitals were exempted from the DRG approach. Studies sponsored by the Department of Health and Human Services as well as those done by the National Association of Private Psychiatric Hospitals and the APA all indicated that only about 7% of the cost of inpatient psychiatry can be accounted for the DRG methodology. For that reason, the Department of Health and Human Services have continued the exclusion of psychiatric hospitals under the DRG system.

DRGs, as proposed, would have a profound effect on all psychiatric treatment but could virtually eliminate long-term hospital treatment. Early information suggests that the norms for the cost of the psychiatric conditions may be established largely through experience in the psychiatric units of general hospitals. Lengths of stay under the DRG currently used in general hospital units are exceedingly short (*Federal Register*, 1983): depressive neuroses, 9.4 days; psychoses, 10.8 days; alcohol dependence, 8.1 days; and childhood disorders, 15.4 days. Absolute cutoff dates range from 24 to 35 days. For practical purposes, treatment beyond 30 days would be eliminated.

Although doubtful, it is possible that the present norms could reasonably be applied to general-hospital psychiatric units. Symptom relief and stabilization can be accomplished for most patients, even those who are severely ill, in a relatively short period of time. Long-term reconstructive and rehabilitative treatment takes longer. The goal of such treatment is to help the patient achieve an increased capacity to tolerate stress, improved impulse control, and greater coping abilities. Such improvement can result from improved intrapsychic and interpersonal functioning. Such changes require more than the resolution of a specific conflict; they require ego growth and psychological maturation achieved through learned interpersonal and adaptive skills. The psychotherapeutic relationship essential to reconstructive, rehabilitative treatment programs takes time. Typically, this type of treatment is available in psychiatric hospitals offering intermediate to long-term care.

The prospective pricing approach utilizing DRGs has strong economic incentives to provide less intensive and shorter term care. Such a system relies heavily on the commitment of professionals to preserve quality care despite economic disincentives. Given the varied opinions about what constitutes quality psychiatric care and the inherent methodological difficulties in evaluating it, psychiatrists are concerned that the DRG approach, if applied to general hospital units, will seriously undermine quality.

It is of interest, that in the late 1970s, Medicare permitted the Maryland HSCRC to develop a guaranteed inpatient revenue program (GIR) that is similar to the DRG methodology. The HSCRC never applied this approach to psychiatric hospitals because it concluded that there was, at best, a poor correlation between diagnosis and length of stay.

At least two alternatives to the current DRG techniques have been proposed for psychiatric hospitals: first, a modification of the DRG approach has greater flexibility and takes into account variables other than diagnosis, for example, age, complication (drug and alcohol dependence, potential dangerousness), severity at admission, history of prior psychiatric illness and treatment, intensity and combinations of treatment, and discharge potential, along with availability of community-based treatment and psychosocial support; second, the development of norms for the costs on a per diem basis for programs designed to treat specific problems such as chronic psychosis and alcoholism, and specific patient populations, such as adolescent, geriatric, and so forth.

In an attempt to use variables other than diagnosis, Richard E. Gordon and Catherine K. Gordon (1983) developed a functional-level equation designed to distinguish between patients who need short-term inpatient care and those who require longer-term treatment. This equation brings together interacting variables in which aggravating stress, biomedical impairment, coping skill, directive power, and environmental support are used to determine functional levels. These authors found that, in general, patients having high scores on aggravating stress and biomedical impairment require hospital care. Of patients requiring hospitalization, those with higher scores on coping skill, directive power, and environmental support need shorter term hospitalizations than those with lower scores in these factors. Physicians and nurses have been found to be best able to make good judgments about the aggravating stress and biomedical impairment. Rehabilitation personnel are frequently experienced in judging coping skill, social workers in assessing environmental support, and virtually all clinicians in judging directive power.

It is noteworthy that the functional-level approach makes no reference to diagnosis, supporting the contention of psychiatrists that diagnosis alone does not indicate the length of hospitalization needed for a given patient. Further study and validation of this and other approaches based on similar principles are urgently needed so that viable alternatives can be proposed to the present psychiatric DRGs now used.

The National Association of Psychiatric Hospitals (NAPH), the American Psychiatric Association (American Psychiatric Association, 1985), NIMH through a contract with Macro Systems (Health Economics Research, Inc., 1985), and other organizations have conducted extensive studies on psychiatric DRGs. Using different data bases and different methodologies, all of these studies have arrived at the same conclusions.

The existing psychiatric DRGs explain a very small percentage (3%–7%) of the variance of length of stay and resource use for the treatment of psychiatric patients. Introducing additional elements into the system can improve the predictability of resource use, but only marginally.

Application of the present psychiatric DRG system would result in a systematic redistribution of resources. Specialized psychiatric facilities, larger general hospitals with discrete psychiatric units, and teaching hospitals would be penalized. There is no evidence to suggest that application of the psychiatric DRGs to all hospitals would improve the psychiatric delivery system. Indeed, in my judgment, the studies all indicate that universal application of the DRGs would cause a serious erosion in the quality of care.

Whatever approach is ultimately used, it will be essential to have quality assurance programs that can truly measure whether the pressure for cost containment erodes quality. The assumption that erosion of quality will be prevented by the clinical judgment of individual practitioners is absurd. Such an assumption ignores the complex interplay of variables outside the control of the treating physicians, as well as the economic and political dynamics of the health care system.

Although these DRGs apply only to payments under Medicare, efforts are already under way to extend them, to third-party payers. These efforts are understandable because Blue Cross and other commercial carriers fear that they will be expected to subsidize losses incurred under Medicare. We must assume that all inpatient psychiatric care will be affected by prospective pricing using DRGs. Indeed, there is likely to be a spillover to outpatient and partial hospitalization treatment.

CONCLUSION

Quality assurance must take into account therapeutic effectiveness, compliance with standards of practice, accessibility of services, costs, safety, and patient satisfaction. Forces external to the treatment process focus on cost (third-party payers, cost review, utilization review, and DRGs), on safety (the various regulators and litigation), and on patient satisfaction (right-to-treatment and advocacy groups).

The concerns about cost, safety, and patient satisfaction are legitimate. There is a danger, however, that the most important question may be ignored: Can patients get the treatment they need, when they need it, by competent professionals, in accordance with current standards of practice? These are urgent issues in these times when resources are being withdrawn. Quality care cannot be provided without adequate resources.

It is difficult to defend against cuts in funding. Court decisions, once made, can have profound effects extending over a long period of time. Regulation typically takes years to change. Experience has shown that the impact of these external forces on the quality of care is far greater than that of formal quality-assurance programs.

An awareness and understanding of the external forces that impinge on quality will help in designing quality assurance programs that can respond to and, when necessary, defend against challenges by forces outside the health care system. Even more important, properly designed quality-assurance programs can be used proactively to document the therapeutic effectiveness of psychiatric treatment and to support the increased commitment of resources needed for quality treatment of the mentally ill.

REFERENCES

American Hospital Association. (1983). *Managing under medical prospective pricing.* Chicago: Author.

American Psychiatric Association. (1985). *Findings and conclusions of the American Psychiatric Association Study & Evaluation of the Medicare Prospective Payment System—Diagnosis related groups and psychiatric patients.* Washington, DC: Author.

Authorization Act. (1984). 97 Stat. 648, Part C., Section 931.

Blue Cross and Blue Shield of Maryland. (1985). *Blue Cross and Blue Shield of Maryland Utilization Control Program for Psychiatric Facilities.* Baltimore, Maryland: Author.

Department of Health, Education, and Welfare. (1975). *Forward plan for health FY-1977–81.* Washington, DC: U.S. Government Printing Office.

Federal Register. (1983, September). *Rules and Regulations, 48*(171). Washington, DC: U.S. Government Printing Office.

Gordon, R. E., & Gordon, K. K. (1983). Predicting length of hospitalization with diagnostically related groups of psychiatric patients. *Bulletin of Southern Psychiatry, 2*(4), 131–206.

Health Economics Research, Inc., Health Data Institute, Inc., Macro Systems, Inc. (1985). *A study of patient classification systems for prospective rate-setting for Medicare patients in general hospital psychiatric units and psychiatric hospitals.* Washington, DC: Author.

HEW, too, says PSROs don't save money. (1977, November 28). *Medical World News, 18*(24), 15–16.

Joint Commission on Accreditation of Hospitals. (1976). *Accreditation manual for hospitals.* Chicago: Author.

Joint Commission on Accreditation of Hospitals. (1983). *Consolidated standards manual for child, adolescent, and adult psychiatric, alcoholism, and drug abuse facilities.* Chicago: Author.

Joint Commission on Accreditation of Hospitals. (1984). *JCAH AMH/84 accreditation manual for hospitals.* Chicago: Author.

Lahar, E. (1973). *Psychiatric care manual.* Chicago: Blue Cross Association. Maryland Laws. (1971). Vol. 627.

Laws of Maryland. (1971). (Volume 1, Chap. 627, pp. 1311–1318). Baltimore: King Brothers (State printers).

Maryland State Department of Health and Mental Hygiene, Title 10, Subtitle 07 Hospital, 10.07.01 Acute General Hospital and Special Hospitals. (1985). Emergency action regulations. *Maryland Register, 12,* Issue 18.

National Association of Private Psychiatric Hospitals. (1983). *A proposal on prospective reimbursement for psychiatric hospitals.* Washington, DC: Author.

The National Association of Private Hospitals' Prospective Payment Study, Rockburn Institute. (1985). Washington, D.C.: Lewin & Associates.

Westlake, R. J. (Ed.). (1976). *Shaping the future of mental health care.* Cambridge, MA: Ballinger.

Wyatt v. Stickney. 344 F. Supp. 73 (M. D. Alabama 1972).

12

Quality Assurance in Outpatient Psychotherapy

SHARON A. SHUEMAN AND WARWICK G. TROY

INTRODUCTION

This chapter is concerned with the application of quality assurance (QA) mechanisms to the planning, delivery, and monitoring of outpatient mental-health services. Although the content embraces the technology used to assess and ensure adequate quality of services in organized care settings, the particular emphasis is on strategies that show promise for use in the independent practice setting.

Although the role of QA in organized care settings has received considerable attention (see especially Mattson, 1984), a comprehensive review of those methods that may usefully be adapted for independent practice settings is, by and large, lacking because, until the late 1970s, external pressures or accountability to sanctioners of mental health services (third-party insurance programs, federal and state governments, self-insured programs, and consumers) had barely reached the private practice setting. Secondarily, given the essential entrepreneurial and market-driven aspects of *independent* practice (the adjective here is itself significant), few independent practitioners had developed a commitment to the systematic use of formal QA methods. Until these external demands materialized, quality assurance methodology came almost exclusively from agencies and institutions, and those procedures that came to be used in independent practice were actually variations of mechanisms developed in these larger scale organized care settings.

Changes came, of course, and the resultant pressures in the mental health marketplace caused independent practitioners to confront the reality of QA (although unwillingly). Even in the absence of these pressures, however, professional practitioners in any setting have an obligation to demonstrate that their services meet professionally acceptable standards. Hence, although acknowledging organized care settings as the prime source for the development of QA technology,

---

SHARON A. SHUEMAN • Shueman Troy and Associates, 246 North Orange Grove Boulevard, Pasadena, California 91103.   WARWICK G. TROY • California School of Professional Psychology, 2235 Beverly Boulevard, Los Angeles, California 90057.

this review focuses on independent practice and those strategies that appear to have utility for this setting.

ORGANIZATION OF THE CHAPTER

The first section of the chapter sets the context for QA in independent practice. We discuss the changes in practice and in socioeconomic climate that contribute to the need for accountability procedures (of which QA is only one manifestation). We then present definitions of quality assurance, quality assessment, and the related concept of utilization review, and we discuss the role of criterion development in QA. Definitions are followed by an overview of select strategies for monitoring the process and outcome of services rendered in independent practice. The particular QA mechanisms included here were chosen because they are both useful and capable of implementation outside an organized care setting. Much of the discussion focuses on *document-based case review* of specific episodes of care, the method that currently seems to hold the greatest promise for assessing and raising the overall quality of outpatient mental-health care.

The final sections of the chapter are based on the authors' experiences with a variety of quality assurance programs and contain our projections for the future of quality assurance in independent practice. We conclude with a brief discussion of what independent practitioners providing outpatient psychotherapy services need to know in order to confront legitimate accountability demands and to implement QA principles and technologies in their practice.

THE CONTEXT OF ACCOUNTABILITY

Within a period of approximately 10 years, mental health professionals have been forced to learn to deal with significant demands for accountability from sanctioners of mental health services. With these demands has come a corresponding decrease in the "independence" of independent practice. Although there are a number of reasons for these changes, the most obvious is the spiraling costs of health care.

Staff members in health care agencies and institutions have generally been aware of budgeting and funding constraints and bureaucratic needs for accountability that make it necessary for service providers to account for their time and their practices. Increasingly, however, these same constraints are being felt by the independent practitioner. It no longer suffices for the psychiatrist, psychologist, or social worker who is providing outpatient therapy services to place a diagnosis and a procedure code on a claim form and proceed to meet with the patient until both provider and patient decide the patient is "cured." Sanctioners of mental health services, as well as consumers of these services, are asking for more—and more specific—information, for predictions of treatment length and outcome, and, to the dismay of many traditionally trained practitioners, for specific modifications in the usual ways of delivering services (short-term, problem-focused treatment rather than open-ended, insight-oriented, "uncovering" psychotherapy).

In view of many recent findings in empirical research on the efficacy of psychotherapy, these demands may not themselves be unreasonable, regardless of whatever degree of reluctance practitioners may have about accommodating them. For example, there is increasing research support for (a) time limits with a wide variety of patients (Budman & Gurman, 1983); (b) specificity in treatment plans (Singer, 1981); (c) relatively active and direct therapy (Budman & Gurman, 1983; Malan, 1976); and (d) the use of alternate treatment modalities such as specific psychoeducational programming designed to increase coping and self-management skills (Singer, 1981). There is also a growing consensus about the importance of patient involvement in the development of treatment plans and in the evaluation of treatment (Bent, 1982; "Letting Patients Help," 1985; Mattson, 1984).

It is to these developments that the practitioners, many trained yesterday, are being asked to adapt today. And for the independent practitioner who has had neither the sanctions nor the support of an agency to cajole, force, or induce him or her to use the new strategies, who has not learned the relevent concepts or methods in graduate training, and who may see the new expectations as an intrusion or infringement on his or her usual way of doing business, the adjustment may be exceedingly difficult. As a consequence, the real challenge in implementing a quality assurance system is less technical (because the technology, as we shall see, is not conceptually complex) than it is "political": the challenge of ensuring acceptance among mental-health service-providers of the basic tenets of accountability.

The context depicted above helps explain why quality assurance in independent practice has come to assume, sadly, a somewhat adversarial aspect. The overt demands for QA outside organized care settings have come from third-party payers, most significantly from CHAMPUS, the Civilian Health and Medical Program of the Uniformed Services (Claiborn, Bisken, & Friedman, 1982). Many service providers have seen these demands as peremptory, as the third-party payers' saying, in effect, "Prove to us that what you are doing is worth our paying for!" The general posture for the third-party payer has too frequently been seen as one of coercing treatment information from the reluctant, sometimes hostile, provider. For the provider, the response has been to give as little case-management information as he or she could provide without actually threatening reimbursement itself.

These adversarial overtones have had an effect on the development of QA procedures. That is, to a certain extent, developers of the systems have looked for "foolproof" procedures, ones that are likely to reflect inconsistencies in the providers' reporting of treatment. Mental health providers, on the other hand, have been reluctant to participate actively in the development of new systems that might, because they would more directly reflect practitioner input, be more "provider-friendly" (Shueman & Penner, 1987).

WHAT IS QUALITY ASSURANCE?

We shall use the term *quality assurance* to refer to *any formal activity implemented within the service delivery system to improve the outcome of mental health care.* (It should be noted that, in a situation where care is judged to be of adequate quality, the goal would be to *maintain* rather than *improve* outcome.) The qualifier "within the service delivery system" allows us to eliminate the consideration of criteria related to credentialing, such as licensing or certification and type and level of professional education, indicators that may be seen to reflect quality that, although significant, are peripheral to the thrust of this chapter. Ensuring quality via credentialing is analogous to evaluating the quality of a library on the basis of the number of books or credentials of the librarians: Although not trivial considerations, they do little to advance directly the assurance of quality.

Quality assurance ranges from ensuring accessibility to care to improving the ability of providers, individually and collectively, to adhere to the procedures (standards) thought to be correlated with better outcome. The foremost theoretician on QA in health care, Avedis Donabedian (1980), defined the quality assurance process as consisting of:

- Activities intended to determine whether standards of care are being adhered to (quality assessment)
- Additional activities intended to eliminate any deficiencies identified during quality assessment

As will become clear, most "QA" efforts that have occurred in the independent practice setting are more appropriately called *quality assessment* (the first set of activities, above). That is, such efforts do not usually extend to educational or remedial activities directly intended to raise the level of performance of providers (the second set, above). Agencies frequently have or can acquire adequate material and human resources to enable them to develop programs for using quality assessment findings. This may also be true of integrated group practices. It is the unaffiliated (solo) practitioner providing outpatient psychotherapy services, however, who, even if he or she has the motivation, may have neither the expertise nor the organizational structure to implement such programs.

UTILIZATION REVIEW

Closely related to QA is the concept of *utilization review* (UR), the *collection of activities used for monitoring the way in which the available service resources are being allocated*. UR concerns can be articulated in the two questions:

- Is treatment *necessary* for this person at this time?
- Are the services being provided in this case *appropriate* for the specified problems?

UR programs are generally undertaken because the supply of services or the money to pay for the services is limited, and *judgments* need to be made about who should get how much of what treatment. The content area of utilization review embraces, to a great extent, cost-effectiveness issues (Levin, 1983). For example, a UR activity might be aimed at determining which patients coming to an agency will be placed in therapy groups rather than being assigned to individual therapy. The decision involves both cost considerations (i.e., group therapy is less expensive than individual therapy) and efficacy considerations (i.e., it is generally agreed that certain types of clients may benefit more from group than from individual therapy).

It is commonly argued that quality assurance activities need to be implemented along with UR programs. The rationale is that, when cost considerations influence the allocation of services (as is true in UR activities), the quality of services is invariably compromised. The implication is that the best care can occur only in an environment where the patient and the service provider make an independent decision about the necessity for and the continuation of treatment.

A somewhat related argument is that, in the workaday world of health care, programs referred to as quality assurance are no more than disguised utilization-review programs. This particular position holds that third-party payers and government agencies are concerned about cost almost to the exclusion of quality, and that what they do is motivated by cost-containment considerations. On the other hand, calling an activity QA makes it more acceptable to the mental health professionals who must assist in its implementation.

Despite these arguments, experience has shown that it is difficult to separate quality and cost considerations. Donabedian (1980) maintained that cost is an important dimension of quality if the needs of society are a consideration. More important, perhaps, is the empirical evidence that more expensive mental-health care is not necessarily better care (Schultz, Greenley, & Peterson, 1983).

THE ELEMENTS OF QUALITY ASSURANCE

Donabedian (1980) delineates three treatment parameters with which QA activities are concerned: *structure, process,* and *outcome.* He defines *structure* as the *human, physical, and financial resources necessary to provide care: process relates to what goes on within and between providers and patients;* and *outcome* refers to the *consequences of treatment.*

Table 1 provides, for each of these parameters, representative questions that would be addressed in efforts to assess quality, as well as the mechanisms commonly used to answer the questions.

The focus of the discussion that follows is on process and outcome, those aspects that have most frequently been the focus of quality assurance efforts in mental health care. These particular parameters are directly relevant to independent practice and are aspects of case management about which sanctioners such as third-party payers usually seek to obtain information.

Process. When the focus of quality assurance activities is on the process of health care, there is an implicit assumption that there are certain operations that, if done by service providers, will result in a better outcome. The concern is the provider's case-management behaviors: what he or she *does* in therapy. The assessment of process in treatment requires an analysis of the nature of the operations and how they may be translated into *criteria,* or *standards of performance,* against which the adequacy of the provider's work with the client may be measured.

Outcome. To focus solely on process does not provide sufficient information to enable questions relating to quality to be adequately answered. It is necessary in addition to focus on outcome, what happens "down the road" as a result of the treatment. It is conceivable, for example, that a provider may satisfy the criteria of pro-

TABLE 1. Examples of QA Questions and Methods:
Structure, Process, and Outcome of Treatment

Treatment parameters of QA	Questions asked	Evidence
Structure	Is facility accessible (i.e., adequate parking; on bus line)?	Survey of patients
	Are service providers well qualified?	Review of credentials
	Are services available (in a timely manner)?	Survey of patients Record review
Process	Do providers deliver services according to agreed-upon criteria?	Record review Observation of treatment process Survey of patients
Outcome	Do patients improve as a result of treatment?	Objective measures (pre- and posttreatment) Patient self-report Significant-other report Provider evaluation of outcome

cess (i.e., exhibit the right behaviors in managing a particular case) but may still not succeed in helping the patient achieve the desired changes; in other words, the provider may fail to meet the preestablished criteria for the outcome of interventions. Hence, when the focus of QA is on outcome in mental health treatment, there is an assumption that there are certain client changes in thinking, feeling, and behaving that can be specified in advance and that should come about as a result of treatment. An assessment of outcome requires that there be a definition of the desired outcomes (goals) and an agreement on what indicators (criteria) would reflect the attainment of the goals.

THE USE OF CRITERIA

Criteria are standards or rules that are used to make decisions about treatment quality in both the process and the outcome phases. It is useful to consider two types of criteria:

1. Those that are objective, public statements, the application of which requires little professional judgment
2. Those that are more subjective and that usually involve judgments of professionals[1]

Examples of the former include such prescriptive statements as that the treatment plan for a child patient should normally include regular consultations with the parents (a process criterion), or that the patient would achieve a 50% reduction in the frequency of fights with her husband (an outcome criterion). Whether or not the

---

[1]For an exhaustive but highly technical discussion of criteria for quality assurance in health care, see Donabedian (1982).

parent is consulted or the frequency is decreased is easily determined by looking at the written record or by asking the provider or the patient. Thus, these can be seen as explicit, or objective, criteria.

An example of a subjective criterion would be the statement that, by the 10th session, there should be evidence that reasonable progress has been made in treatment. The judgment of what is "reasonable progress" would need to be made by a professional reviewer whose own experience provides the basis for the judgment. To the extent that this reviewer's judgments are based on his or her expert knowledge of the science of practice, then, the judgments may approach objectivity. On the other hand, one would certainly expect this person's indicators of good care to be somewhat idiosyncratic, that is, to reflect his or her interpretation of professional consensus or the research literature. Clearly, such a process is not "objective"; neither, however, is it entirely "subjective." It is perhaps more accurate to depict the process of using the judgments of professional peers as fulfilling the demands of *intersubjectivity.*

Donabedian (1982) stated that criteria that depend on professional judgment require much time and training to implement; they are not amenable to large-scale QA programs, or to programs that require a great deal of control. It is in these situations that the more easily applied objective criteria should be used, but only as "screening" mechanisms, not as full representatives of quality. In other words, explicit or objective criteria are useful for making relatively crude discriminations regarding quality of care, whereas finer discriminations may require the closer scrutiny of other professionals using intersubjective criteria. It is according to these principles, in fact, that most review programs implemented by third-party payers operate. These payers are seeking to increase the sophistication of their criteria sets so that greater numbers of decisions can be made without incurring the expense of using the judgments of professionals. (From the independent practitioner's point of view, this thoroughly pragmatic approach has some negative aspects, but consideration of this issue is beyond the scope of this chapter.)

For the general mental-health practitioner to restrict his or her QA efforts exclusively to objective criteria would still do much for the advancement of QA in the independent practice setting. The development of formal treatment plans, for example, is an objective QA criterion, and one that is relatively simple to implement. Although the use of formal treatment planning does not "assure" quality, when it is followed by the use of a subjective criterion (such as peer review), QA does begin to assume a harder edge. Thus, certain QA activities in the independent practice setting may be simple in and of themselves, but when they are sequenced with other, more complex QA activities (peer review), the assurance part of QA is given real substance. We will return to this issue in the last section of this chapter.

THE PROBLEM OF SPECIFICITY

The most significant problem facing developers of QA programs is the lack of specificity displayed by providers of mental health care in the description of their clients' problems and goals (desired outcomes). This is the same problem that has traditionally plagued those involved in formal research designed to investigate the efficacy of psychotherapies, and it is no less a problem for those attempting to improve outcome.

The difficulty stems in part from a basic and long-standing difference of opinion about the proper focus for psychotherapeutic interventions. The source of this contention is firmly grounded in the notion of theroretical orientation. For example, problems may be defined, on one hand, behaviorally (that is, operationally) or, on the other, in terms of intrapsychic functioning. The goals under these two approaches would most likely be, respectively, very specific (e.g., reducing the frequency of fighting between spouses) and very general (e.g., helping the client deal with his or her internalized rage). Indeed, if a psychodynamic therapist and a cognitive behavioral therapist evaluated an individual for treatment, one would expect their assessment of the person's problems and the appropriate goals to differ significantly.

It is apparent that, if providers cannot delineate in clear, observable terms what it is they are trying to accomplish in therapy, little can be done to ensure its achievement. Independent practitioners thus have the significant challenge of identifying ways of expressing problems and outcomes that make the process of psychotherapy more amenable to documentation and evaluation.

We suggest, then, that the most significant contributions to quality assurance in outpatient mental-health care have been related to procedures for *documenting* treatment, that is, for conceptualizing and expressing what is being done in the therapeutic relationship (the contract). Such documentation procedures have provided the mechanisms for breaking through the semantic barriers that those of particular theoretical orientations tend to erect around themselves, thereby making mental health services less understandable (and hence less accessible) to consumers, the public, and third-party and other sanctioners. The documentation systems described below provide means of making mental health services more objective and amenable to evaluation (Singer, 1981), a prerequisite in any approach that aspires to incorporate tenets of accountability.

THE PROBLEM-ORIENTED RECORD

Traditionally, judgments about the quality of mental health care (most of which have been restricted to inpatient settings) have been centered on the psychiatric diagnostic system, the revised third edition of the *Diagnostic and Statistical Manual* (DSM-III-R), a system that evolved directly from the medical model of health care delivery (American Psychiatric Association, 1987). Criteria defining necessary conditions for admission to, or continuation in, treatment have typically been based on these diagnoses (American Psychiatric Association, 1985). Empirical findings suggest, however, that this type of system has major shortcomings for mental health care as well as for general medical care (Horn & Horn, 1985; S. Horn, personal communication, April 1985). In particular, the system generally requires very little, if any, client information relating to the level of functioning, the support system, or the history of mental health problems—all of which have been shown to have significant implications for treatment, day-to-day case management, and prognosis.

In the outpatient setting, especially in independent practice, not only does the diagnosis seldom contribute significantly to the understanding of a patient's problems, it also frequently confounds understanding. For example, the practitioner may use a false or inaccurate diagnosis to "protect" the patient from the stigma of the accurate but more serious diagnosis; he or she may select a diagnosis because a certain condition is known to be covered by the particular third-party plan.

Much of the progress that has been made in inpatient and outpatient mental-health QA since the mid-1970s is to a great extent built on the early work of Weed (1964) with the problem-oriented medical record (POR). Originally thought to be helpful in multidisciplinary settings because it replaced the previous discipline-specific set of records, the POR is a patient-focused document requiring specific problem identification and intervention planning.

Rather than focusing on diagnosis, the POR focuses on *problems* (or dysfunctions) and intervention strategies aimed at the amelioration of the particular problems of deficits. The common adaptions of this unified record for mental health care also require that service providers speak a "generic" language (the language of behavior) rather than the language of the particular psychotherapeutic orientation.

These two characteristics of the POR—the focus linking specific dysfunctions and interventions, and the requirement that the document be written in "generic" terminology—have contributed significantly to recent progress in quality assurance in independent practice settings. The POR has provided practitioners with a means of *specifying* what they are doing, thereby giving them and others the vehicle for determining whether what they are doing is effective. In other words, it is the POR's degree of specificity that simplifies greatly the task of deciding to what extent case management is appropriate. Thus, in simplifying the application of criteria to the service being delivered, the QA process is greatly facilitated. That the POR model permits cross-disciplinary exchange is a significant additional benefit.

DOCUMENT-BASED REVIEW OF TREATMENT

Currently, outpatient QA rests on document-based review of a problem-oriented treatment plan. There are those who would argue that a desirable alternative is a face-to-face review, in which the provider of services meets with the person evaluating the care. Indeed, if one considers the educational potential of face-to-face interactions, the process seems promising. On the other hand, it would appear that there is valuable educational potential in requiring service providers to develop treatment plans. In fact, anecdotal evidence from service providers and peer reviewers involved in the peer review programs of the American Psychological Association, the American Psychiatric Association, and the National Association of Social Workers attests to the professional development function of providers' completing such treatment plans. In addition, it is costly and time-consuming to conduct face-to-face reviews involving professionals; this is true in organized care settings and obviously even more true in independent practice.

The third reason for favoring document-based review is related to the argument frequently advanced in defense of face-to-face review, namely, that it is impossible to explain the complexity of the therapeutic situation in a brief document. There is some empirical evidence, however, that a professional caregiver who cannot produce a coherent treatment plan would have difficulty providing coherent treatment (Mattson, 1984; Sechrest, 1987). Finally, the interpersonal dynamics of the face-to-face review process may result in either the service provider's using personal influence to support his or her treatment strategy or the meetings taking on the characteristics of the traditional case conference. Neither occurrence would necessarily contribute to improved quality of care.

THE TREATMENT PLAN

It is generally agreed that a treatment plan needs to contain (a) a clear description of the patient's problems (expressed in terms of difficulties in functioning in various areas of life) and (b) specific intervention strategies (something more specific than "individual psychotherapy"). For the former, what is required are specific goals, the accomplishment of which would lead to a diminution of the problems; the latter would require strategies defined to effect the desired changes. If the review process occurs at a time other than at the beginning of treatment, then it is helpful, if not necessary, to indicate in the documentation the extent to which the treatment has resulted in progress. The rationale for requiring functional descriptions of mental "illness" is that, no matter how problems are conceptualized (as intrapsychic, behavioral, or any other), one would expect them to be manifested as dysfunctions in some areas of the patient's life. Thus, we expect to find in an adequate documented treatment plan at least the following:

- Statements of client problems
- Treatment goals expressed as functional objectives
- A set of specific interventions
- Statements describing progress toward goals

Variations of this treatment plan have been endorsed by four major professional associations (the American Psychiatric Association, the American Psychological Association, the National Association of Social Workers, and the American Nurses' Association) and, in the authors' experience, constitute the most common format for conducting utilization review or quality assurance activities. Table 2 contains examples of problem and goal statements expressed in terms of patient functioning such as those one might find on a POR as well as other statements expressed in the more traditional language of specific theoretical orientations. As can be seen, the statements expressed in terms of client functioning can be understood without familiarity with the language of a particular theoretical orientation. In fact, the statements could—and, we would argue, should—be understood by laypeople.

A major purpose of requiring treatment plans of the POR format is to obtain a degree of standardization in the documentation of treatment. Without this modest standardization, reasoned judgments about the quality of care are next to impossible; that is, the application of criteria is very difficult. Experience has shown, however, that mental-health service-providers are often ill equipped to provide this kind of documentation. Those responsible for QA programs, therefore, have sought

TABLE 2. Examples of Problem and Goal Statements

Problem statements expressed as observables:
  Patient gets 2–3 hours' sleep per night.
  Patient is isolated and has no friends.
  Patient repeatedly provokes fights with boss.
Goal statements expressed as observables:
  Patient will report an improved ability to sleep.
  Patient will report increased personal comfort and show reduced incidence of arguing.
  Patient will follow up on five job-availability notices by submitting résumés and calling employers.
Problem statements expressed in language of theoretical orientation:
  Patient has tendency to project anger onto others and to become involved in angry confrontations.
  Patient retains unresolved anger toward authority figures in his/her past.
  Patient typically practices splitting: viewing others as either idealized or degraded.
Goal statements expressed in language of theoretical orientation:
  Patient will report being better able to understand the relationship between guilt and displaced
    aggression.
  Patient will come to terms with unresolved anger toward his/her parent.
  Patient will learn to integrate self- and other-perceptions to allow for more consistent view of
    others.

---

to identify additional, clearly delineated mechanisms for obtaining relevant information about the patient and the treatment, mechanisms that would depend less on the provider's ability to document adequately. The types of data collection mechanisms that have, in fact, been used to good effect fall into two general categories: level-of-functioning and severity scales, and structured procedures for defining problems or goals.

LEVEL-OF-FUNCTIONING AND SEVERITY SCALES

These two types of scales or measures represent opposite ways of viewing how well a person is functioning in various areas of life. Level-of-functioning scales focus on how well the person can conduct the activities of everyday living (a more-or-less positive, "wellness" perspective), and severity scales focus on how serious the dysfunctions are (a more negative, "illness" perspective). Commonly used scales range from highly standardized measures that have been subjected to careful validation and reliability testing to nonstandardized instruments having little more than face validity.

Typical examples of the range of standardized scales include the multidimensional Severity of Illness Scale developed by Horn and her colleagues at Johns Hopkins University (Horn & Horn, 1985), the more well-established Global Assessment Scale (Endicott, Spitzer, Fleiss, & Cohen, 1976), and the Level of Functioning (LOF) Scale developed in the Ravenswood Community Mental Health Center in 1978 (Sherman & Gomez, 1982). In general, scales of this type provide some descriptive information (often in vignette format) that is to be used by the rater in assigning a numeral indicating the level of the client's functioning or severity of dysfunction. The descriptions have been carefully developed to reflect factors that have been shown empirically to be significant in determining a person's need for treatment. The instruments guide the service provider in making an assessment of the patient's ability to function, and they help ensure that the ratings for a given client will be similar across a spectrum of different providers.

Scales that are less technologically developed include, for example, simple severity (and progress) ratings of functioning in several areas of the client's life (home/family, job/school, interpersonal relationships, bodily functions, personal comfort, and protection of self or others). This type of scale was developed by the American Psychological Association and is used in the nationwide CHAMPUS Peer Review Program (Stricker, 1983). Severity ratings range from 1 to 5, with 1 reflecting no problem (least severe) and 5 reflecting significant dysfunction. Progress ratings also range from 1 (no progress) to 5 (complete progress—no remaining problem). In the author's experience, this particular type of scale is very commonly used in relatively small-scale QA activities, such as would occur within a mental-health preferred-provider organization.

A similar scale is found in the multiaxial diagnosis system of DSM-III-R (American Psychiatric Association, 1987). Axis V (Highest Level of Adaptive Functioning Past Year) is thought to have considerable prognostic significance because "usually an individual returns to his or her previous level of adaptive functioning after an episode of illness" (p. 20). In a QA review program, however, this same scale can be used as a rating of *current* level of adaptive functioning. Although it is not a standardized measure in the sense of the Global Assessment Scale, it has the advantage of having carefully developed descriptive anchors, and (perhaps more important) it is a familiar instrument to many independent service providers.

PROBLEM AND GOAL CHECKLISTS

The second type of instrument, and one that is sometimes combined with the rating principle discussed above, provides standardized descriptions of the patient's problems or goals. That is, rather than depending on the service providers' ability to create a "free-form" description of the patient's dysfunctions, the provider is given lists to choose from. These lists may be developed intuitively and by consensus (for example, by a group of practitioners considering the range of problems that they usually confront or the types of goals that they usually set with clients) or through a formal methodology such as the factor analysis of a series of very specific problems or goal statements.

As with severity scales, problem or goal lists may also range from relatively global to highly specific. An example of a simple global problem list would be the six areas of life functioning developed for the CHAMPUS Peer Review Program. An example of a relatively specific instrument would be the well-established Mooney Problem Checklist.

A commonly used variation on the above is Goal Attainment Scaling (Kiresuk & Sherman, 1968). In this method, the therapist and the client specify the problems that will be chosen as the subjects of the therapy and, for each problem, develop a set of possible outcomes that may result from treatment. Each of the possible outcomes is assigned a numeral from (for example) −5 to +5, with −5 reflecting the worst possible outcome (usually a decrease in the level of functioning) and +5 reflecting the best outcome that the client could have. The rating of zero reflects the pretreatment level of functioning. This method not only allows the client and the therapist to work together toward the clear identification of problems and the establishment of goals, but it also provides a quantitative outcome measure that can be used in a QA activity. Finally, the technique permits aggregating across clients within a program to yield objective indices of efficacy for incorporation in program evaluation research.

The types of instruments and methods mentioned above accomplish several things beyond conveying meaningful information about treatment that might not be found on the treatment plan. First, by causing the patient and the provider to focus more clearly on the nature of the specific problems and the desired outcomes, the procedures facilitate the case management process. Second, they provide a ready mechanism for evaluating the outcome or progress resulting from interventions aimed toward specific goals. Finally, in the eyes of external sanctioners, they objectify a commitment to accountability.

THE FUTURE OF QA IN INDEPENDENT PRACTICE

Two parties will most influence the future of QA in independent practice: those providing the services and those purchasing the services, whether they be direct consumers (patients), third-party payers, employers, or governmental agencies. To date, the greatest influences have come from the latter groups, primarily third-party payers. It remains to be seen whether service providers will assume some of the responsibility for shaping the QA programs of the future. No matter who shapes them, however, their specific forms will depend on three factors: (a) empirical research on the efficacy of psychotherapy; (b) the increasing use of computers and automated data-processing systems; and (c) developments in professional training. We will discuss these three issues and relate them to those who will, in turn, influence, and be affected by, the climate of accountability in mental-health service-delivery: the parties who provide and those who consume. It is clear that, whatever their particular manifestations, QA mechanisms in general will shape the work of the individual provider every bit as much as they have shaped the form and structure of service delivery in large organized care settings.

RESEARCH ON EFFICACY OF TREATMENT

As we have seen, there has been significant consistency between what many third-party payers are demanding and what the research literature suggests is good mental-health practice. In particular, it appears that the development of treatment plans, specificity in problem identification, therapeutic time limits, more activity on the part of the therapist, and patient participation in treatment planning are some of the factors that may be related to good case management and hence to efficacy. It is obvious that demands from payers and consumers for service approaches with these characteristics will not diminish in the near future. Instead, there will be an increasing number of settings in which these types of provider behaviors will be required as those involved in mental health care seek improved fiscal control and evidence of increased accountability from providers. The preferred provider organization (PPO), the health maintenance organization (HMO), and the independent practice association are currently the most common examples of these "highly accountable" organizations. There will be others, however, as the new principles of service delivery expand to affect all mental health practitioners and (a trend of great significance) as the results of research are increasingly used in designing benefits and service delivery systems.

We anticipate that research will serve to identify a wider range of patients for whom highly structured, brief therapy is appropriate. In addition, it will generate more sophisticated ways of specifying problems, goals, and treatment strategies and of defining level of functioning or severity of problems. Progress in these latter areas will yield more precision in the treatment process and more effective ways to monitor treatment.

This increased precision and the ability to monitor treatment will result in attempts to develop treatment protocols in which the appropriate (i.e., most efficacious) modes of intervention for particular problems will be identified. Increased specificity will also allow the development of more sophisticated sets of objective criteria, which may be used in large-scale UR and QA activities and which may be adapted for computer applications.

CONTRIBUTIONS OF THE COMPUTER

What follows is a depiction of a fantasy frequently expressed by both third-party administrators and administrators of mental-health service-delivery systems: All service providers belonging to a particular delivery system or serving beneficiaries of a particular third-party program will be linked to a single computer network. The provider will develop a treatment plan, enter it into the terminal in his or her office, and receive rapid feedback concerning the appropriateness of the plan as well as a decision relating to reimbursement. The decisions may be made *by the computer* on the basis of explicit criteria (protocols) defining what type of treatment of what length and intensity is appropriate for what type of problem. In certain cases for which the protocols are inadequate, the decision may be made by a professional reviewer who reviews the plan from an office terminal and enters an opinion (judgment) into the system. All the while, the system aggregates data on patients, providers, and services that may be useful for making quality-of-care and management decisions about benefits and services. For example, the data base may include routine data on treatment outcome that would allow decisions to be made about the most appropriate providers for particular types of patients or particular problems. It may also allow the identification of especially cost-effective providers, those patients who appear to be overusers of mental health services, or problems that seem to be especially prevalent among specific subgroups of patients.

Although the above depiction may seem somewhat futuristic, it does afford a glimpse of aspects of the QA process that are available *now*. It is merely an extension and integration of elements of QA that currently prevail. It is the responsibility of the independent practitioner to become familiar sufficiently with current concepts so that the pace of future developments will not leave him or her professionally disadvantaged and consumers inadequately served. There is no doubt that there will be computer networks, there will be treatment protocols, and service providers will become increasingly accountable to systems outside their consulting rooms.

THE IMPORTANCE OF TRAINING

The importance of training graduate students in the mental health disciplines explicitly in the principles and practices of accountability has been addressed in detail elsewhere (Shueman & Troy, 1982). Our experiences and those of professional colleagues engaged in training (Bent, personal communication, August 1985; Barr, Wolstadt, & Kinast-Porter, 1979) have borne out the postulate advanced in the Shueman and Troy article, namely, that students who are introduced to the concepts and technology of accountability and who are educated about its importance are more likely to believe that being accountable in formal ways is both a natural and an appropriate aspect of a service provider's professional role.

To the extent, then, that the attitudes of therapists influence patients and clients, one might also expect patients to be more accepting of the demands for accountability. Providers will be able to ensure that their clients will understand fully that treatment that is underwritten by third parties carries with it certain obligations for both patient and provider, and that these obligations are in no way sinister: they may even serve to advance the therapeutic outcome.

The ultimate goal of introducing accountability into professional training should be to change the professionals' perception of QA activities, from requiring unexamined and reactive compliance to contributing to good-quality care for the consumer as well as financial stability for the service provider. It is even to be imagined that, with this change, more mental health professionals will find themselves anxious to contribute to research in QA procedures and to their collegial incorporation into the everyday world of the service provider.

A QA ROLE FOR THE INDEPENDENT PRACTITIONER

We conclude the chapter with a brief review of those QA activities that are both appropriate to the mission of the independent provider of outpatient mental-health services and capable of implementation within the constraints of an independent practice. The review summarizes QA mechanisms previously considered and concludes with a brief analysis of the advantages to consumer and provider alike of familiarity with the context and substance of QA in mental-health service-delivery.

EASILY IMPLEMENTABLE QA MECHANISMS

The reader is referred to an observation made earlier in this chapter to the effect that, if the independent provider did little else in the way of QA other than to focus on a few activities commonly explicated as objective criteria, the cause of QA and accountability in mental-health service-delivery would be significantly advanced. We set out in Table 3, therefore, a selection of such activities, whose implementation by practitioners is neither difficult nor particularly consuming of resources. The activities are subsumed within a classification scheme whose parameters are explained in this chapter, and they are phrased prescriptively, as befits their status as QA criteria.

QA and the Professional Role

To hold oneself accountable and to lead an examined life are only two of the tenets that many agree embody the notion of what it is to be a professional. A *license* may be a credential that allows one to function professionally in specific ways; the term in its modern, technical usage does not, however , connote *license* in the sense of unbridled freedom. Rather, it invokes a set of *obligations* frequently incorporated in some normative code of professional conduct (such as in the *Ethical Principles of Psychologists* of the American Psychological Association, 1981). There are, however, at least three other reasons that the truly professional caregiver should hold himself or herself "accountable." First, the provider's services are sanctioned by society (indirectly), by the consumer (directly), by the third-party payer (directly), by the statutory licensing authority (indirectly), and by the caregiver's own profession (in-

TABLE 3. Quality Assurance Activities in Case Management

Intervention phase	Activities incorporated as objective (explicit) quality-assurance criteria
Process of therapy	1. Develop formal *treatment plan* in problem-oriented record format.
	2. Incorporate treatment plan in a formal *service contract* with the client.
	3. Ensure the direct *participation of client or patient* in development of treatment goals and in review of progress.
	4. Establish *periodic review* of service contract, sessions in which progress toward goals is formally reviewed with client or patient in context or preestablished, mutually negotiated criteria for desired change.
	5. Systematically and explicitly use *treatment plan as case management tool,* whereby progress toward goals is monitored.
	6. Regularly *consult professional colleagues* with the focus on active case management rather than diagnostic or etiological issues.
	7. Develop competence in *strategic, focused methods of intervention.*
Context of therapy	1. Become informed about *new developments in mental-health service-systems* (HMO, PPO, EAP,[a] etc.).
Outcome of therapy	1. Following termination, seek client's or patient's compliance in completing *questionnaire of posttreatment status* and *degree of satisfaction.*
	2. Make use of available *standardized and nonstandardized* scales for assessing level of client functioning, pre- and posttreatment.

[a]Employee assistance programs.

directly). Without the "blessing" of such groups, the provider will not long be able to continue to offer services. There is a clear obligation. Second, caregivers whose mode of professional conduct is designed to make explicit their direct accountability to their clients or patients serve this constituency well. Third, accountable providers tend to be effective in their search for techniques that embody this sense of obligation.

The use of QA mechanisms is, to be sure, somewhat removed from the lofty, aspirational tone of the previous paragraph. Nonetheless, familiarity with and appropriate use of such procedures do reflect a commitment to certain functional aspects of accountability in the provision of mental health services. The evidence is mounting that an independent provider who is able to make appropriate use of some of the QA mechanisms listed in Table 3 will indeed serve his or her client or patient constituents well, that is, will be *efficacious*. Thus, to the extent that the systematic and examined use of QA procedures eventually becomes part and parcel of the independent provider's professional armamentarium, it is good professional practice. To the extent that such mechanisms are precisely the kind of professional behaviors that third-party payers require for reimbursement, it is, at the very least, good business.

REFERENCES

American Psychiatric Association. (1985). *Manual of psychiatric peer review* (3rd ed.). Washington, DC: Author.

American Psychiatric Association. (1987). *Diagnostic and statistical manual of mental disorders* (3rd ed, revised). Washington, DC: Author.

American Psychological Association. (1981). *Ethical principles of psychologists.* Washington, DC: Author.

Barr, D. M., Wollstadt, L. J., & Kinast-Porter, S. (1979). Will physicians learn to like quality assurance? The effect of a curriculum on medical student attitudes. *Journal of Medical Education, 45,* 649–650.

Bent, R. J. (1982). The quality assurance process as a management method in psychology training programs. *Professional Psychology, 13,* 98–104.

Budman, S. H., & Gurman, A. S. (1983). The practice of brief therapy. *Professional Psychology: Research and Practice, 14,* 277–292.

Claiborn, W. L., Biskin, B. H., & Friedman, L. S. (1982). CHAMPUS and quality assurance. *Professional Psychology, 13,* 40–49.

Donabedian, A. (1980). *The definitions of quality and approaches to its assessment.* Ann Arbor, MI: Health Administration Press.

Donabedian, A. (1982). *The criteria and standards of quality.* Ann Arbor, MI: Health Administration Press.

Endicott, J., Spitzer, R., Fleiss, J. L., & Cohen, J. (1976). The Global Assessment Scale: A procedure for measuring overall severity of psychiatric disturbance. *Archives of General Psychiatry, 33,* 766–771.

Horn, S. D., & Horn, R. A. (1985). *A study of the reliability and validity of the severity of illness index.* Baltimore, MD: Johns Hopkins University, Center for Hospital Finance and Management.

Kiresuk, T. J., & Sherman, R. E. (1968). Goal attainment scaling: A general method for evaluating comprehensive community mental health programs. *Community Mental Health Journal, 4,* 443–453.

Letting patients help plan treatment could reduce hospitalization, costs. (1985, August). *Psychiatric News,* p. 21.

Levin, H. M. (1983). *Cost-effectiveness: A primer.* Beverly Hills, CA: Sage.

Malan, D. H. (1976). *The frontier of brief psychotherapy.* New York: Plenum Press.

Mattson, M. R. (1984). Quality assurance: A literature review of a changing field. *Hospital and Community Psychiatry, 35,* 605–616.

Schultz, R. I., Greenley, J. R., & Peterson, R. W. (1983) Management, cost, and quality of acute inpatient psychiatric services. *Medical Care, 21,* 911–928.

Sechrest, L. B. (1987). Research on quality assurance. *Professional Psychology: Research and Practice, 18,* 113–116.

Sherman, P. S., & Gomez, M. (1982). Quality assurance uses of level of functioning ratings. *Quality Review Bulletin, 8,* 40–48.

Shueman, S. A., & Penner, N. R. (1987). Administering a national program of mental health peer review. In G. Stricker & A. R. Rodriguez (Eds.), *Quality assurance in mental health.* New York: Plenum Press.

Shueman, S. A., & Troy, W. G. (1982). Education for peer review. *Professional Psychology, 13,* 58–65.

Singer, J. L. (1981). Clinical interventions: New developments in methods and evaluations. In L. T. Benjamin, Jr. (Ed.), *The G. Stanley Hall Lecture Series* (Vol. 1). Washington, DC: American Psychological Association.

Stricker, G. (1983). Peer review systems in psychology. In B. D. Sales (Ed.), *The professional psychologist's handbook.* New York: Plenum Press.

Weed. L. L. (1964). Medical records, patient care and medical education. *Irish Journal of Medical Science, 6,* 271–282.

IV

ADMINISTRATIVE STRUCTURE

13

The Role of the Federal Government in Peer Review

PATRICK H. DeLeon, Joan G. Willens, J. Jarrett Clinton,
and Gary R. Vandenbos

INTRODUCTION

This chapter reviews two roles that the federal government has adopted in the area of peer review: that of a purchaser of care and that of a provider of care. The authors note that, depending on the particular federal health-care program involved, those who have enacted the governing statute and those who currently administer the program may, or may not, have envisioned a substantial role for peer review. The issues of cost containment and assurance of quality care are intimately intertwined. The development of peer review has genuinely been an evolutionary one, with the federal government frequently using its "purchasing power" to guide the system toward an envisioned but elusive goal. Within the U.S. Department of Defense, where the federal government serves as a provider of care, a number of innovative approaches have recently been adopted to systematically institute objective standards of performance and quality care.

It is unfortunately all too rare for those who are intimately involved in our nation's mental health programs to view the federal government as a true partner in a collaborative effort. This is the case whether one considers the views of mental health advocates, consumers, or professionals. Instead, most individuals and professional associations conceptualize the federal presence in the mental health arena to date as being either adversarial or insufficient. In our judgment, this attitude reflects both a fundamental misunderstanding of the appropriate role of the federal government and the relative youth—in a public policy sense—of the mental health expertise in our nation (DeLeon & VandenBos, 1984).

Mental health professionals really do not appreciate the extent to which our nation's health policies, including our nation's mental health policies, are for

PATRICK H. DeLEON • Office of Senator Daniel K. Inouye, 722 Hart Senate Office Building, Washington, D.C. 20510.   JOAN G. WILLENS • Private practice, Beverly Hills, California 90212. J. JARRETT CLINTON • Office of Health Affairs, The Pentagon, Department of Defense, Washington, D.C. 20310.   GARY R. VANDENBOS| • American Psychological Association, 1200 17th Street, N.W., Washington, D.C. 20036.

mulated by politicians (elected and appointed public officials). Further, most mental health professionals do not appreciate the extent to which being a successful politician is in itself a unique professional career (DeLeon, 1983, DeLeon, Frohboese, & Meyers, 1984). When one reviews the professional educational backgrounds of our nation's elected officials, one finds that most are attorneys, especially at the federal level. In fact, during the 99th Congress (1985-1986), slightly more than 60% of the members of the U.S. Senate were attorneys, whereas in sharp contrast, only one senator possessed any formal health-care expertise *per se*, and he was a veterinarian. Comparable figures exist for the members of the House of Representatives. Further, even within the ranks of congressional staff, there is a dearth of health care expertise. Grupenhoff (1983) reported that only 2% of the professional staff members assigned to health care issues had any formal training in any of the health professions. He also noted that their turnover rate was high, with 40% having worked less than 2 full years in Congress and 79% having worked 5 years or less. Of congressional health staff members, 60% were under the age of 29, and 95% were under the age of 40.

These statistics are relevant to several different but related issues. First, the vast majority of those who establish our nation's health programs and/or health care priorities, as well as their staff, do not personally have the technical (professional) expertise necessary to be truly comfortable with "micromanaging" any aspect of the health delivery system. Instead, they are considerably more comfortable with following the legislative public-hearing procedure under which complex issues are "flushed out" in an admittedly sometimes adversarial manner. The underlying philosophy of this approach is very close to that of trial work; that is, objective and reasoned decisions can best be made when parties with adverse interests are given the opportunity to present their "best cases" before an impartial judiciary. Whereas scientists or health care practitioners might, as a result of their professional training, attempt to personally learn all that they can about a particular disease entity or promising clinical technique, legislators, by contrast, would strive to arrange a public forum in which those with genuinely different (preferably economic) interests can "argue out" their position on what should be the preferred course of action. From a legal or legislative perspective, this latter procedure should result in reasoned and appropriate decisions, not to mention insights into what subsequent actions and priorities should evolve.

From a public policy perspective, there are at least three distinct approaches to conceptualizing the federal government's appropriate role in our nation's health care system, including its approach to peer review. And depending on the approach taken, one's view of the issue of peer review can be quite different.

First, one can view the government as a major purchaser of health care. Federally financed health expenditures rose from $5.5 billion, or almost 5% of total federal expenditures in 1965, to over $93 billion, or 12%, in 1982 (GAO, 1985). Currently, the federal government is responsible for paying nearly 30% of our nation's personal health-care expenditures. This coverage is second only to private health insurance plans, which account for 31% of the coverage of personal health-care expenditures. The 1984 federal expenditure for personal health care was $101 billion.

Second, one can view the federal government as an actual deliverer of care. Within the U.S. Public Health Service, the Department of Defense, and the Veterans Administration, there are approximately 135,000 federally employed physicians, nurses, psychologists, and other health-care practitioners. The beneficiary population is also quite large and diverse. For example, the Indian Health Service regards its eligible beneficiary population as including 909,000 Indians and Alaskan natives. The Veterans Administration's Fiscal-Year 1986 budget-request projected 1.4 million inpatients and 18.7 million outpatient visits.

Third, one can also attribute to the federal government a special societal responsibility for setting objective standards in an area as intertwined with the public health and safety as the delivery of health care.

In this chapter, we focus on the first two approaches to conceptualizing the federal government's role in peer review. It is our hope, however, that, in reviewing the material provided, the reader will gain an appreciation of the different approaches that the government has taken to date and how, accordingly, it would be extremely difficult at this point in the development of peer review for the government to attempt to be overly directive or restrictive.

THE FEDERAL GOVERNMENT AS A PURCHASER OF HEALTH CARE

In the federal government's role as a primary *purchaser* of health care, the Congress has, over time, developed a number of different programs. Each of these has been established to address the needs of a different beneficiary population, and each has viewed peer review in a different frame of reference. Although, in the abstract, and perhaps in a public policy frame of reference, it would appear that the federal government should actively seek to develop a uniform benefit program (and a uniform cost-containment/quality-assurance mechanism) for all its beneficiaries—and thus, at a minimum, use its considerable purchasing power (i.e., size) and experience to ensure the most cost-effective benefit package possible—in reality, this is not what has occurred.

At present, there are four fundamental programs under which the federal government purchases health care. These are the Federal Employees Health Benefit Program (FEHBP), the medical expense provisions of the Internal Revenue Code, Medicare and Medicaid (Titles XVIII and XIX of the Social Security Act), and the Department of Defense Civilian Health and Medical Program of the Uniformed Services (CHAMPUS) (DeLeon & VandenBos, 1980).

Each of these programs comes under the jurisdiction of different congressional (sub)committees, and as a result, each is considerably different in both philosophical orientation and overall benefit structure. It is a legislative fact of life, the significance of which, unfortunately, most mental health professionals do not appreciate, that members of Congress actively seek out those committees that will have jurisdiction over areas (or programs) that either were of personal concern to them during their previous professional lives, or that will directly affect their individual constituents. Thus, if one reviews the membership, for example, of the House and Senate Veterans Affairs Committees or the Ways and Means and

Finance Committees, one will find that the individual interests and philosophy of the elected officials serving on those committees very closely parallel those of the people directly affected by their actions. Stated more concretely, the elected officials who have been involved in developing our nation's various federal health-care initiatives have traditionally made a conscious effort to establish programs that would be specifically aimed at the perceived needs of those directly affected (DeLeon, Forsythe, & VandenBos, 1986). On the one hand, this makes inherent and intuitive sense. However, accompanying this approach has also been the significant institutional reluctance of elected officials to become involved in programs that come under the jurisdiction of committees on which they do not personally serve (i.e., those serving other beneficiary populations).

Since the mid-1970s, our nation's health care costs have risen faster than almost any other segment of our economy. As a nation, we now spend $387.4 billion, or 10.6% of our gross national product (GNP), on health care; this is among the highest amounts, if not the highest amount, spent by any Western industrialized nation. Further, credible projections have been made that this amount will increase to $660 billion, or more than 11% of our GNP, by 1990 (Arnett, Cowell, Davidoff, & Freeland, 1985). Concerted efforts have been made under each of the federal health initiatives to curtail the rate of escalation. At the same time, increasing concern has been expressed among providers and in the popular media regarding the extent to which the various cost-containment proposals may also be causing significantly diminished quality of care.

It is important for the reader to understand that, as a practical matter, the congressional committee structure described above has meant that, even when viewing the federal government as the purchaser of health care, issues such as cost containment and peer review are treated in distinct and quite different ways under each of the four governmental programs. These differences will be described in more depth below.

Federal Employees Health Benefit Act

At present the federal statute governing FEHBP (5 U.S.C. Section 8901-8913) neither requires nor even directly encourages the use of peer review. FEHBP provides access to necessary health care for approximately 10 million federal employees, annuitants, and dependents. The most recent figures available indicated that, in fiscal year 1983, the total revenues for FEHBP were $5.9 billion, of which the federal share was $3.5 billion.

Each year the director of the Office of Personnel Management (OPM) negotiates a range of program options for this particular beneficiary population. During the last "open season," there were approximately 120 different plans offering options, although nearly 79% of the beneficiaries were enrolled in one of six major plans.

It is of interest that nearly 10% of the enrollees participate in one of approximately 100 health maintenance organizations (HMOs) (GAO, 1983). This fact is significant in terms of peer review because Section 1301 (c) (7) of the U.S. Public Health Service Act states that, in order for any HMO to become federally qualified, it must

have organizational arrangements, established in accordance with regulations of the Secretary, for an ongoing quality assurance program for its health services which program (A)
stresses health outcomes, and (B) provides review by physicians and other health professionals of the process followed in the provision of health services.

The Department of Health and Human Services estimates that approximately 65%–
70% of all HMOs are federally qualified. Further, as a practical matter, nearly every
HMO, including those that have chosen not to seek federal qualification, has
developed an aggressive internal peer-review mechanism. They have done this, if
for no other reason, in order to obtain a handle on their professional and clinical expenses (i.e., as a basic cost-containment mechanism).

As described earlier, policy discussions surrounding the issue of quality
assurance and peer review often become intertwined with the issue of cost containment, and this has also been the case for FEHBP. 1981 FEHBP experienced nearly a
half-billion-dollar shortfall, and as a direct result, OPM instituted a substantial premium increase and two rounds of benefit reductions. The expanding cost of the program and the expressions of concern by providers and beneficiaries about OPM's
decisions led to congressional hearings in December 1983 by the Senate Committee
on Governmental Affairs, which has jurisdiction over the program. At those hearings,
the director of OPM reported that his efforts to date had been successful in generating
a savings of $.5 billion during the initial year and savings of $2 billion since that time
(Devine, 1983). Further, by the end of fiscal year 1985, the administration was to request legislation, which has now been enacted into public law, that authorized a rebate
from the program's fiscal intermediaries of approximately $1 billion.

At the time of the 1983 Senate hearings, one of the legislative proposals that
was being given serious consideration would have mandated that OPM establish a
formal system for peer review of the utilization and quality of health care being furnished under the program. This proposal (S2027) had been introduced by the
chairman of the Senate subcommittee and, as drafted, would have expressly required that OPM establish a system similar to that of Title XI of the Social Security
Act (i.e., the one being used for Medicare). The driving force behind the chairman's
proposal was an effort to curtail the then ever-escalating costs of health care. During
the hearings, the administration expressed its preference for being granted greater
administrative flexibility but was also clearly dedicated to controlling costs. Because
OPM was able to demonstrate convincingly that its efforts were being successful,
the proposed peer-review requirement was not enacted into public law. Thus, as of
this writing, with the exception of the HMO statute provisions, peer review *per se* is
not legislatively nor administratively mandated under FEHBP. However, as the
Senate hearings demonstrated, requiring peer review has been considered by the
committees with jurisdiction, and there is every reason to assume that it may be proposed again in the future, especially if costs continue to escalate or if a concerted effort evolves to integrate (or coordinate) the various federal health-care initiatives.

INTERNAL REVENUE CODE

The medical-expense-deduction provisions of the Internal Revenue Code (26
U.S.C. Sections 162, 213) also do not in any manner address the issue of peer

review. The individual deduction provision (Section 213) is quite similar to the one originally adopted as an amendment to the Internal Revenue Act of 1942 (P.L. 77-753). In essence, it allows a taxpayer to take a deduction for the extent to which his or her total medical expenses are not covered by insurance and exceed 5% of adjusted gross income. In 1965, the percentage was modified to 3% (P.L. 89-97), but in 1982, the 5% floor was again reinstated (P.L. 97-248). It is of historical interest that the original provision had imposed a limitation on the total maximum deduction allowable, but this limitation is not currently in effect.

The Congress has never attempted to legislatively define what would constitute a "medical expense." However, the legislative history makes clear that, from the inception of the individual deduction provision, only unusual and extraordinary *medical expenses* were to be covered, in contrast to personal (or growth) expenses (DeLeon, 1981). In implementing the provision, the Internal Revenue Service has attempted to draw a functional distinction between medical treatment that is for the prevention or alleviation of physical or mental defect or illness (deductible) and treatment that may be merely beneficial to one's general health or sense of well-being (nondeductible). Having the care "ordered" or provided by a health care practitioner *per se* is not sufficient to ensure its deductibility. Similarly, the business expense provision (Section 162) uses the general phrase "ordinary and necessary expenses" and does not attempt to legislatively define "medical care."

The staff of the Joint Committee on Taxation have estimated that, in fiscal year 1986, the individual provision will cost the U.S. Treasury $3.5 billion, and that the business provision will cost an additional $23.7 billion (GPO, 1985). Given the substantial dollars involved, serious consideration has been given by the Reagan administration to significantly modifying what is essentially a "health subsidy" (Dorken & DeLeon, 1986). However, as is the case with the Federal Employees Health Benefit Program, their efforts have focused exclusively on ways of recouping funds to the Treasury, rather than on legislatively defining "medical expenses" or exploring the role that peer review might play in either ensuring quality of care or controlling costs.

THE SOCIAL SECURITY ACT

Medicare and Medicaid constitute the largest share of the federal government's commitment to financing quality health care to its citizens (DeLeon, VandenBos, & Kraut, 1984). At present approximately 40 million Americans receive "medically necessary" care under these two programs. In 1984, this care accounted for an expenditure of approximately $82.7 billion, or nearly 82% of the federal government's expenditures on personal health care.

The Social Security Amendments of 1965 (P.L. 89-97) established Medicare and Medicaid, and although the mental health benefit under each of these two programs has historically been meager at best (GPO, 1978), there can be no question that peer review as we know it today owes its very existence to the magnitude of the concern about the federal government's investment in these two programs.

The Social Security Amendments of 1972 (P.L. 92-603) authorized the then Department of Health, Education, and Welfare to establish independent professional standards review organizations (PSROs) nationwide (Title XI of the Social Security Act). Even at that time (and before), escalating health care costs were of significant congressional concern. In recommending the enactment of this new program, the Senate Finance Committee maintained that the proposed PSRO program would serve at least as a partial solution to the problem of rising health-care costs. The committee even then noted that the economic impact of the overuse of services was significant (GAO, 1980).

The committee was also quite concerned about the issue of quality assurance and similarly noted that there was an unacceptably high incidence of medically inappropriate services being rendered to Medicare and Medicaid patients (O'Sullivan, 1981). In February 1970, the Senate Finance Committee staff had issued a report entitled *Medicare and Medicaid—Problems, Issues, and Alternatives* (GPO, 1970). This document resulted in 12 days of Finance Committee hearings that dealt extensively with the failure of utilization review under Medicare and Medicaid. Utilization review was described as frequently being of a token nature and as essentially constituting merely a *pro forma* activity. In summary, there was clear documentation of a history of extensive, widespread, costly, and inappropriate utilization of health care services. As eventually enacted into public law, the PSRO program was to review all health services provided under the entire Social Security Act, including Medicare, Medicaid, and the Maternal and Child Health programs. The last category in particular, however, was never implemented because of budgetary constraints.

As initially conceptualized, the goal of the PSRO program was to ensure that federal funds would be spent only for services that were "medically necessary" and that were provided in accordance with locally determined professional standards. When the services were provided in an institution, there was to be a further determination that the care had been rendered in an appropriate setting. From the beginning, cost containment and quality of care were viewed as being intimately intertwined. Also in existence from the program's inception was the underlying philosophy that these clinical determinations could best be made by professional peers and, further, that the active involvement of physicians *per se* was absolutely crucial to the success of the program.

In response to growing congressional concern, in early 1970 the American Medical Association (AMA) had brought a professional review organization (PRO) plan to Senator Wallace Bennett, who served on the Senate Finance Committee. This plan ultimately turned out to be the genesis of the eventual PSRO legislation, even though the AMA decided that it could not accept a proposal that did not mandate that state medical societies would be the administrative mechanism (Constantine, 1981). Nevertheless, in arguing for the passage of the PSRO legislation, Senator Bennett, who was its chief sponsor, stated:

> I want to reiterate that my amendment is firmly based on the principle that only physicians are capable of deciding whether a service is medically necessary or meets proper quality standards. Therefore, peer review must mean just that—only physicians should review physicians. (*Congressional Record*, S1019, January 25, 1972)

In an evolutionary public-policy frame of reference, the federal government was using its purchasing power to insist that our nation's health care providers (i.e., physicians) should be accountable for providing objective evidence of quality care.

In administering the program, the Health Care Financing Administration (HCFA) eventually established 195 separate PSROs nationwide. Each of these organizations represented substantial numbers of physicians and, as a practical matter, had very little behavioral-science input—especially as the mental health expertise was generally provided by psychiatrists.

During its early stages, the PSRO program focused primarily on providing concurrent review (review at admission, with periodic rereviews) of short-term general hospital services. Eventually, there evolved a steadily increasing emphasis on targeted (or focused) review of known or suspected problem areas. Although the original legislation provided authority for the review of noninstitutionalized care, when requested by the PSRO and approved by the Secretary of the Department of Health and Human Services (DHHS), as a practical matter this aspect of the program was never implemented. The reader should appreciate, however, that the ultimate goal of the PSRO legislation was to ensure the review of the full range of health care services being delivered under the Social Security Act. Of historical interest to the mental health community were the staff recommendations that grew out of the 1978 Senate Finance Committee hearing on the scope of the mental health benefit under Medicare and Medicaid (GPO, 1978). One of the recommendations was to establish a separate national mental-health PSRO system that would have paralleled that already in existence, but that would have focused exclusively on mental health treatment. This proposal was to be directly related to ensuring that only mental health services that were deemed "safe, effective, and appropriate" would be reimbursed; unfortunately, primarily because of concerns of mental health practitioners, this provision was never enacted into public law (Inouye, 1983).

As the 1970s came to an end, budgetary constraints generally precluded any significant expansion of the PSRO program outside the hospital along the lines that it was originally envisioned, although some PSROs did carry out ancillary service review and/or health-services research-projects (Lohr, 1985). As Lohr (1985) noted:

> Most significant, however, is probably the change in the attitudes of the medical community during the 1970s. Physicians eventually were motivated to band together in the interests of improving quality of medical care. It was shown that they could, through peer review agencies, identify quality-of-care problems and effect measures to overcome them. The average physician came to accept the idea that length-of-stay review, medical record audit, and other peer review activities were here to stay. Perhaps most remarkably, this progress occurred in an environment almost wholly concentrated on controlling the costs of medical care. (p. 13)

With the 1980 election of the Reagan administration and the control of the U.S. Senate simultaneously going to the Republican Party for the first time in nearly 30 years, a consensus evolved that the PSRO program had to be dramatically modified if it were to be continued. The initial Reagan administration budget proposed a total phaseout of the PSRO program over the Fiscal-Year 1981–1983 period, and accompanying legislation was submitted to the Congress to eliminate the requirement for utilization review committees in institutions that were not covered by PSRO review (O'Sullivan, 1981).

From a historical frame of reference, it is important to appreciate that, even before the election, there had been growing concern that the PSRO program was not as cost-effective as its advocates had projected. In 1979, the Department of Health and Human Services (HHS) estimated that the program had resulted in a net savings for Medicare of approximately 20% over the review costs. Nevertheless, the Congressional Budget Office (CBO) reviewed the same data and concluded that it would be very difficult to credit PSROs with producing any substantial savings, and CBO further suggested that, because of the fixed nature of various hospital costs, any savings generated by the PSRO program for the Medicare population were partially offset by the practice of the hospitals of subsequently transferring these costs to private patients (CBO, 1979).

Why CBO decided to apply this "shifting thesis" to this particular program is unclear, and the former staff director of the Senate Finance Committee's health subcommittee has stressed that applying this doctrine to other federal expenditures would leave virtually no room for the government ever to reduce expenditures or to be a prudent buyer (Constantine, 1981). Nevertheless, CBO (1979) did conclude that "although PSROs seem to be effective in reducing Medicare utilization, it is doubtful that they produce a net savings" (p. ix). The General Accounting Office (GAO) also reviewed certain aspects of the PSRO program and concluded that there were definite ways in which the HCFA could improve the program (GAO, 1980).

Regardless of whether the PSRO program was truly cost-effective, its long-term significance was undoubtedly in getting those concerned with our nation's health policies to appreciate that objective evaluations of care were possible. This success is underscored by the fact that the program and the various policy issues surrounding it were reviewed in depth by the Institute of Medicine (IOM)—which in many ways serves as a health policy "think tank" for the Congress—in at least two of its publications: *Advancing the Quality of Health Care: Key Issues and Fundamental Principles* (IOM, 1974) and *Assessing Quality in Health Care: An Evaluation* (IOM, 1976).

The Omnibus Budget Reconciliation Acts of 1980 (P. L. 96-499) and 1981 (P. L. 97-35) generally narrowed the scope of the activities that PSROs were required to conduct in order to receive full designation by HCFA. Further, the secretary of DHHS was directed to assess the relative performance of each PSRO and was authorized to terminate up to 30% of the PSROs, based on the results of this review. The most significant modification in the federal peer review program, however, was the provisions of the Tax Equity and Fiscal Responsibility Act of 1982 (TEFRA) (P. L. 97-248), which repealed the PSRO section of Title XI and established in its place a new program of Utilization and Quality Control Peer Review (O'Sullivan, 1982).

TEFRA proposed a fairly straightforward transition from the PSRO program into one based on peer review organizations. This new legislation required the secretary of the HHS to enter into performance-based contracts with physician-sponsored or physician-access organizations known as *peer review organizations* (PROs). The secretary was required to designate geographical areas that were to be served by an individual PRO, each state generally being designated as a single area. At present, there are 54 PROs nationwide. Although there are considerable similarities between the two programs, the most significant difference is the performance-based nature of the PRO contracts, which include negotiated objectives against which the organization's performance is to be judged. Further, there is an improvement in the PRO's sanction authority, and at present, all hospitals that receive reimbursement from the Social Security Trust Fund are required to be reviewed by PROs. There are also changes in the program's funding arrangements; PROs are financed with 2-year fixed-price contracts, the total contract amount to be derived from a set price per review. The statute also provides for an expansion of PRO eligibility to include for-profit groups and payer organizations (i.e., insurance companies and fiscal intermediaries), although priority was still given to physician groups for the 1st year. It was congressional expectation that the PROs would be less regulated by HCFA and thus would be able to be more creative. The standards for review are quite similar to those of the original PSROs, with PROs to review, subject to the provisions of the contract, the professional activities of physicians, other health practitioners, and institutional and noninstitutional providers who render services to the Medicare population. Their focus is on the necessity and reasonableness of care, the quality of the care, and the appropriateness of the setting.

The Social Security Amendments of 1983 (P. L. 98-21) authorized the establishment of an entirely new way of providing reimbursement under Medicare based on a prospective payment approach, rather than the traditional fee-for-service (retrospective) method. This new approach is based on the use of diagnosis-related groupings (DRGs), psychiatric hospitals being exempt for an initial period. The statute requires hospitals to enter into agreements with PROs as a condition for receiving Medicare payments, and one of the issues to be addressed by the PROs is the validity of the diagnostic information being provided by the hospital. The Deficit Reduction Act of 1984 (DEFRA) (P. L. 98-369) made several further modifications in the PRO program, generally in the program's underlying structure, rather than in its function. For example, up to 20% of the members of a PRO governing board can now be affiliated with providers (Committee on Finance, 1985).

Finally, the Consolidated Budget Reconciliation Act of 1985 (COBRA), which at the time of this writing was still being debated by the Congress, would require the mission of the PROs to include services provided in Section 1876 ("risk contract") HMOs. The original PSRO statute included authority for the review of all Medicare services (including those in HMOs). However, as we indicated, inpatient care became the prime focus of the earlier program. TEFRA established a new Section 1876 "risk-sharing" reimbursement provision, thus making the traditional comprehensive approach of an HMO eligible. The regulations published by HCFA to implement TEFRA also made such care subject to review under the new PRO program. The Office of Management and Budget (OMB), however, would not allow any funds to be spent for this additional review. Accordingly, the COBRA provision was enacted in order to get around OMB's objections and to statutorily mandate PRO review of HMO care. We would note that if this provision becomes public law, the secretary of DHHS will undoubtedly use his or her broad statutory authority to ensure that all targeted HMO services, not merely those provided to Medicare patients, will become reviewed by PROs.

As the federal government has moved to tighten up its payment system under Medicare and Medicaid, growing expressions of concern have become evident in the provider community, among beneficiaries, and ultimately, in the Congress regarding the fundamental issue of quality assurance. The Senate Special Committee on Aging, for example, held three hearings in the fall of 1985 specifically addressed to the issue of quality oversight and enforcement in order to ensure that elderly Americans would not be subjected to poor-quality medical care. The Prospective Payment Assessment Commission (ProPAC), which was established under the Social Security Act Amendments of 1983 to advise and assist the Congress and the secretary of DHHS in maintaining and updating the Medicare prospective-payment system, has indicated that the issue of quality care will be one of its high-priority areas (Young, 1985). The commission has submitted several reports to the Congress describing its observations and recommendations (Pro-PAC, 1985, 1986). Similarly, the Office of Technology Assessment (OTA) has also completed an extensive evaluation of the Prospective Payment System (U.S. Congress, 1985b), as well as two reports on the ProPAC (U.S. Congress, 1985a, 1986). Partly in response to these concerns, HCFA issued a special contract in late 1985 to establish a medical evaluation team for the PRO program, or what might be considered a "super PRO." This new entity is to be an independent, professionally recognized organization that will assess the accuracy of the medical determinations made by the PROs. It is to focus on admissions review and DRG validations (i.e., the overall quality of the PRO decision-making process itself).

Clearly, the underlying question that is evolving is whether the PRO program can adequately address the "quality-of-care" issue. One concrete example of how the new PRO program might be viewed as having obtained respectability from the administration was the testimony of the acting HCFA administrator before the Senate Appropriations Committee during the Fiscal-Year 1987 hearings. In response to a question about why HCFA was not requiring second opinions on Medicare surgical procedures, as was recommended by the DHHS Inspector General and was projected to save up to $135 million annually, the administrator indicated that the department felt that PROs could best address the problem of unnecessary surgery (Desmarais, 1986). Similarly, the *Washington Post*, in a January 22, 1986, editorial comparing the PSRO and PRO programs, stated:

> Between 1973 and 1984 these organizations (PSROs) disciplined 70 doctors and hospitals, temporarily barring them from treating Medicare patients because of inferior treatment or billing for unnecessary procedures ... In 7½ months these agencies (PROs) have begun proceedings against 950 doctors and 183 hospitals. Most of the cases involving physicians were brought because of poor care ... Wisely, the reorganized review boards are taking this responsibility seriously and stepping up the pace of review. (p. A-10)

Hopefully, these favorable prognostications will turn out to be correct.

CHAMPUS

For our nation's mental health providers, the Department of Defense CHAMPUS program has undoubtedly been the most significant federal initiative in the peer review arena. Each of the four core mental-health disciplines (psychology, psy-chiatry, psychiatric nursing, and clinical social work) is expressly enumerated (and given professional parity) in the governing federal statute (10 U.S.C. Section 1071-1089), and the CHAMPUS mental-health benefit has historically been one of the most generous of any federal or private insurance program.

Perhaps of even more importance, however, has been the fact that it has been around issues involving CHAMPUS that each of the nonphysician mental-health disciplines has now developed a significant federal legislative presence. At present, 70%–75% of all Department of Defense beneficiary care is provided in military facilities; however, approximately 7.6 million dependents, retired members, and their families and survivors are eligible to receive reimbursement under CHAM-PUS for health care that is received in the private sector. The present annual CHAMPUS budget is approximately $1.5 billion. Of historical interest, the Veterans Administration has developed a similar, although substantially smaller, program (CHAMPVA), which is projected to cost the federal government $74.1 million in fiscal year 1986, when it is expected to provide reimbursement for some inpatient care and approximately 145,000 outpatient visits. The regulations for CHAMPUS and CHAMPVA are identical.

In January 1977, CHAMPUS promulgated detailed regulations specifying utilization and peer review requirements. In July of that same year, CHAMPUS en-tered into contracts with both the American Psychological Association (for out-patient care) and the American Psychiatric Association (for inpatient and outpatient care) in order to allow the two professional organizations to develop appropriate "criteria sets," so that comprehensive peer-review procedures could be instituted nationwide (Clarborn & Zaro, 1979).

In retrospect, the decision to take this innovative approach may very well have allowed CHAMPUS the administrative flexibility to continue its liberal mental-health benefit for a number of years. The House and Senate Appropriations Com-mittees were becoming increasingly concerned about the ever-escalating costs of CHAMPUS, including its mental health benefit.The year following the issuance of the peer review contracts, the House of Appropriations Committee, during its deliberations on the fiscal year 1979 Department of Defense Appropriations Bill (P. L. 95-457), stated in its report that

> even though there has been some improvement in this area, utilization of psychiatric serv-ices by CHAMPUS beneficiaries is too high and additional control measures are needed. Limits should be placed on current CHAMPUS benefits for psychiatric services, i.e., num-ber of days, number of visits, and total dollars. Further, it does not appear that any amount of utilization, medical and peer review effort will bring this area into control. Therefore, all admissions to psychiatric facilities, whether or not hospitals, should require preauthoriza-tion. (House Report no. 95-1398, p. 153)

The Senate report expressed its concurrence with the House view; however, it also expressed its support for the two APA contracts and stated that

> CHAMPUS is urged to implement the review mechanisms resulting from these contracts at the earliest practicable date, and to continue to work with the two associations in mon-itoring the effectiveness of the two peer procedures. (Senate Report no. 95-1264, p. 20)

CHAMPUS's efforts in the peer review area have continued to be expressly addressed by the Senate Appropriations Committee in its annual committee reports. For example, directives were included to ensure that appropriate attention would be also given by the Department of Defense to the growing interest in peer review of the American Nurses' Association and the National Association of Social Workers. The two initial contracts with the psychological and psychiatric associations were continued until CHAMPUS agreed to accept the further recommendations of the Senate Appropriations Committee, made during its deliberations on the Fiscal-Year 1983 appropriations bill (P. L. 97-377), that the time had come to establish an integrated approach. Bids were then solicited for a unified contract, and in 1985, this was awarded to the American Psychiatric Association.

In a very real and tangible sense, CHAMPUS has had a monumental impact on the delivery of mental health care in our nation. For example, as a direct result of their experiences with CHAMPUS, both the American Psychological Association and the American Psychiatric Association have now developed their own internal offices for peer review and now provide such professional services for a number of private insurance companies including Prudential, Aetna, Connecticut General, and Blue Cross and Blue Shield.

One of the authors of this chapter had the opportunity to serve on the Psychology CHAMPUS Peer Review National Advisory Panel from its inception until the members began to rotate off in 1980. For a number of years, she had also been very active in lobbying at the national level on behalf of psychology and, in particular, in support of its efforts to obtain independent recognition under the programs of the Social Security Act (i.e., Medicare and Medicaid). In her judgment, the awarding of the initial CHAMPUS peer-review contract came at a particularly politically opportune time (Willens & DeLeon, 1982). As a profession, psychology was just beginning to achieve respectable credibility among federal legislators, although its actual legislative successes were still rather minimal. What was new, however, was a climate of greater respect for the profession, and for mental health in general. This was the era of the President's Commission on Mental Health. Congressional staff— and particularly that of the health subcommittee of the Senate Finance Committee, which has jurisdiction over Medicare and Medicaid—had just begun to offer concrete suggestions working toward psychology's eventual inclusion as an autonomous profession under the Social Security Act (Bent, Willens, & Lassen, 1983). In the House of Representatives, Representative Henry Waxman, who was chair of the health subcommittee of the then Interstate and Foreign Commerce Committee, has similarly begun to demonstrate a rather consistent support for psychology.

The awarding of the CHAMPUS contract by the Department of Defense unequivocally placed the profession of psychology on equal footing with organized psychiatry, particularly in the eyes of federal legislators and their staff. Thus, this award presented a significant political opportunity. Those psychologists who served on the original CHAMPUS panel (Drs. George Stricker, chairperson; Russ Bent, Anna Rosenburg, Lee Sechrest, Joan Willens, Harl Young, and Bill Claiborn, staff) were quite aware of the longterm legislative implications for organized psychology, and they further felt that they also possessed a truly unique opportunity to significantly influence the fundamental practice patterns of the profession. It was believed that, as psychiatry was the "establishment" profession within the mental health community, its peer review would have less flexibility (i.e., less support from its membership) to develop a truly innovative system. Psychology, on the other hand, primarily as its practitioners had not historically been as accepted politically, possessed the exciting opportunity to experiment with considerable deviance from tradition.

It was the panel's hope that the same forces within the Congress that were actively encouraging fundamental change in the overall health-care system—for example, in the form of sunsetting laws and required consumer participation—might also be receptive to other, more substantive innovations in mental health care. CHAMPUS appeared to be the only governmental or private system to recognize psychology's potential contribution to altering the entire mental-health delivery-system. The panel, with this vision in mind, concluded that the CHAMPUS contract was psychology's best opportunity to date to systematically introduce a creative, nonmedical model that might conceivably propel psychology into inclusion in future national health systems. Consequently, the panel viewed a major element of its charge as being to devise a functional, flexible system that would eliminate major abuses and that, at the same time, would also determine both the quality and the necessity of the psychological care that was being provided.

The seven members of the panel came together, each with his or her own particular experience and theoretical biases. The authors are aware that, both individually and collectively, the psychology panel members have been accused by their colleagues of being overly psychoanalytic, overly behavioral, and antifamily. In fact, of the ultimately six members and project director, three had a psychodynamic-psychoanalytic orientation, two had a behavioral orientation, one's background was primarily in research, and the last member specialized in family therapy. Of the original seven, six had spent most of their professional lives engaging in full or part-time private practice. Contrary to what many who were not intimately involved in the development of the system may have conjectured, the psychology criteria grew out of arguments, negotiating, compromise, use of "outside" experts, and, above everything else, evolution. The underlying principle guiding the panel's deliberations was that the development of national standards should be an evolving process, with the same dialogue occurring within the profession as had occurred among the panel members.

As the psychology panel went through its conceptualization process, several guiding concepts eventually emerged, which it believed would distinguish psychology from the traditional practice of psychiatry, and which also had the potential for significantly modifying the basis of most mental-health reimbursement-systems. They were the following:

1. That professional judgments, once established, could be carried out by trained nonpsychology personnel, and that a psychologist need not be involved at every level of clinical decision-making.
2. That the traditional psychiatric diagnosis was not a valid basis for the development of criteria to determine the adequacy of mental health care. The panel proposed instead the institution of a "treatment plan" concept that delineated the stated problems (or functional impairments), the goals of

treatment, the procedures to be used to get there, and the level of progress made. In this way, the criteria by which the determination of "adequacy of care" was to be made would be related to the treatment plan, rather than to a diagnosis.

3. That the patient was to be involved, unless circumstances suggested the contrary, in the development and execution of his or her treatment plan.

4. That the criteria to be used in evaluating the "adequacy of care" should be publicly disseminated so that there would be no "guessing" about what aspects of treatment would be evaluated, and so that public professional debate about such criteria could be fostered.

For the most part, these underlying premises were a departure from both psychology's and psychiatry's "usual and customary" procedures in dealing with third-party payers. They were based on the panel's collective belief that the only way that the government and private insurance carriers would continue to reimburse mental health professionals (and similarly, that consumers would ever trust the psychology profession's integrity) was if the profession itself became more visibly accountable for what was being done in the privacy of its members' offices. The panel took this stand fully understanding the vital need to maintain high levels of patient confidentiality. Far from conceiving of this project solely as a way of "policing" private practice (or, as some have accused the panel members, of "selling out" the profession), the members of the psychology panel believed that the time had come to admit formally that "quality service" would not always be synonymous with "usual and customary" practice, especially when, in fact, the actual practice of members of the psychology profession often lagged considerably behind more contemporary considerations of consumer protection and participation, as well as the changing views of the integrity of the profession.

One of the major concerns raised by the critics of the psychology CHAMPUS peer-review system has been that the criteria developed did not exclusively represent "usual and customary" practice. This and other criticisms were predicted by Kiesler, Cummings, and VandenBos as early as 1979 (pp. 343–346). As we indicated, to some extent, this criticism is true. Rather than simply accepting that what professionals were currently doing in their practices, was by definition, "the state of the art," the panel had instead attempted to spell out operational definitions of "usual and customary" *quality* service as postulated by a combination of expert opinion, current literature, and general usage. The panel was aware, for example, that what was considered "general usage" might vary from one geographical region to another, as well as among the varying treatment orientations. Where differing practices did emerge, the panel felt that the proposed model of activity review by one's peers would provide the opportunity for a practitioner to explain the need for his or her proposed services and would at the same time serve as an institutional check and balance on any fundamental errors of judgment that might have been made by the panel.

In essence, the system as proposed allowed for continual modification through open dialogue and the gathering of actual data from treatment reports. The panel members never thought that their particular criteria would be imbedded in cement. Instead they expected two things to happen: first, that, over time, the actual data that were reviewed would result in substantial and appropriate modifications of the criteria in the direction of high-quality usual practice, and second, that the continuous surveying of professional psychologists about what they felt constitutes "quality care" would also result in appropriately modifying the criteria.

The issuance of the two peer-review contracts to the American Psychological Association and the American Psychiatric Association also reflected a fundamentally new venture for the Department of Defense (Rodriguez, 1983). Thus, there was no real track record or guidelines for how to accomplish the underlying objective. In a real sense, the department was taking a chance that ultimately the two professions would be able to evolve what, in retrospect, would be considered a reasonable approach to addressing highly complex issues.

At the same time, CHAMPUS had already developed an ongoing relationship with its fiscal intermediaries and had in place a formal system involving federal regulations and contract requirements, which unfortunately contained a number of mental health provisions that had not been developed by the mental health professions, and which appeared to the peer review panel to be without any clear rational or appropriate theoretical basis. For example, sex therapy and Gestalt therapy were excluded from reimbursement as representing "growth" procedures, rather than "clinical" procedures, and CHAMPUS prohibited the use of psychological assistants by psychologists, but not by psychiatrists. Similarly, a policy had been established through prior regulations—and again, not at the recommendation of the peer review panel—that formal peer review would be required at 8, 24, 40, and 60 sessions. The national panel was unsuccessful in its efforts to modify a number of these preestablished regulations, as it appeared that, once established, they had taken on a life of their own. To further complicate matters, as the program initially got under way, one of the larger fiscal intermediaries appeared to be holding up and/or denying claims and, at the same time, passively and inaccurately implementing the new peer-review system. Needless to say, these types of "internal battles" were never really appreciated by the psychology profession as a whole, and especially not by those practitioners who were having difficulty in obtaining reimbursement for services they had already rendered.

To be successful, the panel clearly had to engage in a delicate balancing act between the conflicting forces of the political considerations within the psychology profession and the need for responsiveness to a broader federal legislative agenda.

Although they received substantial criticism from certain elements of the practitioner community, the members of the psychology CHAMPUS peer-review panel continued to believe strongly that psychology as a profession must accept the underlying premise that its primary responsibility is to the consumer (and not merely to the economic well-being of the practitioner). Further, the panel was convinced that the profession (and the public) would be best served by ensuring that practitioners would be held accountable and would excel not only in their therapeutic relationships, but also in their public relations efforts.

From the panel's frame of reference, it was important to recognize and acknowledge the political reality that psychology's inclusion in any health-care program would cost the government dollars, no matter how reasonable to psychology such inclusion might seem. And as the dollars got tighter and there was more and more competition, that there would be increasing resistance to psychology from organized medicine and psychiatry in particular. Both national and local health-policy makers still found it easiest to assume that one can measure the upper limit of mental health costs by simply multiplying the number of psychiatrists in any given area (or throughout the entire nation), by the number of possible hours in each of the working days, by the average hourly charge. Simply stated, if psychiatrists remained the only independent providers of mental health care, the upper limit of their cost could easily be calculated. From the panel's vantage point, given the political and economic realities, it simply made sense to develop a national system that was truly creative.

It was also the psychology panel's distinct perception that, at the time, there was a significant chance to effect a fundamental change in the status quo. Certain pockets within the political system were beginning to recognize that solely limiting mental health reimbursement to psychiatry did not guarantee quality care or even cost-effective treatment (VandenBos, 1983). As we indicated, it was the panel's judgment that it would ultimately be innovative proposals such as the panel's original plan that would eventually demonstrate that psychology should be a true partner in our nation's mental-health-care programs. Nevertheless, if those within psychology who resist change succeed, and if the profession's primary concern becomes the protection of its practitioners from perceived enemies such as insurance companies and the federal government, it will become increasingly difficult to offer innovative approaches. That does not suggest that the profession should "sell out" in any way. Rather, the underlying question is whether the members of the psychology profession can bring themselves to appreciate the fact that something other than their own individual "usual and customary" approach might be in the best interest of their patients.

The CHAMPUS peer-review project brought out a healthy and vigorous tension within the profession between those who wanted truly to take into account the changing nature of our society, with its different expectations of professional responsibility, as well as those who appreciated the political reality that requires an innovative approach to the delivery of health care, and those within the profession who felt that the status quo could continue for an indefinate period of time. The resolution of these fundamental differences is crucial to the very future of the profession.

At present, the Department of Defense is giving serious consideration to the possibility of contracting out the entire CHAMPUS program to, possibly, three private regional contractors covering the United States (Project IMPRINT). The department plans to issue a request for information (i.e., a concept paper) in the spring of 1986, to be followed by a request for proposals for actual implementation in the fall of 1986. The projected implementation date is October, 1987. Preliminary projections are that, by taking this approach, the department will be able to contain the cost of CHAMPUS at the fiscal year 1986 level and, further, that the eventual successful bidders may be able to make a profit (or offer new benefits) ranging from 20% to 25% of their contract price. If this proposal is put into effect, the fate of the current peer-review mechanisms will be very much up in the air.

THE FEDERAL GOVERNMENT AS A PROVIDER OF HEALTH CARE

The Department of Defense efforts to ensure that quality health care is provided to all military beneficiaries provide an excellent example of what the federal government can do regarding such review activities. However, this approach will not necessarily be adopted by the U.S. Public Health Service or the Veterans Administration. One of the authors currently serves as Deputy Assistant Secretary for Health Affairs (Professional Affairs and Quality Assurance) of the Department of Defense and is intimately involved in the department's efforts.

The three military departments (Army, Navy, and Air Force) staff and operate 168 hospitals, 520 free-standing clinics, and 394 dental clinics. Of the 168 hospitals, 38 are outside the United States, for example, at U.S. military installations in Germany, Italy, the United Kingdom, the Philippines, and South Korea. Of the total 168 hospitals, 109 have fewer than 100 beds, and 6 have more than 500 beds. On an average day, approximately 15,000 beds are occupied by inpatients, and 50,000 outpatient visits occur. There were nearly 1 million inpatient admissions during 1985. Staffing for this system includes nearly 13,000 physicians and approximately 35,000 additional health-care providers.

In response to adverse publicity regarding specific medical-care cases and the professional assessment that followed, the Department of Defense (DOD) initiated in 1982 an intensive effort to strengthen the quality of medical care standards (and quality-of-care monitoring) throughout the military health-care system.

QUALITY ASSURANCE POLICY DEVELOPMENT

This 1982 review (and the development of the quality assurance program) became the primary function of a Quality Assurance Office within the Office of the Assistant Secretary of Defense (Health Affairs), and similar groups were established within the offices of each of the three surgeons general. Subsequently, quality assurance positions were created at the hospital level (and at both regional and central commands for health care services).

Policy guidelines within the department are issued in the form of a DOD Directive (DODD). These are translated by the three military departments into regulations that describe how a particular policy "directive" is to be implemented. These regulations (and the "service instructions") form the matrix by which a local hospital, for example, organizes and implements its quality-assurance functions. Although the initial policy decisions in 1982-1983 often responded to specifically identified deficiencies, subsequent policy directives have created a coherent framework by which the quality of care can be monitored, assessed, and, through appropriate corrective activity, improved.

QUALITY ASSURANCE INITIATIVES

The department issued a "directive" in 1982 requiring each military department to report to the Federation of State Medical Boards those medical officers whose clinical practice privileges are suspended, limited, or withdrawn at the time the individual is separated from the military. This directive was an initial response to the issue raised in the adverse publicity surrounding inadequate health-care providers' moving from the military to civilian positions. In addition, if a medical officer seeks training at a civilian institution, the training institution must be notified. Subsequently, similar reporting requirements were issued for nurses and dentists, such reporting being to the appropriate profession-specific national clearinghouse for disciplinary actions.

As a first step in beginning to gather systematic information about quality assurance issues, the military services, also in 1982, were required to report quarterly the details of malpractice claims and of credentialing sanctions against physicians to the Quality Assurance Office established within the Office of the Assistant Secretary of Defense (Health Affairs).

In 1983, two major quality-assurance directives were developed and issued. The first of these 1983 directives, "Standards for DOD Health Care Provider Performance," required the calculation of mortality rates for 26 surgical procedures and the use of an 18-point "quality-of-care checklist" with each inpatient case. The data generated by these two processes provide an excellent means of monitoring provider performance and of assessing trends (so that appropriate actions can be taken when any of the rates is either outside predetermined parameters or indicates that an individual's performance requires immediate intervention or more complete review).

The second of these 1983 directives focused on issues surrounding nonphysician health-care providers. This policy directive increased the amount and quality of physician supervision exercised over departmental nonphysician health-care providers, such as physician assistants and nurse practitioners, who are given the authority to alter, suspend, or terminate a regimen of medical care. The directive further established minimum requirements for nonphysician providers in the areas of supervision and review, duties and responsibilities, and education and training.

Slow organizational response to serious medical-care incidents or mischances continued, and a 1984 directive was developed to address this problem. It required that timely action be taken to remove from patient care activities any departmental health-care provider found to be involved in improper, unethical, or unprofessional conduct (or substandard patient care). Beyond providing for adequate review, and for due process, the directive specified that all reasonable efforts must be made to protect the identity of a person who alleges misconduct by a departmental health-care provider that could lead to suspension or withdrawal of that provider's credentials, until the action to suspend or withdraw credentials is initiated. At that point, the release of the identity of that person is to be made only to those parties legally authorized to have this information.

The keystone of the Department of Defense quality-assurance program emerged, in 1985, as a DOD directive governing the credentialing of health care providers. This policy requires that all department health-care providers given the authority and responsibility for making independent decisions to initiate, alter, or terminate a regimen of medical or dental care being provided to a patient within the military health-care system be credentialed. Although medical and dental interns, residents, and fellows are not fully credentialed (because they are in training programs), a process was developed for defining and evaluating the professional and clinical activities of these professionals that was equivalent to that required for fully recognized physicians and dentists (which is administered by the department's committees on graduate medical education and training). In addition, as a part of a precredentialing process, all prospective departmental health-care providers evaluated for possible active duty, employment, or contractual relationships to provide medical or dental care within the department's health care system are now reviewed in terms of education, training, and performance documentation.

This 1985 policy further specifies that a provider activity profile (which documents each departmental health-care provider's clinical and professional activities) must be maintained. At a minimum, the provider profile must contain the number and type of medical and surgical procedures performed, the number of inpatients discharged, the number of deaths that do not meet specified criteria, the number of validated occurrences from the inpatient quality-of-care checklist, the number of drug-use and blood-transfusion variations that do not meet specified criteria, the number of surgical patients with normal tissue removed other than in those instances specified, the number of validated patient complaints, the number of provider-specific medical-record deficiencies, the number of malpractice claims filed, the number of medical record delinquencies, continuing medical-education credits, license renewal date, date of last training in cardiopulmonary resuscitation (or advanced cardiac life support or advanced trauma life support), and the date of the last credentials review. These profiles are updated at 6-month intervals, are kept in the provider's credentials file, and must cover the provider's most recent 3-year profile history.

The Department of Defense health-care providers included in the credentialing process are physicians, dentists, podiatrists, nurse anesthetists, nurse practitioners, nurse midwives, physician's assistants, optometrists, and clinical psychologists. Clinical social workers, clinical dieticians, clinical pharmacists, physical therapists, occupational therapists, audiologists, and speech pathologists are incorporated into the credentialing process if the individual has the privilege of initiating, altering, or terminating a regimen of medical care.

This extensive credentialing and profiling activity for military health-care providers is one of the most extensive in the United States. It establishes a powerful, explicit record of the high standards required by the Department of Defense.

Coupled with this credentialing requirement, the Department of Defense also established in 1985 a minimum licensure requirement for four groups of health care providers: physicians, dentists, clinical psychologists, and nurses. Such a requirement had been recommended by the Senate Appropriations Committee, and the department's requirement was incorporated into federal statute by the Department of Defense 1986 Authorization Act (P.L. 99-145). This policy directs that health care providers entering the armed forces (either on active or reserve duty or as civilian employees or contract personnel) possess and maintain a valid, current license at the independent practice level. Exceptions are provided for those individuals entering either the first or second year of departmental graduate-health-education programs, and those few health-care providers in these categories who had not yet sought a license were required to do so within 3 years of the issuance of the DODD.

To place greater limitation on the use of off-duty employment of physicians and other health-care providers (commonly called *moonlighting*), the department also issued in 1985 more stringent guidelines with a particular emphasis on the geographic and time limitations of off-duty employment.

In early 1986, the department announced details for a bold initiative to use an external (that is, civilian) physician-peer-review process to assess the quality of medical care in departmental hospitals. This intense focus on the quality of medical care further augmented the vigorous departmental quality-assurance program already in place by that date.

The concept was sparked by the Health Care Financing Administration's (HCFA) multiple contracts with state-based professional review organizations (PROs) to monitor the utilization and quality of hospital-based medical care for Medicare patients. This defense program, however, differs from the HCFA PRO contracts by establishing a standardized process for all reviews at military hospitals around the world rather than allowing state-based contractors to select specific problems. Further, the department's program focuses almost exclusively on the quality of medical care being provided, rather than on the additional elements of hospital utilization and cost containment, which are dominant functions in the HCFA PRO contracts. During its deliberations on the Fiscal-Year 1986 appropriations bill (P.L. 99-190), the Congress appropriated $7.5 million for this peer review effort and further urged the Office of Health Affairs to work with the General Accounting Office (GAO) in its efforts to improve the quality of care being provided within the various military facilities.

The Department of Defense external-peer-review program provides a common framework for review of medical and surgical cases representing approximately 15% of the nearly 1 million admissions per year to departmental hospitals. These reviews are accomplished by measurements against explicit criteria developed by an outside contractor in conjunction with departmental health-care leadership. The information gained by these external reviews assists the military hospitals in identifying quality-of-care issues requiring attention and (to the extent that similar data are available elsewhere in America) provides a data base for comparison between military medicine and other elements of American medicine.

In the area of general inpatient care, the reviews focus on all readmissions within 14 days of discharge from prior hospitalization, all surgically related deaths, all cases of neurological deficit related to anesthesia but not present on admission, all cases of organ failure (for example, kidney failure) that were not present at admission, and all postoperative or postprocedure complications (for example, infection complications following a leg artery graft). Emergency service reviews examine all unplanned returns to the emergency room within 48 hours of previous treatments, all patients dead on arrival at the emergency room when emergency transportation was the responsibility of the military hospital, all patients who die in the emergency room, and all patients who die in the hospital following inpatient admission through the emergency room.

To focus on the use and quality of surgical services, the program reviews 20% of all primary cesarean sections, hysterectomies, cholecystectomies (removal of the gall bladder), transurethral resections of the prostate, coronary-artery-bypass grafts, cardiac catheterizations, carotid endarterectomies (removal of fatty deposits from the carotid artery), femoral and aortic grafts, total joint replacements, and laminectomies (to relieve pressure on nerves emerging from the spinal cord).

Of all cases transferred to major military hospitals from referral hospitals, 10% are reviewed to assess the timeliness and appropriateness of the hospital transfer. This is of particular importance because of the extensive referral system within the Department of Defense hospital system.

To examine medical cases associated with high risks of complications, reviews examine all of the following cases: primary diagnosis of hypertension, primary diagnosis of cancer, congestive heart disease, atrial fibrillation (an abnormal heart-rhythm condition), diabetic ketoacidosis (a severe complication of diabetes), status asthmaticus (a sustained asthmatic condition), meningitis, ruptured appendix, acute myocardial infarction (heart attack), bone nonunions (following fractures), cerebrovascular accidents (strokes), neonatal and maternal deaths, preeclampsia and eclampsia (a severe complication of pregnancy), and the delivery of any infant weighing less than 2,500 grams (about 5 pounds). A review of the associated outpatient medical records, both before the hospitalization and following the usual posthospitalization checkups, is conducted to assess the quality of care provided before and after hospitalization in these high-risk cases.

In the event that the above cases are less than 10% of the prior month's patient discharges, supplementary reviews are made from among the more commonly seen medical cases. In this instance, the local military hospital can select medical conditions for review from the list, which includes injuries, gastrointestinal hemorrhage, pneumonia, pulmonary embolus (blood clot in the lung), kidney and other urinary-tract infections, gastrointestinal obstruction, septicemia (blood poisoning), and electrolyte imbalances (high or low levels of sodium or potassium in the blood).

The program contractor uses civilian nonphysician health-care personnel (with appropriate medical-record-review experience) to perform the preliminary screening of the medical records under review. Board-certified physicians later assess the subset of cases identified during this preliminary screening as having possible quality-related problems.

Information derived from all of the above-mentioned reviews is provided to each of the hospital commanders, the military department's surgeon general, and the Department of Defense Health Affairs office.

The PROs conducting review for HCFA differ from the Department of Defense in several aspects. By early 1986, each state had a separate PRO, and that PRO had considerable freedom to select problems for review, particularly regarding the quality assurance issues. Under the HCFA-PRO contracts effective through mid-year 1986, the PRO was required to develop one quality objective from each of five areas: reducing unnecessary hospital readmissions resulting from substandard care provided during the prior admission; ensuring the provision of medical services that, when not performed, have significant potential to cause serious patient complication; reducing avoidable deaths; reducing unnecessary surgery or other invasive procedures; and reducing avoidable postoperative or other complications.

Thus, the HCFA-financed PROs select their quality-assurance review-issues, and the Department of Defense has explicitly stated the problems to be examined in these areas. The HCFA PROs have additional reviews focusing on admission criteria, procedure utilization criteria (for example, the placement of cardiac pacemakers), and DRG validations that are not part of the defense review.

CONCLUSIONS

Very few of our nation's health care providers truly appreciate the evolutionary nature of the peer review process. The federal government, which has been the critical catalyst in developing peer review, has done so primarily in its role as a purchaser of health care. Thus, the issues of containing costs and ensuring quality of care have been intimately intertwined. Our nation's health policies are established primarily by our public (elected) officials. The vast majority of those elected to office, however, are not personally familiar with the nuances of the health delivery system. Accordingly, our nation's health providers, and particularly those in the mental health community, must become active in the public policy process in order to ensure that the evolving programs will really reflect the realities of practice.

A review of the role of peer review under the four major federal-health-care initiatives indicates that a truly unified approach has not yet evolved. To some extent, this diversity reflects the lack of sophistication of the Congress, but it also reflects the lack of essential agreement within the health-care-provider community itself. Within the Department of Defense health-care system, where the government serves as a provider of care, an exciting new movement is evolving toward ensuring objective standards of care and data-based definitions of quality assurance. One can only hope that, eventually, the data obtained from the various peer-review and quality-assurance efforts will, in fact, become integrated and will will reflect the most up-to-date scientific and clinical knowledge available.

ACKNOWLEDGMENTS

We would like to express our appreciation to Jay Constantine and Bob Hoyer, formerly of the Senate Finance Committee staff, for their assistance in reviewing the Social Security section of this chapter.

REFERENCES

Arnett, R. H., III, Cowell, C. S., Davidoff, L. M., & Freeland, M. S. (1985, Spring). Health spending trends in the 1980's: Adjusting to financial incentives. *Health Care Financing Review, 6,* 1–26.

Bent, R. J., Willens, J. G., & Lassen, C. L. (1983). The Colorado clinical psychology/expanded mental health benefits experiment: An introductory comment. *American Psychologist, 38,* 1274–1278.

Claiborn, W. L., & Zaro, J. S. (1979). The development of a peer review system: The APA/CHAMPUS contract. In C. A. Kiesler, N. A. Cummings, & G. R. VandenBos (Eds.), *Psychology and national health insurance: A sourcebook.* Washington, DC: American Psychological Association.

Committee on Finance Background Paper. (staff, 1985). *The peer review organization (PRO) program.* Prepared for the use of the Members of the Committee on Finance.

Congressional Budget Office. (CBO). (1979). *The effect of PSROs on health care costs: Current findings and further evaluation* (background paper). Washington, DC: U.S. Government Printing Office.

Constantine, J. (1981). *Proposed phaseout of PSROs and utilization review requirements.* Statement for the Subcommittee on Health, Committee on Finance, on Professional Standards Review Organizations. Washington, D.C.: U.S. Government Printing Office.

DeLeon, P. H. (1981). The medical expense deduction provision—Public policy in a vacuum? *Professional Psychology, 12,* 707–716.

DeLeon, P. H. (1983). The changing and creating of legislation: The political process. In B. Sales (Ed.), *The professional psychologist's handbook.* New York: Plenum Press.

DeLeon, P. H., & VandenBos, G. R. (1980). Psychotherapy reimbursement in federal programs: Political factors. In G. R. VandenBos (Ed.), *Psychotherapy: Practice, research, policy.* Beverly Hills, CA: Sage.

DeLeon, P. H., & VandenBos, G. R. (1984). Public health policy and behavioral health. In J. D. Matarazzo, J. A. Herd, N. E. Miller, & S. M. Weiss (Eds.), *Behavioral health: A handbook of health enhancement and disease prevention.* New York: Wiley.

DeLeon, P. H., Frohboese, R., & Meyers, J. C. (1984). Psychologist on Capital Hill: A unique use of the skills of the scientist/practitioner. *Professional Psychology: Research and Practice, 15,* 697–705.

DeLeon, P. H., VandenBos, G. R., & Kraut, A. G. (1984). Federal legislation recognizing psychology. *American Psychologist, 39,* 933–946.

DeLeon, P. H., Forsythe, P., & VandenBos, G. R. (1986). Federal recognition of psychology in rehabilitation programs. *Rehabilitation Psychology, 31,* 47–56.

Desmarais, H. (1986). Testimony before the Senate Appropriations Subcommittee on Labor, Health and Human Services, and Education. Fiscal Year 1987 Budget Request for HCFA. (Senate Hearings 99-850, Pt. 2, pp. 949–1032). Washington, D.C.: U.S. Government Printing Office.

Devine, D. J. (1983). Testimony before the Senate Committee on Governmental Affairs, Subcommittee on Civil Service, Post Office, and General Services (Senate Hearings 98-688). Washington, DC: U.S. Government Printing Office.

Dorken, H., & DeLeon, P. H. (1986). Cost as the driving force in health care reform. In H. Dorken & Associates (Eds.), *Professional psychology in transition: Meeting today's challenges.* San Francisco: Jossey-Bass.

General Accounting Office. (GAO). (1980). *Department of health and human services should improve monitoring of professional standards review organizations* (HRD-81-26). Washington, DC: U.S. Government Printing Office.

General Accounting Office (GAO). (1983). *Financial and other problems facing the federal employees health insurance program* (GAO/HRD-83-21). Washington, DC: U.S. Government Printing Office.

General Accounting Office (GAO). (1985). *Constraining national health care expenditures: Achieving quality care at an affordable cost* (GAO/HRD-85-105). Washington, DC: U.S. Government Printing Office.

Grupenhoff, J. T. (1983). Profile of congressional health legislative aides. *Mount Sinai Journal of Medicine, 50,* 1–7.

Inouye, D. K. (1983). Mental health care: Access, stigma, and effectiveness. *American Psychologist, 38,* 912–917.

Institute of Medicine. (IOM). (1974). *Advancing the quality of health care: Key issues and fundamental principles* (IOM Publication 74-04). Washington, DC: National Academy of Sciences.

Institute of Medicine. (IOM). (1976). *Assessing quality in health care: An evaluation* (IOM Publication 76-04). Washington, DC: National Academy of Sciences.

Kiesler, C. A.., Cummings, N. A., & VandenBos, G. R. (Eds.). (1979). *Psychology and national health insurance: A sourcebook.* Washington, DC: American Psychological Association.

Lohr, K. N. (1985). *Peer review organizations (PROs): Quality assurance in medicare.* Working paper prepared for the Office of Technology Assessment.

O'Sullivan, J. (1981). *Professional standards review organizations and utilization review: FY82 administration budget proposals* (Library of Congress mini brief number MB81232).

O'Sullivan, J. (1982). *Medicare and Medicaid provisions of the "tax equity and fiscal responsibility act of 1983"* (P.L. 97-248) (Library of Congress Report No. 82-173 EPW).

Prospective Payment Assessment Commission (ProPac). (1985). *1986 Adjustments to the Medicare Prospective Payment System: Report to the Congress.* Washington, DC: Author.

Prospective Payment Assessment Commission (ProPac). (1986). *Medicare Prospective Payment and the American Health Care System: Report to the Congress.* Washington, DC: Author.

Rodriguez, A. R. (1983). Psychological and psychiatric peer review at CHAMPUS. *American Psychologist, 38,* 941–947.

U.S. Congress, Office of Technology Assessment (OTA). (1985a). *First Report on the Prospective Assessment Commission (ProPac).* Washington, DC: U.S. Government Printing Office.

U.S. Congress, Office of Technology Assessment (OTA). (1985b). *Medicare's Prospective Payment System: Strategies for Evaluating Cost, Quality, and Medical Technology* (OTA-H-262). Washington, DC: U.S. Government Printing Office.

U.S. Congress, Office of Technology Assessment (OTA). (1986). *Second Report on the Prospective Assessment Commission (ProPAC).* Washington, DC: U.S. Government Printing Office.

U.S. Government Printing Office (GPO). (1970). *Medicare and Medicaid—Problems, issues, and alternatives* (report of the staff to the Senate Committee on Finance). Washington, DC: Author.

U.S. Government Printing Office (GPO). (1978). *Proposals to expand coverage of mental health under Medicare-Medicaid* (hearings before Subcommittee on Health, Committee on Finance, U.S. Senate). Washington, DC: Author.

U.S. Government Printing Office (GPO). (1985). *Estimates of federal tax expenditures for fiscal years 1986–1990.* Prepared by the staff of the Joint Committee on Taxation (JCS-8-85). Washington, DC: Author.

VandenBos, G. R. (1983). Health financing, service utilization, and national policy: A conversation with Stan Jones. *American Psychologist, 38,* 948–955.

Willens, J. G., & DeLeon, P. H. (1982). Political aspects of peer review. *Professional Psychology, 13,* 23–26.

Young, D. A. (1985). Address at the American Psychological Association Public Policy Forum. Washington, DC.

14

Legal Considerations in Quality Assurance

Donald N. Bersoff and Kit Kinports

INTRODUCTION

Like other forms of professional conduct, those functions that can be subsumed under the heading of *quality assurance* have come under increasing judicial scrutiny. Although peer review has borne the brunt of this scrutiny in the 1980s, peer review is not equivalent to quality assurance, nor is it the only such activity to evoke the concern of the courts. In psychological training and practice, the process of enhancing the probability that professional psychologists will provide services of good quality encompasses many activities. It begins much earlier in time than peer review, involves examination by persons and groups who may be external to psychology itself, and includes review not only of individual performance but of institutional performance as well.

At the institutional level, quality assurance may begin with regional accreditation of the university or college in which the psychology department is housed. It continues with the university's grant of permission to the department to develop and offer a doctoral training program in one or more of the branches of professional psychology. The process becomes increasingly rigorous when the training program seeks accreditation by the American Psychological Association (APA).

At the individual level, the process is more complex and may be lifelong. At the preprofessional level, quality assurance begins when students apply for entrance into professional training programs and take the Graduate Record Examination, continues when they seek admission to candidacy as doctoral students and undergo internship training, and ends when they defend their dissertation immediately before the awarding of the doctorate itself. But as all practitioners are well aware, examination of ability does not stop when the degree is granted. The potential practitioner faces scrutiny and testing by licensure or certification boards, and by the APA itself, should he or she seek membership and eventually fellowship status in the association. Both the novice and the experienced practitioner who belong to APA are bound by a set of standards whose purpose is to specify acceptable levels of quality assurance and professional and ethical performance. These standard-

DONALD N. BERSOFF and KIT KINPORTS • Ennis, Friedman, & Bersoff, 1200 17th Street, N.W., Washington, D.C. 20036.

setting documents include the *Standards for Providers of Psychological Services* (APA, 1977; now *Guidelines for Providers of Psychological Services*), the "Specialty Guidelines in Clinical, Counseling, Industrial/Organizational, and School Psychology" (APA, 1981b, c, d, e), and the "Ethical Principles of Psychologists" (APA, 1981a). In addition, several states require that practitioners who wish to retain their license maintain their proficiency through formal continuing-education activities. Experienced practitioners who wish advanced credentials must endure evaluation by such non-APA organizations as the American Board of Professional Psychology, the American Board of Forensic Psychology, and the National Register of Health Service Providers.

Thus, peer review is at the end of a long list of quality assurance activities that the practitioner must face, and it is certainly not the only one that may be challenged in legal proceedings. All of the functions just described are vulnerable to judicial scrutiny. In fact, it is surprising that some of these activities have virtually escaped legal challenge, for example, the use of licensure examinations of questionable psychometric soundness. By and large, however, most activities that attempt to protect the public and enhance competent professional performance have been upheld by the courts.

There is no doubt that, at the present time, peer review is the most controversial mechanism designed to help assure the public that it is receiving reasonably effective and necessary psychological services. It is increasingly being challenged by practitioners who find the process anticompetitive, distasteful, time-wasting, and invasive both to themselves and their clients. Thus, we will devote the majority of this chapter to an analysis of recent and significant litigation concerning peer review. Nevertheless, to give a broader context for the overall subject matter of the book of which this chapter is a part, we will begin by discussing the status of the law with regard to other quality-assurance activities.

PRE-PEER-REVIEW ACTIVITIES

From the possible pool of quality assurance activities unrelated to peer review, we have selected three that may serve as exemplars: (a) assessing the competency of students in preprofessional training; (b) admitting applicants to professional associations; and (c) granting licenses to those who wish to engage in private practice. None of these will receive the attention that they may deserve, but readers may glean some general principles that can lead to a broader understanding of how courts adjudicate challenges to quality assurance functions.

PREPROFESSIONAL TRAINING

The most significant case in this area is *Board of Curators of the University of Missouri v. Horowitz* (1978). A student in her 4th year of medical training was dismissed by a state university even though she had successfully completed all her academic courses up to that time. The university contended that the student's clinical performance was substandard and that, on the basis of observation by her supervisors, the medical school had determined that she would not be able to care properly for patients. Among other contentions, the student asserted that she had been entitled to a hearing before expulsion so that she could contest the university's actions. The university responded by proving that, for 2 years, the student had been counseled concerning her poor patient care. It argued that a court should not become involved in what was essentially a matter of academic discretion, and that the creation of formal procedural safeguards in that context would waste valuable time that should be devoted to training and that would subject the university to harassing litigation each time it sought to dismiss incompetent students. The U.S. Supreme Court upheld the university's position and ruled that no hearing was necessary. In so holding, the Court said:

> The decision to dismiss [Ms. Horowitz] ... rested on the academic judgment of school officials that she did not have the necessary clinical ability to perform adequately ... Such a judgment is by its nature more subjective and evaluative than the typical factual questions presented in the average disciplinary decision.
>
> The determination whether to dismiss a student for academic reasons requires an expert evaluation of cumulative information and is not readily adapted to the procedural tools of judicial or administrative decisionmaking.
>
> We decline to further enlarge the judicial presence in the academic community and thereby risk deterioration of many beneficial aspects of the faculty–student relationship. We recognize, as did the Massachusetts Supreme Judicial Court over 60 years ago, that a hearing may be "useless or even harmful in finding the truth as to scholarship." (pp. 89–90)

Thus, assuming that a training program has provided notice during supervision to students who are having serious problems in providing competent professional services and has acted in good faith, the program appears relatively immune to challenges to its academically based determinations.

ADMISSION TO PROFESSIONAL ASSOCIATIONS

Most professional organizations, like the APA and its state affiliates, are nonprofit, voluntary associations or corporations established under state and federal law. Their purposes and powers are set forth in articles of incorporation and in bylaws, rules, and regulations that the associations formulate and enforce. In this sense, professional associations are part of a broad category of private entities that the judicial system has traditionally treated with deference.

Insulating private, voluntary associations from judicial scrutiny is justifiable for a number of reasons. First, there is the deep-rooted belief that individuals should be allowed to associate with whomever they please. Second, it is feared that public supervision of private groups could lead to excessive governmental power and bureaucracy, creating a society in which diversity and individual initiatives are replaced by orthodoxy. Third, courts often lack the expertise to review the judgment of private associations, particularly regarding credentials, and although courts could be educated in these matters, the cost of ensuring the propriety of every membership decision would be, on balance, a waste of judicial resources.

The policy of judicial noninterference has been more pronounced in cases involving exclusion from membership than in those involving expulsion. In expulsion cases, the courts have justified intervention on the grounds that the expelled member has been deprived of property rights as well as existing and valuable personal relationships, or that the association's actions breached a contractual agreement between the member and the association. Because neither of these grounds is relevant in the case of applicants or nonmembers, the courts have distinguished expulsion cases and have generally held that professional associations have unlimited discretion to grant or refuse admission to membership.

Nevertheless, there have been some exceptions to this rule. The first case to break completely from the traditional deferential approach of the judiciary toward professional associations was *Falcone v. Middlesex County Medical Society* (1961). It was also the first case to use a monopoly power theory to justify compelling admissions; this theory is that admission should be compelled because the association exercises monopoly power over the profession, and that membership thus becomes a matter of economic necessity. In *Falcone,* a licensed physician sought entrance to a county medical association but was denied admission because he had not been trained at an AMA-approved medical school. Because he did not belong to the local association, Dr. Falcone could not obtain staff privileges at any local hospitals. Thus, the association was perceived as wielding monopoly power affecting the ability of Dr. Falcone to practice his profession. In that context, the court held that when a private association possesses monopoly control, that power

> should be viewed judicially as a fiduciary power to be exercised in a reasonable and lawful manner for the advancement of the interests of the ... profession and the public generally .... When its action has no relation to the ... evaluation of professional standards and runs strongly counter to the public policy of our State and the true interests of justice, it should and will be stricken down. (pp. 799–800)

The medical society in *Falcone* can be distinguished from national and state associations composed of psychologists. Membership in the APA, for example, is not a prerequisite for licensure, hospital privileges, or access to such economic advantages as third-party reimbursement. Thus, membership is not an economic necessity.

Nevertheless, exclusion from a professional association, it could be claimed, results in substantial harm not measurable in pure economic terms. Membership in an association may be necessary to obtain malpractice insurance, social and professional contracts necessary for referral, professional advancement, and access to publications, libraries, and meetings that help practitioners keep abreast of developments in their field. Insofar as the association is influential in monitoring and supervising professional practice, exclusion also deprives disappointed applicants of the important right to participate in decision making concerning regulation of the profession.

The noneconomic arguments described here were used by the plaintiff in the only reported case on this subject involving a psychologist, *Salter v. New York State Psychological Association* (NYSPA) (1964). Salter was a practicing psychologist denied membership in NYSPA because he did not meet its criterion that members must have done graduate work in psychology. Salter had graduated from college and had gone directly into practice before New York passed a licensure law requiring an advanced degree in psychology. The law contained a "grandparenting" provision that allowed those in practice for a certain number of years to continue practicing without obtaining advanced training. NYSPA, however, made no exception in Salter's case. Salter claimed, among other arguments, that NYSPA exercised monopoly power over his profession. He emphasized the noneconomic interests that were jeopardized by exclusion, chiefly his inability to be heard in the forum most concerned with the regulation and development of the practice of psychology in New York. Further, he claimed, membership itself was a credential that signaled superior skill, and it was possible that he would have access to fewer patients because of the public recognition that NYSPA had refused to endorse him, that he would receive fewer referrals from members of the association, and that he would be forced to charge lower fees for his services.

Ultimately, Salter's arguments were to no avail. New York's highest court took a narrow view of the role of judicial intervention, holding that deprivation of anything short of economic necessity, the association's exercise of monopoly power over the profession, or the imposition of arbitrary or unreasonable standards did not warrant intrusion into the association's affairs. It found that Salter was a successful practitioner and could not show direct economic injury; any harm resulting from exclusion stemmed from the respect the public gave to the judgment of NYSPA rather than from any monopoly power it exercised.

Some courts, particularly those in California, have, however, taken a much more expansive view of the judicial system's obligation to review a private association's admission decisions. California courts have held that intervention is proper not only in situations where an association has monopoly power, making membership an economic necessity, but also where membership is a "practical necessity" without which an individual cannot "realize maximum potential and recognition" in his or her chosen profession (*Pinsker v. Pacific Coast Society of Orthodontists*, 1969, p. 499; see also *Ascherman v. Saint Francis Memorial Hospital*, 1975). In jurisdictions like California, therefore, it may be more difficult for professional associations to escape challenges to denial of admission. But, as *Falcone* indicates, association standards that are reasonable, that are applied fairly, and that are genuinely designed to protect the quality of the profession and of professional practice are likely to be sustained.

LICENSURE

Until peer review, licensing was the most frequently challenged of the activities whose aim is to ensure the provision of quality psychological services. When licensing began, the majority of litigation was brought by unsuccessful applicants challenging state-licensure-board interpretations of "grandparenting" provisions. When that issue waned in importance, later applicants began to challenge the legitimacy of the requirement that candidates have a doctorate degree from a program that is substantially psychological in content. In all but five states, successful completion of the doctorate is the minimal educational prerequisite for licensure.

The requirement that applicants for licensure who are not yet in practice have a doctorate in psychology (or its equivalent) has not been successfully challenged in the handful of cases brought by psychologists since the mid-1970s. Educational prerequisites have been seen as reasonable statutory provisions designed to implement the state's authority to regulate professions to protect the public. For example, in *Cohen v. State* (1978), an applicant who had a master's degree in psychology and a doctorate in social science failed to convince a state supreme court that he should be permitted to take the certification examination. In holding that the requirement of a doctorate in a program that is primarily psychological in nature was not unconstitutional, the court acknowledged that the "right to pursue a profession is subject to the paramount right of the state ... to regulate business and professions in order to protect the public health, morals, and welfare" (p. 303).

A recent case concerning licensure indicates, however, that the determination of which programs are to be deemed primarily psychological cannot be vested in the university; rather, the responsibility lies with the state. In *Charry v. Hall* (1983), an applicant was refused permission to take New York's licensure examination for psychologists because he had obtained his Ph.D. degree from a program designated as "Human Relations and Social Policy." New York required "a doctoral degree in psychology, granted on the basis of completion of a program of psychology registered with the department [of education] or the substantial equivalent thereof." The university officer charged with making the relevant determination, who was a psychologist, concluded that the human relations program did not meet the statutory definition. The applicant asserted that it did.

In reviewing the competing arguments, the court first concluded that the right to take a licensure examination was not as substantial as the right to be protected against unfair revocation of the license itself. Nevertheless, the court held the right to take an examination as a prerequisite to licensure was an interest protected by the due-process clause of the Fourteenth Amendment to the U.S. Constitution and, therefore, could not be denied arbitrarily by the state. The court then suggested that the determination of which programs satisfy the educational criterion is best left in the hands of state officials rather than university officials, who may be motivated by intramural disputes between competing programs. It was the state that should establish standards for substantial equivalency and should independently investigate the facts with respect to each individual's application. This procedure, the court held, would better protect the applicant's property right to take the licensure examination.

But neither *Charry* nor prior cases pose a significant danger that the doctoral-level requirement will be abrogated in the foreseeable future. That requirement appears to be established as one that, among others, helps to ensure the provisions of quality psychological services.

PEER REVIEW

The imposition of licensure requirements since the mid-1960s evoked claims by excluded practitioners that licensure was only a sham to protect the profession rather than the public. The initiation of peer review has evoked similar complaints from diverse sources, including psychologists, that it, in effect, inappropriately uses psychologists to help lower the costs of insurance companies and thus has little or nothing to do with quality assurance. In addition, opponents of peer review argue that peer review may lead to anticompetitive activity among competing practitioners, may endanger the confidentiality of the client–clinician relationship, and may increase the liability of psychologists in a number of areas. The debate about whether peer review is merely cost-cutting utilization review on behalf of insurance companies or another positive means of ensuring quality services is beyond the scope of this chapter, though this subject is discussed elswhere in this book. But the other concerns are relevant, and we will now address the major aspects of each of them.

ANTICOMPETITIVE CONDUCT

A core assumption underlying American economic philosophy is a belief in the value of competition. This strong faith led to the passage, beginning in the latter part of the nineteenth century, of federal antitrust laws prohibiting certain conduct that was judged to unreasonably restrain trade. Thus, in 1890, Congress passed the Sherman Act, forbidding monopolies and conspiracies in restraint of interstate commerce; in 1914, it passed the Federal Trade Commission Act, prohibiting unfair methods of competition and creating the Federal Trade Commission and later, through amendment, barring deceptive business practices; in 1914, Congress also passed the Clayton Act, barring several forms of anticompetitive activity and permitting those plaintiffs who ultimately prove that they have been economically harmed by anticompetitive conduct to recover threefold (treble) damages. In addition, many states have enacted their own antitrust statutes to regulate anticompetitive activities that do not involve commerce across state lines.

Until the 1970s, activity by health care providers and institutions was considered beyond the reach of the antitrust laws. The health care profession was widely thought to be protected by an exemption created by both federal and state antitrust laws for the "learned professions" (Bersoff, 1983; Overcast, Sales, Pollard, 1982; Rich, 1980). However, beginning with *Goldfarb v. Virginia State Bar* (1975), the U.S. Supreme Court has made it clear that professions are not exempt from the antitrust laws. In a decision holding that a local bar association violated the antitrust laws by establishing minimum fee schedules for real-estate closings, the Court stated, "The nature of an occupation, standing alone, does not provide sanctuary from the Sherman Act . . . nor is the public-service aspect of professional practice controlling in determining whether § 1 [of the act] includes professions" (p. 787).

Goldfarb did include a footnote indicating that a different standard might apply to determining whether a restraint imposed on a particular profession violates the antitrust laws than would apply in the usual commercial context: "The public service aspect, and other features of the professions, may require that a particular practice, which could properly be viewed as a violation of the Sherman Act in another context, be treated differently" (*Goldfarb*, pp. 788–789). However, the continued applicability of that statement to professional activity was severely undercut, if not destroyed, 3 years later by another Supreme Court decision, *National Society of Professional Engineers v. United States* (1978), in which an engineering association sought to evade the antitrust laws by asserting that its challenged activity (i.e., prohibiting competitive bidding) was necessary to ensure public safety. But the Court held that a restraint on the economic aspects of a profession is not necessarily justifiable even if it is claimed to be in the public interest:

> The fact that engineers are often involved in large-scale projects significantly affecting public safety does not alter our analysis. Exceptions to the Sherman Act for potentially dangerous goods and services would be tantamount to a repeal of the statute . . . . The judiciary cannot indirectly protect the public against . . . harm by conferring monopoly privileges on manufacturers. (pp. 695–696)

Two decisions by the Supreme Court in 1982, *Arizona v. Maricopa County Medical Society* and *Union Labor Life Insurance Co. v. Pireno*, clearly indicated that the antitrust laws would be applied vigorously to the activities of peer review committees. At issue in *Maricopa County* were the activities of a nonprofit state corporation, an adjunct of the local medical society, composed of licensed physicians, osteopaths, and podiatrists engaged in private practice. Those activities included, among others, establishing a schedule of maximum fees that participating practitioners agreed to accept as payment under plans approved by the Maricopa Foundation for Medical Care and reviewing the medical necessity and appropriateness of treatments provided by its members to patients insured by the foundation. The state of Arizona, suing on its own behalf and on behalf of its citizens, brought suit against the society and the foundation. The state alleged that the setting of maximum fees constituted price fixing in violation of Section 1 of the Sherman Act, which prohibits contracts, combinations, or conspiracies in restraint of trade.

In a 4–3 decision (Justices O'Connor and Blackmun did not participate), the Supreme Court held that an agreement among competing physicians setting fees that they may claim for services provided to certain insurance policyholders violates the Sherman Act. Not only did the court rule that maximum, as well as minimum, price-fixing violated Section 1, but it refused to permit the defendants to argue that such an activity might have procompetitive effects that would protect the public. Specifically, the Court said that "price-fixing agreements are unlawful *per se* under the Sherman Act and that no showing of so-called competitive abuses or evils which those agreements were designed to eliminate or alleviate may be interposed as a defense" (*Arizona v. Maricopa County Medical Society*, 1982, p. 345). Practices subject to the *per se* rule—like the price-fixing scheme at issue in *Maricopa County*—are conclusively presumed to be unreasonable and therefore illegal; the courts do not engage in an elaborate inquiry into the precise harm they have caused or the business excuses for their use.

Both *Goldfarb* and *Professional Engineers* left open the theoretical possibility that some marginally anticompetitive activity that could properly be held unlawful in another industry might be upheld if it occurred in one of the learned professions. Similarly, in *Maricopa*, the Court alluded to the possibility that an agreement "premised on public service or ethical norms" (p. 349) might escape *per se* treatment. But, the Court held, "the price fixing agreements in this case . . . are not premised on public service or ethical norms . . . . The claim that the price restraint will make it easier for customers to pay does not distinguish the medical profession from any other provider of goods or services" (p. 349). The *per se* rule was applicable because the price restraint at issue tends to provide the same economic rewards to all practitioners regardless of their skill, their experience, their training, or their willingness to employ innovative and difficult procedures in individual cases. Such a restraint also may discourage entry into the market and may deter experimentation and new developments by individual entrepreneurs. It may masquerade for an agreement to fix uniform prices, or it may in the future take on that character. (p. 348)

Although the dissent characterized the plan as an attractive consumer-protection measure that "seems to be in the public interest" (p. 357), its opinion did not prevail. For the first time, the *per se* liability standard was expressly applied to a restraint involving a learned profession. The Court did leave open the question whether, and to what extent, other professional practices may be immune from treatment under the *per se* approach and may instead be subject to a full-scale evidentiary analysis of whether they are in fact reasonable: "Restraints involving ethical rules designed to prevent deception or to protect the public in some way may yet be determined under the rule of reason if there is no effect on price ... " (Taylor, 1983, p. 239). But it is now clear that the *per se* rule will apply if the restraint relates to the commercial aspects of the professional practice, such as setting fees for services. It should be noted in this regard that the APA's peer review activities, both for private insurance companies and for CHAMPUS, do not involve scrutiny of the fees that psychologists charge; they are confined solely to determining whether the psychological treatment being provided is warranted (Rodriguez, 1983).

The Supreme Court rendered another significant decision 10 days after it decided *Maricopa*, concerning peer review in *Union Labor Life Insurance Co. v. Pireno* (1982). Pireno, a chiropractor who was a licensed practitioner in New York, alleged that the peer review practices of the Union Labor Life Insurance Co. (ULL) and the New York State Chiropractic Association (NYSCA) constituted a conspiracy in restraint of trade. Pireno had been subjected to frequent peer review under ULL insurance policies that required the company to reimburse only reasonable charges for necessary medical care and services. These examinations were conducted on several occasions by NYSCA's Peer Review Committee (PRC), a group of 10 practitioners who served on a voluntary basis, Pireno claimed that ULL and NYSCA had used the PRC to conspire to fix prices. As a result, he alleged, he had been restrained from providing his services freely and fully, and potential patients had been deprived of the benefits of competition. Pireno claimed no damages; he wished only to enjoin continued use of the PRC.

The defendants claimed that their actions were exempt from antitrust scrutiny under the McCarran-Ferguson Act (15 U.S.C. § 1001 *et seq.*). That act creates an exception to the Sherman Act for any activity that can be characterized as the business of insurance as long as that insurance activity is regulated by the state and does not involve a boycott, coercion, or intimidation. In *Group Life and Health Insurance Co. v. Royal Drug Co.* (1979), the U.S. Supreme Court had articulated three factors to be considered in determining whether a practice constitutes the business of insurance:

1. Whether the practice has the effect of transferring or spreading the policy holders' risks
2. Whether the practice is an integral part of the policy relationship between insurers and insured

3. Whether the practice is limited to entities within the insurance industry

The Supreme Court applied these criteria to the facts in *Pireno* and held that the activities of the PRC were not exempt under the McCarran-Ferguson Act on any of the three grounds. Specifically, the Court held that the PRC was not involved in transferring risk from insured to insurer; it was unrelated to the parties to the insurance contract; and the use of a state professional-peer-review committee involved "third parties wholly outside the insurance industry—namely, practicing chiropractors" (p. 132). The dissent agreed with the NYSCA and asserted that the PRC served the same function as claims adjusters and, as such, constituted "a critical component of the relations between an insurer and an insured," thus making it "part and parcel of the 'business of insurance'" (p. 139). That position failed to persuade the six-member majority, however, which held that if the insurance company was free to disregard the PRC's judgments, those judgments were "merely ancillary to the claims adjustment process" (p. 134). Moreover, even if the peer reviewers' decisions were binding, it is doubtful that the PRC's activities would be considered an integral part of the policy relationship between the insurer and the insured, and it is even more doubtful that those activities would be judged to have the effect of transferring or spreading the policyholders' risk. In support of this conclusion is the 1983 decision of a U.S. court of appeals in *Ratino v. Medical Service of the District of Columbia*, which held that peer review of the usual and customary charges paid to participating physicians by Blue Shield was not protected from antitrust scrutiny under the McCarran-Ferguson Act.

It should be noted that in none of these cases did the U.S. Supreme Court impose liability. Rather, it merely said that the case could go to trial because the defendants were not permitted to short-circuit full review of the charges by claiming that their activities were exempt under the business-of-insurance exemption. The dissenters in *Pireno* did express concern that the majority's "decision will vastly curtail the peer review process [as] few professionals or companies will be willing to expose themselves to possible antitrust liability through such activity" (p. 140). But as subsequent cases appear to imply, if peer reviewers do not engage in activities that concern fees and if they participate in the peer review process in good faith (i.e., do not intend to harm colleagues they review), they will very likely escape any liability for money damages, especially in those jurisdictions where states have passed statutes protecting reviewers (see, e.g., *Quasem v. Kozarek*, 1983).

VIOLATION OF CONFIDENTIALITY

Beyond the concerns generated by the esoteric dictates of the antitrust laws, psychologists subjected to peer review have raised more easily recognizable issues. Perhaps the most important is whether clients who authorize the release of treatment information to insurance companies and other third-party payers, such as Blue Shield, for the limited purpose of peer review thereby waive their claim to confidentiality for all purposes. As a corollary, if the release of information creates a general waiver, can the client sue the psychologist for breach of confidentiality or of any other obligation?

In the usual case, the insurance company asks the psychologist to complete a report form and asks the client to sign an information release form. The release form specifies that treatment information will be used solely for the review of the necessity and quality of care by professionals and will not be used for any other purpose without the patient's consent. Although the form contains the client's name when it is submitted to the insurance company, the name is not included in the information sent to those who perform peer review for the APA.

The oldest and perhaps most important right guaranteed to clients seeking mental health intervention is confidentiality. Several states have passed statutes preventing the disclosure in a legal proceeding of confidential information that is learned in a professional relationship between client and psychologist. This right may be embedded in a state's rule of evidence creating a psychotherapist–inpatient privilege or may be found in statutes providing for the licensure or certification of psychologists (Bersoff & Jain, 1980). Some state statutes go further and protect treatment information from being disclosed to any third party, whether in a legal proceeding or not. In addition, some states have passed mental-health-information acts that recognize a client's right to prevent a psychologist from disclosing confidences to third parties outside a legal proceeding without the client's permission, and to sue for damages if the psychologist breaches that obligation. Finally, there is a growing recognition of a federal and state constitutional right to privacy that includes a limited right of privacy between therapist and client.

But all these privileges belong to clients, not to practitioners, and they can be waived in several ways. A client can waive the privilege expressly, either orally, as in courtroom testimony, or in writing. Waiver may also be implied from the client's conduct. Waiver can be implied even when the client may not actually have intended to waive the privilege, but when it would be unfair to the opposing party to permit retention of the privilege. Thus, where clients initiate legal actions that raise the issue of their mental condition (as in civil claims involving the infliction of emotional harm or criminal actions where the defendent asserts an insanity defense), the courts have held that the client's right to protect information concerning that issue has been waived.

The release form used by insurance companies to obtain information relevant to claims review can be considered an express waiver by contract. But to be effective, such waivers cannot be coerced or given under undue influence. Similarly, agreements entered into by a party who has little choice and bargaining power (i.e., contracts of adhesion) will not be legally enforced. Further, any waiver must be a voluntary and knowing act done with sufficient awareness of the relevant circumstances and the likely consequences, and an individual release form might therefore be invalidated if the client was not fully informed of the consequences of releasing treatment information. Thus, in some jurisdictions, a release of information might not be a valid waiver of the privilege if the client did not know that, by releasing information to the insurer, the client thereby waived protection of the privilege as to the insurance company as a whole. Or if the client was incorrectly informed that the disclosure of treatment information for the limited purposes of insurance and peer review would preserve the privilege for all other purposes, the waiver might be invalidated because it was not knowing and voluntary. Although specific releases may be challenged under one or more of these theories, most courts have not regarded the release of information authorized for insurance or peer review purposes as necessarily invalid because it is coerced, is a contract of adhesion, or is involuntary.

Assuming that the release-of-information form used by insurance companies for peer review is valid, what is its effect? The release form itself expressly limits the persons to whom the information may be disclosed, the purposes for which the information may be used, and the length of time for which the form is valid. Are these limitations legally enforceable? Or by releasing treatment information to the insurance company, has the client completely waived the protections of the psychologist–patient privilege and lost the right to prevent employers and others from seeking to introduce that information in court?

Unfortunately, the answers depend on applicable state law, and readers should consult the statutes and case law of their particular jurisdiction. There are, however, a number of arguments that can be made to support a limited reading of the waiver effected by insurance release forms and to justify refusing to disclose treatment information submitted to insurance companies to other persons. But it should be noted that some courts have held that, once partial disclosure of protected information is made to a party, that party has the right to complete disclosure (*Sicpa North America, Inc. v. Donaldson Enterprises, Inc.,* 1981). For example, if the client authorizes the release of treatment information to an insurance company, the client probably waives the privilege at least as to the insurance company. In a suit between the client and the insurance company, the company may compel the client's therapist to testify. But if another party seeks to introduce the information at trial, the client can argue that the privilege has not been waived as to that person. Some courts have, in fact, taken this approach, refusing to find an unlimited waiver of the privilege when the client has made only a limited disclosure of confidential information. But at least one court in California has held that the entire privilege is waived if the holder of the privilege, without coercion, has disclosed a significant part of the communication or has consented to such disclosure made by anyone else (*People v. Pic'l,* 1981).

Except in jurisdictions like the one just noted, it can be argued that release of confidential information to an insurance company is only a limited waiver of the psychologist–patient privilege for a very specific purpose. The argument is supportable on a number of grounds. First, some courts have held that the privilege is to be construed broadly in favor of the client. Second, the courts have held that waivers of the privilege are not lightly to be presumed. Third, the courts held that consent forms will be strictly construed against the insurance company.

In addition to a limited-waiver theory, clients and their therapists may argue that the insurance company is a necessary third party to whom confidential communications may be disclosed without any loss of privilege. For example, California's Evidence Code, Section 912, provides that the therapist-patient privilege is not waived by a "disclosure in confidence of a communication that is protected by a privilege . . . when such disclosure is reasonably necessary for the accomplishment of the purpose for which the . . . psychotherapist was consulted."

One California court has applied this provision to the disclosure of information to an insurance company. In *Blue Cross of Northern California v. Superior Court of County of Yolo* (1976), the court held:

> The [insurance] carrier's participation transforms the dual medico-economic relation between physician and patient into a tripartite relationship. Anticipated payment is a prerequisite of medical care in all cases involving financial resource to a prepaid health plan. Coverage determinations ineluctably call for disclosure of the patient's name and ailment. The information's disclosure to accomplish payment is reasonably necessary to achieve the consultation's diagnostic and treatment purposes. (p. 801)

Although we have found no other cases that have taken this approach, it is supported by language in the Supreme Court's opinion in *Whalen v. Roe* (1977), where the Court noted that "disclosures of private information to doctors, to hospital personnel, to insurance companies, and to public health agencies are often an essential part of modern medical practice" (p. 602). The argument that disclosure of information to insurance companies is reasonably necessary to achieve the purposes of treatment should likewise be applicable to disclosure for purposes of peer review.

Thus, even when state law does not expressly protect disclosures of confidential treatment information for insurance or peer review purposes, the client may argue that a limited disclosure for such purposes does not constitute a general waiver of the testimonial privilege because the privilege was waived only as to the insurance company and peer reviewers, or because the disclosure was reasonably necessary for the accomplishment of the purposes for which the therapist was consulted.

Many of these issues were played out in a very direct way in the case of *Doe v. Prudential Insurance Co.* (1985). In mid-1984, the APA was named as a defendant along with nine third-party payers who contracted with the association to provide professional services review. In a complaint filed in the Superior Court of New Jersey, the plaintiffs were four anonymous clients of psychologists permitted to practice in New Jersey and four licensed psychologists, all of whom are members of the APA. In the course of this case, eight of the insurance companies were dismissed as plaintiffs for a variety of technical reasons, leaving only the APA and the Prudential Insurance Company as defendants. Later, three of the patient-plaintiffs and three provider-plaintiffs were dismissed as well, leaving only one psychologist and one patient, the latter proceeding anonymously as Jane Doe.

The complaint had two counts, the second of which is relevant to this chapter. The plaintiffs alleged that the peer review contract between the APA and the third-party payers violated the New Jersey privileged-communication statute, was motivated by income and personnel needs, and was irreparably harmful because it violated the trust and confidentiality necessary for successful treatment. The plaintiffs sought no damages but asked the court to preliminarily and permanently enjoin the APA from "promoting and marketing" the peer review program and accepting treatment report forms and to enjoin the third-party payers from requesting, obtaining, and forwarding treatment report forms to the APA for peer review purposes.

The APA board of directors unanimously voted to vigorously defend the suit. The APA's law firm (of which the author is a part), along with local New Jersey counsel, represented the Association, but after over a year of preliminary motions and the initiation of discovery (e.g., the taking of depositions and the answering of written interrogatories), the case was dismissed by agreement of all the parties on October 30, 1985.

The basis of the dismissal was the passage of a statute restructuring the nature of peer review in New Jersey by a unanimous state legislature and its enactment by the governor. This statute provides for a collegial form of peer review rather than the previous format. If an insurer has reasonable cause to believe that the patient's treatment may be neither usual, customary, nor reasonable, the insurer may request peer review. But the review will be performed by an "independent professional review committee," that is, a group of experienced licensed psychologists established and appointed by the New Jersey Board of Psychological Examiners. On the basis of preliminary information, if there is reasonable cause to believe that the treatment is neither usual, customary, nor reasonable, the committee can request additional information from the treating psychologist, pursuant to a valid authorization signed by the patient. The information is prohibited from being disclosed to the insurer or any other person outside the committee and the two or more members of the committee who will conduct the peer review itself. The statute explicitly states that the authorization and disclosure of the confidential information needed for the review are not to be construed as a waiver of the psychologist–patient privilege.

This is not the proper space for a lengthy critique of the statute, but it places peer reviewer colleagues in the unenviable position of coming close to making a claims benefit determination, the proper role of the third-party payer. And, even more detrimental to the interests of psychologists, statutes such as this one may lead insurance companies that believe they should receive the information contained in mental-health treatment-reports and make the determination about whether the treatment is psychologically necessary to exclude mental health benefits altogether or to severely restrict access to mental health providers by sharply increasing copayments or reducing the number of allowable visits.

The APA contended in *Doe* that no waiver of the psychologist–patient privilege resulted from the disclosure of confidential information to insurers. Some psychologists, however, were concerned that disclosure would lead to their own liability, especially if it were ultimately held to be no longer protected as a result of disclosure for peer review purposes. Even if a court were to hold that information released for peer review purposes was discoverable in a lawsuit by a party other than the insurance company, and if the client suffered some actual damage as a result of the disclosure of that information, it is still difficult to find a legal basis for holding the psychologist liable for divulging the information pursuant to court order. Courts have held specifically that physicians who have been ordered by a court to divulge confidential information cannot be held liable in actions brought by their patients. In addition, many state statutes now protect peer reviewers against money damages liability for actions taken without malice and within the scope of their function as members of peer review committees.

If a psychologist failed to inform a patient that the express limitations on a release form would prevent the disclosure of the information for other purposes, a court might subject the psychologist to liability on the theory that he or she was in a better position than the insured to know the effects of the release form and was under a duty to inform the client of those effects. Or if the failure to inform the client invalidated the waiver, the psychologist could be held liable for unauthorized disclosure of confidential information. But we have found no cases on that question, and thus, it is difficult to tell whether psychologists face liability on either of those grounds. However, clients, like everyone else, are held responsible for knowing or finding out the legal effects of their acts. Nevertheless, the better solution, in the interests of client trust, is to amend release forms to reflect accurately the state of the law in the relevant jurisdiction.

Readers should have received the impression that the law of confidentiality in the context of peer review is still uncertain at this time. Thus, as a precaution, in the light of extant case law, we would suggest the following:

1. Information provided in treatment reports should be as limited as is consistent with the purposes of disclosure.

2. Client release forms should indicate what effect the release will have on the client's right to confidentiality and should specify that access to the information will be limited except when otherwise required by court order.

3. Contracts among insurance companies, their insureds, and peer review entities should provide that the insurance company shall use information in treatment report forms only for peer review purposes.

4. Psychotherapist–patient privilege statutes should be amended to make clear that the disclosure of treatment information solely to insurance companies and peer reviewers waives the privilege only as to those entities and does not waive the privilege generally—or, alternatively, that such disclosure is reasonably necessary for the accomplishment of the purposes for which the psychologist was consulted and does not constitute a waiver at all.

5. Statutes should be enacted that severely restrict the breadth of information that insurance companies can force clients to disclose and that specify that disclosure may be made only when the client has given written authorization for such disclosure.

ABANDONMENT

A final issue surrounding peer review concerns the potential liability of therapists and peer reviewers for "abandonment," a civil wrong (tort) for which money damages may be recovered. The problem would arise in the following manner. A psychologist, seeing a client in psychotherapy, undergoes peer review. The peer reviewers recommend to the third-party payer that it no longer reimburse the client for the cost of therapy. The carrier agrees with the recommendations and decides to discontinue reimbursement. The client is unable to otherwise pay the psychologist's fee. Although they mutually agree that therapy should continue, the psychologist tells the client that because the client can no longer pay for therapy, the therapist will no longer be able to treat the client.

Under these conditions, two questions arise. First, are psychologists who act as peer reviewers and whose recommendations are instrumental in an insurance company's decision not to reimburse clients for psychotherapy subject to liability for abandonment? Similarly, are treating psychologists who refuse to continue therapy with clients who are no longer reimbursed by the insurance company and can no longer pay liable for abandonment?

The law of abandonment is relatively clear and does not vary significantly among the various jurisdictions. A psychologist has no duty to undertake the treatment of all clients who desire treatment and may refuse to treat any person for any reason or even for no reason at all. But once a psychologist agrees to treat a client and a therapist–patient relationship is established, the psychologist has a duty to continue treatment and not to abandon the client except under certain circumstances (McIntire, 1962).

Psychologists are justified in discontinuing treatment if they are discharged by the client, if they agree with the client that treatment will be ended, or if the client no longer needs treatment. But in determining whether the need for treatment has ended, psychologists must exercise reasonable and ordinary care and skill; that is, they must not act negligently. Some courts have also suggested that psychologists cannot be held liable for abandonment if they refuse to treat clients who are uncooperative or who do not comply with their practitioners' treatment recommendations. However, the client's inability to reimburse the psychologist for the cost of therapy is not, by itself, an adequate justification for discontinuing treatment. In addition, psychologists may unilaterally withdraw from a case if they give the client enough notice to enable him or her to find an alternative source of treatment or if they make available a qualified substitute therapist.

Nonetheless, even a psychologist who does abandon a patient will not be subjected to liability unless the patient suffers injury as a result of the psychologist's discontinuing treatment. Moreover, psychologists may limit their potential liability for abandonment by expressly limiting the nature of the services to be provided to the client. For example, they may agree to see the client for only a specified number of visits, rather than for an unlimited period of time or until the client is treated successfully. Psychologists will not be exposed to liability if they refuse to provide services beyond the scope of their contract with the client.

The refusal to continue treating a client whose insurance company will no longer provide reimbursement for the costs of treatment can be construed as a form of abandonment. But in such circumstances, the psychologist will typically be able to avoid liability in a number of ways.

First, practitioners are permitted to withdraw from a case as long as they give adequate notice to the client and afford the client a reasonable amount of time to secure another source of therapy. In almost all cases involving peer review (as in the APA's system), the insurance company gives significant notice before discontinuing reimbursement. Thus, the psychologist will be able to give the client notice and time to find another therapist without having to provide free treatment.

Second, the psychologist may withdraw from the case if the client no longer needs therapy. The psychologist does not have limitless discretion in deciding that therapy is not necessary; that determination must be a reasonable one, and the psychologist must exercise ordinary skill and care in reaching that decision. If the insurance company's decision to terminate reimbursement is based on the evaluation of peer reviewers, their assessment that treatment is no longer necessary will prove that the practitioner's decision was a reasonable one.

Third, psychologists may withdraw from a case if they arrange for a qualified therapist to treat the client in their place. Admittedly, it may be difficult to find another private practitioner who will agree to see a client without reimbursement. In those cases, the psychologist might be able to refer the client to a community mental-health clinic or to another therapist who will treat the client at little or no cost. If the client is seriously disturbed and also presents a danger to him- or herself or to others, the therapist may have a duty to seek civil commitment.

Fourth, psychologists may be permitted to withdraw from a case if they make clear from the outset that they will continue to treat the client only if they are paid for their services by an insurance company or some other source. By expressly limiting the agreement with the client in this way, psychologists are arguably free to discontinue treatment when reimbursement is not forthcoming. The judicial decisions permitting health care providers to expressly limit their services, however, have not involved restrictions based on financial considerations. Rather, they involve limits based on the type of services to be provided, the number of treatments to be given, and the times and places at which services are offered. Given that a client's inability to pay ordinarily does not justify abandonment, it may be that the courts will refuse to enforce agreements that limit a psychologist's services to those clients who are able to pay. But an agreement to that effect at the start of therapy may provide a partial, if not total, defense to a claim of abandonment.

Finally, a psychologist who abandons a client after the insurance company terminates reimbursement will not be subjected to liability unless the client suffers injury as a result. If the client secures another therapist almost immediately or is not damaged by the psychologist's withdrawal from the case for some other reason, the psychologist will not be required to pay damages. Injury itself will not lead to liability; liability will be imposed only when the client can show that the injury was a reasonably foreseeable result of the termination of therapy.

With regard to the liability of peer reviewers for abandonment, the resolution is much simpler. Peer reviewers will not be subjected to liability bacause they have not entered a therapist–client relationship. Therefore, they have no duty to provide treatment, and the theory of abandonment is not applicable.

In fact, the peer review process is not likely to increase materially psychologists' potential liability for abandonment. In one sense, peer review is unrelated to questions of abandonment. Even if peer review is not used, a practitioner may confront the problems we have discussed if the client is unable to pay for therapy or if the client's insurance company decides on its own, without benefit of peer review, to discontinue reimbursement for the costs of therapy. If peer review is used, however, the role of the peer reviewers in the insurance company's decision to discontinue reimbursement will not increase the chances that the practitioner will be held liable for abandonment. If anything, peer review may diminish the likelihood of liability by providing proof that therapy was no longer necessary and that the treating psychologist was therefore justified in withdrawing from the case. Moreover, the use of psychologists as peer reviewers may make an insurance company more likely to discontinue payment unreasonably, than if the company's decisions about reimbursement are made only by insurance agents or nonpsychologists.

CONCLUSION

This chapter does not exhaust all the possible legal challenges to quality assurance practices. For example, as accreditation of doctoral programs and internship sites becomes more important to those who wish to enter practice, accrediting agencies may find themselves increasingly subject to litigation. Similarly, as postdoctoral credentials, such as listing in the National Register of Health Service Providers, are used as criteria for employment or reimbursement, credentialing bodies may also discover that they have become defendants in cases brought by disappointed applicants. Certainly, the issues we have discussed, especially peer review, have not been fully resolved by the courts or the legislatures. We hope, however, that readers have been both informed by this chapter and aided in their attempts to increase the quality of the services that psychologists provide their clients.

REFERENCES

American Psychological Association. (1977). *Standards for providers of psychological services* (rev. ed). Washington, DC: Author.

American Psychological Association. (1981a). Ethical principles of psychologists. *American Psychologist, 36*, 633–638.

American Psychological Association. (1981b). Specialty guidelines for the delivery of services by clinical psychologists. *American Psychologist, 36*, 640–651.

American Psychological Association. (1981c). Specialty guidelines for the delivery of services by counseling psychologists. *American Psychologist, 36*, 652–663.

American Psychological Association. (1981d). Specialty guidelines for the delivery of services by industrial/organizational psychologists. *American Psychologist, 36*, 644–669.

American Psychological Association. (1981e). Specialty guidelines for the delivery of services by school psychologists. *American Psychologist, 36*, 670–681.

Arizona v. Maricopa County Medical Society, 457 U.S. 307 (1982).

Ascherman v. Saint Francis Memorial Hospital, 45 Cal.App.3d 507, 119 Cal.Rptr. 507 (1975).

Bersoff, D. N. (1983). Hospital privileges and the antitrust laws. *American Psychologist, 38*, 1238–1242.

Bersoff, D. N., & Jain, M. (1980). A practical guide to privileged communication for psychologists. In G. Cooke (Ed.). *The role of the forensic psychologist.* Springfield, IL: Charles C Thomas.

Blue Cross of Northern California v. Superior Court of County of Yolo, 61 Cal.App.3d 800, 132 Cal.Rptr. 635 (1976).

Board of Curators of the University of Missouri v. Horowitz, 435 U.S. 78 (1978).

Charry v. Hall, 700 F.2d 139 (2d Cir. 1983).

Cohen v. State, 121 Ariz. 6, 588 P.2d 299 (1978).

Doe v. Prudential Insurance Company. No. C-2307-84 (New Jersey Supreme Court, 1985).

Falcone v. Middlesex County Medical Society, 34 N.J. 582, 170 A.2d 791 (1961)

Goldfarb v. Virginia State Bar, 421 U.S. 773 (1975).

Group Life & Health Insurance Co. v. Royal Drug Co., 440 U.S. 205 (1979).

McIntire, L. (1962). The action of abandonment in medical malpractice litigation. *Tulane Law Review, 36*, 834–842.

National Society of Professional Engineers v. United States, 435 U.S. 679 (1978).

Overcast, T., Sales, B., & Pollard, M. (1982). Applying antitrust laws to the professions. *American Psychologist, 37*, 517–525.

People v. Pic'l, 114 Cal.App.3d 824, 171 Cal.Rptr. 106 (1981), *rev'd on other grounds*, 31 Cal.3d 731, 646 P.2d 847, 183 Cal.Rptr. 685 (1982).

Pinkster v. Pacific Coast Society of Orthodontists, 1 Cal.3d. 160, 460 P.2d 495, 81 Cal.Rptr. 623 (1969).

Quasem v. Kozarek, 716 F. 2d 1172 (7th Cir. 1983).

Ratino v. Medical Service of the District of Columbia, 718 F.2d 1260 (1983).

Rich, J. (1980). Medical staff privileges and the antitrust laws. *Whittier Law Review, 2,* 667–681.

Rodriguez, A. R. (1983). Psychological and psychiatric peer review at CHAMPUS. *American Psychologist, 38,* 941–947.

Salter v. New York State Psychological Association, 14 N.Y.2d 100, 198 N.E.2d 250, 248 N.Y.S.2d 867 (1964).

Sicpa North America, Inc. v. Donaldson Enterprises, Inc., 179 N.J. Super. 56, 430 A.2d 262 (1981).

Taylor, C. (1983). Doctors' maximum fee plan is unlawful per se under Section 1 of the Sherman Act, *Brigham Young Law Review, 10,* 217–240.

Union Labor Life Insurance Co. v. Pireno, 458 U.S. 119 (1982).

Whalen v. Roe, 429 U.S. 589 (1977).

15

The Role of the State

JUDY E. HALL

This chapter summarizes state governments' influence on mental health services, giving examples of and providing a rationale for the methods that the states have used. Because the state's mechanism for affecting change is through legislation and the accompanying rules and regulations, we will focus on examples of legislation, the intent of which is to affect, directly or indirectly, the provision of mental health services. This legislation may be in the form of a licensure or a certification bill that regulates the mental health professions. Once licensure is in place, legislation may establish a mechanism for determinating the provider of services, the payment of services, or the evaluation of services (Kelvorick, 1981).

In order to understand how legislation directly affects quality assurance, we must first define quality assurance. Sechrest and Hoffman (1982) provided a definition that we can keep as a frame of reference as we discuss the role of the state:

> Quality assurance is concerned both with assessing the availability, adequacy, and appropriateness of health care resources, processes, and activities by applying professionally determined standards and criteria and with implementing methods to institute corrective actions. (p. 14)

The next reference point to establish is: Who are the mental health professionals? In a majority of states, they may include the following groups of licensed or certified service providers:

- Physicians (in particular, psychiatrists)
- Nurses (in particular, psychiatric nurses)
- Psychologists (in particular, clinical, counseling, and school psychologists)
- Social workers (usually clinical social workers)

In each of these professions, those practitioners who provide mental health services are a subset of the profession and often perform other services, such as teaching, research, and industrial consulting, in addition to activities in the mental health arena.

Furthermore, there are other practitioners who may provide services in some

JUDY E. HALL • New York State Education Department, Cultural Education Center, Albany, New York 12230.

states without state recognition and regulation and whose titles are derived from the techniques used:

- Counselor (such as rehabilitation, mental-health, marriage, or family counselor)
- Psychotherapist (which may also be a title used by one of the licensed mental-health professionals)
- Psychoanalyst (which may denote a physician, a social worker, or a psychologist)
- Therapist (which may include marriage, family, or sex therapists)

State government not only determines which of these groups will be licensed but also who will be eligible to receive direct or indirect payment from third-party insurers. Those same governmental agencies may apply other standards to those providers and the services provided. Let us look at the type of standards adopted by state governments, by the professions, and/or by third-party payers.

STANDARDS

There are four types of standards which affect mental health care: (a) clinical standards, (b) practitioner standards, (c) programs or facilities standards, and (d) payment system standards (Nelson, 1979).

CLINICAL STANDARDS

Clinical standards are usually developed by practitioners of the profession and relate to specific aspects of clinical care. Examples are (a) the criteria sets developed by the American Psychological Association and the American Psychiatric Association for peer review of outpatient psychological and inpatient and outpatient psychiatric services reimbursed by the Civilian Health and Medical Program of the Uniformed Services (CHAMPUS) and used by the professional standards review committees (PSRCs) of the state psychological associations in the review of care; (b) the *Standards for Providers of Psychological Services* of the American Psychological Association (1974, 1977); and (c) the National Association of Social Workers (NASW) *Standards for Social Work in Health Care Settings* (1981).

PRACTITIONER STANDARDS

Practitioner standards are applied by agencies of government (usually states), professional associations, and third-party payers to the individuals providing the services to determine qualification. Examples of the first include regulations that define professional staff in a state facility (e.g., civil service) or those that list providers in that state who are qualified for reimbursement by federal or private funds. For instance, in New York there are state standards that recognize certified social workers eligible for third-party reimbursement for psychotherapy services. Examples of standards developed by professional associations include the *National Regis-*

ter of Health Service Providers in Psychology (Council for the National Register of Health Service Providers in Psychology, 1985b; Wellner & Zimet, 1983) and the *Register of Clinical Social Workers* (NASW, 1982a). The Texas State Board of Examiners in Psychology developed a procedure for determining which licensed psychologists are eligible for third-party reimbursement. The standards are the same as those developed by the Council for the National Register of Health Service Providers in Psychology (Council, 1985b), which includes psychologists who are licensed and have met certain experiential requirements. Some third-party payers may find it convenient to use licensure only as a basis for reimbursement, except in managed health-care systems that require willingness to accept a set payment (see discussion to follow on health maintenance organizations, preferred provider organizations, and individual practice associations).

PROGRAM AND FACILITY STANDARDS

Program and facility standards may be developed by (a) states, to review, approve, or license facilities; (b) program or institutional associations, to ensure the quality of the programs or facilities they represent; (c) third-party payers, in the form of requirements for licensure, certification, or accreditation; or (d) teaching or research institutions, for the purpose of determining who is eligible for financial aid (Nelson, 1979). It is likely that more reliance will be placed on institutional licensure than on individual licensure, especially because a major portion of health care activities are being carried out by a variety of nonphysician health-care providers in institutional settings, and the malpractice insurance costs are paid by hospitals and other institutions rather than by the individual providers. As a result, the monitoring of institutional standards will become even more important and critical than it is now.

PAYMENT SYSTEM STANDARDS

Third-party payers, such as private insurance companies and federal insurers like Medicare and Medicaid, have developed standards for payment. Rather than quality of care being the primary concern, the objective is most often cost containment, often at the expense of quality. Although some insurance policies may provide adequate coverage for mental disorders and substance abuse problems, most limit coverage to physical disease. This is also true under the Medicare and Medicaid programs. In some states, the concerns over these limitations have led to legislation mandating minimum mental-health coverage (MMHC) in health insurance policies. More recently, states have intervened in the battle between cost containment and quality service provision by passing legislation to allow the establishment of health maintenance organizations (HMOs), preferred provider organizations (PPOs), and individual practice associations (IPAs). All these types of managed health systems have significantly affected the provision of mental health services, especially since the mid-1970s, and will continue to do so.

Given the above standards that regulate the provision of mental health services, let us examine what the role of the state is in the overall scheme and how the state may affect quality assurance in licensure and other types of legislation.

LICENSURE IN THE MENTAL HEALTH PROFESSIONS

The state legislatures pass licensure laws for the purpose of protecting the general health, safety, and well-being of the public. Licensure laws are based on the assumption that the public must be protected from unqualified, unscrupulous, and/ or unethical practitioners. Especially in situations where the service provider knows more about the quality and effectiveness of the care than does the consumer, the state steps in and provides regulation. This unbalanced situation involves "asymmetrical information and characterizes the relationship between the consumer and the provider of professional services" (McGuire & Weisbrod, 1981). In addition to the protection of the public, the goals of regulation may include cost containment, ensuring quality standards, and perhaps ensuring that some minimal care will be available to all (Kelvorick, 1981).

The authority for the credentialing of individuals and the regulation of practice is vested by law in a state board of examiners, professional association, or other equivalent body. This authority to regulate admission to the practice of a profession is well documented in case law. A landmark decision is *Dent v. West Virginia* (1889), which held that

> the nature and extent of the qualifications required (to practice the profession) must depend primarily upon the judgment of the State as to their necessity. If they are appropriate to the calling or profession, and attainable by reasonable study or application, no objection to their validity can be raised because of their stringency or difficulty. (Reaves, 1982, p. 1)

These boards are also empowered to develop rules and regulations to administer the provisions of the licensing act, including specialty identification, continuing education, professional discipline, and so on. Therefore, the laws, regulations, and rules of state regulatory boards are the major mechanism by which the state ensures that quality services will be provided (Kelvorick, 1981; Nelson, 1979; Reaves, 1982).

Of course, the criticisms of licensure have been many, especially in recent years (Simon, 1983; U.S. Department of Health, Education, and Welfare, 1967, 1971, 1977); the complaints allege that licensure is counterproductive because competition is curtailed, the cost of services is increased, the supply of services is less, and innovation in training and practice is stifled (cf. Herbsleb, Sales, & Overcast, 1985). However, none of these complaints have persuaded the state legislature that a more effective approach exists (see Theaman, 1982, for a comprehensive response to the critics). The few empirical studies (rather than simply opinion) often attack the question of whether licensure is necessary and whether the inputs into the licensure process relate to effectiveness in practice. Carroll and Gaston (1983) responded by saying that there are situations in which licensure is necessary but that asking that question is nonsensical. For them and others, the question is an empirical one that must be addressed profession by profession and restriction by restriction.

After acknowledging that there appear to be two main areas of the effect of licensure, price and quality, Carroll and Gaston looked at the measures used by three professions that have evaluated quality: lawyers, pharmacists, and optometrists. The measures used by lawyers were peer ratings, malpractice insurance rates, and disciplinary actions. For each measure, greater quality control was associated with a higher average quality of the licensed practitioners. Similar measures used in the evaluation of pharmacists and optometrists produced a similar outcome: there was a direct association between the quality of service delivered and the stringency of licensure standards, higher licensure standards "yielding enhanced quality of practitioner" (p. 141). These authors went on to distinguish between quality of service delivered and received, to examine the effect of restrictive measures on the number and quality of practitioners, and to measure the quality received by the use of proxy measures (Carroll & Gaston, 1983).

These restrictive procedures in state regulation are termed *inputs* by those attempting to determine how to evaluate mental-health service-regulation, noting that experts disagree on whether it is easier to evaluate the inputs or the outputs (Kelvorick, 1981). In the sections that follow we examine these inputs and the way in which each is regulated by the state, through professional licensing laws and through other state legislation. The inputs leading up to licensure include: (a) education, both the definition of basic knowledge, skills, and abilities and the approval of education and training programs; (b) experience; (c) examination; and (d) moral character. The inputs following licensure include; (e) continuing education; (f) requiring relicensure; and (g) professional discipline. Together, these constitute regulation by the state of the licensed professions. Malpractice will be considered briefly as one output.

All states regulate the practice of medicine, nursing, and psychology. Thirty-six states, the District of Columbia, and two other U.S. jurisdictions credential in social work.

STATE REGULATION BEFORE LICENSURE

INPUT 1: EDUCATION—DEFINITION OF THE REQUISITE KNOWLEDGE, SKILLS, AND ABILITIES

Each of the mental health professions has literature stating what constitutes the knowledge, skills, and abilities that every practitioner must have to be minimally competent. Typically, the knowledge, skills, and abilities are generic; that is, all those licensed to practice should be able to demonstrate readiness for a wide range of activities in a variety of settings (Kane, 1982). Once the knowledge, skills, and abilities are identified, a method of assessment is developed that evaluates these components for practice. As pointed out by Loveland (1977), "a test is really a set of stimuli designed to evoke a representative sample of some specified behaviors under controlled conditions which permit classification and/or evaluation of those behaviors" (pp. 5–6). In order for any test to be constructed, the domain of professional practice must be defined by experts and supplemented by empirical studies of the actual conduct of practice (Kane, 1982). In some cases, the knowledge assessed on the licensure examination has been empirically tied to practice through a role delineation study and/or a job analysis (Hall, 1987a; Richman, 1982; Rosenfeld, Shimberg, & Thornton, 1984; Wohlgemuth & Samph, 1982;) or to another method of content validation that uses a critical incident technique followed by other forms of content validation (Dvorak, Kane, Laskevich, & Showalter, 1982; Hubbard, 1978).

As all these professionals may practice in areas besides the mental health area, and as licensure is theoretically granted for a lifetime, the entry level for each of the four professions is evaluated on a generic level; an approach that provides for flexibility in career development as the individual and the field matures, and that leaves the identification of specialists to the private certification organizations. Licensure tells the consumer that minimal requirements have been met but makes no statement about specialty qualifications, except in the few states that license (or certify after licensure) health care providers in psychology or that license clinical, school, and other psychologists (Hall, 1985a), and except in a few states that recognize the additional title of *clinical social worker* (Georgia, Massachusetts, Oregon, and Virginia). More typically, the licensure and certification laws provide for the recognition of the generic titles of *physician, psychologist, nurse,* and *social worker,* although the two latter professions have multiple levels of recognition.

Typically, specialization takes place after licensure. In social work, no established specialties are recognized either by statute or by the National Association of Social Workers, although the NASW has approved standards for practice in schools (NASW, 1978) and in health care settings (NASW, 1981). In medicine, nursing, and psychology, specialty identification is an option for those who apply and who meet the standards of the specialty boards. As specialty certification is available but not required for entry into practice, this process is labeled as being voluntary (American Medical Association, 1983; American Nursing Association, no date; Hall, 1985b).

Input 1: Education—Approval of Programs

Three of the four professions—medicine, nursing, and social work—have developed a national consensus on what constitutes an appropriate curriculum and where the education must take place. By requiring that candidates for licensure graduate from an approved professional school, they ensure that those graduates will meet the educational requirements for licensure and also facilitate interstate mobility, as a person licensed in one state will meet the educational requirements for licensure in another state.

In 1977, representatives of a number of educational, professional, and credentialing organizations defined the core education necessary for a psychology program (Wellner, 1978). A similar policy was adopted by the American Association of State Psychology Boards (AASPB, 1977), by the Accreditation Committee of the American Psychological Association in 1979, and by the Council of Representatives of the American Psychological Association in 1985 (APA, 1985c). However, in the absence of a mandatory accreditation system for psychology, there is no guarantee that graduates will meet those criteria. The accreditation process for psychology programs is voluntary but is currently open only to programs in the specialty areas of clinical psychology, counseling psychology, and school psychology. As of this writing, approximately 230 (APA, 1986) programs are approved. This means that the state psychology boards must assess each applicant's education to determine whether the core educational requirements have been met and whether the applicant has graduated from the required program in psychology. Thus, the states may implement or interpret those criteria differently.

As an outgrowth of these concerns, the National Register for Health Service Providers in Psychology, a private credentialing body in psychology, established a procedure for designating which of the programs purporting to train psychologists actually meet the criteria (Council for the National Register of Health Providers in Psychology, 1985a; Wellner & Zimet, 1983). The American Association of State Psychology Boards has agreed to cooperate formally with the National Register by establishing a joint review system for use by state and provincial boards.

State-level accreditation does occur, also. New York has a mandatory accreditation system called *registration*, which is supervised by the New York State Board of Regents, administered by the New York State Department of Education, and used by the New York state boards to determine admission to licensure. For any doctoral program to be initiated in New York, the proposed program must first meet certain structural standards specified in the Commissioner's Regulations, must be evaluated by peers from outside the state, and must be given a positive recommendation, approved by the State Board of Regents. This unique approach applies to any degree-granting program, whether in the area of a profession or not, and is a responsibility of the state department that also administers the licensure of individuals graduating from those programs. A similar but less intensive effort has occurred in some other states (e.g., Ohio and Pennsylvania), where a list of approved programs in psychology has been developed by the state psychology licensing board, based on a paper review of the individual program requirements.

Where a national accreditation system does exist, such as in medicine, nursing, and social work, state boards participate through representation on the accreditation council, the executive committee, or the council of the national associations of state licensure boards. The state representatives reserve the right to disagree with a decision made by the accreditation council, thereby maintaining control over what is really the state's responsibility, that is, admission to licensure. In New York, where a strong program registration system exists (i.e., accreditation) and where the New York State Board of Regents is a regional accrediting agency recognized by the U.S. Office of Education and the Committee on Postsecondary Accreditation, occasionally these disagreements have focused on the quality and the extent of educational preparation. For instance, some graduates of APA-approved training programs have been ascertained not to meet the core educational requirements, namely, completing the basic educational requirements specified in APA accreditation criteria and/or meeting the regulations of the state licensure boards.

Input 2: Experience—A Requirement for Licensure

The assessment of a candidate's performance in training or on the actual job is a critical component of the state definition of minimal competence of physicians, psychologists, and social workers but is not required for nursing. A nonempirical study conducted by Cathcart and Graff in 1978 is of interest (in Herbsleb *et al.*, 1985). The authors developed a model evaluating 30 health professions' need for experience as a prerequisite for licensure. Using seriousness of impact, the degree to which practitioners are required to exercise independent judgment, and the amount of practical training in the educational program, the authors found physicians and osteopathic physicians needed to experience for licensure the most of all the 30 professions. Psychologists received the next highest rating.

In reality, those findings reflect state requirements. Experience for medical licensure is required in 41 states. Those jurisdictions that have no such requirement are Arkansas, Indiana, Louisiana, Maryland, Massachusetts, Missouri, New Mexico, Ohio, and Tennessee (AMA, 1983). Most of the psychology statutes require some supervised experience before licensure at the independent practice level. The exceptions are Alabama, Arizona, Puerto Rico, and Wyoming, as well as three Canadian provinces (Alberta, Quebec, and Saskatchewan) (Hall, 1987). Nursing has no experience requirement for the entry level of licensed practical nurse or registered nurse. All states with statutory recognition of social workers require appropriate experience for the independent (or highest) practice level except for Illinois, New York, and South Carolina. Although New York does not require experience for entry into the private practice of social work, certified social workers in New York State are required to demonstrate that they have 3 or 6 years (depending on the insurance policy) of appropriate and supervised experience for third-party reimbursement for psychotherapy services in New York State.

INPUT 2: EXPERIENCE—TRAINING INSTITUTES

Another example of state standard-setting is the review and approval process for psychotherapy training institutes. In New York, the State Board of Regents and the Commissioner of Education must review and approve (technically, charter) any institute intending to incorporate for the purpose of education and training. The review is an extensive one focusing on faculty credentials, student admission requirements, content of curriculum, administration, and space and is performed by representatives of the four mental-health professions. The Board of Regents in New York has approved 75 institutes (R. Madrazo-Peterson, personal communication, January 21, 1986), which provide clinical experience to an unknown number of students, many of whom are also enrolled in or are graduates of schools of social work.

California has a research-psychoanalyst-registration law for graduates of four California psychoanalytic institutes (or their equivalent). The research psychoanalyst title allows the registrant to engage in psychoanalysis as an adjunct to teaching, training, or research. However, only a few people (fewer than 50) have applied for registration as a research psychoanalyst since the inception of the law in 1977 (Chapter 5.1, Section 2529 of the Business and Professions Code). Although applicants for registration are required to have the equivalent of a doctorate, members of the California licensed professions of medicine, psychology, social work, and marriage, family, and child counseling are not required to register in order to practice research psychoanalysis.

INPUT 3: EXAMINATION

One of the major roles accorded state licensing boards in the statutes is selecting or developing and approving the licensing examination. These choices are an individual state matter, and even if the examination is developed by an outside examination committee in collaboration with a testing agency, each state board is responsible for the exam. The degree to which the examination may conform to the *Standards for Educational and Psychological Tests* (APA, 1985b) and the *Uniform Guidelines for Employee Selection Procedures* (EEOC, 1978) are of concern to all states. As a result, the national organizations of state licensure boards in the four professions have funded extensive research programs assessing the degree to which the exams are content-valid, then using that information to modify the examination (Dvorak *et al.*, 1982; Hall, 1987; Hubbard, 1978; Richman, 1982; Rosenfeld *et al.*, 1984 Wohlgemuth & Samph, 1983).

The professions of medicine, nursing, and psychology have a national examination used in the review of all candidates for licensure in every state (with the exception of Michigan, which does not use the Examination for Professional Practice in Psychology). The American Association for State Social Work Boards (AASSWB) has completed a job analysis and developed a new examination for social work (Wohlgemuth & Samph, 1983), although the exam derived from that research is not in use in all states. In fact, some states do not require an exam for the independent practice level in social work (Michigan, New Hampshire, Oregon, South Carolina, and Tennessee) but may require an exam at a lower level of recognition (Idaho, North Dakota, and Texas).

In some of the states, there is a limit on the number of times an applicant can take the licensure examination without having to wait a specified time period or number of test administrations, or to engage in some type of educational remediation. According to a survey by the Federation of State Medical Boards (1983), 37 states put a limit on the number of times an applicant can take the Federation Licensing Examination (FLEX) without waiting and/or engaging in additional study or training. Those states without the requirement are Arizona, Colorado, Connecticut, the District of Columbia, Florida, Nebraska, New Jersey, New York, North Carolina, Ohio, Rhode Island, Tennessee, and Washington. The following 22 states do not put a limit on the number of times an applicant for a psychologist's license can take the examination: Arizona, California, Colorado, Connecticut, Florida, Hawaii, Illinois, Kansas, Maryland, Massachusetts, Minnesota, Nebraska, New Mexico, North Carolina, North Dakota, Oklahoma, Pennsylvania, South Carolina, South Dakota, Tennessee, Vermont, and Wisconsin. As indicated, in the profession of social work, there is no uniformity in whether an examination is given nor in the choice of the examination for licensure. Some states do place a limit on the number of times the registered-nurse examination can be taken by a failed applicant; however, some states do not.

Finally, the last responsibility of the state is determining the pass point on the examination. This decision may be reached by a variety of methods, all of which focus either on the candidate population's performance or on the examination content, or both (Hall, 1987). The pass point should be located at the score that maximizes the probability that those licensed will be minimally competent and that those not licensed will not be minimally competent. In nursing and medicine, the pass point is determined for all jurisdictions at once; for psychology and social work, the pass point is determined by each state board. The procedures used in determining the pass point are as critical as the procedures ensuring a content-valid examination.

INPUT 4: MORAL CHARACTER

The determination of an applicant's character and fitness for practice is often subsumed under the term *moral character*. For instance, if a person has been convicted of a felony or a misdemeanor, has a history of mental illness or substance abuse, or has misrepresented facts on a licensure application, his or her current behavior and character will be carefully considered before he or she is granted a lifetime license to practice a profession. If the behavior occurred in the past, the candidate is given an opportunity to document evidence of rehabilitation. Additionally, the courts have interpreted this requirement of "good moral character" to mean that there must be a relationship between past behavior and the practice of the profession. For instance, in New York, a statute forbids denial of a license because of a previous criminal conviction, in the absence of the "direct relationship between one or more of the previous criminal offenses and the specific license sought" (NYS laws Chapter 931, 1976—Section 752 of the Correction Law). To determine whether there is a basis for questioning the moral character of an applicant and denying the license, New York provides a hearing before a panel of the professional board. This decision regarding moral character may, on appeal by either party, be presented to the administrative review body of the State Education Department. Similar patterns exist in other states (Hall, 1986).

STATE REGULATION AFTER LICENSURE

The methods used by state governments to regulate the individual practitioner after licensure currently include (a) mandating continuing education, (b) requiring relicensure, (c) restricting the licensee to specialty practice, and (d) disciplining licensees. If the practitioner is judged incompetent or negligent, the tort system may enter the picture through a charge of malpractice.

INPUT 5: CONTINUING EDUCATION

In a number of states for a variety of professions, continuing education requirements have been mandated by statutes that either specify the hourly requirements or provide authorization for the board to set requirements. The continuing education concept originated in the 1930s with a concern that physicians can be required to keep abreast of important developments in medical science. Initially, it was assumed that practitioners would voluntarily maintain competence by taking advantage of various educational activities sponsored by state and local medical societies, and there would be no need for states to require continuing education for the renewal of a license. However, when reports were published by the U.S. Department of Health, Education, and Welfare in 1967 and again in 1971 that promoted relicensure and/or required continuing education as the solution to obsolescence in the profession, state medical associations instituted continuing professional education (CPE) as a condition of membership. In 1971, New Mexico became the first state to mandate CPE for physician relicensure, and by 1977, 16 state medical societies had adopted such a requirement for the maintenance of membership. Although the 1977 report by the U.S. Department of Health, Education, and Welfare concluded that continuing education "is often invalidated and of questionable relevance" (p. 17), the same report stressed the necessity of exploring a variety of approaches to continuing competence. However, no systematic studies have evaluated the effectiveness of the various approaches to the maintenance of skills in a profession. The Michigan legislature passed a law in 1978 requiring a pilot study in which the boards were encouraged to experiment with a variety of procedures for competence assurance (reexamination, continuing education, peer review, and practice audits), but no funding was provided to evaluate the mechanisms that worked best (Shimberg, 1982). Over the period 1985–1986, the New York legislature agreed to support a CPE bill for certified public accountants contingent on a required 3-year evaluation of the effectiveness of the various activities, funded by a $35 fee paid by licensees.

Requiring a systematic evaluation of continuing education was one of the recommendations in a draft report examining current efforts at competency assurance by the National Commission for Health Certifying Agencies (NCHCA, 1985). The National Commission was established in 1977 because the federal government felt that an outside organization should help set standards for the credentialing of health professionals by private organizations. The NCHCA's guidelines require as a condition of membership that certifying agencies present a plan for the development of a recertification program or a plan for research and development on continuing competence. As a part of their mission, that organization completed a study of the status of continued competence activities in the health professions, providing recommendations for research to address the defined issues and problems (NCHCA, 1985).

Although the number of states that require CPE in each of the four mental-health professions has increased since the 1970s, it remains a controversial topic even for the medical profession, which has the longest history with CPE. In 1983, Illinois and Colorado repealed their mandatory CPE laws for physicians; other states have never implemented their law. There are five states that require CPE to maintain a license as a practical nurse, nine states with requirements for registered nurses, and two states mandating CPE for nurse practitioners. In addition, Tennessee requires CPE following a 5-year period of inactivity. Three jurisdictions have not implemented CPE for the nurse (Alaska, the District of Columbia, and Idaho) (NCHCA, 1985). Psychology has been slow to adopt mandated continuing education but now has 21 states with requirements for relicensure (see Table 1) (AASPB, 1983; APA, 1985c).

TABLE 1. Mandated Continuing Education by Profession[a]

State	Medicine	Nursing	Social work	Psychology
Alabama			X	
Alaska		X	NA	X
Arizona	X		NA	
Arkansas			X	X
California	X	X		
Colorado		X	X	
Connecticut				
Delaware			X	X
District of Columbia				
Florida		X	X	X
Georgia			X	X
Hawaii	X		NA	
Idaho				
Illinois				
Indiana			NA	
Iowa	X	X	X	X
Kansas	X	X	X	X
Kentucky		X	X	
Louisiana				X
Maine	X		X	
Maryland	X		X	X
Massachusetts	X	X	X	
Michigan	X			X
Minnesota	X	X	NA	
Mississippi		X	NA	
Missouri			NA	
Montana			X	
Nebraska		X		
Nevada		X	NA	X
New Hampshire	X			
New Jersey			NA	
New Mexico	X	X	NA	X
New York				
North Carolina			X	
North Dakota			X	
Ohio	X		X	
Oklahoma			X	
Oregon		X		X
Pennsylvania			NA	
Rhode Island	X			
South Carolina				
South Dakota	X	X	X	X
Tennessee				
Texas	X		X	X
Utah	X	X	X	X
Vermont				X
Virginia			X	X
Washington	X	X	NA	X
West Virginia			X	X
Wisconsin	X		NA	
Wyoming		X	NA	X

[a]The information in Table 1 was supplied by the American Medical Association (1983), the National Commission of Health Certifying Agencies (NCHCA, 1985), the National Council of State Boards of Nursing (1984), the National Association of Social Workers (1984), and the American Psychological Association (1985c). This information reflects the elimination of a few laws and the fact that some states have CPE provisions for physicians (Alaska, Arkansas, Kentucky, Nebraska, and Nevada) and for nurses, but have not implemented them. Only those states with mandated and implemented CPE are included in the table.

INPUT 6: CONTINUING COMPETENCE THROUGH RELICENSURE

Although renewal of the licensure is automatic on payment of fees and/or demonstration of having met the mandated continuing-education requirement, occasionally individuals and organizations raise the issue of requiring a reexamination of licensees in place of or as evidence of CPE. In spite of that debate since the mid-1970s, no state has required reexamination of licensees in any mental-health profession. (Florida has legally mandated that certified public accountants pass an examination on Florida accountancy law and the boards' rules before license renewal.)

One of the difficulties with reexamination is the fact that the licensing exam is intended to assess entry-level knowledge, skills, and abilities. Presumably, the mature practitioner would have different knowledge and skills and would be competent to practice but might have difficulty with a generic, entry-level licensure examination. As Loveland (1977) pointed out, "the variables which distinguish between successful and unsuccessful junior members of the profession may not be the same as those which differentiate among senior practitioners" (p. 11). The other difficulty is that with experience, practitioners tend to narrow their focus and to specialize, some voluntarily restricting their practice.

Another reason for the strong resistence to relicensure by examination is the recognition that there are other and perhaps more appropriate methods of regulation that assess a mature practitioner's current capabilities, such as peer review and specialty certification boards (e.g., through simulations and audit of private practice). Although reexamination may make sense from a public protection viewpoint (i.e., all practitioners should possess similar knowledge, skills and abilities), the mature practitioner often does poorly on certain types of exams (e.g., multiple choice) but well with interactive methods, such as simulation problems. Other approaches, such as chart audits or practice audits, have not been thoroughly assessed. As a substitute for continuing education, licensed psychologists and social workers in Virginia may elect to have their private practices evaluated.

Clearly, more systematic research is needed before the states should embrace the concept of relicensure. The recommendations of the NCHCA should provide suggestions on how to assess the effectiveness of different approaches, as well as the cost effectiveness and professional acceptance of each. Hopefully, funds will be secured to implement the research proposed (NCHCA, 1985).

INPUT 7: PROFESSIONAL DISCIPLINE

The state also discharges its responsibility for ensuring and maintaining the quality in service provision by requiring that licensed practitioners adhere to standards of practice and professional conduct codes. In some instances, these codes were developed by the profession and endorsed by the state; in other instances, 15 states such as New York and Minnesota have developed their own professional conduct codes (Hall, 1986). The failure to practice according to these standards may result in a complaint filed with the licensing board or the regulatory agency responsible for the investigation of such complaints. If a complaint is brought against a practitioner, an investigation ensues; if there is merit to the complaint, a disciplinary hearing is held before a panel of peers and members of the public. Based on the weight of the evidence, the professional may be disciplined, and the action may range from a simple reprimand to the revocation of his or her license with or without a fine.

That is the theory of professional disciplinary action. The reality is different, for few professionals are ever disciplined. The reasons for the inactivity vary from state to state but include weak laws, an inadequate range of sanctions, and limited resources (Shimberg, 1982). If the client is not sophisticated enough to detect a violation, nothing may be reported. Even when persons from outside the profession are included in the disciplinary process, the number of practitioners affected is very low.

Do we have any assistance in understanding the lack of success in disciplinary activity? Dolan and Urban (1983) studied empirically the effectiveness of discipline by medical boards during the period 1960–1977. Their results suggest that the principal determinant of the boards' vigilance was the degree to which they were not dominated by physicians. However, there was no trend over time showing an increase in effectiveness—defined as more disciplinary actions taken—even though physician dominance has decreased (see the discussion to follow on public representation on boards). Other than suggesting an empirical approach to this issue, we are still left with the question of how to increase the effectiveness of the current disciplinary process.

In 1982, the first full year after the implementation of a New York statute aimed at streamlining the disciplinary process, the number of practitioners disciplined by the State Board of Regents was 470, out of a total of 488,012 licensed to practice. The number of mental health practitioners disciplined in that figure was 176, plus 22 given administrative warnings, out of a total of 294,743 licensed mental-health practitioners (As of April, 1987, the figure had increased to 310,543.) How many complaints were filed during that same period for the four professions? We can easily examine those figures for psychology, social work, and nursing, but not for medicine, as medical complaints are handled by the State Health Department instead of the State Education Department (the administrative arm of the State Board of Regents). The number of complaints in 1982 for psychologists, for social workers, and for registered nurses was 107, 45, and 291, respectively. The number of cases closed with no action taken was 97, 43, and 165, respectively (New York State Education Department, 1982; W. Wood, personal communication, July 6, 1983). In 1986, the parallel figures were 144, 96, and 799; for medicine, 530; those that were closed with no action taken were 136, 72, 479, and 799.

Presumably, there are legitimate reasons for an action to be closed, and the number of Board of Regents' actions did increase to over 600 in 1985 and to over 900 in 1986 (J. Fisch, personal communication, April, 1987), but each figure indicates that, even when persons from outside the profession are responsible for investigating, prosecuting, and charging unprofessional conduct, the number of practitioners affected is very small. Hence, the criticism cannot be that the profession is reluctant to act.

Attempts to strengthen the disciplinary system by increasing the number of complaints reaching a licensing board include required reporting of malpractice actions and criminal convictions, providing immunity from civil action for those who provide information to regulatory boards, and educating the public. More than 25 states have passed legislation to require reporting to medical boards by medical societies when actions are taken. At least 10 states require reporting by other physicians when a colleague violates the conduct code (Derbyshire, 1983). Minnesota has a similar provision that applies to psychologists. Reporting by hospitals of violations remains a problem, according to Derbyshire (1983), and some hospitals prefer to "export their problem" rather than to actively self-regulate, despite the 45 laws that require hospitals to report such occurrences. Although these laws are seen as ineffective because little reporting actually occurs, there is still a backlog of cases awaiting investigation and prosecution in many states.

Perhaps the backlog could be eliminated by adequate staffing. California's medical board employs 50 full-time investigators, and with a budget of $11 million, the state spends about $200 a physician. In contrast, New York spends about $67 and had one of the lowest disciplinary rates: in 1983, .49 serious actions per 1,000 doctors; in 1984, .6 actions per 1,000. The maximum figure for any state in 1984 was in Nevada, where 18.3 actions per 1,000 physicians took place (Brinkley, 1985).

To prevent hopping from one state to another after being disciplined by a state board, there now exists a centralized reporting procedure for reporting violations by physicians, nurses, and psychologists developed by the organizations of state boards for those professions and a state-financed national disciplinary data bank organized by the Clearinghouse on Licensure, Enforcement, and Regulation. However, only 15 states have legislation providing for suspension or revocation of a physician's license based on discipline in another state (Derbyshire, 1983). The absence of this provision and the lack of information on previous offenses have led to state hopping and to the necessity for multiple convictions and wasting of state resources. Another loophole exists for those convicted: a physician may continue to practice under another title, for instance, *psychotherapist*; such an individual is outside the usual mechanism for ensuring quality of service.

Another legislative approach enacted in over 20 states is the impaired-professional law, which attempts to provide rehabilitation and to forestall punishment (Derbyshire, 1983; Hall, 1986). The number of practitioners in such a program is very small, and the emphasis is on restoring the practitioner to competent practice after rehabilitation.

Clearly, discipline after licensure has been the most difficult input mechanism to implement effectively and to evaluate. Are there many more practitioners who should be disciplined? If there are, what is the current number of errant, incompetent, negligent, and unscrupulous practitioners? Apparently, no one has the answer to those questions.

Various educational efforts have attempted to provide the consumer with sufficient information to judge the service provided and, if not satisfied, to know where and how to complain. Clearly, it is desirable to educate the consumer on what constitutes minimally competent practice. In fact, one measure of success might be the proportion of complaints coming from consumers rather than from professionals or from other reporting mechanisms (e.g., required reporting, and licensure boards). In New York, approximately one-third of the complaints against the four mental-health professions in 1982 and 1983 were classified as individual complaints (from private agencies individual complaints, and the newspapers or other media sources).

MALPRACTICE

Malpractice is simply professional negligence, and as such, it is an indication of the failure of quality assurance. Therefore, no examination of those activities that the state engages in to ensure proper service provision is complete without examining the failures and why they occur. It is entirely likely that the recent crisis in malpractice costs to physicians and other health personnel and the significant increase in action against psychologists are due to a failure in enforcement of the professional conduct codes, either by the state or by the professional organizations, as well as the oft-cited over-supply of attorneys. In 1978, there were 3 malpractice claims per 100 physicians; in 1984, the ratio was 16 per 100. In 1985, in specific states, such as California, Florida, and New York, the number was 20 per 100 (Brinkley, 1985). If one looks at the relationship between disciplinary activity and the increase in malpractice actions, it becomes apparent that part of the fault lies in ineffective discipline and/or in a decline in professional competence. Assuming that the latter might be true, Florida intends to investigate any licensee with three malpractice claims over a 5-year period. Recently, there have been reports of special information systems being developed to provide doctors with the names of patients who have a history of bringing malpractice actions; in retort, another organization intends to provide consumer information on physicians' experience with malpractice actions. The premium costs to physicians and to hospitals are in the billions. Needless to say, the malpractice crisis is here.

Although the numbers are smaller, even psychology has had a significant increase in successful malpractice actions. However, little information is available on the psychologist who is sued. In fact, malpractice cases that result in an award to the consumer are cloaked in secrecy by the American Psychological Association's Insurance Trust. None of these cases are reported to the APA's Ethics Committee or to the state boards. Self-reporting to the state board is mandated by only one state law (Hall, 1986). Because cases that are settled out of court against any practitioner may be sealed by court action, only those that result in damages can be searched in legal sources (e.g., the American Trial Lawyers Association). In general, and for psychology, a significant source of information on the detection of incompetence is unavailable to the public. The small amount of solo practice may be the reason for the smaller problem in social work and in nursing. There is every reason to assume that the same problem will occur in those professions that has occurred in medicine and psychology.

Until states and private and professional organizations work together more effectively to remedy the problems of professional incompetence, it is likely that individuals will seek redress in the courts, and quality assurance will be only partly effective. Other individuals feel that the only effective means for weeding out incompetent practitioners is through managed health-care systems, whose incentive is to hire competent people and keep them competent (Paxton, 1985).

SPECIALIZATION

One of the ways in which professionals maintain competence is through specialization. The licensed practitioner who specializes typically devotes himself or herself to that area of practice to the exclusion of other areas of practice. One reason for specialization is that the field has become too complex for any one person to master all of the information available and to offer all of the different services competently. Because of specialization, the care provided is of higher quality, and the access to the quality care is facilitated through the accurate advertising of those services. Two benefits encountered in the literature are cost containment and professional satisfaction (Schnaps & Sales, 1985).

Medicine has the longest history of specialties and formalized specialty-certification boards, the first originating in 1936. The largest growth in the number of specialties took place in the period after World War II. By 1970, the American Medical Association had recognized 33 different types of specialized medical residencies, multitudes of specialty boards, and over 25 subspecialty boards certifying physicians in over 65 areas of medical practice (Schnaps & Sales, 1985). The profession of nursing developed specialties later, with five divisions identified by 1980; the impetus came from the first area to be recognized as a specialty: the nurse practitioner.

In most states, the statutes regulating psychology provide for the assessment of the core education, for specific training, and for generic examination that, in a few states, may be followed by specialty certification or the restriction of practice to specified areas (Hall, 1985a). Similarly, in a few states, the legislation equates a psychologist to a health service provider (Hawaii, South Dakota, and Michigan are examples); and in others, a specialty certificate is awarded for practicing as a clinical, school, or counseling psychologist (for example, in Virginia, South Carolina, and Nebraska). Psychology has four recognized specialties (clinical, counseling, school, and industrial-organizational psychology), with a certification process administered by the American Board of Professional Psychology (APBB) for those four specialties; The APBB also awards diplomas in two other areas of practice (clinical neuropsychology and forensic psychology) not yet recognized by the APA as specialties in psychology.

However, as the number of these private certification boards grows, one must ask whether the goal of high standards has induced a movement from specialty recognition to specialty monopolization. Have specialists furthered their own self-interests by excluding general practitioners from practicing in the specialty areas? How is that general practitioner evaluated if he or she practices in an area that is within a specialty area of practice?

This second question has found some interpretation in the courts:

> Even for the general practitioners, some courts hold that the appropriate standard of care is to determine on the basis of a scope far wider than that of the practice of a comparable professional in the same or similar community. For recognized specialists or those practicing in a specialty area, the trend among the jurisdictions is to recognize and adhere to a national standard of care referenced to the unique training, education and circumstances of each specialty. (Overcast & Sales, 1985, p. 24)

As a result, it is not surprising to find that, in medical malpractice actions, some form of a specialty standard of care is recognized in 29 jurisdictions (Alaska, Arizona, Arkansas, California, Connecticut, the District of Columbia, Florida, Indiana, Iowa, Kansas, Kentucky, Louisiana, Maine, Massachusetts, Michigan, Missouri, Nebraska, New Jersey, New York, North Carolina, Ohio, Oklahoma, Oregon, Pennsylvania, Texa, Vermont, Virginia, Washington, and West Virginia; Annotations, 1980).

Is a nonspecialist restricted from practicing in a specialty area? The profession of psychology states the following in the introduction to the *Specialty Guidelines* (APA, 1981b):

> Traditionally, all learned disciplines have treated the designation of specialty practice as a reflection of preparation in greater depth in a particular subject matter, together with a voluntary limiting of focus to a more restricted area of practice by the professional. Lack of specialty designation does not preclude general providers of psychological services from using the methods or dealing with the populations of any specialty, except insofar as psychologists voluntarily refrain from providing services they are not trained to render. (p. 3)

However, in a 1980 malpractice case brought in Ohio against a licensed psychologist (*Midwestern Psychological Services, Inc. v. Potts*), the appeals court found that the psychologist trained as an educational psychologist

> had been practicing clinical psychology although he was not a specialist in that field. ... Generally, one who undertakes to practice a given specialty, even on a part-time or limited basis, is held to the standard of care of that specialty while practicing it. (p. 3948)

The court apparently relied on the case law in medicine and ruled that the specialty standard of care applied in psychology as well.

There are other questions that lack empirical verification: Does the specialist overwhelm the generalist in competition for education resources and consumer dollars? Does the consumer choose the specialist over the generalist more frequently? If so, is the quality of care better, more expensive, or both? Does specialization fragment the profession and, as a result, lose its ability to represent members of the profession to the consumer? If so, does this fragmentation have any impact on the quality of care? These questions remain only partly answered.

RESPECIALIZATION

Respecialization refers to the changing of a specialty or the addition of another specialty area of practice. Because there are multiple areas of specialization in three of the four professions, there is the likelihood that an individual practitioner might choose to practice in an area for which his or her education or training has been inadequate. Psychology confronted this issue by including a section dealing with respecialization in the first revision of the *Standards* (APA, 1977), stating that education and training for the new specialty must take place in an accredited university or professional school. Merely obtaining experience is not enough (Hall, 1983). This philosophy corresponds with what the *Ethical Principles of Psychologists* (APA, 1981a) state as competence: verified education, training, and experience. The first grappling with this difficult issue took place in 1976; by 1982, three states (Oregon, Tennessee, and Virginia) had included in law or in regulation the necessity for re-specialization through a formal educational process. Tennessee added another caveat to the use of experience to qualify one in a new area: "functioning as a member of an institutional staff shall not be considered as adequate preparation" (Tennessee Code Annotated Rules of the Tennessee Board of Examiners, Chapter 1180-2).

Even though specialty boards exist in both medicine and nursing to identify proper education and training for those practice areas, the major control measure for the practice of those two professions and their specialties is through the awarding of hospital staff privileges. Only a few jurisdictions have adopted legislation that mandates hospital staff privileges for qualified psychologists.

Certain jurisdictions have incorporated the *Standards for Providers of Psychological Services* (APA, 1977) (Alabama, Alaska, Arkansas, and Indiana) or the *Specialty Guidelines for the Delivery of Services* (APA, 1981b) into law (Ontario and South Carolina). This is similar to what happened in the states that adopted the Ethical Principles (APA, 1981a) as the state's measure of unprofessional conduct. To include these other documents might provide an additional yardstick for determining if a practitioner should be disciplined by the state for practicing outside his or her area of competence. Clearly, psychology has grappled with the issues of practicing outside one's area of competence more than any other mental-health profession.

In social work, there are voluntary standards for social workers in health care settings, schools, child protection agencies, and long-term care facilities. These are not considered specialty areas, and there are no formal recognition and examination procedures, and no restrictions are placed on providers who wish to work in those areas.

PUBLIC REPRESENTATION

A recent movement designed to provide another check on the effectiveness of the licensing process is the placement of representatives of the public on the state licensing boards. The first state to require the appointment of one public member to each licensing board was California in 1961 (Common Cause, 1982). Then, in 1976, the law was changed to require that a majority of positions on many of the boards be reserved for public members and that on the other boards, one third of the positions be set aside for consumer representatives. Since 1961, a number of states have placed varying numbers of public representatives on the boards and hence on the discipline panels that sit in judgment on errant practitioners. Although the inclusion of public representation on state licensing boards has not yet been systematically evaluated (Shimberg, 1982), a majority of the state licensure boards now have public representatives (see Table 2). The 1982 survey by Common Cause of the membership and executive officers of the state licensing boards in California indicated the experience to be neither a success nor a failure. The most frequently mentioned areas of favorable public-member influence were consumer education, disciplinary actions, board legislative programs, and complaint investigations. However, very few board members believed that public members had had much influence on the quality or the cost of professional service.

Shimberg (1982) indicated that one of the problems has been the lack of a clarity of purpose and selection criteria. He asserted that the same patronage system occurred with public members as had occured in some states with the professional members, pointing out that recruitment, orientation, and available support services were the key to success.

PAYMENT LEGISLATION

In addition to the licensure provisions, there are other statutory provisions that affect quality assurance and/or cost containment, as well as choice of mental health providers. Examples of legislation that dictates or provides access to specific provider groups and possible mental health, alcohol, and drug abuse benefits require an understanding of legislation categorized as freedom of choice, vendorship, mandated minimum coverage, and preferred provider.

FREEDOM OF CHOICE

Thirty-nine states and the District of Columbia, representing nearly 92% of the American population, have enacted laws establishing the direct recognition of psychologists for reimbursement purposes (Dorken, 1986).

These laws amend a state's insurance code and require carriers to provide reimbursement for services performed by a psychologist if the insurance contract covers services within the scope of practice and authorized for psychologists by the state's licensure law. Freedom-of-choice (FOC) laws are intended to deal with the eventuality that an insurance or health service plan will elect to pay only physician providers for services, even though other providers (for instance, psychologists) are licensed to perform many of those same services. The laws require third-party payers to make their reimbursement policies consistent with state law. In a number of state codes, these laws appear in conjunction with similar laws recognizing podiatrists, optometrists, and other practitioner groups qualified for independent and autonomous practice and otherwise eligible for direct reimbursement (APA, no date). Most FOC laws specifically name psychologists as eligible providers, and the laws in Florida, Oregon, and West Virginia also permit reimbursement to licensed social workers. Twenty-six states have passed legislation providing for third-party reimbursement of nurses. Only a subset of those states reimburse for mental health services when provided by either psychiatric nurses or registered nurses (e.g., California, Connecticut, Maine, Massachusetts, New Jersey, and New York).

VENDORSHIP LAWS

In the 35 jurisdictions where one or more levels of social workers are recognized, the potential exists for reimbursement of those licensed or certified for independent practice. Although the FOC legislation may provide access to psy-

TABLE 2. Public Representation on State Licensure Boards[a]

State	Medicine	Nursing	Social work	Psychology
Alabama				
Alaska	X	X	NA	X
Arizona	X	X	NA	X
Arkansas	X		X	X
California		X	X	X
Colorado	X	X	X	X
Connecticut	X			X
Delaware	X		X	X
District of Columbia	X		X	
Florida	X	X	NA	X
Georgia	X		NA	X
Hawaii	X	X	NA	X
Idaho	X	X		
Illinois				
Indiana	X		NA	X
Iowa		X	X	X
Kansas	X	X	X	X
Kentucky	X	X	X	X
Louisiana				
Maine	X	X	X	X
Maryland	X		X	X
Massachusetts	X	X	X	
Michigan	X	X	X	X
Minnesota	X	X	NA	X
Mississippi		X	NA	X
Missouri	X		NA	X
Montana	X		X	X
Nebraska	X	X	?	
Nevada	X	X	NA	X
New Hampshire	X		X	X
New Jersey	X	X	NA	X
New Mexico	X		NA	X
New York	X	X	X	X
North Carolina	X		X	
North Dakota		X	X	
Ohio	X		X	X
Oklahoma			X	
Oregon	X		X	X
Pennsylvania	X	X	NA	X
Rhode Island	X			X
South Carolina		X		X
South Dakota	X	X	X	X
Tennessee			X	
Texas	X		X	X
Utah	X			X
Vermont	X	X		X
Virginia	X			
Washington	X	X	NA	X
West Virginia	X	X	X	X
Wisconsin	X	X	NA	X
Wyoming			NA	

[a]Table 2 is based on information obtained from the four national professional associations and the four national associations of state licensure boards and indicates which states by profession require at least on public representative on the state board. This table is current as of June, 1987 (APA, 1985b; Federation of State Medical Boards, 1983; NASW, 1984; National Council of State Boards of Nursing, 1984).

chological services, it typically has no impact on the reimbursement of social workers and psychiatric nurses. The term used to refer to the type of legislation that provides reimbursement to licensed social workers, either with or without referral by a physician or a psychologist, is *vendorship*. Fifteen states have such legislation, four of which require referral. The social worker still must obtain supervision in many states listed in Table 3.

MANDATED MINIMUM MENTAL-HEALTH COVERAGE

Unless linked to legislation mandating mental health or related coverage, freedom-of-choice or vendorship legislation *per se* does not require new or increased benefits or coverage. This legislation does increase consumer choice and promote competition by expanding the number of qualified service providers from whom an individual may choose:

> On the other hand, mandated coverage laws reflect society's understanding that it may be in the general public interest to require universal coverage or to require that such coverage at least be offered to group subscribers to combat what economists call "adverse selection." (This is the negative incentive which arises when good insurance risks are driven out of health plans and those insurance policies providing high levels of coverage as high-risk utilizers sign on to take advantage of the greater range of benefits, causing a cost spiral. The plan either cuts benefits or raise its premiums.) Mandated minimum coverage legislation can restore stability to the health coverage marketplace by making it public policy that mental health costs should be spread out evenly across the board. (APA, no date)

There are three types of laws covering mental health and alcohol and drug abuse. *MBP* (mental health benefits) refers to a law that requires policies to include the specified benefits package. *MA* stands for *mandated availability*, which requires only that the insurance company offer the benefits to the subscriber. To further confuse the matter, some laws are a combination of MBP and MA, that is, mandating certain benefits and making available others. By September 1983, 26 states had some form of mental health insurance mandate. Of these, 13 required insurers to include a minimum set of mental health benefits (MBP), outpatient and inpatient, and 13 required insurers to offer a minimum set of mental health benefits (MA). An additional 13 states without benefits for mental health provided either MBP or MA for alcoholism only, and four of those same states included drug abuse MBP or MA (APA, no date; National Mental Health Association, 1983). Table 3 indicates the states that have FOC and vendorship laws, as well as those states that have passed legislation mandating benefits (MBP) or mandating availability (MA) of the mental health (MH), alcohol (A), or drug abuse (D) services.

PREFERRED PROVIDER LEGISLATION

Whereas freedom-of-choice and vendorship legislation provide access for consumers to a wider range of qualified mental-health providers, usually at a lower cost, and for third-party reimbursement, in a few states legislation has been introduced that supersedes fee-for-service reimbursement and mandates mental-health-coverage requirements. Under this legislation, both individual and in-

TABLE 3. Direct Recognition and Benefits

State	FOC	Vendorship	MBP	MA
Alabama	X			A
Alaska				
Arizona	X			
Arkansas	X		MH	
California	X	X		MH,A
Colorado	X		MH	A
Connecticut	X		MH,A	MH,A
Delaware				
District of Columbia	X			
Florida	X	X		MH,A,D
Georgia	X			MH
Hawaii	X		A,D	
Idaho				
Illinois	X		A	MH
Indiana	X			
Iowa				
Kansas	X	X		MH,A,D
Kentucky				A
Louisiana	X	X		MH,A,D
Maine	X	X	MH,A,D	
Maryland	X	X	MH,A	MH,D
Massachusetts	X	X	MH,A	
Michigan	X		A,D	A,D
Minnesota	X		MH,A,D	A,D
Mississippi	X		A	
Missouri	X		A	A
Montana	X	X	MH,A,D	
Nebraska	X			A,D
Nevada	X			A,D
New Hampshire	X	X	MH	
New Jersey	X		A	
New Mexico	X			A
New York	X	X		MH,A
North Carolina	X			
North Dakota			MH,A,D	
Ohio	X		MH,A	
Oklahoma	X	X		
Oregon	X	X	MH,A,D	
Pennsylvania	X			A
Rhode Island			A	
South Carolina				
South Dakota			A	
Tennessee	X	X		MH
Texas	X			A,D
Utah	X	X		A
Vermont				MH,A
Virginia	X	X	MH,A,D	MH,A,D
Washington	X		MH	A
West Virginia	X			MH,A
Wisconsin			MH,A,D	
Wyoming	X			

stitutional providers are eligible to become the exclusive providers of services for a set fee. This legislation was introduced originally in California in 1982 and was to become effective on January 1, 1983, for Medi-Cal and in the private sector on July 1, 1983. However, a debate in the courts prevented it from being fully implemented. In the meantime, similar proposals were introduced and passed in eight additional states (Florida, Indiana, Louisiana, Michigan, Minnesota, Nebraska, Virginia, and Wisconsin) and is pending in another 15 states. Twenty-eight states have legislation that inhibits the establishment of preferred provider organizations (PPOs). As would be expected, initially, health provider groups strongly opposed this industrialization of health care delivery, which dictates who will provide the service and for how much, as well as, ultimately, what service will be provided (Ginsberg & Buklad, 1983). More recently, these same provider groups have become intent on inclusion, not exclusion. PPOs are certainly in an embryonic stage, and each is different from the other. At present, there is no empirical research that evaluates the ability of PPOs to control costs or to ensure access and quality of care. Only 10 million enrollees have committed themselves to such a concept (Gabel & Ermann, 1985).

STATE-LEVEL QUALITY ASSURANCE

In addition to the provisions of the licensure statute specifying who is qualified to provide services independently, there are other provisions that affect the quality of services, for those clients who are served by state inpatient and outpatient facilities. Another example of the state's influence on services is the clause in the licensure statutes that exempts employees of state (and federal) institutions from having to be licensed, even though they are providing services under a title that would be protected in the private sector. The providers of services who are employed by the state do not have to meet the standards set by the state, although the state may voluntarily adopt the same or similar standards.

EXEMPTIONS AND DOUBLE STANDARDS

Included in almost all statutes specifying the requirements for licensure are sections that list settings (e.g., chartered elementary or secondary schools and degree granting institutions) and/or services (those provided while in training) that are exempt. That means that the persons who provide those services in those settings do not have to meet the standards for licensure. For instance, in New York State, persons employed by municipal, county, or state facilities may use the employment title *psychologist*, in that employment setting only without being required to be licensed. Similar exclusions occur in the statutes of other states.

The degree to which exemption from licensure exists varies by state and by profession (NASW, 1984; Stigall, 1983). However, a series of class-action suits brought on the behalf of institutionalized patients in Alabama had an impact on the criteria used to define professional staff. The landmark case was the *Wyatt v. Stickney* (1972) litigation, brought to the U.S. Middle District Court, which resulted in the first constitutional standards of care and habilitation. National recognition of the in-equity of a dual level of care came when Federal District Judge Frank Johnson ruled that the providers of mental health services in the state institutions must meet the same standard or standards as exist for those in the private sector.

This concern about quality services for those who are served by state facilities set off a shock wave that affected most of the state institutions for the mentally ill and the mentally retarded. The American Psychological Association, as well as the American Civil Liberties Union, the American Orthopsychiatric Association, and the American Association on Mental Deficiency, served as *amici curiae* ("friends of the court"), presented testimony during the deliberations, and participated in the development of the standards in Alabama.

Beginning with this ruling in Alabama and followed by similar court cases (reported as 174 as of January 1979, in Cavalier & McCarver, 1981), there has been increasing interest by county, state, and municipal agencies in requiring their pro-viders of mental health services to meet the state standards. Cavalier and McCarver (1981) noted that the constitutional right to treatment for the mentally ill had been in the federal courts for several years before this case, but the *Wyatt v. Stickney* right-to-treatment case was the first to address the mentally retarded and the first to in-clude a set of standards for the operation of a public facility. The court order pro-vided definitions of a *qualified mental health professional* (QMHP) and added that each "shall meet all licensing and certification requirements promulgated by the State of Alabama for persons engaged in private practice of the same profession" (p. 16). This particular criterion was not upheld, as a federal appellate court held that there was no constitutional basis to compel this requirement (*Wyatt v. Aderholt,* 1974).

From the viewpoint of administrators of state facilities, the major impetus for the increase in quality in the 1970s came from the Joint Commission on Accredita-tion of Hospitals (JCAH) accreditation standards, which, in turn, had been affected by the *Wyatt v. Stickney* staffing-patterns and patient-care concerns. The teams from the JCAH were well trained and took their jobs seriously, and entitlement programs became dependent on accreditation for support (L. Chase, personal com-munication, April 1984).

As a result, when the APA produced its first set of *Standards for Providers of Psychological Services* in 1974, the experience in the U.S. Middle District Court left its imprint:

> The Task Force sees no justification for maintaining the present double standard whereby providers of private fee-based services are subject to statutory regulation, while those pro-viding similar psychological services under governmental auspices are usually exempt from such regulations. This circumstance tends to afford greater protection under the law for recipients of privately delivered psychological services. On the other hand, the recipients of privately delivered psychological services currently lack many of the safe-guards that are available in governmental settings; these include consultation, peer review, records review, and staff supervision. (APA, 1972, p. 2)

In the first revision of these *Standards* in 1977 (the second was approved in 1987 as "guidelines"), a similar section was included but with stronger language on the necessity of change:

> There should be a uniform set of standards governing the quality of services to all users of

psychological services in both the private and public sectors. There is no justification for maintaining the double standard presently embedded in state legislation. (APA, 1977, p. 2)

In the most recent revision of the *Standards* (1987), the position taken was as follows:

When providing any of the covered psychological service functions, in any time, and at any setting, whether public or private, profit or nonprofit, any persons representing themselves as psychologists are expected, where feasible, to observe these Guidelines of practice to promote the best interests and welfare of the users of such services. (p. 1)

Even today in most states, and for all four professions, there still exists the possibility of an exemption from licensure for those employed with an otherwise protected job title. This is the well-known "exemption clause," although the types of practitioners exempted vary from state to state and from profession to profession. According to Stigall (1983), the states of Colorado, Indiana, Kentucky, Michigan, Nevada, North Carolina, Ohio, South Dakota, Utah, and Wyoming have no exemption for state employees with the title *psychologist*.

In 19 of the 35 jurisdictions with credentialing for the profession of social work, there is no exemption for public employees (Arkansas, Colorado, Idaho, Illinois, Iowa, Kansas, Michigan, New York, North Dakota, Puerto Rico, Oklahoma, Rhode Island, South Carolina, South Dakota, Tennessee, Texas, Utah, Virgin Islands, West Virginia, and possibly in Vermont and Connecticut, which only recently passed legislation; NASW, 1984). In a NASW policy statement (1976), the following statement is included:

Regulation must cover all areas or settings in which social work is practiced, including public and voluntary, profit and nonprofit. If the basic justification for legal regulation is the interest of the consumer, there is no justification for limiting the coverage of laws that govern practice. In fact, the primary responsibility may be said to be that of protecting the public from unqualified, incompetent practitioners in publicly supported services. The practice of social work in institutional settings, such as hospitals, does not ensure that the client is afforded the services of practitioners who meet public and professional standards and, of course, offers no accountability that extends beyond the institution. Legal recognition of social work as a profession necessarily involves recognition of professional standards applied by a public authority and should cover all all practitioners of the profession. (p. 7)

In psychology and social work, the statutes may protect only the title. Therefore, there is no necessity for exemption because another title may be used. For instance, in New York, the title *social worker* is not protected, but the title *certified social worker* is. The title *psychotherapist* is not protected in several states, including New York and Colorado.

Exemptions for physicians and nurses employed by state facilities are limited to those working under a limited permit, possibly because the majority of physicians (including psychiatrists) and nurses (including psychiatric nurses) are affiliated with hospitals that themselves require licensure to obtain hospital staff privileges. In fact, 49 medical boards provide for the issuance of educational permits, limited and temporary licenses, or other certificates for the restricted practice of medicine. The terms for the issuance of such certificates vary. Such certificates may be issued (a) for the hospital training of those eligible for licensure, (b) for supervised employment in state or private hospitals, and (c) for full-time private practice until the next regular session of the licensing board. Therefore, the double standard appears to be primarily a problem for psychology and, to a lesser degree, for social work.

The New York Public Mental Health System

The next logical question is: Do the exempt settings have standards that directly or indirectly contribute to the assurance of quality, rather than cost containment? Let me provide as one example the public mental-health system in the State of New York. Within the Department of Mental Hygiene are the Offices of Mental Health (OMH) and of Mental Retardation and Developmental Disabilities (OMRDD), which are the administrative agencies for providing services to those populations. They provide direct services for a total of 34,999 (OMH) and 30,948 (OMRDD) inpatients treated yearly at the state's 31 psychiatric and 20 developmental centers, as well as 9,410 (OMH) and 21,777 (OMRDD) outpatients served yearly at the 249 licensed outpatient OHM programs, the 624 non-state-operated OMH programs, the 359 state-operated outpatient programs, and the undetermined number of facilities serving the mentally retarded and developmentally disabled in the community and in the schools (L. Chase, personal communication, February 1984).

In addition, and more important, these agencies regulated the services provided by both the state and the nonstate providers of services. The regulation by the state is through the recently developed criteria for defining the providers of services in outpatient facilities for purposes of reimbursement (Part 579 and Part 585 of Title 14 of Mental Health Hygiene Law). Approximately 75% of the outpatient programs hold a current operating certificate and meet the standards of Part 579 and 585 (L. Chase, personal communication, February 1984).

In 1978, the New York State Legislature established the Commission on Quality of Care for the Mentally Disabled, giving the statutory responsibility for oversight to an independent body. Most of the commission's energies since 1978 have been devoted to reviewing the organization and operation of the Offices of the Department of Mental Hygiene and scrutinizing various facets of the operations of state facilities and local government, voluntary, and proprietary programs, as well as the workings of the financing and regulatory mechanisms that exist. Such scrutiny has occurred both through individual investigations of complaints or deaths and through broader program reviews, policy analysis, and cost-effectiveness studies (New York State Commission on Quality of Care for the Mentally Disabled, 1982).

This approach to quality assurance may be unique in that the oversight commission is located organizationally within the executive department, with complete access at all times to any mental hygiene facility, its books, and its data (Article 45 of the Mental Hygiene Law). Although there may be a question about whether the commission is focusing on quality assurance, it is clear that its in-depth studies concentrate on the detection of negligence.

The Annual Report from the Commission also comments on the problems that arise when single agencies, such as OMH, retain the roles of service provider and regulator. The first problem is that the agency may be an ineffective regulator and, as a consequence, may lack credibility with agencies under its authority, as programs with deficiencies may go uncertified and still receive state funds. The second problem is that all the programs within OMH and OMRDD do not meet the standards applied to the programs outside those agencies, and as the state needs those agencies to provide services to those discharged or diverted from the state system, there is a reluctance to invoke drastic sanction. This lack of sanction further weakens the respect for the OMH and the OMRDD.

As that Annual Report indicates,

> We must question whether these objectives can be achieved by maintaining the regulatory function within OMH and OMRDD. It is realistic or desirable for these agencies to serve as the primary regulatory bodies when they themselves are major providers of services in need of independent regulation, and when they are inevitably constrained in their regulatory role by considerations of their dependence, as service providers, on the non-State sector?" (p. 11)

In recognition of these diffiulties, OMH is attempting to move out of the direct provider role, first by developing the regulations and then by requiring the nonstate providers to meet the standards. Certainly, this is a policy question, whether a provider can be a regulator and be effective at both, but in all fairness, this issue may also be a political one, and an oversight body may be needed as long as both jobs cannot be done effectively by one agency. The Governor's Select Commission on the Future of the State-Local Mental Health System (July 1984) recommended that OMH have three functions: (a) to regulate, certify, fund, and direct the public mental health system; (b) to stimulate a comprehensive and properly balanced community-based care system; and (c) to provide direct services through state psychiatric centers. The latter function should be segregated from the first two in order to give proper support and resources to the first two functions. Partially in response to the select commission's report, the Office of Mental Health recently issued its 1985–1990 comprehensive plan for services to the mentally ill in New York state.

For the profession of social work in New York, there is another question for those regulated or affected by the standards for professional staff covered in Part 585 of Title 14 of the Mental Hygiene Law, in which a standard exists that contradicts the state definition of a qualified provider of social work services. Although physicians, nurses, and psychologists must be licensed to serve as professional staff in outpatient programs recognized by OMH, the social worker may be licensed as a certified social worker or have a master's degree in social work. As there are over 24,000 certified social workers in New York State, and therefore, supply is not a problem, it would seem appropriate to use the single standards of licensure and thereby to use a standard that is reasonable, practical, and in keeping with the standards for the other licensed professions (P. Johnston, personal communications, August 24, 1983).

Even though these two offices have the responsibility of ensuring that their programs will meet minimal acceptable regulatory standards and of improving the quality of care and treatment derived by these programs, it appears that a separate oversight commission, such as that set up in New York, performs an important role and, given what exists elsewhere, perhaps a unique role.

SUMMARY AND CONCLUSION

Although there are many people who provide services within mental health who remain largely unregulated by the states, the key mental-health providers are physicians, psychologists, social workers, and nurses. These four professions are regulated through licensure, the development of various standards, and the passage of other types of legislation, including the parameters for third-party reimbursement. Although the goals of any state regulation may include cost containment, the major purpose of such regulation is to protect the public, to ensure quality standards, and to a certain extent, to ensure that some level of minimal care will be available to all. Other state and federal regulation is done through the actions of planning agencies for facilities and judicial regulation via common-law decisions in malpractice suits. This chapter focused on the former (licensure and third-party legislation).

To a degree, there is consistency in the licensure requirements within a profession and across states, as well as some commonality between the professions in mental health, but the states have left their unique imprint on these provisions. The idiosyncratic state laws and regulations may cause confusion for the consumer as to labels, consternation for the practitioner as to mobility, and recognition problems for the third-party reimbursers. Some order has been established by virtue of the professional standards adopted by the states. The states have also played an active role in supplementing these standards with program and facility standards, which often require the licensure of staff even though the facility is classified as exempt. The requirement of JCAH accreditation for the federal funding of state as well as private institutions has had a major impact on quality assurance. In addition, voluntary adherence of the profession adds to the assurance of quality.

In addition to regulating the providers through requirements for licensure, standards for staff, and a list of approved providers, the states have instituted programmatic monitoring of quality assurance through the creation of staff positions, special offices, or commissions, such as in the example provided from New York.

Many of these attempts to ensure quality are relatively recent and await through analysis and evaluation. Hopefully, this overview has provided a perspective for initiating that process by focusing on the inputs into quality assurance initiated by the state.

REFERENCES

American Association of State Psychology Boards. (1977, August). *Toward a definition of training in professional psychology.* (Available from AASPB, PO Box 4389, Montgomery, AL 36104.)

American Association of State Psychology Boards. (1983). *Newsletter.* (Available from AASPB, PO Box 4389, Montgomery, AL 36104.)

American Medical Association. (1983). *Continuing medical education fact sheet.* Chicago: Author.

American Nurses Association. (no date). *Certification: Become the professional's professional.* Kansas City, MO: Author.
American Psychological Association. (1974). *Standards for providers of psychological services.* Washington, DC: Author.
American Psychological Association. (1977). *Standards for providers of psychological services* (rev. ed.). Washington, DC: Author.
American Psychological Association. (1979). *Criteria for accreditation of doctoral training programs and internship in professional psychology.* Washington, DC: Author.
American Psychological Association. (1981a). *Ethical principles of psychologists* (rev.). Washington, DC: Author.
American Psychological Association. (1981b). *Specialty guidelines for the delivery of services.* Washington, DC: Author.
American Psychological Association. (1985a). *Standards for educational and psychological tests.* Washington, DC: Author.
American Psychological Association. (1985b). *Summary of state laws regulating psychological practice through licensure.* (Available from APA, 1200 Seventeenth St. N.W., Washington, DC 20036.)
American Psychological Association, Council of Representatives. (1985c, February). *Minutes.* Washington, DC: Author.
American Psychological Association. (1986). APA accredited doctoral programs in professional psychology. *American Psychologist, 40,* 1392–1398.
American Psychological Association. (1987). *General guidelines for providers of psychological services.* (Available from APA, 1200 Seventeenth Street, N.W., Washington, DC 20036.)
American Psychological Association. (no date). *Recognition and reimbursement for psychological services.* (Available from APA.)
Annotation. (1980). *A.L.R.3d, 21,* 953–955.
Brinkley, J. (1985). U.S., industry and physicians attack medical malpractice & medical discipline laws: Confusion reigns. *The New York Times,* (December 2–3), pp. 1, 10, & 1, B6.
Carroll, S. L., & Gatson, R. J. (1983). Occupational licensing and the quality of service: An overview. *Law and Human Behavior, 7,* 139–146.
Cavalier, A. R., & McCarver, R. B. (1981). Wyatt v. Stickney and mentally retarded individuals. *Mental Retardation, 19,* 209–214.
Common Cause. (1982). *Going public: Consumer advocates on California's professional licensing boards.* (Available from Common Cause, 636 S. Hobart Blvd. No. 226, Los Angeles, CA 90005.)
Council for the National Register of Health Service Providers. (1985a). *Designated programs in psychology.* (Available from the Council, 1200 Seventeenth St., N.W., Washington, DC 20036.)
Council for the National Register of Health Service Providers in Psychology. (1985b). *National register of health service providers in psychology.* Washington, DC: Author.
Dent v. West Virginia. (1889). 129 US 114, 32 L. Ed. 623, 9 S. Ct. 231.
Derbyshire, R. C. (1983). How effective is medical self-regulation? *Law and Human Behavior, 7,* 193–202.
Dolan, A. K., & Urban, N. D. (1983). The determinants of the effectiveness of medical disciplinary boards: 1960–1977. *Law and Human Behavior, 7,* 203–218.
Dorken, H. (1986). *Professional psychology in transaction.* San Francisco: Jossey-Bass.
Dvorak, E. M., Kane, M. T., Laskevich, L. A., & Showalter, R. E. (1982). *The National Council Licensure Examination for Registered Nurses.* Chicago: Chicago Review Press.
Equal Employment Opportunity Commission, Civil Service Commission, Department of Labor and Department of Justice. (1978). Adoption by four agencies of Uniform Guidelines on Employee Selection Procedures. *Federal Register, 43,* 38290–38315.
Federation of State Medical Boards of the US, Inc. (1983, January 1). *Membership list.* (Available from the Federation, 2630 West Freeway, Suite 138, Fort Worth, TX 76102.)
Gabel, J., & Ermann, D. (1985). Preferred provider organizations: Performance, problems, and promises. *Health Affairs, 4,* 24–40.
Ginsberg, M. R., & Buklad, W. (1983, January). *Memo to State Psychological Associations.* (Available from APA.)
Governor's Select Commission on the Future of the State–Local Mental Health System. (1984, July). *The*

future of the state–local mental health system (final report). (Available from Patricia Lamphear, Empire State Plaza, Corning Tower Building, No. 1455, Albany, NY 12229.)

Hall, J. E. (1983). Respecialization, licensure and ethical conduct. *The Clinical Psychologist, 36*(3), 68–71.

Hall, J. E. (1985a, August). *Credentials in clinical psychology: Licenses, registries, diplomas and certificates.* Paper presented at the annual convention of the American Psychology Association, Los Angeles.

Hall, J. E. (1985b). *The ABPP diploma privilege: Does it really exist? Professional Practice of Psychology, 6,* 251–265.

Hall, J. E. (1986). Issues and procedures in the disciplining of psychologists. In R. Kilburg, R. Thorenson, & P. Nathan (Eds.), *Professionals in distress: Issues, syndromes, and solutions in psychology.* Washington, DC: American Psychological Association.

Hall, J. E. (1987). Licensure and certification of psychologists. In B. Edelstein & E. Berler (Eds.), *Evaluation and accountability in clinical training.* New York: Plenum Press.

Herbsleb, J. D., Sales, B. D., & Overcast, T. D. (1985). Challenging licensure and certification. *American Psychologist, 40,* 1165–1178.

Hubbard, J. P. (1978). *Measuring medical education: The tests and the experiences of the National Board of Medical Examiners* (2nd ed.). Philidelphia: Lea & Febiger.

Kane, M. T. (1982). The validity of licensure examinations. *American Psychologist, 37,* 911–918.

Kelvorick, A. K. (1981) Regulation and cost containment. In T. McGuire & B. A. Weisbrod (Eds.), *Economics and mental health* (DHHS Publication No. ADM 81-1114). Washington, DC: U.S. Government Printing Office.

Loveland, E. (1977). Alternatives and innovations. In *Proceedings of a national conference for evaluation competence in the health professions.* (Available from Professional Examination Service, 475 Riverside Drive, New York City, 10017.)

McGuire, T., & Weisbrod, B. A. (1981). NIMH Conference on economics and mental health: Introduction. In T. McGuire & B. A. Weisbrod (Eds.), *Economics and mental health* (DHHS Publication No. ADM 81-1114). Washington, DC: U.S. Government Printing Office.

Midwestern Psychological Services, Inc. v. Potts. (1980). Unpublished opinion. Ohio Appellate Court.

National Association of Social Workers. (1976). *Standards for the regulation of social work practice.* (Available from NASW, 7981 Eastern Ave., Silver Spring, MD 20910.)

National Association of Social Workers. (1978). *Standards for social work services in schools.* (Available from NASW.)

National Association of Social Workers. (1981) *Standards for social work in health care settings.* (Available from NASW.)

National Association of Social Workers. (1982a). *NASW register of clinical social workers.* Silver Spring, MD: Author.

National Association of Social Workers. (1984, July), *State comparison of laws regulating social work.* (Available from NASW.)

National Commission for Health Certifying Agencies. (1985, November 15). *The state of the art: Continuing competence assurance* (draft report). (Available from NCHCA, 1101 30th St., N.W., Suite 108, Washington, DC 20007.)

National Council of State Boards of Nursing. (1984, January). *Survey.* (Available from National Council, 303 East Ohio, Suite 2010, Chicago, 60611.)

National Mental Health Association. (1983, September). *Addendum to for ayes only.* (Available from NMHA, 1800 N. Kent St., Arlington, VA 22209.)

Nelson, S. H. (1979). Standards affecting mental health care: A review and commentary. *American Journal of Psychiatry, 3*(3), 303–307.

New York State Commission on Quality of Care for the Mentally Disabled. (1982). *Annual report: 1981–82.* (Available from the Commission, 99 Washington Avenue, Albany, NY 12210.)

New York State Education Department. (1982). *Annual report on the professions.* (Available from the Office of the Professions, Cultural Education Center, Albany, NY 12230.)

Paxton, A. (Ed.). (1985, September). Malpractice crisis: How is it related to competence? *Professional Regulation News,* pp. 1–2.

Office of Mental Health. (1984, October). *Five year comprehensive plan for mental health services: 1985–1990.* (Available from Office of Mental Health, 44 Holland Avenue, Albany, NY 12229.)

Overcast, T. D., & Sales, B. D. (1985). *Antitrust and malpractice implications of specialty recognition.* Unpublished manuscript.

Reaves, R. P. (1982). *Regulating the professions: A legal and legislative handbook.* (Available from author, PO Box 4389, Montgomery, AL 36104.)

Richman, S. (1982). *Final report to the AASPB on the role delineation study for the Examination for Professional Practice in Psychology.* (Available from American Association of State Psychology Boards, PO Box 4389, Montgomery, AL 36104.)

Rosenfeld, M., Shimberg, B., & Thorton, R. F. (1984). *Job analysis of licensed psychologists in the United States and Canada.* (Available from American Association of State Psychology Boards, PO Box 4389, Montgomery, AL 36104.)

Schnaps, L. S., & Sales, B. D. (1985). *Specialization in Psychology: Lessons from other professions.* Unpublished manuscript.

Sechrest, L., & Hoffman, P. E. (1982). The philosophical underpinnings of peer review. *Professional Psychology, 13,* 14–18.

Shimberg, B. (1982). *Occupational licensing: A public perspective.* Princeton, NJ: Center for Occupational and Professional Assessment.

Simon, G. C. (1983). Psychology, professional practice, and the public interest. In B. D. Sales (Ed.), *The professional psychologist's handbook.* New York: Plenum Press.

Stigall, T. T. (1983). Licensing and certification. In B. D. Sales (Ed.), *The professional psychologist's handbook.* New York: Plenum Press.

Theaman, M. (1982). A critical appraisal of Daniel Hogan's position on licensure. *Professional Practice of Psychology, 3*(1), 1–18.

U.S. Department of Health, Education, and Welfare. (1967). *Report of the National Advisory Committee on Health Manpower.* Washington, DC: U.S. Government Printing Office.

U.S. Department of Health, Education, and Welfare. (1971). *Report on licensure and related health personnel credentialing* (DHEW Publication No. 72–11). Washington, DC: U.S. Government Printing Office.

U.S. Department of Health, Education, and Welfare. (1977). *Credentialing health manpower* (DHEW Publication No. [05] 77-50057). Washington, DC: U.S. Government Printing Office.

Wellner, A. M. (Ed.). (1978) *Education and credentialing in psychology.* Washington, DC: Steering Committee of APA, AASPB and National Register.

Wellner, A. M., & Zimet, C. N. (1983). The National Register of Health Service Providers in Psychology. In B.D. Sales (Ed.), *The professional psychologist's handbook.* New York: Plenum Press.

Wolgemuth, R. R., & Samph, T. (1983). *Content validity study in support of licensure examination program of the American Association of State Social Work Boards.* (Available from AASSWB, 6404 Garners Ferry Road, Columbia, SC 29209.)

Wyatt v. Aderholt, 503 F. 2d 1305 (5th Cir. 1974).

Wyatt v. Stickney, 344 F. Supp. 373 (M.D. Ala. 1972).

16

Accrediting Agencies and the Search for Quality in Health Care

MYRENE MCANINCH

In his book *Zen and the Art of Motorcycle Maintenance: An Inquiry into Values,* Robert Pirsig (1974) wrote that Socrates made the following statement in 300 BC: "Quality isn't method; it's the goal toward which method is aimed." This statement is as relevant and important today as it was when Socrates made it. In health care, the goal of providing quality patient care doesn't change; the methods of attaining that goal or of determining whether that goal has been achieved do. The method of attaining quality patient care may vary from patient to patient, depending on the individual's treatment plan, on the particular treatment system or approach taken (for example, a behavior modification or a psychoanalytic approach), or on the hospitalwide or facilitywide quality-assurance programs. Methods change to improve the chances of attaining the goal. The process through which methods change is dynamic and ongoing, based on the identification of problems, the implementation of corrective action, and the achievement of improvement. As progress is made and objectives are met, the process improves, and its focus shifts to other areas in need of attention. This procedure goes on at the patient care level and, through a facility's quality assurance program, at the management level.

Documentation is necessary to the pursuit of quality patient care because there must be evidence that something happened in the staff–patient relationship for quality of care to be measured, studied, and improved. Although quality of patient care is admittedly elusive, in our attempt to define or demonstrate it we need more than a subjective opinion that care and treatment are appropriate. There must be some tangible evidence of the clinical course of treatment that identifies the patient's problems, the various treatment interventions used, and the patient's response to each if we are to assess the appropriateness and effectiveness of patient care. Verbal transmission of this information is insufficient and inadequate because it is subject to misinterpretation. Without documentation, we have nothing to study, nothing that is available to others involved in the care of the patient for direction, participation, and guidance. Without documentation, there is no accountability.

MYRENE MCANINCH • Accreditation Program for Psychiatric Facilities, Joint Commission on Accreditation of Hospitals, 875 North Michigan Avenue, Chicago, Illinois 60611.

Although Socrates helps us to understand the important distinction between the goal of quality care and the methods used to achieve it, Hipprocrates identified the key ingredients in documenting patient treatment. From before the time of Hippocrates, documentation of patient care is found in the caves of France. These pictures, which depict the use of herbs and leaves to bind a warrior's wounds, indicate an early realization of the need to document care for others to review. During the Egyptian period, documentation of care was formalized by Imhotep, the first physician we know of by name (Siegel & Fischer, 1981). Imhotep believed that quality treatment requires, first, diagnosis of the problem through careful observation of the patient. He established detailed requirements for conducting a case history that could then be used to make an accurate diagnosis and to render appropriate treatment. Specifically, he required identifying data, so that one patient could be distinguished from another, and a history of the patient's problems, including a comprehensive physical examination and identification of any emotional factors that might influence the patient's care. Once this initial assessment was completed, Imhotep believed an accurate diagnosis could be made, and appropriate treatment could be determined. He also developed an elaborate system of symptoms that he believed were associated with particular treatment interventions.

Hippocrates contributed the ideas of progress notes that described a patient's response to treatment and a discharge summary that served as a clinical resumé of the patient's response to all interventions and that provided a description of the patient's status at the time of discharge. Because of the contributions of Socrates, Imhotep, and Hippocrates, treatment now involved assessment of the patient, followed by development of an individualized treatment plan based on an analysis of the assessment data, the recording of progress notes that described the patient's response to treatment, and the completion of a discharge summary that indicated the outcome of treatment. With this information, the quality of patient care could be evaluated because a methodology and sufficient documentation existed for care to be reviewed by others.

With the advent of hospitals in the Middle Ages and the development of medical schools, documented care was used to train new practitioners, to bill for services, and to account for staff time. It's interesting that, for the first time, the cost effectiveness of care was addressed in conjunction with the detailed review of care as a teaching technique for certifying quality of care. Henry VIII contributed an indexing system that he had designed for retreiving records and for cataloging various maladies and types of associated treatments. By 1609, all physician orders for medication in England had to be entered into the record. This afforded a more complete chronological review of the patient's clinical course of treatment. With the opening of American hospitals and medical schools during the late 1700s and early 1800s, experimentation with various methods of determining quality of care continued, while the goal of providing quality care remained constant.

Throughout this period in which the medical record was evolving as the primary source of data for determining the provision of quality patient care, the psychiatric record was not differentiated from the medical record. However, as general hospitals began to move psychiatric patients to other buildings on the grounds and then to more distant locales, special institutions or asylums for the insane were developed. Originally, these facilities were quite small, and few records were kept for fear of stigmatizing the patients when they returned to the community. However, as state funds were increasingly used to support patients in these institutions and new accountability issues were raised about the competency of staff and the quality of patient care, statistical data and separate psychiatric records began to be required. Except in a few unusual cases that were well documented for training purposes, these records were rudimentary. State mental-health codes, as well as medical groups such as the New Jersey Medical Society, began to make specific what had to be in the record for reimbursement purposes and for the determination of the quality of care. Despite these efforts, however, record-keeping systems remained incomplete and fragmented during the late 1800s and the only documentation of a cure was discharge from treatment.

In 1890, a New York psychiatrist named Pliney Earl challenged the use of discharge from treatment as evidence of quality care because many patients reentered the system within a brief period and were supposedly "cured" again at the time of their discharge. Dr. Earl believed that progress notes that clearly described the treatment rendered and the patient's response to each treatment were necessary if one were to assess which treatment interventions had worked with a particular patient and which had failed and to determine what the outcome of treatment might be at discharge.

Little actual progress was made, however, as indicated by the concern expressed by both the American Hospital Association and the American Medical Association at their annual meeting in 1902 and 1905, respectively. Medical records were an issue at both conventions. Serious concern was expressed about the prevalence of incomplete, fragmented, and illegible medical records. Medical records were of such poor quality that the level of care being provided could not be determined. At the meeting of the Third Clinical Congress of Surgeons of North America in 1912, concern about the quality of care provided in hospitals was so serious that the surgeons recommended that there should be some system of standardization for hospitals to protect patients from inappropriate or suboptimal care.

When the American College of Surgeons was formed in 1913, one of its goals was to develop minimal essential standards of care for hospitals as a first step toward the provision of quality care in American hospitals. By 1917, a single page of standards had been developed that addressed the medical staff's credentialing, privileging, and monitoring functions; complete medical records; and equipment. The five basic standards were these:

1. That physicians and surgeons privileged to practice in the hospital be organized as a definite group or staff. Such organization has nothing to do with the question as to whether the hospital is "open" or "closed," nor need it affect the various existing types of staff organization. The word *staff* is here defined as the group of doctors who practice in the hospital inclusive of all groups such as the "regular staff," the "visiting staff," and the "associate staff."

2. That membership upon the staff be restricted to physicians and surgeons who are (a) full graduates of medicine in good standing and legally licensed to practice in their respective states or provinces, (b) competent in their respective fields, and (c) worthy in character and in matters of professional ethics; that in this latter connection the practice of the division of fees, under any guise whatever, be prohibited.

3. That the staff initiate and, with the approval of the governing board of the hospital, adopt rules, regulations, and policies governing the professional work of the hospital; that these rules, regulations, and policies specifically provide:

(a) That staff meetings be held at least once each month. (In large hospitals the departments may choose to meet separately.)

(b) That the staff review and analyze at regular intervals their clinical experience in the various departments of the hospital, such as medicine, surgery, obstetrics, and other specialties; the clinical records of patients free and pay, to be the basis for such review and analysis.

4. That accurate and complete records be written for all patients and filed in an accessible manner in the hospital—a complete case record being one which includes identification data; complaint; personal and family history; history of present illness; physical examinations; special examinations; such as consultations, clinical laboratory, X-ray and other examinations; provisional or working diagnosis; medical or surgical treatment; gross and microscopical pathological findings; progress notes; final diagnosis; condition discharge; follow-up and, in case of death, autopsy findings.

5. That diagnostic and therapeutic facilities under competent supervision be available for the study, diagnosis, and treatment of patients, these to include, at least (a) a clinical laboratory providing chemical, bacteriological, serological, and pathological services; (b) an X-ray department providing radiographic and fluoroscopic services.

The first surveys of hospitals using these minimal standards were disastrous. They highlighted the disparities in hospital care. By 1918, however, some hospitals had succeeded in meeting the standards, and voluntary accreditation was launched. The value of the program soon became apparent. The quality of care provided in hospitals and the medical records in approved hospitals were more acceptable. As more and more hospitals sought the educational and consultative benefits of the program, the number of approved hospitals increased. By 1951, some 3,000 general hospitals were accredited by the Hospital Standardization Program of the American College of Surgeons.

The success of the Hospital Standardization Program proved that health care professionals were concerned enough about the quality of care they provided to establish and participate voluntarily in an effective quality-assurance program. It also demonstrated that, once again, the method of attaining the goal of quality of care had changed to include not only documentation of care in the record, but also assessment of the competence of staff through credentials review, assignment of privileges, and peer review of direct patient care, as well as of the availability, adequacy, and safety of clinical laboratory and X-ray equipment.

In 1951, the American College of Surgeons in collaboration with four other national professional organizations—the American Medical Association, the American College of Physicians, the American Hospital Association, and the Canadian Medical Association, all of which had as a common goal the development of national standards as a means of determining and improving the quality of care provided in American hospitals by medical staffs—formed the Joint Commission on Accreditation of Hospitals (JCAH) as a private, nonprofit, voluntary accreditation organization. The Canadian Medical Association withdrew in 1959 to participate in Canada's own national accreditation program, the Canadian Council on Hospital Accreditation.

The minimal essential standards developed by the American College of Surgeons were used by JCAH until 1953. Because of its commitment to evaluating and improving quality of care through its standards and survey process, the JCAH developed standards that specified new methods of determining the goal, including an examination of the qualifications and competence of departmental directors and evidence of appropriate supervision of staff, of sufficient numbers of staff to carry out particular responsibilities, and of monitoring of quality of care through clinical management conferences. Standards addressing these issues were published in 1953 as the *Standards for Hospital Accreditation*.

The 1960s were an era of change in health care. *Public accountability* became buzz words, and the belief that every American was entitled to a reasonable degree of quality emerged from the activities of consumer advocates, such as Ralph Nader, who questioned the quality of care provided by health care professionals. To be responsive to these changes, the JCAH, in 1960, added requirements for medical record review, so that it could assess the completeness of records; for patient care evaluation studies, so that it could determine the effectiveness of different treatment regimes; and for utilization review, so that it could determine the appropriateness of care. Again, the methodology of achieving the goal of quality care was in a state of flux.

The 1965 Medicare Act was a realization of the promise that every American was entitled to a reasonable degree of care. When Congress was fashioning the Medicare legislation, it met with great resistance from provider groups, including the American Medical Association and the American Hospital Association. These groups believed that, while the government was insuring health care for the nation's elderly, governmental regulations tied to the Medicare Act would result in undue policing of health institutions and possible interference with the private practice of medicine. Congress addressed that issue by recognizing JCAH hospital accreditation standards as being equivalent to the federal conditions of participation. Thus, a hospital accredited by the JCAH, an organization that represents the health professions' quality assurance efforts, was "deemed" to meet eligibility requirements for participation in Medicare. Subsequently, when the government insured health care for the poor or the mentally ill, JCAH accreditation again became a quality-of-care criterion for reimbursement.

In response to this confidence, the JCAH Board of Commissioners voted to review and revise the standards in order to raise them from minimal essential to optimal achievable. By 1970, the standards had expanded from 10 to 152 pages of indicators of the quality of hospital care. The new standards further elaborated on clinical practice. For example, medical-care-evaluation studies were required to determine the efficacy and appropriateness of practice; related therapeutic services, such as occupational and physical therapies, were made requisite to habilitation services; and documentation of patient care was emphasized both for quality and reimbursement reasons.

During the 1970s, business, industry, federal and state government agencies, and third-party payers became increasingly concerned about the utilization as well as the quality of services provided by health care facilities. Even the JCAH strengthened its utilization review standards to stress the needs for accessible, appropriate, effective, efficient, and timely hospital services. In 1972, the Social Security Amendments initiated the professional standards review organization

(PSRO) program, through which a hospital could delegate all or almost all of the review process for validating medical and hospital practice. Some hospitals actively embraced the PSRO program to participate actively in the evolving national health-care delivery-system. Other hospitals perceived the PSRO as further governmental encroachment on the physician–patient relationship. Part of the problem for both the PSROs and the JCAH was that, although everyone agreed that quality care should be provided to the public, there was no consensus on how to measure it. Therefore, much of the early effort of PSROs was focused on utilization review because the appropriateness of admissions and lengths of stay could not be assessed (Gordon, Roberts, & Rodak, 1976). The JCAH made an attempt by encouraging the performance of retrospective medical-care-evaluation studies. These studies, however, had only a modest quality-assurance aspect. Everyone wanted to ensure quality, but no one knew how to assess it.

The government and the JCAH continued to try to define methods for assessing quality of care throughout the 1970s. The 1972 edition of the *Accreditation Manual for Psychiatric Facilities,* which reflects JCAH's *Accreditation Manual for Hospitals,* required that appropriately qualified medical staff assume the responsibility for monitoring and reviewing medical care and for supervising other related health professionals who were not clinically privileged to function independently; that medical records be complete; that there be an adequate number of other qualified staff to support the service system as determined by departmental reviews; that utilization review and patient-care-evaluation studies be conducted to influence staff privileges and to identify continuing education needs; and that the physical plant and environment be safe and therapeutic. Patients' rights to treatment and to participation in the planning of treatment were also addressed.

In the mid-1970s, the private and public sectors made several advances in their quality assurance requirements. In 1975, the JCAH added an audit requirement for acute hospitals as a systematic retrospective-review process intended to identify potential problems in patient care. In the 1974 edition of the JCAH's *Accreditation Manual for Child and Adolescent Psychiatric Facilities,* individual case review was highlighted as a systematic concurrent-review process for identifying and resolving patient care problems while the patient was still in the hospital. In 1975, the JCAH's *Accreditation Manual for Alcoholism Programs* added a requirement of program evaluation: it was to be conducted at least annually at the departmental or program level to ensure that the goals of the facility were being met. In 1976, to ease the fragmentation occurring in the field as a result of the different requirements of the JCAH and the PRSO program, the PSRO program adopted the JCAH's quantitative and qualitative audit requirements. The JCAH also added other monitoring activities to existing medical monitors of surgical case review, tissue review, and pharmacy and therapeutics review. Infection control requirements that focused on the control of nosocomial disease were adopted, and utilization review, having received so much emphasis by both the JCAH and PSROs, became a separate chapter in the JCAH standards manual.

By 1978, it was becoming increasingly evident that the audit requirements were inappropriately overshadowing other essential quality-related activities that were practiced by hospitals. For example, other required quality-assessment and quality-related activities, such as the delineation of privileges, individual case review, the monitoring of clinical practice, and the review of support services, were not being coordinated with audit activities or incorporated into an overall quality-assurance program. For some hospitals, adherence to the numerical requirements and the methodology of audit was mistakenly taken to be synonymous with quality care. In other instances, the rigidity of the audit methodology, called the *performance evaluation procedure* (PEP), limited the amount and scope of care evaluated. At best, audit was a quality assessment technique that measured the quality of care provided at a given point in time, apart from any effort to change or improve that care. At its worst, audit was a paper exercise that failed to focus on identified or potential problems in patient or clinical performance.

Quality assurance activities needed a broader perspective. The 1979 edition of the JCAH's *Consolidated Standards for Child, Adolescent, and Adult Psychiatric, Alcoholism and Drug Abuse Programs* endeavored to provide that perspective, first by indicating the need for a quality assurance program that linked and integrated all quality-related activities and, second, by requiring a written plan that described that quality assurance program. The standards in the "Quality Assurance" chapter cited such activities as program evaluation, audit, individual case review, utilization review, and staff growth and development (continuing education) as quality-related activities. Although these standards encouraged hospitals to include quality-related activities other than audit in a hospital-wide quality-assurance program, they neither addressed other monitoring activities, departmental reviews, credentials reviews, and clinical privileging nor specified a problem-focused approach. Hence, hospitals still had to conduct a number of quality-related activities that were not specifically identified as being part of the quality assurance system.

Accordingly, the JCAH Board of Commissioners took two major steps in 1979. First, the board eliminated the numerical requirements of audit. It should be noted this action was meant neither to diminish the value of audit as a means of assessing and improving patient care nor to lessen the responsibility of medical and other professional staffs for reviewing and evaluating care. Second, the board approved a new quality-assurance standard that encouraged hospitals to implement an overall quality-assurance program designed to ensure the delivery of quality patient care. This quality assurance program was to link all committees, functions, and activities concerned with quality assurance and to integrate all quality-related activities in such a way that existing data might be fully used in quality assessment activities. The quality assurance system emphasized the need to identify problems in patient care, to take corrective action, and to monitor that action to determine if the problem has been successfully resolved. In developing this standard, the JCAH continued to focus on improvement in the quality of care as the central purpose of the accreditation process. Once again, the goal of quality patient care remained constant while the methods of attaining the goal changed. And so closed another decade in the search for quality in patient care.

QUALITY ASSURANCE IN THE 1980s

The 1980s opened on a note of hope: a brand new quality assurance standard was in place. The new standard appeared in a chapter by itself to emphasize its importance. But hospitals and programs that read only the quality assurance chapter, thinking this was where all quality-assurance requirements would be found, made a mistake because quality assessment and quality-related requirements were woven into the very fabric of other related standards throughout the JCAH standards manuals.

In the revised 1981 *Consolidated Standards Manual,* for example, the quality assurance standard was essentially an umbrella standard drawing together standards addressing other quality-related activities and functions specified throughout the manual. These included standards that addressed the role of the governing body and administration and their responsibility to support and provide sufficient resources for a quality assurance program, as indicated in the opening chapters of the manual; that addressed the organization and competency of the staff, as specified in standards concerning credentials review and privileging mechanisms by which the essential services in the facility are maintained through the coordinated effort of the organized staff, administration, and governance of the facility; that addressed the patient management issues documented in patient records; and that addressed the adequacy and safety of the buildings in which patients are housed or receive treatment, as defined in the last chapters of the manual. Specific quality-assurance activities, such as utilization review, patient care monitoring, program evaluation, and staff growth and development, were highlighted in separate chapters, whereas other monitoring functions, such as those that concerned the issue of restraint and seclusion, incident reporting, or review of adverse responses to medications, were embedded in appropriate chapters and cross-referenced in the quality assurance standard to ensure linkage of information.

From 1981 through 1984, the JCAH refined and refocused its quality assurance standard (Caffeldt, Roberts, & Walczak, 1983). The new focus of the standard was on the identification and resolution of problems in patient care, apart from an emphasis on a particular number of studies or audits to be conducted in each clinical area or programmatic component. The standard required evidence of an organized program based on an ongoing objective assessment of important aspects of patient care through which problems or opportunities to improve care could be identified and then resolved. A major purpose of the standard was to ensure the accountability of the medical and professional staff for the care they provided and to define and strengthen their roles and functions in identifying and correcting potential problems in patient care. The range of sources of data from which potential problems in patient care could be identified was broadened from the 1979 standard. Because the assessment evaluation process was to be ongoing, it encompassed and subsumed numerous periodic evaluations and monitoring activities to ensure that the system would be a continuous dynamic process.

The standard did not, however, indicate any particular way in which potential problems were to be identified. This lack of clarity appeared to be a problem for many hospitals and facilities. For many, quality assurance activities became a "problem of the month club." Each clinical department and program component thought it had to identify and resolve a certain number of problems to demonstrate that the department or service had fulfilled its quality assurance responsibilities. In addition, JCAH survey results indicated that many hospitals and other facilities were focusing on issues that could be readily resolved, or that the problems did not reflect the major clinical activities of the department or service, in terms of either volume or high-risk potential. What appeared to be missing was a rational approach to identifying potential issues that would impact on patient care.

However, the quality assurance standard was specific in requiring the use of clinically valid criteria for assessing identified problems. The inclusion of clinically valid criteria in the assessment process expanded the range of clinical concerns that could be addressed in the assessment of real or potential problems beyond the original indicators (for example, diagnosis procedures used in the audit process). The standard also required that appropriate action be taken to resolve or alleviate the issue and that the corrective action be followed up or monitored so that it could be determined whether the action had been effective in producing the desired changes. This was a significant improvement.

Another new concept introduced into the JCAH quality-assurance standards during this period was the assignment of responsibilities for the coordination of all quality-assurance activities conducted throughout the organization, as well as oversight responsibilities for determining not only that the quality assurance activity was conducted but that it was conducted effectively. The intent here was to integrate all ongoing quality-assurance activities into the system to ensure the communication of information that could affect multiple clinical areas and to reduce duplicative quality-assurance efforts.

The goal of the JCAH quality assurance standard during the early 1980s was to broaden the range of issues that would be addressed in quality assurance programs. Efforts were also made to assist health care facilities to move from a series of fragmented or poorly integrated quality-assurance activities toward a facilitywide system that was effective in identifying and resolving problems in patient care and that provided for the integration of all quality-related activities.

It became increasingly evident, however, that several problems continued to exist in the quality assurance standard. The field told us that we needed (a) to provide a framework within which a facility could identify potential problems or opportunities to improve patient care; (b) to clarify the role of the governing body in quality assurance activities; and (c) to clarify the scope of quality assurance so that a facility would know what was expected of it. The JCAH also wanted to strengthen its approach to quality assurance in terms of managing patient care and to close the loop by asking facilities to incorporate quality assurance information into credentialing and privileging activities. In essence, the JCAH believed that quality assurance information should serve as an important determinant of the current competency of staff.

The revised 1985 quality assurance standard in the *Consolidated Standards Manual* (CSM) and the *Accreditation Manual for Hospitals* (AMH), both of which are responsive to psychiatric and substance abuse hospitals and such program components within a general hospital, follows basically the same format. It is noteworthy that the "Quality Assurance" chapter in the 1985 CSM is located in the section on hospital or facility management. This placement reflects the JCAH's current focus on helping facilities to implement in clinical areas the same kind of management systems that they already have in place for the management of finances and other resources.

What do the standards call for? Standard 1 in the AMA and Standards 9.1 through 9.4 in the CSM ask for facilities to implement an ongoing facilitywide system for monitoring and evaluating the quality and appropriateness of clinical care. It is important in a period of limited resources for facilities to consider the appropriateness as well as the quality of care. The appropriateness of care involves determining whether the patient received the correct level of care in a timely manner. In surveying the standard, the JCAH expects to find, in the aggregate, evidence of an organized approach to quality assurance. In addition, quality assurance activities should be an integral part of the day-to-day operations of the facility. Paper compliance will not be sufficient to meet the intent of the JCAH standards calling for a planned and systematic process that focuses quality assurance activities on clinically relevant areas versus random attention to quality assurance through periodic studies. To meet the intent of this standard, a facility would have to do its homework first. To establish a process of continuous monitoring of clinically important aspects of care, the facility must first identify which elements of care are important and must then implement a system for regularly checking the actual care provided against established criteria. Now, facilities are expected to have a system in place from which departures from the expected or from opportunities to improve patient care can be readily identified. When established, this planned and systematic process for monitoring and evaluating important aspects of patient care should help facilities to identify the clinical issues requiring improvement.

The standard goes on to identify the responsibilities of the governing body and the clinical administrative staffs in the quality assurance process. The standard asks the governing body both to require and to support a facility-wide quality-assurance program. The governing body does so not only by providing for sufficient resources, but also by delegating responsibility to the facility's administration and medical and professional staff for "monitoring and evaluating the quality and appropriateness of patient care and clinical performance." This delegation of responsibility does not decrease the legal responsibility of the governing body for the provision of quality care in the facility. Litigation such as that in the *Darling* case (*Darling v. Charleston Community Memorial Hospital*, 1963) has made it clear that a governing body is responsible for knowing what is going on in the organization and particularly for knowing about any problems in patient care and any action that the staff has taken to resolve them. Consequently, it's important for the professional and management staff to report the results of their quality assurance activities to the governing body. The standard does not spell out the frequency or format for reporting such information, but it is obvious that it must be of such frequency and in such form that the governing body can fulfill their responsibilities. Rather than giving the governing body computer printouts or statistical data, it might be more helpful to summarize or aggregate the quality assurance information in such a way as to identify trends or patterns of performance and to provide discussion of what the data indicate about problems in patient care and how the problems are being addressed and resolved by the staff. The intent of the JCAH in delineating the responsibility of the governing body and the responsibility of the management and professional staffs in carrying out the various quality assurance activities, as well as reporting the results back to the governing body, was to close the loop in the responsibilities between the governing body and the hospital management and the professional staff.

Planning activities for quality assurance should culminate in a written plan that clearly describes the purpose, scope, and process of monitoring and evaluating the quality and appropriateness of patient care and clinical performance. It is important to remember that the plan is only a blueprint for action. It serves to delineate what the facility intends to do, who will do it, when data will be collected, what will be included, how it will be evaluated, and so on. The plan should be revised if problems in patient care are either not being identified or not being adequately resolved by the system.

Standard 9.5 in the 1985 *Consolidated Standards Manual* and Standard 11 in the 1985 *Accreditation Manual for Hospitals* represents an attempt to identify the minimal requirements for the scope of quality assurance (QA) that should be addressed in the QA plan and by the program. The scope includes facilitywide functions such as infection control and utilization review, as well as individual quality-assurance activities for direct patient care and support services. The scope of quality assurance in the AMH includes the traditional medical-staff functions relating to the monitoring and evaluation of care, including monthly departmental review of the care provided, medical staff monitoring functions, and the monitoring and evaluation of patient care provided by various support services. Similarly, the scope of quality assurance in the CSM includes the monitoring of care provided by all departmental services and units, and a variety of monitoring activities including, but not limited to, review of incident reports, use of restraint and seclusion procedures, treatment plans, and adverse responses to medications, as well as the quality and appropriateness of patient care provided by various support servces. New to the psychiatric and substance abuse field is the requirement of both individual and aggregate analysis of information derived from various quality-assurance activities. Psychiatric and substance abuse facilities are quite familiar with the monitoring of patient care case-by-case, such as in patient care monitoring, but they are not familiar with analyzing aggregate data to detect trends or patterns in performance.

Additionally, the importance of reviewing the care provided by all who come in contact with patients, including those with individual clinical privileges and those who work under supervision, was highlighted in the revised standards. In the 1985 AMH, this provided the necessary link to the governing body requirement that there be a mechanism in place to review the care provided by those who are supervised. In the 1985 CSM, it strengthened the requirements in the chapters on "Staff Organization" and "Written Plan for Professional Services" pertaining to the review of patient care provided by staff not privileged to function without supervision.

But the most important requirement in the section of the QA standard pertaining to scope is the requirement that facilities incorporate the results of QA activities into the process for reappraising and reappointing medical and professional staff members, renewing or revising clinical privileges, and appraising the competency of all practitioners who do not have individual clinical privileges. This requirement for incorporating quality assurance information into the appraisal and privileging process represents a closing of the loop and is designed to assist facilities to incorporate QA activities into their day-to-day operations. This requirement may be difficult and time-consuming for hospitals to achieve, but in the long run, it will help them to incorporate quality assurance into their management processes.

Standard III in the 1985 *Accreditation Manual for Hospitals* and Standard 9.6 in the *Consolidated Standard Manual* reiterate the familiar elements of the monitoring and evaluation process. These include identification of the major clinical activities, based on the type of diagnostic and therapeutic procedures used, the clinical characteristics of the patient populations served, and the intensity of care provided. Major clinical activities or functions should include, minimally, those activities and functions that have the greatest volume, and that are of high risk potential or are most problem prone. The next step is to identify indicators of both the quality and the appropriateness of patient care associated with these clinical activities. It is important to differentiate between the quality of a particular dimension of care that is associated with the competency of a practitioner and the appropriateness of the task or function being provided to a given patient. The latter relates to whether the right patient received the right treatment or procedure at the right time.

Once the indicators of care have been delineated, the staff must then identify clinically valid criteria to be used in the assessment of actual practice versus expected practice. Then the familiar process of data collection, periodic review of data against predetermined clinical criteria, evaluations of variances from the expected results, followed by corrective action and further monitoring to determine the effectiveness of the actions taken, should be carried out. For example, having a high percentage of schizophrenic patients who have not improved and who also have not received therapeutic doses of neuroleptic drugs is neither an expected result of appropriate treatment nor one that can be readily justified and, hence, will require corrective action. On the other hand, if the hospital sets the criteria for preventing all suicidal attempts at 100%, this goal may not be within staff capability, although it may be possible to decrease the frequency of attempts below the level originally identified as problematic. Therefore, the criteria against which patient care data are reviewed should not be set so low or so high that opportunities to improve care go unrecognized or cannot be resolved. In either instance, the requirement is that a system be in place by which, through an evaluation of the information collected, there can be a determination of whether a problem exists. If a problem or an opportunity to improve care is identified, action should be taken through appropriate channels, whether through administration, medical or professional staff, supervisory staff, or other personnel. The main issues are that responsibility for action should be clearly assigned and that action to correct or improve some aspect of patient care be taken.

Standard IV in the 1985 *Accreditation Manual for Hospitals* and Standard 9.7 in the 1985 *Consolidated Standards Manual* address the overall administration of the quality assurance program. A major issue here is the oversight function of the QA program, which is not intended to create conflict between departmental authority or responsibility and that of the QA program, but to ensure that there will be some means of verifying that quality-related activities are being carried out and that they are being performed effectively. If tasks are being performed without the desired results, then this deficiency would indicate failure of the oversight function. In some instances, however, the quality assurance program may collect data for certain departments and/or services that for whatever reason, cannot do it themselves. However, the department or service should always develop clinically valid criteria that will be used to evaluate the quality and appropriateness of care it provides and should always conduct the actual evaluation itself.

A related function of the QA program has to do with communication of information, as appropriate, between the departments and services, to the medical and professional staff, to management staff, and to the governing body. Here, the emphasis is on ensuring that the result of QA activities will get to the appropriate sources so that they can lead to meaningful changes. The failure of the QA program to detect problems that have broad implications or that affect more than one department or service may result in duplication effort and cost inefficiency. Similarly, if the results of QA activities don't get to the place where they are needed, the process becomes meaningless. The QA program also has a responsibility to follow up on problems through various stages toward resolution, or in essence, to develop a tracking system that cuts across departments, services, and program components and medical, professional, or administrative staff. Finally, there is a requirement for an annual reappraisal of the quality assurance program itself. This is an opportunity to reassess whether the plan that was implemented has done what the facility desired, whether the indicators have captured the important aspects of patient care, whether information discovered has been used appropriately, whether duplication of effort has been reduced or eliminated, and so on. This reassessment provides another opportunity for making changes where the goals to be accomplished have not been achieved or where ways to improve the system are identified.

Thus, the revised quality-assurance standard clarifies what is expected in the conduct of quality assurance activities, provides a framework for the identification of problems in clinical areas by requiring an ongoing systematic process of monitoring and evaluation of the quality and appropriateness of patient care, and spells out the relative responsibilities of the governing body and the clinical and management staff, as well as the responsibilities for the overall administration and oversight of QA activities.

The JCAH's quality assurance standard was designed to be equally responsive to hospitals and facilities that vary in size, organization, structure, procedures, and functions. The standard affords a hospital considerable flexibility in developing and implementing an effective quality-assurance program. Flexibility allows the medical and professional staffs to tailor the quality assurance effort to its organizational structure and patient care activities. New and even innovative approaches to quality assurance and to the identification and resolution of problems in patient care are encouraged. Improvement in the patient care provided in psychiatric hospitals and facilities continues to be the emphasis of the JCAH, but it now believes that the methods used to achieve quality should reflect the unique characteristics of a given hospital or facility and should not be regulated through rigid standards. Facilities have been most responsive to this freedom to design quality assurance systems that meet their unique needs.

The JCAH, Medicare, and the peer review organizations (PRO) program still differ, however, in how they determine the presence of quality of patient care, and these differences confound hospitals. Utilization review remains the major focus of the PROs. For psychiatric facilities, Medicare focuses on 2 special conditions of participation (that is, staffing and medical records), in addition to the original 16 conditions. These 2 conditions were added to make sure that sufficient numbers of qualified competent staff would be available to deliver services and that there would be documentation of individualized, active treatment for each patient. During 1983, an unpublished validation study was conducted jointly by the Health Care Financing Administration and the JCAH to determine the degree of equivalency between the *Consolidated Standards* and survey process. The results of the study suggested a high degree of concurrence in the survey process and standards. Such efforts may decrease the confusion that hospitals are experiencing in how best to develop an effective quality-assurance system that meets the varying requirements of the regulating and accrediting agencies.

MURPHY'S LAW: "WHAT CAN GO WRONG, USUALLY DOES"

When the JCAH's Board of Commissioners initially approved the quality assurance standard in 1979, they realized that to ensure orderly and effective implementation, they would have to phase in the standard by identifying specific areas to focus on and by gradually expanding that focus until all areas were encompassed. They realized that some complex organizations might need several years to implement a fully operational and effective quality-assurance program. All facilities needed time to gather all the information available within the organization and to organize it in such a way that critical problems in patient care could be identified and resolved. Consequently, the JCAH's Board of Commissioners selected compliance assessment elements to focus the initial efforts to implement the standard. Facilities were asked to (a) assign authority, (b) develop a written plan, and (c) make some progress toward the integration of quality related activities.

Although these elements seem simple to comply with, this did not prove to be the case when the JCAH actually began to survey the new standards. First of all, in relation to the assignment of authority, the standard allowed much latitude in how that could be handled by a facility. Responsibility could be assigned to an individual, a group, or a committee. Some hospitals created a committee, a department, or a division of quality assurance and appointed a director or coordinator in an effort to comply with this element. All too frequently, however, there was no clear assignment of authority or responsibility for making decisions concerning which problems most critically impacted on patient care or concerning what action needed to be taken in order to address the identified problems. Sometimes, the director functioned as a clearinghouse, and the committee received only those problems that had been screened by the director. The director or coordinator often lacked the clinical expertise to make this type of decision. Other facilities created a committee, but efforts to assure comprehensiveness resulted in some committees of more than 50 people. Problems sent to such committees never saw the light of day again because no consensus could be achieved. Still other facilities, in an effort to involve the governing body in quality assurance matters, developed a committee of equal numbers of staff and board menbers. The result was that board members became involved in the day-to-day operation of the agency, an activity that had the potential of undermining their policy-setting role. Furthermore, because board members were reluctant to deal with sensitive issues, many serious problems were not even addressed. Other facilities developed a committee of workable size, but the results of this committee's work had to be screened by the executive committee of the professional staff, the medical director, or the chief executive officer before any action could be taken. Again, the tendency to divert those problems that were most sensitive or those issues that an administrator did not wish to fund to individual staff members limited the effectiveness of the committee. As a result of this bureaucratic screening process, facilities again avoided issues that were critical to patient care and thus limited the effectiveness of their quality assurance committee. There are many other similar examples of the difficulties that facilities wrestled with in trying to initiate a quality assurance program.

To implement a quality assurance program, the second implementation element required that there be a written plan of action, a plan that reflected a comprehensive approach to problem identification an resolution. Plans ranged from a page description with an attached table of organization that depicted lines of communication and authority to a 100-page discourse on quality assurance. In some instances, there appeared to be an inverse correlation between the number of pages in the plan and the staff's understanding of their quality-assurance program. However, apart from the extensiveness of this plan, discrepancies were frequently observed between the description of the program in the plan and the actual program. Many plans were not comprehensive enough. They failed to recognize the possible importance of information from areas other than direct patient care and, hence, omitted information from such areas as safety, therapeutics, infection control, incident reports, and claims denials.

Progress toward the integration of related quality assurance activities was the third element of implementation. Many facilities found this task too difficult, and as a result, information remained fragmented. Committees and coordinators were frequently inundated with the data from a variety of sources and were unable to organize the information well enough to determine the persuasiveness of certain problems or to identify common elements between studies undertaken, by, say, the nursing and the dietary departments. Unnecessary duplication of effort prevailed, and significant issues affecting patient care remained buried in the deluge of data.

The JCAH Board of Commissioners, wanting to strengthen the quality assurance standard while still allowing hospitals more time to implement the complete standard, revised the compliance elements after the first 2 years of experience to emphasize the critical importance of problem solving in quality assurance activities. In January 1983, the three compliance elements became (a) to assign authority and responsibility in accordance with the description in the written quality-assurance plan; (b) to integrate and/or coordinate information from quality-related activities in accordance with the description in the written quality assurance plan; and (c) to resolve identified patient care problems in the conduct of individual quality assurance functions (organized staff-monitoring activities, departmental reviews, and support service review and evaluation). Here again, the JCAH was trying to encourage the facilities to establish systematic quality review mechanisms throughout the facility and to focus them on the identification and resolution of problems in patient care or clinical performance.

An examination of the accreditation results from mid-1981 to mid-1983 indicates that only 15% of all contingencies involved the three compliance elements described above. This figure is significantly increased, however, if related patient monitoring functions and credential and privileging activities are added. A contingency represents a proviso placed on accreditation indicating that there are areas impacting on the quality of care or the quality of the environment that requires attention before the next survey in 3 years. Contingencies associated with the three quality-assurance compliance elements (plan, authority, and integration) appeared to be fairly evenly distributed, although problems in achieving the integration and coordination of quality assurance information and activities appeared to be slightly greater. The addition of problem correction and resolution as an element for quality assurance compliance in 1983 resulted in a greater number of deficiencies in the area, plus continuing integration and coordination problems. Plan and authority issues seemed to have been resolved. When related quality-assurance activities, such as utilization review, patient care monitoring, program evaluation, and credentialing and privileging, were added to the data on quality assurance contingencies, there seemed to be some evidence of a relationship, as might be anticipated, between these activities and the overall quality-assurance score. Thus, if the facility was experiencing problems in utilization review, it might be anticipated that the facility would have lower quality assurance scores. Patient management, in terms of the adequacy of assessment and treatment-planning documentation, did not evidence any relationship with the overall quality assurance program, although the adequacy of the patient management documentation may have influenced the results of utilization review and patient-care monitoring in particular. On the other hand, program evaluation, which had been developed primarily as a management tool to ascertain the appropriateness and effectiveness of the departments and services in achieving the desired goals, had the strongest relationship to quality assurance deficiencies. Hence, if the hospital was noncompliant in program evaluation, it was usually noncompliant in quality assurance, and vice versa.

Differences between small and large facilities (large being defined as having more than 100 beds) were also examined. The related quality assurance activities did not seem to have any influence on how well small hospitals fared in developing appropriate quality-assurance programs, perhaps because many of these activities are conducted by a committee of the whole, so that the same staff that reviewed utilization review and quality assurance activities also served as the executive committee and reviewed credentials, privileges, and so on. Thus, integration and problem correction and monitoring were vested and operationalized within a single group. On the other hand, with the use of a corrected contingency-coefficient score, a significant relationship was found between utilization review, program evaluation, and all three compliance elements of quality assurance.

Additionally, there was a significant relationship between patient care monitoring and the quality assurance element of integration and coordination ($\hat{c} = .65$) (JCAH, 1981c). Privileging and patient management as a whole bore no relationship to, or exerted no influence on, the adequacy of the quality assurance program. In regard to the privilege delineation process, although the standards indicate that this function should include quality assurance information, facilities apparently infrequently used quality assurance data in their privileging decisions, and hence, no significant relationship was found. In essence then, the larger hospitals and facilities with the complexities of multiple committee structures, combinations of levels of care (inpatient and residential), and services to more than one age and disability group experienced more difficulty in pulling information together into an organized and effective system of quality review and problem resolution.

By mid-1983, with few exceptions, most facilities had developed a written plan that designated authority and responsibility for quality assurance functions and had identified internal and external sources of quality assurance information that were to be integrated and/or coordinated in the review process. Communication across departments and throughout the facility also seemed to have improved as a result of efforts to integrate and coordinate quality assurance activities. In spite of these advances, facilities continued to have problems with patient management and with such quality assurance mechanisms as credentials review and privileging, utilization review, and patient care monitoring. It could be assumed that facilities had developed new mechanisms to achieve compliance with the quality assurance standard rather than strengthening their existing review and monitoring system. There was an obvious discrepancy between what the facility planned to do in reviewing quality of care, as indicated in the plan, and the actual activities of the quality assurance program. Also, in many facilities, there was a tendency to deal with easy problems that could be corrected by a change in policy, an improved system, or a new medical records format, rather than dealing with important problems that affected patient care and clinical performance. Review tended to focus on information about individual patients rather than on aggregate information for detecting patterns of care. Rarely was it found that the quality assurance committee had evaluated the effectiveness of the quality review process itself to determine if problems with little effect on patient care were the focus of quality assurance functions rather than the identification of significant trends and patterns of professional practice. In a number of instances, the quality assurance program had degenerated into a witch hunt or "problem-of-the-month club" in an effort to find and correct any problems, real or imagined, in patient care. These quality assurance activities had a negative, debilitating effect on staff.

It became apparent that, when a quality assurance committee or coordinator was the locus of identification, assessment, and evaluation of patient care problems, few clinical departments or services other than the medical staff were involved in the review of important patient-care issues. As a result, systematic, facilitywide review processes that involved all professional staff and encompassed all clinical disciplines were not developed. To ensure the involvement of all clinical departments in the quality assurance program, quality and appropriateness review standards were developed for 8 chapters of the CSM, and 14 chapters of the AMH. By emphasizing the responsibility of the department or service in the verification of the appropriateness of its services, as well as in the review of the quality of care, a balance between quality assurance and cost effectiveness was defined. Although the

JCAH standard maintained its focus on quality protective functions, it also required facilities to look at whether the services were appropriate to patient needs. In doing so, it allowed the hospital to more easily define the costs associated with quality patient care—an important issue in an era of cost containment.

The quality and appropriateness review function is seen as a gestalt in which appropriateness review is a form of quality review that also addresses cost effectiveness. The review of the quality and appropriateness of patient care encompassed in major clinical activities within the purview of the department is a departmental responsibility, though the routine collection and analysis of data could be conducted through the quality assurance program. The quality and appropriateness review function, however, is an integral part of the hospital's quality assurance program. The standard was designed to promote flexibility while strengthening quality assurance efforts in patient-care service areas that had not been consistently monitored or well integrated into the overall quality-assurance program in the past. It is of paramount importance that the facility have a planned and systematic process for monitoring the quality appropriateness of patient care and for resolving the identified problems. Through this process, assessment of both the quality of a given procedure, lab result, or treatment regime and the appropriateness of the clinical judgment made in ordering that procedure or requesting that lab report, based on a particular patient's needs, can be reviewed.

Information in the aggregate from the required support services is to be linked to the facility-wide quality-assurance program and the review of the quality and appropriateness of patient care throughout the organization. In this way, trends or care patterns in clinical performance across the hospital can be identified.

As indicated earlier, problems in the QA standard resulted in a revised 1985 QA standard that clearly linked the support service quality and appropriateness monitoring and evaluation standard into the facility-wide, systematic review of patient care. The revised standard also endeavored to close the loop between information gleaned from QA activities and the critical functions of credentialing and privileging of staff. The JCAH Board of Commissioners revised the compliance assessment clusters and elements to incorporate these activities and made them effective as of January 1, 1985.

However, by mid-1985, it was apparent that facilities were having problems in using QA information in the reappointment process, in the assignment of individual clinical privileges, and in defining the scope of supervision necessary for staff who were not privileged to act independently. A second major area of difficulty related to efforts to develop an ongoing monitoring and evaluation process in clinical departments or services. Failure to identify major clinical functions and activities precluded the collection of relevant data as well as the review of care against predetermined criteria. These difficulties were borne out by a large number of hospitals and facilities receiving contingencies in these areas. Accordingly, the JCAH Accreditation Committee, at its August 1985 meeting, decided that because of their flexibility, the standards relating to the monitoring and evaluattion of care provided by clinical departments and support services and to the use of QA information in the determination of the competency and privileging of staff, were more complex for an organization to implement. Because these standards required new systems be implemented, they necessitated a greater degree of organizational change. In the committee's judgment, facilities needed more time to achieve compliance with these standards and additional educational assistance. A new procedure, called *implementation monitoring*, was created by which status for predetermined periods would be surveyed as usual, but noncompliance would not result in a contingency. Instead, the JCAH asks the facility to submit an interim monitoring report that described the progress that the facility has made toward compliance with the standards. Additionally, the JCAH has developed a generic model for monitoring and evaluation to clarify the required QA functions and to assist facilities in putting this process into a perspective that is unique to their own service system.

In adopting this approach, the JCAH has in no way taken a step back from its original requirements for a facility-wide, systematic process that involves review of the quality and appropriateness of care. Rather, the JCAH has recognized that some standards require more time to implement fully. Quality assurance is so critical an area in a hospital's operation that latitude must be provided.

FUTURE DIRECTIONS

Revision and clarification of the JCAH quality-assurance standards are most timely in light of recent legislative changes that will affect hospitals. Proposed revisions to the Medicare *Conditions of Participation* include the addition of a section on quality assurance. The regulations, much as the JCAH standards, require an effective, hospitalwide quality-assurance program encompassing all practicing hospital staff. The requirements are silent on the involvement of nonclinical staff, such as dietary, housekeeping, and safety personnel. The proposed *Conditions* require the review of nosocomial infections, medication therapy, and appropriateness of surgery. They also require verification of credentials and reappraisal of medical staff members. The *Conditions* do not, however, indicate the necessity for credentials review or reapproval of privileges for professional staff who may not be members of the medical staff, nor do they indicate the need to monitor medical records or the medical responsibility inherent in pharmacy reviews.

Because of our troubled economy, cost reimbursement systems have been reviewed and revised nationally. DRGs (diagnostic-related groups) have been identified for medical categories and for chronic alcohol and substance abuse. No psychiatric classifications have been agreed on, but they undoubtedly will be in the near future. A prospective payment system has been designed using historical and regional average costs for a given diagnostic-related group. There is speculation that the system, in rewarding hospitals that keep costs below the fixed payment level, will lead to early discharges. Underutilization of needed services may result in multiple admissions for a given patient. Hence, quality of patient care may be jeopardized.

Similarly, the Tax Equity and Fiscal Responsibility Act of 1982 (TEFRA) introduced changes in the reimbursement structure of hospitals. The legislation creates a professional review organization (PRO) that, by contractual arrangement with the Health Care Finance Administration (HCFA), will validate the diagnostic information provided by hospitals, will determine the appropriateness of admissions and discharges, and will determine the appropriateness, completeness, adequacy, and quality of care provided.

What about future directions for accrediting agencies? The JCAH and other accreditation bodies, such as the Commission for the Accredition of Rehabilitation Facilities, continue to emphasize the quality of patient care . But the public has also made it evident that standards that address organizational structure and process may answer the question "Can the agency provide good quality care?" but fail to answer today's consumers' question: "Does the agency provide quality patient care?" The JCAH board is looking at the possibility of using clinical process and outcome indicators as a technique of determining better whether the organization is, indeed, providing quality patient care. Currently, the JCAH's Professional and Technical Advisory Committees (PTAC) are in the process of identifying the clinical activities that lead to quality care in the mental health field. The PTAC found that review of the quality of patient care, apart from a consideration of the resource expenditures associated with that care, may be inappropriate in light of today's environment. A major task may involve developing a more comprehensive definition of quality patient care, one that encompasses an evaluation not only of the care but of the benefits and risks of that care. It is not yet clear how the clinical process and outcome indicators, once identified, will be incorporated into future JCAH standards or survey processes. It is clear, however, that the JCAH is open to looking at how to modify its current process to address quality of care more directly and to measure the risks, costs, and benefits involved.

In today's economy, therefore, a balance must be struck between cost containment mechanisms and the JCAH major thrust for quality patient care. Quality care must be viewed broadly as encompassing the professional judgment exercised in the provision of health care as well as the appropriateness of resource use. Review of quality of care must go beyond general evaluation of care to a specific, objective, and systematic review process of quality protection in which problems in patient care are either prevented or identified and resolved. There are multiple dimensions within such a quality assurance system, including a problem-focused review of facilitywide concerns, an in-depth review of the critical indicators or a single parameter of care, and the development of indicators that focus on the appropriateness of the use of resources. The governmental agencies and professional associations that license, certify, or accredit health care facilities must work together to ensure that review will address resource utilization and cost containment, but not at the expense of quality of care. A number of methodologies have been developed in an effort to improve the quality and utilization of health services. Documenting the identification of instances of poor quality or inappropriate service use is relatively easy. Measuring aspects of patient care that require changes in professional practice patterns or patient outcomes is much more complicated. The review of quality of care separately—that is, independently of utilization and cost—is the most difficult.

Quality assurance programs have at least created professional and public awareness of issues concerning the appropriateness and level of care provided. But it is still early to tell whether quality assurance programs have led to behavior change and improved levels of quality and appropriate service utilization. Greater emphasis must be placed on using quality assurance information in making routine management decisions about patient care and services. A convergence of methodologies may be requisite to influencing and assessing changes in patient care.

REFERENCES

Affeldt, J. E., Roberts, J. S., & Walczak, R. M. (1983). Quality assurance: Its origin, status, and future direction. *Evaluation and the Health Professions, 6*(2), 245–255.

American College of Surgeons (1924). The minimum standard. *Bulletin of the American College of Surgeons, 8*(1), 4.

Code of Federal Regulations, 42 CFR 482.1–482.66

Darling, v. Charleston Community Memorial Hospital, 33 Ill. 2d 253 (1965), *cert. denied* 383 H.S. 946 (1966).

Goran, M. J., Roberts, J. S., & Rodak, J. (1976). Regulating the quality of hospital care—an analysis of the issues pertinent to national health insurance. In R. H. Egdahl & P. M. Gertman (Eds.), *Quality assurance in health care.* Germantown, MD: Aspens Systems Corporation.

Joint Commission on Accreditation of Hospitals. (1953). *Standards on hospital accreditation.* Chicago: Author.

Joint Commission on Accreditation of Hospitals. (1971). *1970 Accreditation manual for hospitals.* Chicago: Author.

Joint Commission on Accreditation of Hospitals. (1972). *Accreditation manual for psychiatric facilities.* Chicago: Author.

Joint Commission on Accreditation of Hospitals. (1974). *Accreditation manual for alcoholism programs.* Chicago: Author.

Joint Commission on Accreditation of Hospitals. (1974). *Accreditation manual for psychiatric facilities serving children and adolescents.* Chicago: Author.

Joint Commission on Accreditation of Hospitals. (1979). *1979 Consolidated standards for child, adolescent, and adult psychiatric, alcoholism, and drug abuse programs.* Chicago: Author.

Joint Commission on Accreditation of Hospitals. (1981a). *Accreditation manual for hospitals.* Chicago: Author.

Joint Commission on Accreditation of Hospitals. (1981b). *Consolidated standards manual for child, adolescent, and adult psychiatric, alcoholism, and drug abuse facilities.* Chicago: Author.

Joint Commission on Accreditation of Hospitals. (1981c). *Quality assurance guide for psychiatric and substance abuse facilities.* Chicago: Author.

Joint Commission on Accreditation of Hospitals. (1984). *1985 Accreditation manual for hospitals.* Chicago: Author.

Joint Commission on Accreditation of Hospitals. (1984). *1985 Consolidated standards manual for child, adolescent, and adult psychiatric, alcoholism, and drug abuse facilities serving the mentally retarded/developmentally disabled.* Chicago: Author.

Pirsig, R. M. (1974). *Zen and the art of motorcycle maintenance: An inquiry into values.* Toronto: Bantam Books.

Platt, K. A. (1976). Inpatient quality assurance from the viewpoint of the private physician. In R. H. Egdahl & P. M. Gertman (Eds.), *Quality assurance in health care.* Germantown, MD: Aspens Systems Corporation.

Siegel, C., & Fischer, S. (Eds). (1981). *Psychiatric records in mental health care.* New York: Brunner/Mazel.

17

The Role of the Professional Association
Psychology in Quality Assurance

WILLIAM J. CHESTNUT, NANCY LANE-PALES, AND ELIZABETH MEID

Quality assurance, in its broadest sense, is a traditional concern of the mental health professions, professions strongly committed to promoting human welfare. Psychology, as a profession, has historically demonstrated its concern with quality assurance in mental health in a number of ways (see Young, 1982, for an overview of the history of quality assurance). Many past activities have kept pace with the profession and are still operating; other means of ensuring quality assurance in mental health have developed more recently. Current forces in the mental health professions and in society in general are operating to suggest the development of future mechanisms to ensure quality.

As professions have grown, the scope and sophistication of activities relating to quality assurance engaged in by professional associations have increased. Psychiatry, psychology, social work, and nursing all work at the national association level to develop and implement policies and programs to promote and ensure quality services.

In this chapter, several types of quality assurance activities developed by professional associations are described; the experience and activities of the American Psychological Association (APA) are used as an example. Many of the current activities (e.g., licensure, and peer review as exemplified by the APA/CHAMPUS program) are touched on only briefly here, because they are covered in depth in other chapters. This chapter focuses in particular on other quality-assurance activities closely identified with the American Psychological Association, such as promulgation of the *Standards for Providers of Psychological Services* (APA, 1977) and the "Ethical Principles of Psychologists" (APA, 1981b), the process of accreditation for psychology training programs, and continuing education activities of the association. Examples of the activities of other associations of mental health professionals are also included.

WILLIAM J. CHESTNUT • Indiana University, Student Health Center, Bloomington, Indiana 47405. NANCY LANE-PALES • 3112 Westover Drive, S.E., Washington, D.C. 20020.   ELIZABETH MEID • 10 Minot Avenue, Auburn, Maine 04210.

Five different approaches to quality assurance developed in professional associations are presented and discussed:

1. Accreditation of training programs and facilities
2. Credentialing of providers
3. Standards and ethical principles for providers of services
4. Continuing education for professionals
5. Peer review of professional services

Each of these areas concerns a particular facet of professional development and functioning. Each also affects professionals at different points in their careers as they advance from being students to being beginning professionals, and then mature professionals, teachers, and trainers.

We also discuss the historical development of quality assurance mechanisms from their beginnings as conceptual frameworks encompassing what was considered necessary for quality training and practice to operational systems that actively involve professionals, willingly or not, in the quality assurance process. A this progress toward operational systems has continued and as the impact of quality assurance mechanisms has been felt by larger numbers of professionals, controversies have arisen in some professional associations regarding the proper role of an advocacy organization in quality assurance. In quality assurance programs, associations traditionally involved in advocacy for the professions take on standard-setting functions, some of which may be used by other organizations that seek to regulate or differentially reimburse mental health professionals. The impact of this tension will be noted as the quality assurance activities that have touched it off are discussed. Using the example of psychology, we will briefly discuss these issues.

We focus on the American Psychological Association's activities because, as the national organization for psychologists, the APA has developed most of organized psychology's activities in the area of quality assurance to date. In the current economic and political climate, private quality-assurance groups are forming and are beginning to market their services. Credentialing and peer review are the areas of greatest activity for non-association-related programs and activities.

A final section of this chapter discusses future possibilities in quality assurance activities for professions. Trends with regard to psychology's role in quality assurance in mental health are also considered.

ACCREDITATION OF PROFESSIONAL TRAINING

The origins of psychology's involvement in quality assurance activities lie in the historical and traditional antecedents of quality assurance mechanisms in the professions generally. Harl Young's history (1982) of quality assurance summarizes the historical development of quality control in professional training. The existence of professional schools directly linking training to practice and custom of subsequent credentialing by receipt of a diploma following a successful course of study and by licensure after a period of professional practice were the first steps, well established by the beginning of the nineteenth century. A subsequent development was the setting of minimum standards for facilities in training institutions, resulting from the recommendations in 1910 of the physician Abraham Flexner regarding reform of medical education.

Psychology's concern with standards for facilities and faculty in professional practice has been demonstrated in a number of ways. In 1947, the APA began a program of accreditation of training programs in clinical psychology. In 1950, criteria for the accreditation of counseling-psychology doctoral programs began, and in the late 1960s accreditation of doctoral programs in schools of psychology was instituted. In 1956, the American Psychological Association began the accreditation of predoctoral internship programs. At present, the APA accredits 216 psychology training programs (138 clinical, 44 counseling, 29 school, and 5 combined) and 286 internship programs (APA, 1985a,b).

The APA's accreditation process provides for assessment in seven areas of concern: institutional setting; cultural and individual differences; training models and curricula; faculty; students; facilities; and practicum and internship training. As stated in the *Accreditation Handbook* (APA, 1983):

> The purpose of accreditation is to promote excellence in programs designed to educate and train professional psychologists and to provide a professional and objective evaluation of these programs as a service to the public, prospective students and the profession. (Appendix A, p. 1)

For example, the accreditation criterion for "Facilities (VI)" states:

> Training in professional psychology requires adequate facilities. Although the specific facilities may vary depending upon the program specialty (e.g., clinical, school, counseling) and the geographic area (e.g., rural, urban), adequacy of the following will be assessed in relation to program goals. (Appendix B, p. 14)

The criterion then specifies such areas as teaching facilities, library, office space, practicum and internship facilities, equipment, and facilities for handicapped students.

The criterion for "Faculty (IV)" states: "A quality faculty is essential to the development and maintenance of an excellent program." The criterion goes on to note that besides diversity in specialty areas, the faculty should demonstrate their dedication to the profession and "must have acquired professional competencies and experience which enable them to train students about particular settings and problems" (Appendix B, p. 10).

Criterion I-F applies to supervision, a process begun at the professional training level that may well continue throughout a professional career. This criterion states that "the program must include supervised practicum, internship, field or laboratory training appropriate to the practice of psychology" (Appendix B, p. 3). Personal clinical supervision has become essential to the training of psychologists since early in the century. In the context of a training program, the supervision purpose is clearly educational. It is a process in which

> a more knowledgeable and experienced clinician assists the less experienced, developing professional in the expansion and application of clinical knowledge of skills. Contained within this purpose is the expectation that the novice will be helped to establish an appropriate professional identity. (Schaefer, 1981, p. 51)

Although supervision is essentially educational in purpose, it is also clear that the client is the preeminent responsibility of the supervisor and of the supervised clinician. And further, it is the task of the supervisor to see that the supervisee knows the ethical principles and applies them to practice. As a consequence, supervision not only is a viable means of teaching clinical skills and socializing student psychologists into the profession, but also protects the consumer and ensures quality of service.

In this criterion regarding supervision, we see a policy statement that appears in different forms in many areas of quality assurance for professionals: in the requirements for credentialing, in the concept of continuing education, and in standards for providers and ethical principles. The Council for the National Register of Health Service Providers in Psychology (1985) also specifies, "two years of supervised experience in health service in psychology, of which at least one year is in an organized health service training program and one year is post doctoral" (p. xvii) in its criteria for listing in the register. Finally, almost all licensing and certification boards require supervision before awarding independent practice status.

Again, this concept of quality assurance (clinical supervision) recommends a process that may carry on throughout a professional's career, as a trainee, as a journeyman practitioner, and as an independent practitioner engaged in peer supervision.

Policies developed regarding desirable settings and curriculum for professional training have become further operationalized through the development of professional schools of psychology. Professional schools of psychology granting the doctor of psychology degree were first established in 1968 at the University of Illinois and in 1971 at Baylor University. The freestanding California School of Professional Psychology was established in 1969. As is generally known, before that time doctoral-level professional training in psychology was traditionally conducted in graduate schools of arts and sciences to produce graduates fitting a scientist-practitioner model. Such programs result in a Ph.D. degree based on a research dissertation, among other requirements. Following a conference in Vail, Colorado, in 1973 on professional training in psychology, the philosophical basis for the existence of professional schools of psychology was confirmed. The philosophy behind the establishment of professional schools of psychology has a complex history, but it basically arose from a perceived need for training programs in psychology with more of a clinical emphasis than an academics–research orientation for those students intending to become practitioners. According to Caddy and Lapointe (as quoted by Peterson, 1985), in 1982 there were 44 practitioner programs operating, 20 in universities and 24 in freestanding professional schools.

The success of professional schools in developing a "better mousetrap" continues to be debated (Peterson, 1985), and it is unclear how professional schools will develop in the future, as well as how or if their graduates will be regarded differently from graduates of more traditional programs. In this area as in many others, further research is needed to provide empirical information on the costs and effectiveness (in terms of the future performance of trainees) of various types of training programs for professional psychologists.

The APA is not the only source of opinion regarding adequate training and education for professionals. The Council for the National Register of Health Service Providers in Psychology has published its lists of "Designated Doctoral Programs in Psychology" (1985). The list of programs meets the guidelines established at the 1977 Education and Credentialing in Psychology meeting and includes all specialty areas, basing its criteria on the principle: "The foundation of professional practice in psychology is the evolving body of knowledge and the discipline of psychology" (p. 29). This principle neatly ties the "Designation List" into the *Standards for Providers* (1977) as well as the *Specialty Guidelines for the Delivery of Service* (APA, 1981a). It should also be noted that the designated-doctoral-programs project was completed after collaborative efforts with other groups in psychology, including the American Association of State Psychology Boards and the Council of Graduate Departments in Psychology.

A factor affecting psychology's future and its quality assurance efforts in the area of training and accreditation is the issue of differentiating psychology programs from nonpsychology programs. As we shall see in other areas, the development of operational systems (here for designating psychology programs) as opposed to conceptual statements of desirable goals for training and the impact of this process on an individual's access to full participation in the activities of the profession has generated controversy. This has arisen out of a number of developments over the past decade: (a) the growth and influence of state psychology licensure boards; (b) court challenges concerning eligibility for licensure as a psychologist; (c) the emergence of various programs not clearly identified as psychology that enroll students who ultimately apply for licensure or certification; (d) the National Register that noted the wide range of credentials of persons seeking to be listed as health service providers; (e) the development of professional schools with faculties and educational resources that vary widely; and (f) the debate among psychologists about what constitutes the basic nature of graduate education in psychology. The APA's own "Designation Project" (1985c) appeared to be a viable, if controversial, approach to this issue when developed. It was tabled for any further action in 1985, and its future at this point appears to be uncertain.

Accreditation of training programs for mental health professionals (e.g., nurses, psychiatrists, and social workers) is also an important component of the quality assurance activities of these professions. Nursing depends on the National League of Nursing and social work on the Council for Social Work Education. These organizations exist in addition to the membership organizations for these professions. Segmenting of areas of quality assurance activity is common throughout the mental health professions, and no one professional association is responsible for all of the areas of activity described here.

CREDENTIALING

The early custom of credentialing has continued virtually unchanged since it first evolved. A diploma is received following the successful completion of a course of study in psychology at the doctoral level. Licensure (or certification) as a psychologist is mandatory in all of the states and the District of Columbia. (Connecticut was the first state to have mandated licensure and certification. This occurred in 1946. In 1977 Missouri became the last state to achieve mandated licensure and certification. In many states, licensure or certification depends in part on the satisfaction of experience requirements presupposing successful practice, often supervised, for 1 to 5 years. (See the discussion of licensure in psychology in Chapter 15.)

Besides a diploma and licensed status, other credentials are available to psychologists that testify to their standing in and identification with the profession. These are available manily through independent, private accrediting organizations. The American Board of Professional Psychology, established in 1947, examines and certifies advanced competence in the five specialty areas of professional psychology: clinical, clinical neuropsychology, counseling, school, and industrial-organizational. Following successful completion of a rigorous examination, the candidate is awarded diplomate status. This credential is considered by many the mark of a psychologist with a high level of professional expertise.

In 1974, the Council for the National Register of Health Service Providers in Psychology was created to help establish standards for the provision of health services in psychology by identifying, through the publication of a list (the "Register"), psychologists who apply for listings and who meet established criteria regarding training and experience. At present, these requirements are current licensure or certification by the state board of examiners of psychology at the independent practice level of psychology; a doctoral degree in psychology from a regionally accredited educational institution; at least 2 years of supervised experience in health service, of which at least 1 year is postdoctoral and 1 year is in an organized health service training program (Council for the National Register of Health Service Providers in Psychology, 1983). The "Register" is available to consumers of health services, health service organizations, health and welfare organizations, government agencies, and the general public and has become an important credential for eligibility for reimbursement by carriers and third-party payers for mental health services.

In the current marketplace for psychological services, the role of independent credentialing organizations is still being clarified. Application to credentialing organizations is voluntary, and lack of the credential cannot be taken to mean that a professional does not meet the standards for inclusion. Here we see an example of the stresses that begin to manifest themselves as a system that has a strong conceptual base becomes operational and affects the members of a profession.

The use of these credentials by employers, third-party payers, and other organizations that provide reimbursement for psychological services has led to controversy. Although the National Register was established as an independent organization, its use by insurance carriers to identify qualified health-service providers in psychology and its alleged indirect links with the American Psychological Association have caused controversy within psychology. This controversy led to the filing of an unsuccessful lawsuit involving the National Register and the American Psychological Association brought by individuals who alleged that their professional careers were being adversely impacted by their inability to obtain a listing in the "Register." Although the plaintiffs in the suit did not prevail, the suit represented an action brought against an organization (the National Register) and a professional association (the APA) that develop policies and set standards for quality assurance that, by their nature, exclude some practitioners.

One assumption of the credentialing process is that the possession of credentials results in the provision of competent services. Credentialing mechanisms of quality assurance assess providers, not the specific services they deliver.

As Claiborn (1982) indicated, the assessment of competence is an area requiring much greater effort on the part of psychology. Many of our efforts to date define minimum levels of competence (e.g., standards for providers, state licensing and certification boards, and the National Register). Perhaps, at this time, the American Board of Professional Psychology, which has languished in psychology's efforts to date, may be an appropriate mechanism for defining and assessing competence. Its conference on competence assessment, held in October 1982, certainly appears to be a hopeful sign for future effort.

In other mental-health fields, credentialing is again distinguished by the licensing of practitioners at the state level and voluntary advanced credentialing by either the national professional association (the National Association of Social Worker's Academy of Certified Social Workers [ACSW] program and the American Nursing Association's generalist and specialty certification program) or by independent bodies (the American Board of Psychiatry and Neurology).

STANDARDS AND ETHICAL PRINCIPLES

Standards for providers represent a further refinement of quality assurance activities. Here, associations develop statements and policies regarding their conceptions of the ideal practice of the profession. Standards typically encompass a wide variety of areas, including facilities, procedures, and standards for conduct and for the employment of associates. The American Psychological Association, through its governance structure, has demonstrated its commitment to accountability and quality assurance in documents such as the "Ethical Principles of Psychologists" (APA, 1981b), *Standards for Providers of Psychological Services* (APA, 1977), and the *Specialty Guidelines for the Delivery of Services by Clinical, Counseling, School, Industrial/ Organizational Psychologists* (APA 1981a).

In 1977, the American Psychological Association adopted its *Standards for Providers of Psychological Services*. The introduction of the 1977 revision of these standards states that the intent of the standards is to "improve the quality, effectiveness, and accessibility of psychological services to all who require them" (p. 1). It was expected that the standards would serve the respective needs of users, providers, and third-party purchasers and sanctioners of psychological services. Hence, the standards were established by organized psychology as a means of self-regulation and as a protection of the public interest.

The standards uniformly specify the minimally acceptable levels of quality assurance and performance on the part of "psychological service units" and individual providers in all settings. The standards cover providers, programs, accountability, and environment. The standard related to providers defines qualifications, areas of professional functioning, professional roles in organizations, and relationships with other professionals. Standards pertaining to programs define staffing, organization, services, and functioning relevant to the delivery of psychological services. The accountability standard relates to the accessibility, appropriateness, and effectiveness of psychological services. The standard pertaining to environment specifies the psychologist's responsibility for concern with environmental factors as they relate to quality of service.

Psychology's *Standards for Providers* are comprehensive and, like the standards for accreditation, cover facilities as well as operating principles. In the *Standards for Providers*, Standard 4, "Environment," 4.1 states, "All providers of psychological service promote the development in the service setting of a physical, organization and social environment that facilitates optimal human functioning." Similarly, the *Specialty Guidelines* devotes one section in each of the specialty areas to the environment and to the promotion of optimal human functioning through the development of appropriate physical, organizational, and social environments. Here we see another overlap of ideas: the acceptable training environment (discussed in the accreditation process) should lead to the acceptable service delivery environment (as described by both documents).

Whereas the *Standards for Providers* are generic, applying to all psychologists delivering psychological services, the *Specialty Guidelines* apply to the delivery of services by the four recognized specialties: clinical, counseling, school, and industrial-organizational psychology. Hence, there are four sets of guidelines. They are modeled on and parallel to the generic standards but have been modified to meet the unique needs of the four specialties.

With the *Standards for Providers* and the *Specialty Guidelines*, the APA has articulated it official policy with respect to delivery of services by each of the four specialities. However, it should be noted that there is no formal investigative or enforcement mechanism associated with either the *Standards for Providers* or the *Specialty Guidelines*. Although violations may be referred to state or national ethics committees, a state licensing board, or a professional standards review committee, any action would be contingent on a particular behavior's being in violation of the state or national ethical code or of the state licensing or certification law.

Ethical principles are a set of normative "prescriptions" and designate behaviors that are seen as intrinsically desirable or as valued by the psychological profession. In 1953, the "Ethical Standards of Psychologists" were published, and the APA psychologists were required to abide by a code of professional ethics. This code was revised and updated in 1981 and is now called the "Ethical Principles of Psychologists":

> The revised Ethical Principles apply to psychologists, to students of psychology, and to others who do work of a psychological nature under the supervision of a psychologist. They are also intended for the guidance of nonmembers of the Association who are engaged in psychological research or practice." (p. 633)

There are 10 principles; responsibility, competence, moral and legal standards, public statements, confidentiality, welfare of the consumer, professional relationships, assessment techniques, research with human participants, and care and use of animals. An example of the principles, Principle 2, "Competence," follows:

> The maintenance of high standards of competence is a responsibility shared by all psychologists in the interest of the public and the profession as a whole. Psychologists recognize the boundaries of their competence and the limitations of their techniques. They only provide services and only use techniques for which they are qualified by training and experience. In those areas in which recognized standards do not exist, psychologists take whatever precautions are necessary to protect the welfare of their clients. They maintain knowledge of current scientific and professional information related to the services they render. (p. 634)

In addition to publishing the "Ethical Principles of Psychologists," the APA has established its Ethics Committee to investigate alleged violations and to enforce the principles with disciplinary actions. These may range from reprimand to expulsion from the association. Ethical violations may be referred to state-association ethics committees or state licensing or certification boards. Many states have written into their licensing or certification laws that a psychologist must abide by the APA ethical principles. Hence, a psychologist in violation of the ethical principles may also be in violation of state law and may face disciplinary action by the state licensing or certification board or the state's attorney general.

In the past several years, the APA has developed a new version of the *Standard for Providers*. This has also been a period when litigation against mental health professionals has increased. Before final approval of the new version, questions were raised concerning potential legal liabilities for psychologists if the standards were taken as prescriptive statements of what constitutes acceptable psychological practice. Many psychologists felt that the version should be viewed more as aspirational statements—goals that psychologists strive for but cannot be expected to meet totally.

To help protect psychologists and the association in the future, the APA developed a policy in 1985 regarding the legal review of all APA standards, guidelines, principles, protocols, and scientific and professional policy. Before approval or adoption by the APA Council, all such documents must be reviewed for legal aspects such as potential risks for and effects on the APA and on its members. This increasing sensitivity to legal vulnerability reflects the general climate in the culture regarding litigation and the risks inherent in setting standards that may be used inappropriately. Again, the act of an association in setting standards that may be used by other in ways that were not anticipated or intended by the authors, or that may not represent the will of all of its members, can cause tension within the advocacy organization.

Finally, it should be noted that an explicit assumption related to the "Ethical Principles," the *Standards for Providers*, and the *Specialty Guidelines* is that those psychologists who abide by them deliver services of an acceptable quality. Standards and ethical principles also affect professionals' activities over the life of their careers. Knowledge of standards and ethics is assessed by licensing boards at the time of licensure, and current knowledge in these areas is an ongoing professional responsibility.

Nursing and social work also have general and specialty standards of practice. As there are many varied types of social work and nursing practice, standards have been developed that cover these various areas. As a practice-oriented profession, nursing, for example, supplies its standards to other groups and organizations that are striving to define acceptable nursing practice.

CONTINUING EDUCATION

As Young (1982) noted in his history of quality assurance, continuing education is another mechanism for ensuring the quality of professional services. Continuing education as a means of ensuring professional competence is advanced for four basic reasons, as pointed out by Vitulano and Copeland (1980). First, the knowledge base in psychology is increasing at an astonishing rate. Dubin (1972) estimated that the half-life of competence for psychologists averages 10–12 years. Second, our ethical responsibility demands high standards of competence. Third, lifetime education should be an important goal for psychologists. Fourth, the maintenance of competence helps to maintain public confidence in psychology.

Three models of continuing education, which appear to be appropriate for psychology, have been identified (Welch, 1976). The first and most popular model is continuing education, in which credits are awarded for participation in approved workshops, seminars, or other ongoing training. A second method is the examination model. Here, psychologists would take periodic examinations as a prerequisite for relicensure or recertification. Finally, peer review has been proposed as a model. This would entail a system for comparing the practitioner's record of clinical practice with an established set of standards of care for the profession as a whole.

At present, the first model, continuing education for credit, is the method required in a number of states mandating continuing competency demonstrations. The American Psychological Association and its Council of Representatives endorsed a continuing-education sponsor-approved system, and so far, approximately 235–240 sponsors have been approved. A central registry service established for psychologists to centrally record their participation is currently being discontinued.

Continuing education is an active concern in all mental health professions. Credentialing and licensing often depend on fulfilling continuing education requirements, and professional association sponsorship is a desirable characteristic of any continuing-education offering.

There has been less controversy surrounding continuing education than in some other areas of quality assurance. Controversies are most likely to arise in the areas of who may become an approved sponsor, how lucrative continuing education may become as a business, and how organizations such as licensing boards may choose to use continuing education requirements.

PEER REVIEW

Peer review is a quality assurance mechanism focused on the outcome of the professional psychologist's work. Basically, peer review is a review of the services provided by one psychologist by his or her peers in psychology. Although peer review is relatively recent in the service provision area, it is a long-standing tradition in other areas of psychology. For example, in the academic arena, peer review occurs in areas such as review of journal articles, grant proposals, and the use of human subjects. Peer review may occur face to face, or it may be based on documents or some combination of the two. It should be noted here that peer review, unlike credentialing, focuses on the particular services or products of psychologists, not on their training or credentials. This is an important conceptual distinction.

With the advent of rising health-care costs, diminishing resources, and the phenomenon of third-party payers (including the federal government through Medicare), the concept of peer review was implemented as a means of ensuring the quality of the services rendered.

Peer review as a mechanism used to ensure the quality of service can be based on criteria such as whether the service is necessary, appropriate, and effective. However, peer review may have other purposes, such as cost containment, pattern-of-utilization review, "policing," and education all with their own standards or criteria. These varying and growing uses of peer review are reflected in recent policy developments in the APA.

Peer review of professional services is the most recent area of quality assurance activities developed by the American Psychological Association. It has the potential to directly affect large numbers of practicing professionals and is a concept that has led to the development of an operational system.

Although peer review may be traced back to long-standing customs of training and practice such as case conferences, supervision, and problem-oriented record-keeping, its more recent impetus came from the federal government. In 1972, Congress established professional standards review organizations (PSROs) to provide comprehensive, ongoing, federally supported review of medical care. The goals of the PSRO program were both cost containment and quality assurance. At about this time, the American Psychological Association was developing a review system for outpatient psychological services. In 1975, the APA issued guidelines for the establishment and functioning of professional standards review committees (PSRCs) to work in conjunction with each state psychological association to review questions about fees and quality of service (APA, 1975).

At this time, the PSROs vary widely in their level of activity and their influence on the practice of psychology in their jurisdiction. Some, such as California's, are well-developed and active, whereas others have been inactive. Efforts to localize review services are also uneven. Some local associations have sought contracts with organizations such as Blue Cross and Blue Shield. It is still an open question whether review activities are more efficiently carried out on a local level, or if the growing sophistication of the review field requires a national level organization.

In 1977, the American Psychological Association contracted with the Civilian Health and Medical Program of the Uniformed Services (CHAMPUS) to develop a quality assurance program with a peer review mechanism. This program became operational in 1980, resulted in the development of written criteria for selecting cases for review, and established a roster of APA-approved psychologist peer reviewers. A parallel CHAMPUS project was developed by the American Psychiatric Association. Much later, the National Association of Social Workers and the American Nurses Association became involved in CHAMPUS review as well.

In 1983, CHAMPUS sought a single administrative entity to house the peer review projects of these mental health groups. Following an unsuccessful attempt by the psychiatric and psychological associations to develop a joint venture, CHAMPUS solicited bids for the combined project, and the American Psychiatric Association became the program administrator in 1985. Controversy regarding the CHAMPUS project has been intense within the American Psychological Association, as individual psychologists struggle with CHAMPUS policies and decisions that exclude some practitioners and make payment of benefits dependent on peer review of services. (This program is discussed in greater detail in Chapter 13.)

Beginning in 1979, the APA contracted with private insurance companies for peer review services. The program developed out of the CHAMPUS Project and initially followed similar procedures. The controversy surrounding the CHAMPUS Project within the membership of the APA spread to the private peer-review activities, and the program underwent substantial revision during 1984–1985. A new system for professional services review was designed and approved by the APA Council of Representatives in 1985. Under this new system, peer review no longer addresses the questions of whether the treatment is necessary and appropriate for the patient or client but whether the treatment itself is usual and customary or reasonable practice. Other changes in the program involved new criteria for the selection of peer reviewers, procedures for matching providers and reviewers according to theoretical orientation, the number of reviewers assigned to each case, and greater client involvement in providing information to insurance carriers for peer review. In late 1985, the APA began a marketing effort to sell the newly revised program to insurance carriers and third-party administrators.

Peer review activities have aroused bitter and vigorous debate within the psychological community. From late 1983 until the APA Council of Representatives approved the revised peer review system in 1985, the APA placed a moratorium on any new contracts with insurance carriers. In 1984, four psychologists became plaintiffs against the APA and the insurance carriers when they brought a legal action concerning the information requested in mental health treatment reports. At this writing, APA has been dismissed from this suit.

It is unclear at this time if a viable peer-review business can be contained and operated from within a professional association, which must, by its very nature, be responsive to the needs and wishes of its membership. The APA is first an advocacy organization, and the addition of activities such as the peer review of professional services for commercial insurance carriers and the federal government has been viewed as divisive by some. Indeed, the recent changes in the peer review system were the partial result of membership pressure and dissatisfaction with the peer review system developed out of the CHAMPUS project. From a philosophical point of view, it is also unclear what the "proper" role of the professional association in operational quality-assurance systems might be. Should professional associations take on the challenge of evaluating the professional activities of their members, or should professional associations reserve for themselves the traditional role of advocacy? Should they resist efforts to enlist them in activities that serve the purpose of other organizations and that may be antithetical to the goals of some members, or should they assume that their involvement in peer review activity provides greater long-range access to the organizations that recognize psychologists and reimburse consumers for psychological servces? Furthermore, can business operations that involve setting and enforcing standards thrive in the political atmosphere of the professional association, where program directions can be substantially influenced through membership pressure? The answers to these questions are not yet clear, and the area will continue to be one of vigorous debate over the next several years.

It is in this area that the experiences of the various professions vary most widely. Psychiatry, social work, and nursing do not report wide dissension among their membership regarding peer review, although some unrest has been recorded. Nursing, for example, reports no controversy regarding peer review. Nurses are trained in hospital settings, where peer review committees are prevalent and written notes and treatment plans the norm. This background and the desirability of nurses' reviewing the work of nurses may make peer review more palatable to nurses. In these programs, the peer review of individual treatment remains focused on the necessity and appropriateness of treatment, and possible expansion is considered desirable. Social work and psychiatry operate peer review (of their own profession's services only) for private insurance carriers, and nursing and social work provide peer reviewers to the American Psychiatric Association for use in CHAMPUS reviews. State efforts in these areas exist alongside of the national programs, and greater involvement at the state level is encouraged.

PSYCHOLOGY'S FUTURE IN QUALITY ASSURANCE IN MENTAL HEALTH

As can be seen in this chapter, psychology has a long history of concern about involvement in various aspects of quality assurance. It is also evident that the concern about and involvement in various quality assurance activities are becoming increasingly robust. This state of affairs has obviously resulted from a multiplicity of factors. Consumerism and an enlightened public and legislation in the health care areas, as well as the maturation of the profession, are all involved. As psychology seeks greater acceptance for its products and services from the community at large, it becomes more necessary that the profession ensure that its conduct and practices will embody the highest professional standards. To be considered a mature profession, psychology must be accountable and must ensure the quality of its services, necessarily implying that, in the future, psychology will refine and increase its self-regulatory functions.

As was discussed in the section on peer review, quality assurance activities, especially those that are operational systems, can exert great pressure on professional associations. Professional associations will eventually decide whether to continue in the quality assurance business or whether to encourage the development of private quality-assurance organizations that are philosophically compatible with the association. Private, profit-oriented peer- and utilization-review companies are also beginning to appear in the marketplace, and their influence and impact may be considerable.

With the staggering increases in the cost of health care, we can anticipate that there will be a greater demand for quality assurance and cost containment. This demand will most obviously emanate from the public and the insurance industry, but also must be a concern for the profession.

In addition to the demand for cost containment, psychology must be prepared to facilitate quality assurance activities within emerging delivery settings such as health maintenance organizations (HMOs), preferred provider organizations (PPOs), independent practice associations (IPAs), and employee assistance programs (EAPs). Obviously, in these emerging areas, the focus of quality assurance efforts will need to be on organizational group levels in addition to individual providers.

Given that psychology's quality assurance activities will expand, and that psychology will expand, and that psychology will continue to seek greater and expanded acceptance for its product and services, training programs will need to include quality assurance concepts and skills. One example of how this might be accomplished has been provided by Bent (1982). Not only do such programs better prepare psychologists for practice, but they also demystify the process, hence making it less threatening when it is encountered in practice.

A major area for future work is the domain of research (see Chapter 3). At this stage of development, psychology is in serious need of efforts to evaluate the validity and effectiveness of its quality-assurance efforts. Inasmuch as this whole areas is relatively new, this appears to be an ideal time to carefully evaluate our programs to ensure thay will meet their goals of quality assurance, fairness, and efficiency. Of course, research will also assist the profession in building a firm foundation for future developments and in developing a base from which educational efforts for users of quality assurance programs (e.g., insurance carriers) can be designed.

In conclusion, it can be said that psychology has a committed role in quality assurance, as demontrated by the multiple mechanisms now operating to ensure the quality of training programs, individual providers, and psychological services as delivered to consumers. Other associations of mental health professionals share this commitment. In the future, demands for quality assurance will arise from both within and outside professional psychology. Providing adequate quality-assurance systems while maintaining the independence and integrity of psychological practice involves the whole field in a major challenge. To meet this challenge, psychology will need the strong commitment of its members and their willingness to act. Decisions will be made regarding the most favorable location for quality assurance activities (national versus local and professional associations versus independent organizations) and the most responsible role for associations to assume in the process. The interplay of philosophies and politics will be intense. The professional association will need to be prepared to withstand challenges from those who are judged by the system and to continue to judge itself, its policies, and its programs.

ACKNOWLEDGMENTS

The authors wish to acknowledge Norman Penner of the American Psychiatric Association, Judith Browne of the American Nurses Association, and Norma Taylor of the National Association of Social Workers for their assistance. Each provided information regarding quality assurance activities in his or her associations, though none was asked to speak on behalf of that association. Their opinion, and the opinions of the authors, are personal opinions and do not necessarily reflect the official position of any professional association.

REFERENCES

American Psychological Association. (1975). *Procedure manual for professional standards review committees of state psychological associations*. Washington, DC: Author.

American Psychological Association, Committee on Professional Standards. (1977). *Standards for providers of psychological services*. Washington, DC: Author.

American Psychological Association, Committee on Professional Standards. (1981a) *Specialty guidelines for the delivery of services by clinical, counseling, school, industrial/organizational psychologists*. Washington, DC: Author.

American Psychological Association. (1981b). Ethical principles of psychologists. *American Psychologist, 36*, 633–638.

American Psychological Association. (1982). *Approval of sponsors of continuing education for psychologists: Criteria, standards, and procedures*. Washington, DC: Author.

American Psychological Association, Committee on Accreditation. (1983). *Accreditation handbook*. Washington, DC: Author.

American Psychological Association. (1985a). APA-accredited doctoral programs in professional psychology: 1985. *American Psychologist, 40*, 1392–1398.

American Psychological Association. (1985b). APA-accredited predoctoral internships for doctoral training in psychology: 1985. *American Psychologist, 40*, 1380–1391.

American Psychological Association, Task Force on Education and Credentialing. (1985c). *Recommendations for a designated system* (Final report to the APA Board of Directors and the Council of Representatives). Washington, DC: Author.

Bent, R. J. (1982). The quality assurance process as a management method for psychology training programs. *Professional Psychology, 13*, 98–104.

Claiborn, W. L. (1982). The problem of professional incompetence. *Professional Psychology, 13*, 153–158.

Council for the National Register of Health Service Providers in Psychology. (1978, May). *Proposal for a National Commission on Education and Credentialing in Psychology*. Washington, D.C.: Author.

Council for the National Register of Health Service Providers in Psychology. (1983). *The National Register of Health Service Providers in Psychology—1983*. Washington, DC: Author.

Council for the National Register of Health Service Providers in Psychology. (1985). *Designated doctoral programs in psychology—1985*. Washington, DC: Author.

Dubin, S. S. (1972). Obsolescence or lifelong education: A choice for the professional. *American Psychologist, 27*, 486–498.

Peterson, D. R. (1985). Twenty years of practitioner training in psychology. *American Psychologist, 40*, 441–451.

Schaefer, A. B. (1981). Clinical supervision. In C. E. Walker (Ed.), *Clinical practice of psychology*. New York: Pergamon Press.

Vitulano, L. A., & Copeland, B. A. (1980). Trends in continuing education and competency demonstration. *Professional Psychology, 11*, 891–899.

Welch, C. E. (1976). Professional licensure and hospital delineation of clinical privileges: Relationship to quality assurance. In R. H. Egdahl & P. M. Gertman (Eds.), *Quality assurance in health care*. Germantown, MD: Aspen Systems.

Young, H. H. (1982). A brief history of quality assurance and peer review. *Professional Psychology, 13*, 9–13.

18

Third-Party Payment
Psychiatric Peer Review

ROBERT S. LONG

Insurance company personnel, like most other people, assume that medical care is of high quality and strongly favor the concept that the quality remain high. Everyone wants to reduce the cost, not by impairing quality or limiting access, but by increasing efficiency and by getting the treating professionals and providers (health care facilities) to consider cost as they go about providing high-quality care. Insurance companies are also aware that many of their policyholders ask for more care than they need; use more care just because it is paid for (more or less) by some vague third party, which, in their opinion, has inexhaustible resources; and too often take advantage of the insurance system. Claims auditors and management personnel encourage the treating professionals to police the patients, the facilities, and each other other to prevent overutilization and increased costs.

Peer review programs have been supported by the American Medical Association (AMA), most state medical societies, the American Society of Internal Medicine, and other professional associations since the mid-1960s. Until recent years, review of charges for services has been widely available as well as review of utilization of services and facilities.

Although cost control has been the primary motivating factor in the use of peer review by insurers and state and federal medical care programs, the medical associations, physicians, and other professionals have always looked on peer review as an educational process rather than a punitive one. The final goal has always been to provide the best quality of medical care while using all necessary services and facilities in the most economical and efficient way possible. The medical profession generally believes that controlling utilization and quality takes care of cost automatically.

Mutual of Omaha's first awareness of the urgent need for special psychiatric or mental health peer review came in 1976, when one of its medical directors noted in his review of CHAMPUS (Civilian Health and Medical Program of the Uniformed Services—for which Mutual was a fiscal intermediary) claims that there were an in-

---

ROBERT S. LONG • Mutual of Omaha Insurance Company, Omaha, Nebraska 68175.

creasing number for "residential behavior modification" treatment, usually involving teenage or younger dependents of active or retired military personnel.

The "treatment" was provided during a long-term inpatient stay at a residential or ranch-type facility. The stays were very long (1 or more years), and the cost was very high (up to $3,000 a month).

After an on-site inspection of several of these facilities and detailed discussions with consultants and CHAMPUS headquarters personnel, it was decided to decline to pay benefits for such treatment on the basis that the stay was custodial in nature, a home away from home for children with chronic behavior problems and/or an educational experience. It appeared that any psychiatric or mental health treatment really needed could have been provided on an outpatient basis. At that time, specially trained and experienced registered nurses were employed to sort out mental-health treatment claims as well as others for physician review.

After considerable study and consideration by the CHAMPUS National Headquarters Office, an agreement was worked out with the American Psychiatric Association by which the APA would recruit from their membership physicians who would provide peer review for any of the psychiatric cases referred to them by the CHAMPUS intermediaries. Later, the APA signed agreements to provide the same services to private insurance companies.

The CHAMPUS Psychiatric Peer Review Program started in 1979. Soon after that, Aetna Life and Casualty Insurance Company started its program. Mutual of Omaha joined the program with its private insurance claims in late 1981 and has now had 6 full years of experience. We were influenced by the favorable CHAMPUS experience and that of Aetna and believed that we could be as successful. Many other major insurance companies now use the APA peer review program. The same program was initiated with the American Psychological Association in late 1981, again following the favorable experience reported by CHAMPUS and other private insurers.

One of the early problems in any kind of review or audit of psychiatric or mental-health treatment claims had to do with privacy. Representatives of the insurance industry through the Health Insurance Association of America (HIAA) started discussions with the American Psychiatric Association (APA) as long ago as 1970 in an attempt to get more and better information to use in claims processing. Some readers will recall the formation of the National Commission on the Confidentiality of Health Records, formed in 1974 largely by the APA with the help of others, including the insurance industry, the American Medical Association, the Blue Cross and Blue Shield plans, the American Hospital Association, the American Nursing Association, and other professional organizations. The commission did some good and had some input into federal privacy legislation, regarding which there was a lot of action in the early 1970s. The National Privacy Act was passed in 1974. There were extensive studies, bills written, and threats of additional, more restrictive privacy-type legislation for several years, but we have not heard much about it recently. The insurance industry is, of course, ever mindful of the importance of privacy and confidentiality of records, as will be illustrated later.

The development and operation of the peer review programs by the American Psychiatric and American Psychological Association have been a monumental effort and are a testimonial to the dedication, perseverance, and integrity of the membership and staff of those associations. It was not easy to sell this idea to the rank and file of members, who jealously guarded their patients' privacy and their own individuality and independence, probably more than in any other discipline in medicine. The great majority of the members of both associations have now recognized the value of peer review and have cooperated very well with it.

The peer review committees are made up of unusually independent personalities with a strong sense of responsibility to the public and their patients as well as to their colleagues. They are very knowledgeable about insurance mechanisms and why benefits are often limited in the treatment of mental health disorders. They are very patient and understanding, yet very forthright and candid in criticizing those colleagues who do not seem to them to be practicing as well as they should, either in lack of testing and investigating or in overusing treatment methods in frequency and duration.

The committees have constantly sought to improve quality, efficiency, and economy of their operation. We have had no differences of opinion of any kind with the committees' philosophy or activities. We have enjoyed an unusual measure of equal respect and cooperation. In addition, we have had surprisingly few complaints from either the reviewers or those being reviewed. All but a very few of those complaints have been due to a lack of information and understanding of the program. Two psychiatrists opposed the program in principle, and a few psychologists have been hesitant to complete mental-health treatment reports for one reason or another, usually because of privacy and confidentiality.

The first step in joining the programs was the development and signing of an agreement between Mutual of Omaha and the American Psychiatric Association, then soon after with the American Psychological Association. This is a very simple agreement that simply describes the methodology for handling the papers, maintaining confidentiality, keeping records, and paying for services. The last is on a per-case basis. There is no "hold harmless" agreement on either side. The next step was to develop guidelines for insurance claims personnel.

The guidelines for the special claims auditors at the regional office to follow in specific *outpatient* treatment situations are:

I. Outpatient treatment exceeding 40 sessions per episode (an additional 20 can be allowed if there is a one-time change in providers) or 2 years' duration generally require an MHTR (mental-health treatment report). In addition to extending benefits because of policy considerations, the following situations may warrant payment of additional sessions[1]:
   A. The diagnosis is Schizophrenia,[2] Bipolar Disorder, or Depressive Disorder, *and* one of the following situations exists:

---

[1]In either of these situations, benefits may be allowed for up to 60 sessions. After that, an MHTR is required. Benefits are continued while that report is being obtained and reviewed.

[2]DSM-III-R listings refer to the *Diagnostic and Statistical Manual of Mental Disorders* (3rd ed., revised) published by the American Psychiatric Association (Washington, D.C., 1987). See, especially, Chapter 1, "The DSM-III-R Classification," pp. 3–10, for specific classification codes.

       1. Treatment has been decreasing in frequency.

       2. There is a history of previous hospital confinements.

       3. Treatment is provided by an M.D. and includes medication regulation and adjustment.

    B. The treatment being provided is psychoanalysis and the diagnosis is Anxiety Disorder (or Anxiety and Phobic Neurosis).

II. Outpatient treatment exceeding 1 hour of individual therapy per day, 90 minutes of group therapy per day, or two sessions of any therapy per week generally requires an MHTR. Benefits are continued while this report is being obtained and reviewed. The following situations may warrant payment for longer or additional sessions:

    A. The diagnosis is a Psychotic Disorder not elsewhere classified or Delusional (Paranoid) Disorder (297.10), and the treatment is used to avoid hospitalization or as a trial before admission. There may be a need to contact the attending physician to verify this. Treatments usually do not exceed five times a week. If they do, an MHTR is required. The 40- or 60-session guidelines would still apply.

    B. The treatment started at a high frequency but gradually decreased in frequency and was no more than two sessions per week by the sixth week of treatment. The 40- or 60-session guidelines would still apply.

    C. Psychoanalysis is generally provided four to five times per week. This can be allowed without question. The 40- or 60-session guidelines would still apply.

    D. Crisis intervention may be allowed up to 2 hours per session but no more than 8 hours per episode without question or MHTR.

III. When treatment is provided for conditions that affect more than one family member, such as anxiety, depression, or alcohol or drug abuse, the use of marital and/or family therapy, in addition to individual therapy for the primary patient(s), would be appropriate, and benefits may be allowed.

    Marital and/or family therapy provided more than once per week are questioned. An MHTR is requested or the question is referred to the home office. These sessions are combined with individual therapy for the primary patient to determine when the 40- or 60-session guideline is exceeded.

    Only one family member is billed for this type of therapy. There is not a separate billing for each person.

IV. Some claims are seen where multiple providers are treating the same patient or family, and others are seen where multiple therapies are provided to the same patient even on the same day. These are evaluated with the following information in mind. It is recommended that this type of situation be questioned only when there is a significant amount of liability in question on a single or continuing claim:

    A. Individual therapy to the same patient by different providers is not the usual psychiatric practice. Even consultations with another psychiatrist or psychologist are unusual. If this practice occurs for more than one or two sessions, an MHTR is requested from each provider with an attachment stating that the patient is being treated by another provider and asking about the necessity and for documentation of coordination between the providers.

Consultations are questioned if more than one is claimed. The attending physician is asked whether this consultation was requested and why. Consultations by specialists in other fields of medical practice are approved when the diagnosis is appropriate.

B. It is not unusual psychiatric practice for a patient to receive both individual and group therapy on the same day when both treatments are provided on an outpatient basis. A notable exception would be when the patient is in a "day-care" (partial hospitalization) setting. Otherwise, the medical necessity of such multiple therapies on the same day is specifically questioned if it occurs on more than 1 or 2 days.

C. It is not usual psychiatric practice for one physician to give individual therapy to more than one member of the same family. An exception may be allowed when the conditions of the family members are interrelated and one provider sees the different patients in individual therapy for evaluation purposes or to obtain information needed to clarify possible misinterpretations occurring at crisis points in either patient's treatment program. More than six such sessions are questioned.

It is appropriate to approve for benefits those claims in which a single provider is seeing one patient in individual therapy and other family members in marital or family therapy. See heading for further details. The guidelines here apply to different family members receiving *individual therapy* from the same provider.

Additionally, it is not usual to include several members of one family in the same group therapy session. Group therapy is not the same as marital or family therapy. If this occurs, an MHTR is needed for each patient.

D. The administration of medications by injection is not usual, with the exception of medications for schizophrenia. If injection occurs, the specific medical need for such administration must be documented.

E. Outpatient Methadone treatment is not usual and requires an MHTR.

F. A court-ordered outpatient evaluation may not be *medically necessary*, and the patient may not be responsible for the cost. The insurance agent investigates the policy liability, and gets a copy of the court order.

G. Outpatient therapy for sexual dysfunction requires an MHTR. This is true whether sexual dysfunction is the primary or the secondary diagnosis.

H. Marathon therapy is not generally accepted psychiatric practice. If a claim is seen, an MHTR is required.

I. Charges by two providers for the same service may not be covered. In this case, both providers are notified of the other's charge and are asked to clarify why both are providing the same service. Normally, only one charge at the maximum usual and customary (U&C) level may be considered. Any questionable cases are referred to the home office.

V. Outpatient services provided to children are evaluated with the following guidelines in mind:

A. It is unusual to treat a child under 4 years old with individual therapy.

If this occurs, an MHTR is required. Benefits are not provided without clarification.

B. It is unusual to treat a child under 8 years old in group therapy. This treatment also requires an MHTR with benefits delayed.

C. If a child has a diagnosis of Mental Retardation, any more than six individual or group therapy sessions require an MHTR. Psychological testing is usually covered.

VI. Outpatient psychotherapy for a diagnosis of organic brain syndrome is not usual treatment and is generally required only when psychotropic medications are being used. If there are more than six sessions or there is no evidence that psychotropic medications are being provided, an MHTR is required.

VII. There are several other situations that may not be considered standard psychiatric practice. Most of these require an MHTR or some direct input from the attending physician regarding the necessity of the service. Any questionable situations are referred to the home office.

A. The use of hypnosis is not usual unless performed by an M.D. or a Ph.D. If performed by some other provider who is covered by the plan, it must be within that provider's legal scope of practice. If hypnosis is used in conjunction with individual or group therapy, the 40- or 60-session guidelines are used. If it is done alone, an MHTR is required after 20 sessions.

B. Psychological testing in excess of 6 hours may be questionable. If the claim for testing is in excess of 6 hours, the attending physician (not the testing physician, unless they are the same) is asked if she or he ordered all the tests reported and why. A copy of the billing is sent to the physician so that she or he can see which tests were performed and the charge made.

C. Day-care usually requires an MHTR and a discharge summary from the last confinement. Benefits may be allowed for 30 days if the patient is an adolescent (age 14–19), goes to day-care right after a hospital confinement, *and* had a diagnosis of Schizophrenia, Bipolar Disorder, Major Depression, or Psychotic Disorder.

The documentation needed to review *inpatient* claims is somewhat different from that for outpatient claims. Hospital records are very important in these cases. MHTRs are also necessary in some instances, especially when the review process is taking place while the patient is still confined. The following hospital records should always be obtained when the length or necessity of an inpatient stay is in question:

1. Admission history and physical examination
2. Discharge summary (if available) or weekly progress notes or summaries of status
3. Psychological test reports (if testing was done)

Additional daily records may be necessary, but the above, along with an MHTR, will usually be sufficient to determine if the claim is payable. Numerous claims are seen for long-term inpatient psychiatric care. Although these need to be evaluated differently from long-term inpatient nonpsychiatric care, concerns about the necessity for the confinement and a possible custodial care classification are just as important. No one at the regional office may deny any inpatient confinement on the basis that the stay was custodial or that it was not medically necessary. Review by a psychiatric consultant at the home office is required in such cases.

The guidelines for the loss control specialist at the regional office to follow in *specific inpatient* treatment situations are:

I. *Child/Adolescent Hospitalizations*
   A. Any confinement of a child under 6 years old requires review before payment. An MHTR is required in addition to the hospital records listed above. These are referred to the home office.
   B. Confinements of between 30 to 60 days may be approved if three or more of the following criteria are met, as documented by hospital records, and there is no other reason to question the claim:
      1. Patient is over 12 years old.
      2. Patient is a potential danger to self, others, or property.
      3. Other treatment programs, such as outpatient therapy or drug therapy, have failed, and the patient requires skilled observation in an institution.
      4. The patient suffers from some combination of depression, behavior disorder, or thought disorder and has impaired functioning severe enough to require skilled supervision and observation, a structured environment, and therapeutic intervention.
      5. The therapeutic services received by the patient include some combination of group, individual, or family therapy; activity therapy (may be called *occupational* or *recreational therapy*); behavior modification; or collateral therapy, which may include contact with the parents, school, probation officer, social worker, or other involved parties.
   C. Confinements of between 60 and 90 days may be approved if two or more of the following criteria are met, as documented by hospital records, and there is no other reason to question the claim[3]:
      1. Patient is over 14 years old.
      2. Patient is receiving antidepressant medication (such as Desyrel, Ludiomil, Asendin, Tofranil, or Norpramin) or major tranquilizers (such as Thorazine, Mellaril, Stelazine, or Haldol).
      3. Diagnosis is a Disruptive Behavior Disorder, Anxiety Disorder (of childhood or adolescence), Major Depression, Bipolar, or Psychotic Disorders, or Schizophrenia.
      4. The admission summary indicates an anticipated confinement of no more than 90 days and describes plans for discharge or transfer that will be actively pursued as a goal throughout the confinement.
      5. The admission indicates a long-standing history of depression with loss of parenting figures (death, broken home, or separation by distance), intermittent suicidal ideation, or antisocial behavior.

[3]Any confinement of more than 90 days requires an MTHR and/or hospital records and home office referral.

II. *Adult Hospitalization*

    A. Confinements of between 30 and 60 days may be approved if three or more of the following criteria are met, as documented by hospital records, and there is no other reason to question the claim.

        1. Other treatment programs, such as outpatient therapy or drug therapy, have failed and the patient required skilled observation in an institution.

        2. The claim is for the patient's first psychiatric hospitalization. Refer also to heading IVC, "Multiple Admissions."

        3. Patient is a potential danger to self, others, or property.

        4. Treatment plan includes the use of individual psychotherapy.

        5. There is documentation that the patient's medications are being added to or adjusted during the confinement.

        6. Electroconvulsive therapy (ECT) or electroshock therapy (EST) was used after a trial of medication. Refer also to heading IVA on the use of ECT and EST.

    B. Confinements of more than 60 days or more than 30 days that do not meet the above criteria require an MHTR, hospital records, and home office review.

III. *Hospitalization for Specific Diagnosis*

These guidelines may apply to any hospitalization and are used in combination with any others that may apply.

    A. *Alcohol- and Drug-Abuse Hospitalizations*

        1. Those that are for 8 days or less may not involve any treatment and may not be medically necessary. Hospital records are always required for these claims. The following factors are considered, along with any specific policy language, in the evaluation of such claims:

            a. The records must document the need for detoxification and a plan for withdrawal. This need must be supported by documentation that the patient required continuous skilled observation because of coma or stupor and received controlled chemotherapy or some other form of institutionalization type therapy.

            b. Confinements that appear to provide the patient only a place to dry out, sober up, or come down after drug or alcohol intoxication are not usually covered.

            c. When the record indicates a complete lack of detoxification or lack of medication needs, it may be that the treatment could have been handled on an outpatient basis. Such cases are referred to the home office.

        2. Any confinement of more than 60 days for an adolescent (ages 14–19) with a diagnosis of alcohol and/or drug abuse requires hospital records, an MHTR, and home office referral. Cases that are only a very few days over the limit are discussed with the home office. This type of admission may often combine a substance abuse diagnosis with another psychiatric condition that is covered for 60 or 90 days.

        3. Adult confinements of 30 days for the treatment of alcohol and/or drug

abuse may be covered, and up to 6 weeks may be covered if three or more of the following criteria are met, as documented by hospital records, and there is no other reason to question the claim:

    a. Patient is a potential serious danger to self, others, or property.

    b. Outpatient treatment programs have failed or are inappropriate.

    c. The claim is for the patient's first hospital confinement due to this diagnosis.

    d. The records document the occurrence of convulsions during the stay or the existence of two or more of the following: restlessness, tremulousness, disorientation, hallucinations, or delirium tremens.

    e. The records document an ongoing active treatment plan, including psychotherapy, activity therapy, behavior modification, and/or other types of therapy.

4. Any confinement of more than 6 weeks for an adult with a diagnosis of alcohol and/or drug abuse requires hospital records, an MHTR, and home office referral. Cases that are only a very few days over the limit are discussed with the home office. This type of admission may often combine a substance abuse diagnosis with another psychiatric condition that is covered for 60 to 90 days.

5. Claims including unusually high charges for drugs or supplies from the pharmacy, X-ray, or laboratory services require special attention. Usually, very little in these areas is required for the diagnosis or treatment of alcohol and/or drug abuse. The attending physician is asked to document the medical necessity for kinds and amounts of tests, drugs, and supplies that are questionable.

B. *Adjustment Disorders*

These conditions are not usually serious enough to require inpatient confinement. (See DSM-III-R, Chapter 1, for specific codes). Unless there is a different primary diagnosis requiring hospitalization, any confinements for this type of disorder must be identified as soon as possible. Hospital records, an MHTR, and home office referral are required.

C. *Mental Retardation and Developmental Disorders*

Inpatient confinements for one of these diagnoses alone should be handled the same as those for adjustment disorders (see B above). (See DSM-III-R, Chapter 1, for specific codes for Mental Retardation, Pervasive Developmental Disorders, and Specific Developmental Disorders.)

D. *Organic Mental Disorders*

This category includes dementias arising in the senium and presenium, and organic mental disorders associated with known physical disorders or conditions or those whose etiology is unknown. (See DSM-III-R, Chapter 1, for specific codes). All of these patients have a high potential for becoming custodial confinements. Any claim exceeding 30 days requires hospital records, an MHTR, and home office referral. Confinements exceeding the guidelines by only a very few days are discussed with the home office.

IV. *Specific Inpatient Services*

These guidelines address psychiatric services provided to hospital inpatients:

A. *Electroconvulsive Therapy (ECT)*
   or Electroshock Therapy (EST)
   1. This is an acceptable form of treatment, and benefits are provided if the diagnosis is:
      a. Major depression when the risk of suicide is high, when the patient is not taking adequate food or fluids, when the use of drugs or other therapy entails high risk, or when it will take an unusually long period of time to manifest a therapeutic response.
      b. Severe psychosis of any type characterized by behavior that is a threat to the safety and well-being of the patient and others and/or that cannot be controlled by drugs or other means or for which drugs cannot be used because of adverse reactions or because of other risks with their use.
      c. Severe mania or the manic phase of a bipolar disorder when the use of drug therapy would result in unacceptable risks and/or when coexisting medical problems (e.g., recent myocardial infarction) require prompt resolution of the mania and/or make the use of drug therapy unacceptable.
   2. The administration of ECT must be within the following guidelines for benefits to be provided:
      a. Treatment must not exceed three times per week with 1 or 2 days between sessions.
      b. For elderly patients over age 60, ECT is usually provided only once or twice per week.
      c. Usually 6–10 treatments are sufficient for satisfactory response. If more than 15 are utilized, a report from the attending physician and discussion with the home office are required.
      d. When multiple-monitor ECT (MMECT) is used (more than one treatment per day), the usual total of 15 treatments is still applicable. This form of ECT involves, for example, three ECTs per day on 3 days. More than 15 total treatments must be discussed with the home office.
      e. ECT is normally administered under general anesthesia.
   3. A benefit for the use of ECT is not provided for the following diagnoses or under the circumstances described:
      a. Diagnosis of behavior disorder.
      b. Diagnosis of personality disorder.
      c. Diagnosis of anxiety or phobic neurosis.
      d. For feelings of depression related to recent tragic or disappointing life events.
      e. For children under 18 years of age.
      f. When used only to control symptoms of violent behavior.
      g. Benefits for individual psychotherapy are not usually provided on the same day as ECT because of the mental confusion resulting from the ECT.
   4. If any claim involving ECT falls outside these guidelines, it is referred to the home office.

B. *Therapy Utilization and Charges*

Patients typically receive individual psychotherapy, group therapy, family therapy, and other therapies (e.g., activity and recreation) on a concurrent basis while hospital-confined. Below are some guidelines concerning the usual practice in these cases. Claims that exceed these guidelines are investigated further. This is an area where there are either very clear policy limits e.g., (50 doctor visits per year) or none at all:

1. Individual psychotherapy is usually no longer than 1 hour and is provided on a daily basis up to six times per week.
2. Group therapy is usually no longer than 90 minutes and is provided, in addition to individual therapy, on a daily basis up to six times per week.
3. Family therapy is usually between 90 minutes and 2 hours long. It is normally used as a substitute for individual or group therapy once a week. This therapy is usually used when the patient is not an adult or the diagnosis is a substance abuse.
4. Adjunctive therapies may include occupational therapy, recreational therapy, and various kinds of other activity therapies. These are usually performed on a daily basis and multiple therapies each day are not unusual. Some hospitals do not bill separately for these activity-type therapies but include them in the room-and-board charge. There is a wide variance in the charges for these therapies. A charge of $60–$75 per session for any one of them is not unusual. Unusual cases are discussed with the home office.
5. It is not usual for the same physician to bill for both daily hospital visits and individual psychotherapy on the same day. If these are claimed, the physician is questioned, and daily hospital records are obtained to verify both services.
6. If a Ph.D. provides and bills for individual psychotherapy, then it is usual for an M.D. to bill for hospital visits. If one M.D. bills for therapy and another for hospital visits, the billing for daily visits is questioned, and the medical necessity is determined.
7. Refer also to the outpatient guidelines for types of claims and charges that may be applicable to inpatient claims.

C. *Multiple Admissions*

A number of patients receiving psychiatric care are admitted to a hospital several times during their course of treatment. Admissions in excess of 30 days or more than two admissions in 1 year require hospital records, an MHTR, and referral to the home office. However, a second admission for a diagnosis of Schizophrenic Disorders, Affective Disorders, or Conduct Disorders may be provided benefits for 13 days without referral.

D. *Pass Days*

It is not unusual for a patient receiving psychiatric care to be allowed to leave the hospital for part of a day or even overnight. Some hospitals identify these days on the billing. Others identify them only in the nursing records or the physicians' progress notes.

Usually, benefits are provided if the patient is out of the hospital on a daytime pass. However, benefits are not usually provided for any room-and-board charge made for days the patient is on an overnight pass. If the hospital documents that services were provided before the patient left on pass and were not billed for separately, some partial payment may be considered. Payment is prorated based on the services provided or the number of hours the patient was actually in the hospital.

In either situation, daytime or overnight pass, the charges for other services purportedly provided on those days are closely checked.

Numerous pass days may also be an indication that the level of care is custodial. If this is suspected, the policy is checked for applicable language, all the information needed for complete review (hospital records, MHTR, and letter from the M.D.) is obtained, and the case is referred to the home office unless the additional information clearly justifies the medical necessity for and length of hospital stay.

It is now apparent that a considerable number of mental-health treatment reports are required. Each of these is reviewed at the home office by a trained psychiatric nurse, who, after looking at a few hundred, becomes very skilled in spotting the problem cases. If the frequency or length of treatments is outside the usual pattern and the reason is not obvious, the claims nurse refers the MHTR to an experienced, certified psychiatrist consultant for an opinion. Our consultant has had nearly 8 years' experience in the CHAMPUS/APA peer review project and is accustomed to and skilled in this type of evaluation.

If the various treatment factors appear to be reasonable, the claim is paid according to the terms of the contract, and a new MHTR is requested at an appropriate future time.

We have largely abandoned the set 20-, 40-, and 60-day intervals for review of outpatients (or office patients) and use instead a flexible and usually longer time interval, such as 3–6 months or occasionally longer. If it appears that termination or a decrease in the type or frequency of service is indicated, then the file is referred to the medical director for consideration for peer review by the American Psychiatric Association or American Psychological Association Quality Assurance Committee.

A significant percentage of the claims for outpatient or office treatments indicate that the patients have been under regular treatment 1, 2, or even up to 8 years at the time we first see them. This happens because the patient is in an insured group that is new to us and the prior insurers have apparently made no effort at utilization control. Group contracts are made for 1 year, and many groups move around from one insurer to another from year to year. Other insurers apparently have believed that the benefit limits commonly placed in contracts for psychiatric treatment are enough protection, so they have not questioned the kind or amount of treatment and have routinely paid the claims up to the limits of liability. Common limitations are a limited number of visits per year (such as 50); a limited payment of usual, customary, and reasonable (UCR) charges (50%); or a limited aggregate per year ($500–$1,000). Only a very few contracts provide the same benefits for outpatient or office treatment of psychiatric illnesses as for the other types of illnesses.

The treatment of psychiatric illnesses in a recognized hospital is almost always fully covered in the same way as for any other illness. Review of these cases is important to an insurer because the cost is high—$12,000–$15,000 a month (including professional services) at this writing.

Although we would like to review all hospital cases at 60 days, we often see the first billing only after 1 or more months have passed and, not infrequently, after the patient has been dismissed following several months or even 1–2 years of hospitalization. We are amazed that every hospital does not find out who the insurer is, notify the company at the time of the patient's admission, and send a bill every month to the patient or the insurer.

Until recently, we have not been able to get retrospective review on hospital cases. We have some problems with respect to inpatient services provided by the hospitals. For example, a few have a flat daily rate which is easiest to process, but most have a daily room rate and then charge separately for individual psychotherapy, group psychotherapy, recreational therapy, social services, and, commonly, some special services, such as dancing, painting, playing games, and exercising. The attending physician or personal psychologist, of course, charges for a daily visit and/or personal psychotherapy, so the total expense per claim is quite high.

We do not object to the itemization of services if the total of board and room and employee services (professional and semiprofessional) is not excessive compared with the charges of other similar institutions. In some cases, we do question the great number and kinds of these services; we believe that many patients could not tolerate or benefit from so much attention and forced physical or mental responsiveness all in 1 day, on successive days over a rather long period of time.

We now get guidance in helping to control this element of hospital expense from the associations. In the past, we have had guidance only on length of stay.

At a recent meeting of the American Psychiatric Association peer review committee, there was a consensus that there was no reason for most psychiatric patients to be in a hospital more than 90 days for acute care. After that, it is long-term care, which requires a new evaluation of whether it is custodial in nature, and thus excluded from most insurance coverage, or whether there is evidence of continuing long-term active treatment with evident progress month by month.

Dr. Robert Gibson, President and Chief Executive Officer of the prestigious Sheppard and Enoch Pratt Hospital, has made a study of short-term and long-term hospital care. The guidelines he suggested are in Appendix F of the American Psychiatric Association *Peer Review Manual*. In a study of several hundred cases, he found the long-term cases to show a high incidence of schizophrenia, depression, guilt, confusion, social withdrawal, and similar severe disorders. He reported that 92% of this group improved after a median length of stay of 70 days. In the short-term group, there were 82% improved, with a median length of stay of 30 days.

Examples can be found, of course, of patients who have shown a remarkable recovery from very severe mental illness after 1–3 years of in-hospital *active* treatment. Such patients are reported to show a very low rate of repeat hospitalization, particularly as compared with those who had short-term or crisis treatment for a fairly serious mental illness. Most likely patients who are actually receiving active treatment over such long periods of time are in very selected institutions (e.g., Enoch Pratt, Menninger, etc.) and under the care of selected physicians (e.g., Gibson) who from personal experience are convinced they can do some good with very long-term active in-hospital treatment. Most patients who are not responsive to treatment are under long-term *custodial* care. It is custodial care that is not covered by insurance contracts.

Other very long-term in-hospital stays (over 120 days to 180 days) of patients with severe mental illness who are responding little or not at all to fairly routine long-term care programs are commonly identified as custodial care in our evaluations. We quote one of our physician correspondents as follows: "A mental patient may become chronic when the patient gives up hope in himself and the treatment team gives up hope on the patient. The physician's attitude and the patient's attitude about his illness are of crucial importance in the outcome of treatment."

The decision to send a completed mental-health treatment report to a peer review program is based on one of three observations.

1. It appears to those of us who have reviewed it at the home office that the type, frequency, and/or duration of treatment is inappropriate and, therefore, does not fall within the "medical necessity" definition in our contract.
2. The location of treatment or level of care is inappropriate (referring here mostly to hospital cases).
3. We are satisfied that treatment is probably appropriate in all respects, but we want confirmation from the peer review team, and we think they would be interested in seeing examples of some of the types of cases we are approving and not sending to them routinely.

We agreed with the parent associations at the beginning that we would not terminate or reduce benefits without the recommendation of their peer review committees. The reviewers are not always in total agreement. We almost always go along with the majority. On rare occasions, we may use the combination of the opinion of one strong reviewer plus our in-house psychiatrist consultant plus our own claim experience.

During 1985, a change was made by the American Psychiatric Association from three to two reviewers for each case. Each reviews the case separately, and then they confer together by telephone before providing a report and a recommendation. This procedure has been very satisfactory to the insurance companies.

We have had only three or four requests for rereview, always with added information, and we have accommodated them. We have had no serious complaints of any kind from the attending physicians or psychologists or from the patients about the second or later reviews.

With respect to guarding privacy, we have a regular routine to ensure it. All psychiatric files, MHTRs, and correspondence are kept in locked files in the home office with access only to our nurse-auditor and the medical directors. Enough information of a fiscal nature is abstracted for the purpose of claim payment.

MHTRs sent to the associations are sterilized so that the name and address of the patient and the treating professional are not shown. The names and addresses of the peer reviewers are likewise removed from their reports to us. Copies of their reports are sent to the treating professional, along with our statement of intention to continue, change, or discontinue benefits. An appropriate time, usually one to three months, is allowed for the treating professional to effect the desired changes, terminate treatment, or make other arrangements with his or her patients.

Quality control of the process is maintained by the constant watchfulness of the initial screening claims personnel, the psychiatric nurse, the in-house psychiatrist consultant, and the medical directors. On-site visits are made to outlying claims offices by our nurse, and auditors and loss control specialists are brought into the home office regularly for refresher courses. Informational bulletins regarding procedural changes and updating are sent to all claims personnel regularly. The nurse and one or both medical directors meet with the American Psychiatric Association Quality Assurance Committee annually and correspond or talk by telephone with the staffs of both the American Psychiatric Association and the American Psychological Association frequently.

The staffs of the associations and the appropriate committees carefully monitor the work of their reviewers, remind them to be prompt and thorough in reporting their evaluations, and maintain a quality assurance program by monitoring their own reviewers. The programs include utilization review, quality review, and continuing education of members, as well as consultation with intermediaries to improve both availability and appropriate services and cost management. The latter refers to the program, not to individual fees charged by treating psychiatrists and psychologists.

The forwarding of the peer reviewers' reports to the treating professional provides him or her with consultations from highly qualified colleagues at no cost to the professional or the patient. Many peer review reports are detailed, analytical, and explanatory and give valuable advice on how treatment might be changed or improved. This is a tremendous and effective educational program, which we believe has improved the quality of care. We can see it in the kind and quality of MHTRs we are getting now as compared with those we received several years ago. All of us at the insurance company level have also benefited educationally.

Many states have laws that require insurance companies to recognize and provide benefits for the services of certain non-M.D. providers of psychiatric care in the same way as we do M.D.s. Such persons may be identified as psychologists (Ph.D.), licensed or applied psychologists, clinical social workers, registered nurses, marriage counselors, family counselors, or child and adolescent counselors.

In special cases, we may provide benefits for the services of social workers working under the direct supervision of a licensed M.D. or Ph.D. Some policies or contracts are written specifically so that benefits can be provided for the services of these special professionals.

If we receive a claim for services provided by an individual who is not licensed and who is not identified as acceptable under the specific terms of a contract, benefits are denied as provided by the contract.

At this time, the American Psychiatric Association has agreed to review some of the cases of treating professionals, such as the medical social workers and possibly the nurses.

As previously noted, the American Psychological Association has an excellent peer review program, which we use regularly for the Ph.D. psychologist cases in the same way that we use the American Psychiatric Association peer review program for M.D. psychiatrists. Modifications have been made from time to time, as required by experience. All have been beneficial to the program.

All of us have, from the beginning, emphasized that the peer review program is to be an educational device, not a punitive one. We believe that, in the long run, better care will be more economical. Experience has confirmed this opinion. The *rate of increase* in benefits paid out for nervous and mental claims was 34% between 1980 and 1981. The rate from 1981 to 1982 (after the institution of the peer review programs) was 8.7% and is now even lower. Savings to our group policyholders have been substantial.

In the late 1960s, Ross Taylor, then President of the American Society of Internal Medicine, stated, "The quality of medical care cannot be separated from the economics of medical care." Then, as now, the savings are to be found in the proper utilization of professional services and not in lowering their quality. If the diagnosis and treatment are correct in the first place, any seemingly extra cost for quality will be much more than offset in the long run by significantly lower utilization of services. That is why we should always concentrate on quality and proper utilization in peer review activities, for only in this way will the best care be provided at the lowest cost in the long run.

The flurry of activity by the Federal Trade Commission a few years ago in curbing peer review activities with respect to reasonable charges has caused no problem in the American Psychiatric or American Psychological Associations' programs. We scrupulously avoided any exchange of information regarding fees or charges for several years but now frequently request and receive comments on whether charges are reasonable or not.

We are not aware of any legal problems with respect to interference with the doctor–patient relationship or interference with the exercise of the treating professionals' individual judgment in their care of their patients. The program appears to have been widely accepted for what it is, an educational effort to improve the quality of care and the quality of reporting, and to control overutilization of services and facilities.

Some of the future considerations for peer review have been suggested by Alex R. Rodriguez, the former director at CHAMPUS. These may include expanded educational efforts, refined standards, better reporting, increased facility monitoring and surveys, and professional supervision of impaired and incompetent practitioners (*American Psychologist*, August, 1983).

Legal opinions state candidly that peer review is not exempt from antitrust law. Tort allegations have also been made, including libel and slander, interference with the physician–patient relationship, negligence, and invasion of privacy. Although these risks cannot be completely eliminated, steps can be taken to operate the peer review process in a way that will reduce them to a minimum. The two peer review programs described here are good models. Education regarding utilization is stressed, comments regarding charges are carefully worded, and the peer reviewers are instructed to present their opinions in a fair and carefully worded way without overt criticism of the competence or reputation of the treating practitioner or of past and current treatment. There may be suggestions or recommendations for changes. The peer reviewers' comments and recommendations are regarded by the company as advisory only. They are not used as mandates or guidelines.

Any action taken by the company always leaves the door open for reconsideration on presentation of new information. Patients may be told that insurance benefits are to be changed or discontinued, but never that they themselves must or should change or stop treatment.

Rarely, the company may exercise its right, as stated in all contracts, to require an independent medical examination and an opinion of the medical necessity or general medical acceptance of the method of treatment being used. This is always done with the approval of the treating practitioner and with her or his having every opportunity to prepare the patient for examination by a new practitioner.

Physicians and peer review organizations, such as the two described here, are selected on the basis of impartiality and professional standing and competency. Privacy and confidentiality are carefully guarded.

At this time, nearly all non-psychiatric peer review activity has been turned over to specially organized entities in the new federally mandated Professional Review Organization (PRO) program. This does not diminish the need for or the effectiveness of our present mental-health-treatment peer review programs. Psychiatric and psychological treatment, in or out of hospitals, is exempt from PRO review at this writing. The very presence, wide scope, and excellence of these mental-health-treatment peer review programs may prevent the extension of the federal program into the areas of psychiatry and psychology. The federal agencies may ask the present American Psychiatric and Psychological Associations to do all the peer review for psychiatric cases. Certainly, the CHAMPUS program provides a good model with a proven track record.

At this writing, the PRO program for nonpsychiatric hospital care continues its emphasis on cost control, and with considerable success in the way of reducing general hospital admissions, length of stay, and use of in-hospital tests and procedures. It remains to be seen how much this activity will adversely affect the quality of care. Recently the administrators of governmental medical programs have expressed some concern about quality of care and have instructed PROs to monitor quality as well as cost of care.

If and when psychiatric hospitals are brought into the DRG (diagnosis-related groups) and PRO programs, the same serious problems regarding cost versus quality of care will arise.

There are several centers for health-care-technology evaluation, and they are important and useful, but it would also be desirable to have a central, well-organized clearinghouse for the various other disciplines of clinical medicine to which we could send problem cases for an impartial expert opinion and recommendation.

The psychiatrists and psychologists have shown by their peer review activities that they can provide efficient and effective review and reporting of the quality and cost (utilization) of psychiatric illnesses and mental health care. It seems that other specialties could do it, too. It remains to be seen whether they can do it or not in the PRO program as it is now functioning.

The tremendous changes in medical practice and indeed in the provision of all health care services during the past 2 years have made it necessary for insurance companies and self-insured employers to seek an increasingly wide scope of assistance from all professionals who provide health care services. These professionals have come to recognize the importance of their part in the financing of health care services. Their responsibilities include not only providing professional services in the traditional way but considering at the same time how those services can be financed most effectively and economically without affecting their quality.

Peer review is now most commonly known as "quality assurance" and includes assessment of cost as well as quality. Organizations of psychiatrists, psychologists, medical social workers, nurses, and other related health-care providers have been in the forefront of developing internal committees, criteria, and other necessary methodology for helping to control the cost of health care services while maintaining the highest quality of care. Their innovations include preauthorization (as for hospital admission), monitoring and controlling length of stay, utilization of special laboratory and other tests and ancillary personnel; considering alternate methods and locations of treatment (e.g., day-care centers, overnight care, half-way houses, residential treatment centers, and outpatient areas in professionals' offices or elsewhere); monitoring and reviewing the utilization *and* quality of outpatient services; and, most recently, cooperating with insurers and related organizations in *case management*.

Case management is the newest technique for attempting cost control without impairing the quality of services. A specially trained professional employee, usually a nurse or social worker who has earned a master's degree, is assigned by the insurer to a case that appears to be or has proven to be very complex and costly. Case management almost always involves a hospital case. The case manager keeps in telephone contact regularly with the attending professional to keep up to date on diagnosis, treatment plans, and progress; to provide assistance in finding new or alternative *places* for treatment; to discuss plans for the patient's discharge; advising and assisting family and other caregivers; and to inform and remind the attending professional what financial resources are available, how they may be used under the terms of the insurance contract, and what possible exceptions can be made to the usual benefit provisions in order to hold down ultimate costs and stretch the available dollars as far as possible.

All this requires a considerable amount of effort, patience, understanding, cooperation, and caring on the part of the professional health-care providers. Psychiatrists, psychologists, medical social workers, psychiatric social workers, nurses, and other related professionals have not been found lacking in any of these qualities. They are quite aware that professionalism and the quality of health care services must not be lost in the sometimes frantic attempts to reduce the expenditure of health care dollars by those who provide the financing.

ACKNOWLEDGMENTS

We are grateful to the following Mutual of Omaha associates for their assistance in the preparation of this chapter: William B. Long, M.D., consulting psychiatrist;

Kenneth E. Pletcher, M.D., Vice President and Associate Medical Director; Marlene Bohlmann, R.N., psychiatric nurse consultant; M. Lowell Madsen, Assistant Vice President of Group Claims Administration; Mark Taylor, Manager of Group Claim Loss Control; and April McDonald (preparation and editing of manuscript).

19

Quality Assurance Activities for Mental Health Services in Health Maintenance Organizations

MICHAEL J. BENNETT

INTRODUCTION

Currently 13.6 million Americans receive their medical care in 323 health maintenance organizations (HMOs) (Inglehart, 1984). At the current rate of expansion, experts estimate that, by the early 1990s, that figure will reach 50 million. In addition to their direct impact, such organizations have been associated with the current upheaval in the medical community, and have prompted major revisions in practice patterns: a growing emphasis on cost containment, group practice, and reduced hospitalization. The success of for-profit HMOs, as well as a shift in government priorities that has eliminated subsidies, is forcing nonprofit HMOs to consider new sources of funding to finance expansion, and it is likely that such factors will play an increasing role in shaping such organizations as competition for the consumer becomes even more intense in the 1990s. These facts underscore the need to reconcile the economic realities of our times with the traditional values of medical practice, and the maintenance of quality.

The rapid expansion of HMOs, and the recent peaking of interest in them, belies the fact that such organizations have been with us since the early part of the century, when prepaid group practices developed in mining communities in the Mesabi Iron Range, and in the lumber industry in the Northwest. In 1929, Donald Ross and H. Clifford Loos contracted with the Los Angeles Department of Water and Power to provide comprehensive services for its workers and their families; this was the beginning of the Ross-Loos Clinic, which is still in operation. Sidney Garfield began to provide prepaid care to accident victims in 1933, and in 1938, he began to provide similar services for employees of Henry J. Kaiser who were working at the Grand Coulee Dam; this was the beginning of Kaiser-Permanente (Starr, 1982). The Group Health Cooperative of Seattle, Group Health of Washington,

MICHAEL J. BENNETT • Harvard Community Health Plan, Harvard Medical School, Boston, Massachusetts 02215.

D.C., and the Health Insurance Plan of New York, among the largest current non-profit HMOs, began over the next decade.

In essence, an HMO is a health care organization that serves a defined population for a cost that is fixed, prepaid, and periodically renegotiated. There is considerable variation in regard to the relationship that exists between the organization and its providers of care, and among the providers themselves. In a closed panel system, providers either work directly for the HMO on a salaried basis (staff model), or they are organized into a separate entity (e.g., partnership or separate corporation), which then contracts with the HMO to provide necessary services on a per member (capitation) basis. An HMO may contract with one (group model) or several such groups (network model). In an open panel system, the foundation or independent practice association model (IPA), the relationship both among providers and between providers and the HMO are different. Falling halfway between the types of structure described above and a fee-for-service system, the providers in an IPA are loosely affiliated with each other, may practice at separate sites (usually their own offices), and are compensated for services provided to the enrolled and prepaid population on a per visit rather than a per member basis. These differences in structure are likely to influence the pattern of care in a variety of ways, especially in regard to the capacity of the HMO to foster and monitor provider behavior that is consistent with its objectives. This subject will be discussed in detail later in this chapter.

MENTAL HEALTH CARE IN HMOs

The first generation of HMOs described above, as the rest of the health care community, was wary of providing coverage or care for mental disorders. Following Avnet's landmark studies (1962) attesting to the insurability of mental health care, HMOs began in the late 1960s to experiment with the provision of coverage, usually on a supplemental or rider basis for some of their enrollees. Under the influence of large contractor groups (the United Auto Workers and Aerospace Union; union and employee groups such as steel, retail clerks, and state employees; Medicare; and Medicaid), a variety of arrangements were made to provide services, often at sites separate from general medical services, either contracted out or using part-time staff who were not integrated with other providers in the HMO. With the second and successive generations of HMOs, mental health care began to be included as a basic benefit, although with access controlled through a variety of measures designed to cap or control utilization (such as high or graduated copayment, benefit limitation or exclusion, limited budget, and limited staffing). With the growth of larger plans, and with their decentralization, greater integration of mental health with other (especially primary-care and pediatric) services became more common. The HMO Act of 1973 (PL 93-222), which mandated minimal (outpatient) coverage for mental disorders, along with legislation passed in several states, spurred those HMOs that sought federal qualification (with its subsidies and marketing advantages) and state endorsement to make mental health services more readily available, usually inhouse. The tremendous increase in public awareness of, acceptance of, and demand for mental health treatment, along with the failure of public programs to satisfy the need, was a further spur, creating a positive marketing force to balance the cost of including mental health care in the premium.

QUALITY ASSURANCE IN HMOs

Formal requirements for quality assurance are part of both state and federal regulations and are necessary for membership in the HMO parent organization (Group Health Association of America). In addition, both Medicare and Medicaid regulations apply to those HMOs under contract. Federal law requires a quality assurance program that (a) stresses health outcomes to the extent consistent with the state of the art; (b) provides review by physicians and other health professionals of the process followed in the provision of health services; (c) uses systematic data collection of performance and patient results, provides interpretation of these data to its practitioners, and institutes needed change; and (d) includes written procedures for taking appropriate remedial action whenever, as determined under the quality assurance program, inappropriate or substandard services have been provided or services that should have been furnished have not been provided. There are, in addition, provisions for both internal and external monitoring (*Federal Register*, 1979; Office of Health Maintenance Organizations, 1979).

These guidelines are varyingly interpreted and applied. They overlap with managerial and consumer-oriented mechanisms because, in a general sense, concerns about the quality of service and care provided are intrinsic to the well-being and survival of any organization that seeks to attract, satisfy, and retain members in a competitive environment. In seeking to integrate the various levels and components of care, while balancing the interests and priorities of consumers with those of providers and insurers, the HMO inevitably must monitor its activities. Mechanisms and procedures that have been developed to do so, as well as to comply with the various regulations to which the HMO is subject, vary in their coherence, in their relationship to defined and explicit objectives and standards, and in the degree of organizational and staff commitment to them.

The remainder of this chapter describes the range of formal and informal quality-assurance activities in mental health programs of HMOs, found through reports in the published literature and through a national survey of 30 such organizations, conducted by the author. These activities are characterized in regard to classical concepts, and the author then considers the prospects for the future in the light of current trends and the vast system and social transformations in progress.

DIMENSIONS OF QUALITY

The classical categorization of quality-of-care assessment comes from the work of Donabedian (1969, 1980), who formulated three components: structure, process, and outcome. *Structure* refers to the setting in which care takes place and includes the facilities, the provider and support staff, financing, records, and other elements of context. *Process* involves all those elements of the care-giving and care-receiving experience that constitute the provider–patient transaction. *Outcome* refers to the effects of the care, including patient and provider satisfaction, changes in health status, cost, and attitude or understanding, as well as any broader social implications. The assessment of quality in any of these spheres requires definitions of standards (or, in a dynamic sense, objectives), reliable methods for measuring performance, and a set of explicit priorities and values. If the assessment of quality indicates gaps between standards and performance, then the assurance of quality requires some effort to bring performance closer to standards, and to gauge the effectiveness of the intervention. The loop of quality assurance involves a cycle of assessment, selection of interventions to improve performance, and reassessment; where the monitoring of quality fails to include change efforts and their assessment, it is incomplete.

Objectives and standards are neither given nor universal. They rest on values that, in turn, are strongly linked to context. Two contextual features of the HMO are especially germane: (a) the HMO seeks to serve a defined population, a fact likely to influence its relationship to its individual members, and (b) a major objective of the HMO movement is to control the costs of medical care through better management. In regard to the first of these features, the HMO shares the ideals of practice in the public sector, in its attempt to serve some segment of society as well as the individual, and to reconcile and balance personal with social needs. In regard to the second, parallel aims exist: to reduce cost without sacrificing quality. Given the current crisis in financing health care, it is imperative to reconcile cost and quality considerations. As Donabedian has suggested, cost is an important dimension of quality when the needs of society are considered along with the needs or preferences of the individual. If we state the case more strongly, it is possible to dismiss considerations of cost only when the preferences of the patient, his or her family and friends, and society as a whole are overlooked in favor of what Donabedian termed the "absolutist" definition of quality: the practitioner's belief about what is best (Donabedian, Wheeler, & Wyszewianski, 1982). As will be apparent in the following discussion, the value systems that are elaborated into objectives and standards in the various HMOs are greatly influenced by such considerations.

THE EVOLUTION OF QUALITY ASSURANCE ACTIVITIES IN HMOs

Unlike the fee-for-service world or, to a great extent, the public sector, the imperatives for quality assurance activities in HMOs have come from within rather than from outside the system. They are a natural extension of currents and forces already operative: the need to conserve resources, to determine what is optimal care, and to balance the needs of a population with those of the individual. HMO mental-health programs are not likely to share the concerns that predominate in the community at large: overtreatment, excessive third-party costs for the few at the expense of the many, problems of inequitable distribution of and access to necessary services, and excessive use of inpatient care. Rather, the emphasis on brief methodology tends to raise questions about undertreatment and selection of the appropriate methodology for a given patient or problem. Given the interdisciplinary mix of professionals common in such settings, there are likely to be questions about the appropriate division of responsibilities and about the optimal staffing ratio and staff mix. In a system that involves competition for limited dollars, planners and managers are likely to be interested in the relationship between mental health and other services, especially primary care. Because care in the general medical setting is, to a greater or lesser degree, collaborative, there will also be concerns with the efficiency of systems and the adequacy of communication. Member and provider concerns that resources be allocated fairly, and that the contractual obligation to provide necessary services be honored, require attention to such outcome variables as consumer satisfaction (or complaints). Finally, in a system whose viability is based on the competitive status of the premium in the health care marketplace, considerations of cost or its determinants (e.g., hospital days, use of outside resources, utilization rates, and staffing ratios) are likely to be emphasized in any consideration of the structure, process, or outcome of care.

THE STRUCTURE OF QUALITY ASSURANCE PROGRAMS IN HMOs

The structure of quality assurance activities reflects the managerial structure within the HMO. Ranging from the IPAs and network model HMOs, where the clinical and managerial functions are likely to be structurally (and often geographically) remote from each other, to the closed panel systems that engage providers in managerial roles and functions, focus on activities designed to review and monitor performance may be either intrinsic or extrinsic to the delivery system. Using the sophisticated management information systems and computerized records commonly available in HMOs, most organizations monitor utilization and provider and system performance and feed information back to managerial and provider staff in some form. In those organizations where the managerial and clinical functions are not well articulated with each other, providers are likely to view such activities as imposed. At its worst, the disarticulation of clinical and managerial functions may produce a downward flow of numbers, which are perceived by providers as organizational directives designed to control their behavior, disregarding their values and priorities. In those HMOs that integrate service and managerial activities, there is a greater chance that providers can be helped to develop a shared system of values and objectives more closely allied to overall organizational priorities and aims. Under these circumstances, staff are likely to view activities designed to monitor and maintain quality as germane to their practices and as potentially useful to them. Within the mental health area, as in other disciplines, staff who are informed about and allied with organizational objectives (a function of mental health leadership) are likely to accept the need for monitoring activities as a practical need to acquire the data required to address and solve the problems that arise in the pursuit of shared objectives.

The successful integration of mental health programs and staff within the organization produces a climate where scrutiny is allied to the wish for professional growth, as defined in the practice context. Quality assessment and assurance in such systems derives from, rather than being imposed on, clinical practice, and efforts to improve the structure, process, and outcome of care are likely to be viewed as intrinsic to function and to work expectations (Bittker & George, 1981). Formal quality assurance activities require, in addition to such a climate, the information, mechanisms, and staff (or time) to design, implement, and evaluate projects.

Despite legislative requirements, HMOs make varying commitments to formalized quality assurance programs. Such a commitment is reflected in the choice of a director, the allocation of the budget, the provision of adequate staff, and the reporting relationships that either empower or hinder those who operate in such roles. In most settings, the director is a physician, though not necessarily one versed in quality assurance. He or she may or may not have a staff and control of a budget. In those organizations that have a quality assurance department (QA), it may be free-standing or linked to staff education or to research. Most important, the link between QA and provider staff, the crucial element in determining the success of the program, appears to be problematic in most settings. In those programs that seem to function best, provider staff or committees serve to link provider and QA personnel, facilitating collaboration.

THE CURRENT STATE OF THE ART: ZEN AND THE ART OF QUALITY ASSURANCE

In the following discussion, activities are arbitrarily assigned to one of the three categories: structure, process, and outcome. Utilization review, which can be construed in part as a quality assurance device, is primarily concerned with keeping staff aware of cost and cost-equivalent data, in order to enhance practice management and cost containment; for this reason, it has been included under the heading "Outcome." As will be apparent, the full cycle of quality assurance (quality assessment, identification of deficiencies, intervention designed to bring performance in line with standards, and repeat quality assessment) was rarely encountered in the published literature (where there are few reports of mental health quality assurance activities in HMOs) or in the survey of HMOs conducted by the author.

STRUCTURE

Concerns about the setting in which care takes place are ordinarily the province of administrators, health care directors, and, to a lesser extent, medical managers. A number of HMOs use periodic surveys of members to assess satisfaction with facilities (as well as with the staff and services). Most have mechanisms for addressing the problems and complaints of members, and many have prominent consumer input (e.g., on a board of directors or governing body, or on committees with access to administrative leadership). Surveys of patient satisfaction with mental health services have been initiated at Harvard Community Health Plan and at Park-Nicollet (described more fully under the heading "Outcome"), and these have included questions about the structure of care (e.g., facilities, support, and provider

staff,   ambience, scheduling, telephone, and parking). An administrator's performance,
which is likely to be reviewed on a regular basis, includes his or her ability and
willingness to solve problems in these areas, as they are brought to attention.
Although records may be considered indicators of the process of care (to be addressed later), the legal requirement to maintain records, the standards to be met,
the issues of confidentiality and privacy, and the functional state of equipment and
procedures are elements of structure; these are usually monitored by interdisciplinary committees and are subject to periodic on-site review by federal and state
regulatory bodies. When the mental health component is integrated administratively into the health care setting, it participates in planwide monitoring. When
separate records are kept in the mental health department, such records vary in
completeness and quality, a fact that impedes efforts to examine the process of care
and to relate it to outcome; strong concerns about confidentiality are often linked to
such issues.

PROCESS

By far, the most common form of quality assessment and quality assurance activity reported is in this area of process: the various elements of caregiving that involve transactions between the patient and the provider. Activities fall into three
broad categories: retrospective audits of records in order to ascertain provider compliance with agreed-upon standards of care or record keeping; concurrent audits of
the charts of patients undergoing an episode of care, usually inpatient, as a form of
consultation to the therapist or inpatient treatment staff; and face-to-face consultation as a form of concurrent audit, with both inpatients and outpatients.

A retrospective audit of records may be performed in order to gauge the
quality of the record or to allow judgments about the quality of care presumably
reflected in the record. Reviews are usually performed by peers, most commonly
in search of problems or deficiencies in the care process. Specific questions may be
asked, reflecting some parameter of care that has been determined (usually by the
mental health providers themselves) to be desirable or necessary. For example, was
the patient advised about the mental health benefit? Was a depressed patient questioned about suicidal intent? Was a referring doctor or nurse provided adequate response to a consultative request? Prescribing patterns of primary-care and specialty
providers have been analyzed with regard to lithium, benzodiazepines, and
anxiolytics prescribed to known alcoholics.

The following are examples of departmental projects that took place in the
highly integrated mental health component of a staff model HMO. Peers audited
the records at the Harvard Community Health Plan (HCHP) Kenmore facility in
order to determine the completeness of prescribing information for psychoactive
medication prescribed by mental health staff. When the level of compliance was
judged to be below departmental standards, providers were reminded (individually). A follow-up documented compliance in the desired range (95%). In a
study of screening, presciibing, and monitoring practices involving lithium, the
HCHP automated record was used to identify patients newly begun on the drug,
and to assess provider compliance with standards previously agreed on by the

mental-health provider group. Noncompliers were sent computer-generated reminders, and follow-up review of the record demonstrated a significant increase in the rate of compliance. In this project, providers were given a (mandatory) educational session as part of the regular continuing education program and were offered consultation to improve their level of knowledge and competence. Despite these steps, the change in provider behavior was not sustained (Feldman, Wilner, & Winickoff, 1982).

Audits may be concerned with how a given problem is managed by the provider group or by the system as a whole, as determined by review of randomly selected records of patients who present the chosen condition or problem. For example: How soon following discharge from the hospital is a patient seen in the outpatient setting? Patterns of treatment may be examined: Is there any reason to believe that patients who are seen only once (Spoerl, 1975), or who are seen more than 10 times, or who are in continuous treatment for more than 2 years are being mismanaged or are receiving inappropriate care? What problems can be identified in the referral or intake process when patients fail to keep initial appointments, or in the treatment process, when patients terminate unexpectedly?

Concurrent audits usually involve the hospitalized patient. In contrast to utilization data, such audits focus (through record review and, at times, through face-to-face consultation) on the content of utilization. Usually, the focus is on sharpening the clarity of treatment objectives with an eye toward speeding discharge (i.e., through the design of a workable alternative to continuing inpatient care, or through suggestions about resolving treatment impasses). Such consultations are likely to be provided by senior clinician staff, and to have a strong educational (consultative or supervisory) orientation. In some settings, concurrent review of inpatients is mandatory after a certain number of inpatient days. This review differs from extradepartmental scrutiny, in the form of signal reminders sent by utilization control and review staff whose job it is to raise the level of consciousness and to provoke such reassessment. At Park-Nicollet, in conjunction with the emphasis on brevity in outpatient psychotherapy, therapists are required to review treatment with a peer or with a committee of peers after the 10th, 20th, and 30th sessions (Anderson, 1984).

One form of process review concerns the patient's health-care-seeking behavior; the aim is to better understand patterns and needs so that programmatic responses to them can be sharpened and improved. The so-called medical offset literature concerns the interface between mental health and general medical services. Of the many studies that have been done in this area, a number have occurred in HMO settings (Budman, Demby, Feldstein, & Gold, 1984a,b; Fink, Shapiro, & Goldensohn, 1969; Goldberg, Frantz, & Locke, 1970). Beginning with the work of Follette and Cummings at Kaiser (1967, 1968), various investigators have suggested that patients with a mental disorder are apt to overutilize medical services, and that the provision of specialist care reduces such (presumably inappropriate) utilization. The results of such studies tend to be confusing and contradictory, and generally raise more questions than they answer. Nevertheless, they have been used by some to justify the inclusion of mental health services in general health-care settings and are adduced as arguments for the adequate funding of and adequate access to such services. The implications for quality assurance are indirect. One implication is that these studies focus attention on the medical–mental-health interface, and consequently on educating primary-care and other medical providers to recognize covert mental disorder. Another is that they underscore the need for effective liaison of mental health staff with other (especially primary-care) providers. Although offset may or may not prove to be a useful concept, it helps to focus attention on the need for a comprehensive and holistic review of health care, and on the fact that mental health professionals encounter only a fraction of those who present to the health care system because of emotional distress, alone or in conjunction with physical disorders.

At Kaiser of Northern California (L. Rubin, personal communication, 1984), chart audit has been used for a number of years as the vehicle for a comprehensive quality assurance function that links centralized resources with the 23 service facilities. Teams of reviewers, drawn from a pool of 165 doctors and nurses who are employed by the Permanente Medical Group, regularly visit the various centers and, using a highly pragmatic protocol that is based on a search for internal inconsistencies rather than on a set of explicit standards, review charts as the first step in a chain of events designed to identify and correct defects. As reflected in its structure and funding, the program has clear organizational support. It functions under the leadership of the director of education for the Northern California area, who reports to the medical director. Within the centers, the review function is integrated with other, local monitoring devices, through a health center chief of quality, who works closely with departmental leadership and staff. Reviews may be initiated at any level of the system, in response to concerns about quality articulated by members, staff, or internal or external monitoring agents. Emphasis is placed on the congruence of the reviewers with the specific area reviewed; that is, the reviewers are selected to match representatives of the area reviewed, and agreement of the relevant health-center staff with their external peers is required before studies or corrective action can be initiated. A problem is any example of deviation from the ideal: Given the current state of our knowledge, does the record indicate that we are doing what we say we want to do? The reference is to local practice patterns and norms rather than imposed standards. Identified and agreed-upon deficiencies may involve any of the three facets of quality and any component of the health care network. Studies that determine the significance or frequency of the identified defect are followed by recommendations for change. Data are kept on identified problems, and follow-ups are conducted in the same fashion to determine the outcome of the interventions.

Examples of concerns about the process of mental health care include: Has an identified alcoholic been referred to Alcoholics Anonymous? Have an appropriate number of pills been prescribed when a first prescription for an antidepressant has been written? Has suicidal potential been assessed for a patient who is depressed? Have patients who are receiving prescription refills over an extended period of time been adequately followed? Have lithium levels been monitored?

An interesting feature of this approach is the creation of a pool of reviewers that is drawn from active staff and that links with staff at the decentralized level. In the byplay between the reviewers and their peers, there is an opportunity to promote dialogue about variations in practice patterns and norms without imposing rigid standards. It appears to be one thoughtful approach to the problem faced by large HMOs: How much variation in practice norms is optimal, and what methods will prompt physicians to critically consider their professional behavior?

Most attempts to judge the quality or appropriateness of the treatment process are hampered by the difficulty of clarifying the link between process and outcome. A number of writers have suggested that it may be easier to understand process if one starts with outcome, rather than to do the reverse.

OUTCOME

Outcome evaluation of mental health care in HMOs has centered on two parameters—cost and patient satisfaction—with almost no reported efforts to examine other consequences of treatment or to link them with objectives. Although, as Linn and Linn (1975) pointed out, mental health research has produced examples of a multidimensional approach to outcome evaluation, such studies are rare in the setting of the HMO for three reasons: (a) such studies involve the capacity for formalized research, and most HMOs have neither funded nor involved themselves in mental health research programs; (b) as organizations with a highly pragmatic approach, HMOs tend to be less interested in the subtleties of outcome than in its practical consequences; and (c) as service organization, HMOs tend to be wary of "using" their members in the pursuit of knowledge that may benefit members only indirectly.

With its emphasis on cost and cost-effectiveness, the HMO has a wide variety of managerial mechanisms to monitor this dimension of quality, and to feed relevant data back to providers. For the most part, such activities can be characterized as utilization review. The source of such data may be departmental or extradepartmental; these data are usually presented to staff in the form of periodic reports, and they are likely to be a focus of discussion in meetings of mental health staff who are organized as departments. Such reports include information on cost and such cost-equivalent data as the practitioner's use of hospitals and outside resources, laboratory and diagnostic procedures, and consultants. Coleman and Lowry (1983) described an extensive use of management information data to inform managers and providers in an IPA about their practice patterns, their approaches to the treatment of specific disorders, and presumably, their comparison with peers. These authors mentioned mental health providers and data but did not offer details.

In staff model HMOs, medical staff may be subject to performance reviews that address all parameters of function, including the satisfaction or complaints of their patients and colleagues, both within and outside their department. Usually, such reviews, which may be tied to compensation, focus on some element of productivity, defined in system-relevant terms. For the mental health professional, productivity is likely to be defined in terms of the number of patients seen per available clinical hour, the percentage of available clinical hours that are filled, the number or percentage of new patients in the provider's case load, and, in some settings, the provider's willingness to share in nonclinical departmental responsibilities.

A second outcome variable that receives wide attention—although, like cost, not always in concert with other quality variables—is consumer (and patient) satisfaction. Strongest in the cooperatives, consumerism is a key factor in the HMO. Membership surveys are conducted on a regular basis at various HMOs, assessing member attitudes toward systems, services, and providers. These are routinely shared with staff. The same is true with regard to member complaints. Within the mental health department at HCHP, member satisfaction data have generated in-depth reviews of patient expectation and need and system and provider responsiveness. A key finding, supporting observations made in satisfaction studies of general medical care, is the link between patient satisfaction and the interpersonal ambience of the patient–provider interaction (Bennett & Feldstein, 1986). At Park-Nicollet Medical Center, where there is a consistent and long-standing organizational commitment to QA (Batalden, McClain, O'Connor, & Hanson, 1978), consumer attitudes about treatment (including mental health care) are used to guide discussion involving members of the free-standing QA department and staff of the various clinical departments, as these departments meet on a regular basis and seek to identify and prioritize problems. The process begins with a telephone interview, in which a QA nurse, using a protocol, questions consumers about their responses to treatment episodes in the department in question. Findings are fed back to staff with the objective of identifying problems and developing QA projects to formulate, implement, and monitor solutions (Engstrom, personal communication, 1984). In this model, which has also been implemented at Rhode Island Group Health Association, QA staff seek to establish an alliance with practitioners and to act as facilitators, to help staff focus their problem-finding and problem-solving efforts; in this sense, the QA and clinical functions appear to be well integrated.

As service organizations, HMOs are rarely involved in research programs of their own. Prominent exceptions include HCHP, where a number of studies have been done on psychotherapy outcome in brief group, couples, and individual psychotherapy, and where current work concerns the relationship between process and outcome in brief group psychotherapy (Budman, Demby, Feldstein, & Gold, 1984; Budman, Demby, & Randall, 1982). Columbia Medical Plan (CMP) has capitalized on its relationship with Johns Hopkins, and especially with the Health Services Research and Development Center, to examine offset and other aspects of mental health care. A number of studies have been carried out at CMP, Marshfield Clinic (which has both a prepaid and a fee-for-service population), and other sites under these auspices (Burns, Goldberg, Hankin, Hoeper, Jacobson, & Regier, 1982; Hankin, Kessler, Goldberg, Steinwachs, & Starfield, 1983; Hankin & Locke, 1983; Hankin, Steinwachs, & Elkes, 1980; Hankin, Steinwachs, Regier, Burns, Goldberg, & Hoeper, 1982). Although such studies do not originate within the provider groups themselves, the results are shared, and providers have the opportunity to review and consider their practice patterns.

In an ambitious attempt to link psychotherapy outcome with process, with the hope of identifying correctable problems, Hankin studied patients treated at CMP. She assessed problem status via the SCL-90 and therapist ratings (using the Hopkins Psychiatric Rating Scale) and functional status via a new instrument designed to measure job, interpersonal and recreational function. Ratings were done at the beginning of treatment and at 4-, 8-, and 12-month intervals after termination.

Because it proved difficult for staff to agree on what constituted success, she defined it by her measures, as changes sufficient to bring the patient into the normative range; both patients with substandard and successful outcomes were interviewed by a clinician who was not one of the departmental therapists. It proved difficult to link outcome with process, in part because of limitations in records, and it was felt that the program had fallen short of its potential for quality assurance. Nevertheless, satisfaction studies and patient comments indicated a favorable reception by the patients, and there was a suggestion that sensitization of staff to the issues being studied may have improved the service provided (Hankin, Bendit, Koch, & Merrill, 1978). This study exemplifies the problems of research within a busy service-oriented setting, especially when the initiative comes from outside the system. It is, however, a bold step in a necessary direction.

QUALITY ASSURANCE IN HMOs: PROBLEMS, OPPORTUNITIES, AND PROSPECTS

As the HMO concept has evolved, early idealism has given way to an emphasis on pragmatic concerns. Initially viewed as a variant of socialized medicine, the HMO has been discovered by corporate America and the investor, and prepaid group practice is viewed by health planners and economists as a way of controlling spiraling costs through better management of our health care resources (Enthoven, 1984; Manning, Leibowitz, Goldberg, Rogers, & Newhouse, 1984; Starr, 1982). In becoming mainstream, HMO practitioners and administrators find themselves confronted with the same problems that have plagued other health care systems. With the shift away from institutional and hospital-based care, from soft to hard money, and from the public to the private sector, HMOs must develop ways of meeting the cross section of medical demand, and of dealing with the sticky problems involved. In the mental health area, these include problems of chronicity, high demand for service, and the fact that emotional disorder overlaps with medical and social problems in a way that makes it difficult to decide its appropriate boundaries. The HMO may be viewed as a microcosm of the broad transformations taking place in medical care, bringing together the consumer, the provider, and the insurer in a relatively closed system in which the various negotiations required to balance their interests and priorities can be addressed. Viewed as an ideal, the overall quality of the tripartite system is a measure of how well this balance is achieved and maintained. Within the system, any element can be judged by its contribution to this ideal. Such a systems view of quality is, of course, applicable to the community at large; it tends, however, to be most useful where the system is most readily defined and, therefore, is an appropriate way of understanding the telescoped world of the HMO.

In attempting to reconcile the competing agendas of consumers, providers, and insurers, HMOs will be influenced by the relative power and influence of the three components of the system. Because there is no agreed-upon, universal definition of quality, activities designed to assess and monitor it are likely to be shaped by values and priorities relative to the context. Thus, at one extreme, the for-profit staff model HMO is likely to emphasize the monitoring activities most strongly associated with management interests and sound business practice. It is more likely to be concerned about overtreatment than about undertreatment. Consumer-dominated organizations, in contrast, are most likely to emphasize issues of access, an affordable premium, and personalized, responsive service; when physicians are dominant, as in group network model HMOs or IPAs, the yardstick for quality is likely to be the current state of medical knowledge and technology (Vuori, 1980). Donabedian has referred to this position as "the absolutist definition" (Donabedian, 1966). Each perspective, if not balanced with the others, will produce a lopsided set of standards, which are likely to be seen as irrelevant or misguided by those whose interests are underrepresented.

At the present time, quality assurance activities in HMOs tend to focus on the process of care, usually in the form of a retrospective audit that is designed to identify deficiencies. Such activity, usually performed by peers, mirrors peer review in the community and suffers from the same problems: as a form of fault finding, it is unlikely to produce enthusiasm and always runs the risk of becoming an adversarial or imposed function. It is likely to gain, at best, compliance and to produce, at best, normative performance. The outlier who is especially creative is likely to be discouraged and unrewarded, and quality may be maintained, but it is unlikely that it will be enhanced. The same can be said in regard to outcome evaluation that centers on cost and consumer satisfaction, where the priority can easily become pleasing the customer in the least costly way, contributing to the health of the organization but not necessarily to the patient or society. By contrast, when the professional aspirations and pride of the practitioner can be allied with organizational priorities, more creative solutions to old problems are likely. In those HMOs that have strong and cohesive mental-health units or departments, the need to solve practical problems collaboratively forces the professional to come to terms with a core quandary: the perpetual need to reconcile and balance professional and personal values with the practical demands of the setting. It is important to note that, in such settings, influence is likely to flow upward to management as well as downward to the practitioner.

Future trends are likely to be determined by shaping forces currently at work. These include the greater prominence of for-profit HMOs (and for-profit activities in non-profit HMOs), increased consumer influence, growing concerns about cost, and reactive moves on the part of the fee-for-service community. The latter, which is producing an unprecedented flurry of new and varied health care delivery configurations, is most difficult to assess. Spurred in part by the success of HMOs, and in part by regional physician excess (with the prospect of national excess), preferred provider organizations (PPOs), which originated in California in the early 1970s, are springing up in various parts of the country (American Health Consultants, 1982). Most simply put, the PPO, which may be initiated by physicians themselves, by hospitals, by employers, or by third-party payers, involves a selected group of physicians who agree to provide service to a defined population on a fee-for-service basis but at a discount. Although fewer than 100 PPOs are currently operational, it is estimated that several hundred are being organized (Fox, Goldbeck, Kiefhaber, Webber, Spies, Friedland, Tell, & Walsh, 1984). There is little information currently available about the types of services offered in PPOs, and there is little to suggest that quality assurance activities or utilization review is widely practiced within those that are in operation. From the standpoint of the physician, the PPO offers certain advantages: an expanded patient market, often the prompt payment of bills, and no financial risk (the risk is borne by the organization or by the consumer). From the standpoint of the patient, who is permitted to use providers outside the PPO, but who is penalized by having to pay more if he or she does (accomplished by varying mechanisms), this configuration avoids the "lock-in" that is typical of the HMO. From the standpoint of those employers or industries participating, the PPO offers the prospect of reduced costs for employee health care, essentially by forcing providers to compete for their employees as patients.

Although the PPO may initially select its participating physicians on the basis of their willingness to accept lower fees than the competition, such a strategy is likely to produce problems. First, because the physicians are not at risk financially, there is no curb on high utilization unless such curbs are imposed in the form of stringent utilization monitoring and control. Second, it is unlikely that constraints alone will produce the optimal changes in provider behavior. As provider behavior is the ultimate determinant of cost, PPOs will have to address this issue by recruiting with an eye toward favorable practice patterns, rather than simply the willingness to offer a discount. It is likely that the provider groups that can demonstrate high efficiency in conjunction with high quality will compete most successfully in the marketplace of the 1990s. This demand is likely to force PPOs, many of which have developed hastily and with little forethought or preparation, to face the same issues that challenge the HMO: the need to develop, monitor, and sustain cost-effective patterns of medical care. The conflicts inherent in managed health care are likely to be played out in the form of stepped-up utilization review, audits, and other quantitative measures. In a procompetitive environment, with an increasingly aware purchaser, neither quality nor efficiency is likely to be taken for granted. Medical shopping and the influence of "brand names" is one possible consequence of the confusing multiplicity of options likely to be available. As among HMOs, there is likely to be a high failure rate among PPOs, along with a gradual weeding-out process, with a tendency for those groups that succeed to grow and consolidate, as they develop and refine the necessary systems and attitudes. HMOs, as a response to the PPO phenomenon, are likely to develop PPOs of their own, as an additional line of business, capitalizing on their own knowledge and possession of the necessary management-information systems. Such initiatives may involve their administrative and managerial components alone or their provider base as well.

What are the implications for mental health care? First, there will be pressure to make the most efficient use of treatment resources. Given the wide variety of treatment methods available, it will become increasingly necessary to allocate resources fairly and rationally and to determine which method is best for which patient, as well as who needs professional attention and who does not. Second, the growing interest in biological and genetic factors in mental illness calls for better case finding, earlier intervention, and more effective working relationships with medical colleagues, especially those engaged in primary care. Third, mental health professionals are likely to be exposed to scrutiny and their efforts to be judged by the same pragmatic criteria that apply to other services, as decisions are made about cost and service trade-offs. The emphasis on brevity, or on resource-limited inter-vention, will lead to an increasing need on the part of the providers to know that such patterns of care are safe, appropriate, and effective. These concerns favor an at-mosphere of scrutiny and an evolving and balanced notion of quality.

The procompetitive scenario that some experts advocate and predict for the near future (Ellwood, 1971) raises major problems for the mental health pro-fessionals. As Wyszewianski, Wheeler, and Donabedian (1982) suggested, the con-sumer may not always be the best judge of what is the best health-care purchase, and without the proper safeguards, there is pressure on competing health-care-delivery systems to economize through greater cost sharing or through the elimina-tion or curtailing of some services. Historically, mental health has often been one service treated in these ways. The ability of mental health professionals to coun-teract such pressures and trends will rest on their willingness to find more efficient and less costly methods of helping their patients, as well as on a greater readiness to ally with health care planners and with other health care providers in creating a more equitable distribution of their services. There is a need to define and measure the objectives of mental health care, to demonstrate its utility and affordability, and to develop a better working relationship between mental health and other health care services. Mental health providers within the HMO setting, as their brothers and sisters in the community at large, must be willing to subject their work to scrutiny, to engage in treatment planning and the setting of objectives, and to speak out for the "effectiveness" component in the cost-effective dyad. The arguments that are likely to be advanced about the affordability of insured mental-health care in an in-creasingly competitive marketplace must be counteracted by a greater willingness on the part of mental health professionals to accept the challenge of the debate, and to demonstrate that mental health care is an essential component of any com-prehensive program of health care, and that it is its absence that is unaffordable.

It is interesting in this regard to shift the focus from medical care to the history of successful corporate activity, because there is great anxiety within the mental health community (and the medical community in general) about the growth of cor-porate health care and the likelihood that this trend will continue. Despite Starr's bleak scenario (1982), in which medical practice becomes increasingly dominated by corporate enterprise and subject to the transcendent motive of profit, there is no necessary polarity between quality in health care and the economic viability of the organizations that provide it. Peters and Waterman, (1982) in their review of suc-cessful U.S. corporations, pointed to a number of factors that characterize ex-cellence. Among these, consumer orientation and concern with quality are fore-most. In a pluralistic and competitive environment, it is likely that the emphasis on defining, monitoring, maintaining, and enhancing quality that has been associated with the success of well-run corporations will be found as essential in those in-volved in the delivery of health care services. In addition to the influence they exert, there is also reason to believe that those corporations that seek alliance with health care professionals may be influenced by the needs and priorities of those with whom they affiliate (Martin, 1984). For this reason, it is essential to redefine quality in a manner that can then be effectively advocated in the evolving environment. Health care professionals must take the lead in defining excellence in a manner that takes into account the key concerns of our age, and that reconciles the values of the professional with the needs of society, or of the defined populations that he or she serves. Health care has become unwieldy and unaffordable and is a battleground for the broad social issues of our time. Given the need for a blending of clinical and economic wisdom, what looms on the horizon as a shotgun marriage may yet prove a viable partnership.

REFERENCES

American Health Consultants, Inc. (1982). PPOs will proliferate if QA and UR programs improve. *Hospitals Peer Review, 7*(5), 53–64.

Anderson, R. (1984). Individual case review in the outpatient mental health department of an HMO. *Quality Review Bulletin, 10*, 191–193.

Avnet, H. (1962). *Psychiatric insurance.* New York: Group Health Insurance.

Batalden, P., McClain, M., O'Connor, J., & Hanson, A. (1978). Quality assurance in the ambulatory setting: An operating program. *Journal of Ambulatory Care Management, 1*, 1–13.

Bennett, M., & Feldstein, M. (1986). Correlates of patient satisfaction with mental health services in an HMO. *American Journal of Preventive Medicine, 2*(3), 155–162.

Bittker, T., & George, J. (1981). Psychiatric service options within a health maintenance organization. *Journal of Clinical Psychiatry, 41*(6), 192–198.

Budman, S., Demby, A., & Randall, M. (1982). Psychotherapeutic outcome and medical utilization: A cautionary tale. *Professional Psychology, 13*, 200–207.

Budman, S., Demby, A., & Feldstein, M. (1984a). A controlled study of the impact of mental health treatment on medical care utilization. *Medical Care, 22*(3), 216–222.

Budman, S., Demby, A., & Feldstein, M. (1984b). Insight into reduced use of medical services after psychotherapy. *Professional Psychology, 15*(3), 353–361.

Budman, S., Demby, A., Feldstein, M., & Gold, M. (1984). The effects of time-limited group psychotherapy: A controlling study. *International Journal of Group Psychotherapy, 34*(4), 587–603.

Burns, B., Goldberg, I., Hankin, J., Hoeper, E., Jacobson, A., & Regier, D. (1982). Uses of ambulatory health/mental health utilization data in organized health care settings. *Health Policy Quarterly, 2*(3/4), 169–179.

Coleman, J., & Lowry, C. (1983). A computerized MIS to support the administration of quality patient care in HMOs organized as IPAs. *Journals of Medical Systems, 7*(3), 273–284.

Donabedian, A. (1966). Evaluating the quality of medical care. *Milbank Memorial Fund Quarterly, 44*(3), 166–207.

Donabedian, A. (1969). *A guide to medical care administration: Medical care appraisal.* New York: American Public Health Association.

Donabedian, A. (1980). *Explorations in quality assessment and monitoring: The definition of quality and approaches to its assessment.* Ann Arbor, MI: Health Administration Press.

Donabedian, A, Wheeler, J., & Wyszewianski, L. (1982). Quality, cost and health: An integrative model. *Medical Care, 20*(10), 975–922.

Ellwood, P. (1971). Health maintenance strategy. *Medical Care, 291*, 250–256.

Enthoven, A. (1984). The Rand experiment and economic health care. *New England Journal of Medicine, 310*, 1528–1530.

Feldman, J., Wilner, S., & Winickoff, R. (1982). A study of lithium carbonate use in a health maintenance organization. *Quality Review Bulletin, 8*, 8–14.

Fink, R., Goldensohn, S., & Shapiro, S. (1969). Changes in family doctors' services for emotional disorders after addition of psychiatric treatment to a prepaid group practice program. *Medical Care, 7*, 209–224.

Follette, W., & Cummings, N. (1967). Psychiatric service and medical utilization in a prepaid health plan setting. *Medical Care, 5*, 25–35.

Follette, W., & Cummings, N. (1968). Psychiatric service and medical utilization in a prepaid health plan setting, part 2. *Medical Care, 6*, 31–41.

Fox, P., Goldbeck, W., Kiefhaber, A., Webber, A., Spies, J., Freidland, J., Tell, E., & Walsh, D. (1984). *Synthesis of private sector health care initiatives.* Prepaid for the Office of the Assistant Secretary for Planning and Evaluation, U.S. Department of Health and Human Services, by Lewin and Associates, Inc.; Washington Business Group on Health; Center for Industry and Health Care, Boston University.

Goldberg, I., Frantz, G., & Locke, B. (1970). Effect of a short-term psychiatric therapy benefit on the utilization of medical services in a prepaid group practice medical program. *Medical Care, 8*, 419–428.

Hankin, J., & Locke, B. (1983). Extent of depressive symptomatology among patients seeking care in a prepaid group practice. *Psychological Medicine, 13*(1), 121–129.

Hankin, J., Bendit, E., Koch, F., & Merrill, A. (1978). *Developing and testing of methodology to assess quality of mental health care* (Final Report for NIMH Contract No. 278-76-0090 [OP]). Johns Hopkins Medical Institutions, Baltimore, MD: Psychiatry Studies Program, Health Services Research and Development Center.

Hankin, J., Steinwachs, D., & Elkes, C. The impact of utilization of a copayment increase for ambulatory psychiatric care. *Medical Care, 18*(8), 807–815.

Hankin, J., Steinwachs, D., Regier, D., Burns, B., Goldberg, I., & Hoeper, E. (1982). Use of general medical care services by persons with mental disorders. *Archives of General Psychiatry, 39*, 225–231.

Hankin, J., Kessler, L., Goldberg, I., Steinwachs, D., & Starfield, B. (1983). A longitudinal study of offset in the use of nonpsychiatric services following specialized mental health care. *Medical Care, 121*(11), 1099–1111.

Inglehart, J. (1984). Health policy report: HMOs (for-profit and not-for-profit) on the move. *New England Journal of Medicine, 310*(18), 1203–1208.

Linn, M., & Linn, B. (1975). Narrowing the gap between medical and mental health evaluation. *Medical Care, 13*(7), 607–614.

Manning, W., Leibowitz, A., Goldberg, G., Rogers, W., & Newhouse, J. (1984). A controlled trial of the effect of a prepaid group practice on use of services. *New England Journal of Medicine, 310*(23), 1505–1510.

Martin, J. (1984). On the proposed sale of McLean Hospital: Final report of the Faculty Advisory Committee. *Harvard Medical Alumni Bulletin, 58*(1), 25–33.

Office of Health Maintenance Organizations. (1979). *Quality assurance strategy for HMOs.* Washington, DC: U.S. Department of Health, Education and Welfare, Public Health Service.

Peters, T., & Waterman, R. (1982). *In search of excellence: Lessons from America's best-run companies.* New York: Harper & Row.

Public Health Service. (PHS). (1979). Requirements for a health maintenance organization. *Federal Register, 44*(139), 42–70.

Spoerl, O. (1975). Single session psychotherapy. *Diseases of the Nervous System, 36*(6), 283–285.

Starr, P. (1982). *The social transformation of American medicine.* New York: Basic Books.

Vuori, H. (1980). Optimal and logical quality: Two neglected aspects of the quality of health services. *Medical Care, 18*, 975–990.

Wyszewianski, L., Wheeler, J., & Donabedian, A. (1982). Market oriented cost-containment strategies and quality of care. *Milbank Memorial Fund Quarterly, 60*(4), 518–550.

V

MENTAL HEALTH PROGRAMS

20

Administering a National Program of Mental Health Peer Review

SHARON A. SHUEMAN AND NORMAN R. PENNER

In 1977, the Department of Defense (DOD), through the Civilian Health and Medical Program of the Uniformed Services (CHAMPUS[1]), initiated contracts that were to result in revolutionary changes in both practice and attitudes within the nation's two major mental health professions. CHAMPUS contracted with the American Psychiatric Association and the American Psychological Association to develop and assist in the implementation of two essentially independent national programs to monitor the adequacy and necessity of mental health services delivered to beneficiaries of the CHAMPUS program. Called *peer review programs*, these efforts were intended to provide the CHAMPUS claims processors (referred to as *fiscal intermediaries*, or FIs) with the expertise necessary to decide which cases of psychiatric or psychological treatment warranted professional evaluation for adequacy and necessity, and with the mechanism to obtain review of these cases from consultant psychologists and psychiatrists. These consultants would review written treatment plans describing the care delivered by members of their own profession and give the FI written evaluations that could be used by FI personnel in deciding whether or not to pay for the treatment. Those affected by this program would be primarily independent practitioners treating CHAMPUS beneficiaries on a fee-for-service basis.

At the time of this writing, the authors were the directors of the two CHAMPUS peer review projects. The first author held a position for the American Psychological Association, and the second author had equivalent responsibilities for the American Psychiatric Association. We intend in this chapter to describe the tasks and responsibilities that, during the formative years, were part of the day-to-

---

[1]Beneficiaries of the CHAMPUS program include dependents of active duty military personnel, retired military personel and their dependents, and members of the commissioned corps of the National Oceanographic and Atmospheric Administration and the Public Health Service.

SHARON A. SHUEMAN • Shueman Troy and Associates, 246 North Orange Grove Boulevard, Pasadena, California 91103.   NORMAN R. PENNER • American Psychiatric Association, 1400 K Street, N.W., Washington, D.C. 20005.

day administration both of these projects and of the other review activities that developed as the idea of professionally sponsored peer review spread from CHAMPUS to the private insurance industry. Our second purpose is to help the reader understand how administration was complicated by our need to fulfill our legitimate responsibilitites to all constituent groups (providers of services, consumers, peer reviewers, and third parties) while responding effectively to the significant, often negative, reactions to the programs that arose within the professional communities.

The intensity of reactions by psychologists and psychiatrists was somewhat unanticipated. Consequently, the programs originally developed to meet the needs of these groups had to be expanded and new activities instituted as the population of those affected by peer review increased and their needs became more apparent. The responsibility for these programs fell in large part to the directors, and the subsequent demands taxed our resources (and frequently our patience) and became the single biggest challenge to the success of the peer review program.

There were several reasons for the intense reactions to the peer review programs. First, the programs were unprecedented in scope and process and required that therapists do things that had never previously been asked of them. Second, the programs affected a large number of previously unaffected, highly independent (Tryon, 1983) practitioners, many of whom were ignorant of the fact that there was such a thing as review, let alone a review program run by their own professional association. This ignorance remained long after program implementation despite the existence of organized professional activities aimed at educating the professional communities (Shueman & Troy, 1982). Third, the review programs were maintained by two professional associations, both of which had historically played the role of advocates for their members and now were perceived to be working against members while acting as agents of the third-party industry. Fourth, many psychologists and psychiatrists did not understand the principles of health insurance or third-party reimbursement, and peer review became an additional complication in a system about which service providers felt resentment and lack of trust.

In general, then, the peer review programs were a break from tradition that caught the professionals unaware. We will attempt to explain below just how these programs broke from tradition and why this break contributed to the consternation among mental health professionals. We begin with a brief history of the programs and a description of other national and local programs that preceded them.

HISTORICAL PERSPECTIVE

In 1974, in an effort to contain the rapidly escalating costs of mental health care under the CHAMPUS program, the Department of Defense announced the intent to impose a limit on the level of mental health services that would be reimbursed under that program. Until that time CHAMPUS had provided an unlimited mental health benefit that was somewhat unusual among third-party programs, which historically have had serious reservations about being able to control mental health costs without instituting administrative controls (Sharfstein, Muszynski, & Meyers,

1984). The proposed change would allow reimbursement for up to 60 outpatient sessions and 120 inpatient days per year. Before this directive went into effect, however, DOD and a coalition of mental health organizations negotiated an agreement to attempt to control costs through the use of a program of peer review involving the two associations (Rodriguez, 1983). The idea of peer review grew directly out of earlier DOD review activities, which were thought to have been successful in ensuring the adequacy of care and controlling the costs of treatment provided to CHAMPUS beneficiaries in child and adolescent residential-treatment centers (Asher, 1981).

It should be noted that, although the impetus for the peer review program was cost containment, the explicit goals of the program were both to control costs and to ensure that the treatment being paid would be of reasonable quality (Rodriguez, 1983). Had someone within DOD not been concerned about both cost and quality, there would have been little justification for instituting this somewhat complicated and relatively expensive peer review program. Costs could have been contained more simply and with more certainty by enforcing administrative controls.

The compromise agreement between DOD and the mental health coalition resulted in professional review requirements being written into the CHAMPUS regulation in 1977. The contracts between CHAMPUS and the two associations were signed in July of that same year. Over the next 18 months, the two associations independently developed written "screening" criteria (Donabedian, 1982) and procedures that the claims processors would use to select those cases that should be looked at more carefully by professional reviewers. They recruited and trained cadres of psychologists and psychiatrists who would serve as peer reviewers, and they developed procedures for monitoring the entire review process. Through their respective national advisory groups, the associations also provided professional consultation to CHAMPUS on various aspects of mental health policy.

Although the peer review program was ready for implementation by the fall of 1979, the activity was delayed until mid-1980. The delay was due partly to resistance within the fiscal intermediaries to implementing a program that they saw as being of no real benefit to them. They contracted with CHAMPUS only to process claims, and they believed that their existing mechanisms for determining the legitimacy of claims were sufficient and less costly and cumbersome than the proposed system. There was also the sentiment among the DOD personnel that the peer review program was akin to the "fox guarding the hen house."

While CHAMPUS delayed, Aetna Life & Casualty approached the two associations with a request to use a simplified version of the CHAMPUS review program for its own business. Aetna had already placed a limit on mental health care under the policy that it had written for the Federal Employees Health Benefit Program, and the company was considering imposing similar restrictions on other policies. William Guillette, an Aetna medical director and always a strong proponent of the mental health benefit, thought that peer review was preferable to setting administrative limits. Although initially reluctant to work with Aetna, both associations and CHAMPUS eventually agreed to a pilot peer-review program.

This first "commercial" review project, which was actually much simpler in terms of procedures than the CHAMPUS system, was the beginning of national peer review in organized psychology and psychiatry. Soon thereafter, CHAMPUS initiated review, and within another 2 years, many other insurance companies and Blue Cross and Blue Shield organizations had contracted with the associations for peer review services. The mental health professions found themselves in the "business" of peer review in response to requests from third-party payers.

As of this writing, the American Psychological Association administers the CHAMPUS project and also performs review for nine nongovernmental third-party payers. The American Psychiatric Association maintains review programs for 24 payers in addition to its CHAMPUS obligation.

REVIEW BEFORE CHAMPUS

The procedures developed under the CHAMPUS contracts are based on a process of individual *case review*, in which a psychologist or psychiatrist acting in an advisory capacity to a third-party payer evaluates a treatment plan written by the provider of services. The plan describes the problems for which the patient is being treated, the goals for therapy agreed on by the patient and the provider, and the progress being made in treatment. The review focuses on the necessity for treatment and the extent to which reasonable goals have been accomplished and are likely to be accomplished. The process has been referred to as *peer review* because the evaluation is performed by members of the provider's own profession (Sechrest & Hoffman, 1982).

Typical review before the CHAMPUS contracts did not involve peers, nor did it focus on the therapeutic process. For example, it was (and still is) very common for nurses to review treatment provided by psychiatrists by using written criteria defining "quality care." These criteria commonly relate to aspects of care such as the appropriateness of the length of hospitalization or of a certain medication for a particular diagnosis. Nurses applying criteria to evaluate treatment is only one example of the process of *utilization review*. Traditionally, utilization review has been performed principally in three contexts: for private third-party payers by employees of these organizations; under the Medicare and related programs by federally sanctioned review organizations; and for third-party payers, consumers, or providers by the mental health professions themselves through state or other locally based organizations.

Private Third-Party Review

Private third-party payers, including insurance companies and Blue Cross and Blue Shield organizations, frequently had employees or consultants (commonly internists or surgeons, but often nonprofessionals trained in claims review) who advised the company on the payment of mental health claims (Morton, 1982). There is much anecdotal evidence that such review activities did not generally result in any systematic limiting of payment, and it is likely that most mental-health professionals were unaware of the existence of such programs.

Professional Standards Review Organizations

The most notable review program was that of the professional standards review organization (PSRO). The national PSRO program, established by federal legislation in 1972 (Public Law 92-603), was designed to monitor the quality, the medical necessity, and the appropriateness of health services delivered under the Medicare, Medicaid, and Maternal and Child Health programs. PSROs were independent review groups operating outside the organization where care was delivered. They used professionally developed criteria (such as those described above) that defined various dimensions of acceptable care. Although intended eventually to encompass both inpatient and outpatient care, the PSRO focused almost exclusively on the former. It was also true, however, that the level of support for inpatient psychiatric care under the federal programs was so low as to be almost nonexistent, even though the American Psychiatric Association, in hopes of achieving a liberalization of this benefit, became the first medical specialty to cooperate with the intent of the PSRO legislation and to develop criteria for use by PSROs. (The association was hopeful that this obvious commitment to accountability would persuade federal rulemakers to expand psychiatric coverage.) There was no liberalization of the benefit. Hence, except for certain demonstration projects (Lassen, 1982), there was almost no review of mental health care under the PSRO program.

PSRC and PRC

The American Psychological Association and the American Psychiatric Association, since the mid-1970s, supported locally based organizations with the primary function of responding to third-party, consumer, or provider requests for review of mental health services or fees. For organized psychology, the system is that of the professional standards review committees (PSRCs) of the state psychological associations. For psychiatry, it is the system of peer review committees (PRCs) of the district branches. With a few notable exceptions (Chodoff, 1972; Wilson, 1982), underutilization of PSRCs and PRCs has resulted in the systems' having little impact on the professions.

Three factors help account for the underutilization of these review bodies. First, these groups are limited by the financial and other resources available to them and by the general lack of interest among professionals in serving on these committees. These limitations affect the committees' ability to be responsive in a timely manner and to publicize their availability to their target groups. Second, third-party payers, potentially the most frequent utilizers of PSRC and PRC services, have been hesitant to use local groups. Because insurance companies are generally multistate in operation, they would need to deal with up to 50 different review groups for each profession if they were to rely on the PSRC and PRC system. They would rather deal with one. Finally, passing judgment on the professional work of their colleagues is a task that many local groups embrace neither readily nor with enthusiasm. The sensitive nature of the task has often resulted in a review group's inability or unwillingness to articulate its scope, mission, and responsibility. These groups have sometimes been seen as being unwilling to make the "hard" decisions.

There have been other review programs, primarily locally based and operating within organized care settings, a few of which have been quite effective (Newman, 1974). By and large, however, the average psychologist or psychiatrist in independent practice would have had small probability of being directly affected by a review program before the implementation of the two national programs described here.

With this perspective, one can begin to understand why the CHAMPUS and related activities had such a great impact on the members of the two associations. Psychologists and psychiatrists, especially those not affiliated with an organized care setting, had been subject to relatively little monitoring. The private third party payment system conveyed to practitioners a *laissez-faire* attitude. The volume of mental health review under the PSRO program was minuscule, and most psychologists and psychiatrists remained ignorant of its existence. The locally based PSRC and PRC networks lacked the resources and often the commitment necessary to develop effective review mechanisms.

With the implementation of both programs, a large number of psychologists and psychiatrists who had previously been ignorant of and insulated from review became subject to the process. Any provider treating a patient who was covered by a third-party benefit program became a potential subject. For example, small sample surveys of psychologists and psychiatrists conducted in 1982 by the two project offices indicated that 60% and 70% of the respective group members had been asked to submit at least one report under the peer review program sponsored by the associations. Consequently, members of both professions started to scrutinize the programs for the first time.

HOW PEER REVIEW WAS PERCEIVED IN THE FIELD

What many practitioners perceived when they looked closely at the peer review program was their professional association turning against them by supporting an activity that appeared to be antithetical to many of the associations' previous advocacy efforts. They saw their professions as intruding in their practice and acting to limit their incomes. What others perceived, however, and what was promulgated as the official position of both associations, was that the mere existence of the review programs would help ensure maximum third-party coverage for the professions. For example, the American Psychological Association believed that psychologists would never be recognized as independent providers under Medicare or a national health-insurance program unless the profession supported and maintained a viable and visible program to demonstrate that it was willing to be accountable for the professional behavior of its members. Furthermore, it was clear that *someone* was going to make decisions about what was to be paid for, and CHAMPUS offered psychologists their first opportunity to become a legitimate party in the decision-making process (Willens & DeLeon, 1982). It gave them, in a sense, parity with organized psychiatry in this limited arena.

The American Psychiatric Association was concerned not about recognition, but about the market for psychiatric services. It believed that the only way to market its services successfully was to have review as a component of the "package" it was selling (Harrington, 1984). Both associations, then, were willing to support a system that limited reimbursement in individual cases if that system helped widen the market for an entire class of providers, and if they had a role in creating and maintaining the system. These arguments, however, were unpersuasive to rather large groups from both associations, and they did not prevent the controversy and resistance from creating an environment of special challenge for those involved in the administration of the program.

ADMINISTRATION OF REVIEW

We begin this section with a brief description of the process of peer review as it is seen from the administrative offices. As there are significant differences between the CHAMPUS system and that of commercial programs, we describe them separately.

CHAMPUS REVIEW

Using CHAMPUS guidelines, the FI personnel (called *Level II reviewers*) solicit routine treatment reports from psychologists and psychiatrists providing treatment to CHAMPUS beneficiaries. With the help of association-developed criteria, they review the reports and choose the cases that will be sent to peer review. (These decisions are based on factors such as the length of treatment, the frequency of sessions, and the adequacy of the documentation in the treatment report.) After choosing a case to be sent to peer review, the Level II reviewer consults one of the rosters of reviewers that have been provided to the FIs by the associations. She or he chooses three psychologists or three psychiatrists—depending on the discipline of the provider of services—and sends to those reviewers the case materials. Names and all other identifying information have been removed from these documents. The Level II reviewer also sends the appropriate project office copies of these documents and the names of the three reviewers. Each peer reviewer performs the review, returns his or her report and all case documents to the FI, and sends a copy of the review to the project office. The FI consults the three reports before making a benefit decision (i.e., a decision about whether or not to pay for the treatment). She or he sends a record of that decision to the project office. The project office pays the peer reviewer on the basis of his or her completed report and also uses the report as a basis for monitoring the reviewer's work. Selected data from all documents are entered into the computer-based management information system maintained in the project office.

COMMERCIAL REVIEW

Under the commercial review program, each insurance company has its own guidelines for soliciting treatment reports from providers. The company's medical director, using his or her own subjective criteria, decides which of the reports should be sent for peer review. Copies of the treatment reports for these cases (again, with all identifying information removed) are sent to the appropriate association peer-review office, and from there, they are distributed to three peer reviewers. The reviewers send their completed reviews and all case documents back to the association. Staff, in turn, send copies of the three reviewers' reports (with the peer reviewers' names removed) to the insurance company. As in the CHAMPUS system, the associations pay the peer reviewers and enter selected data into the management information system.

THE ELEMENTS OF ADMINISTRATION

The specific functions of the administrative offices that support the CHAMPUS and commercial review programs are presented below. For convenience, we have organized the elements with respect to the dimensions of *structure, process,* and *outcome* of review (Donabedian, 1982).

Structure

The structure consists of the policies, procedures, and documents that have been developed by the associations and that form the basis of the operations of the peer review program. This structure includes the criteria for defining exceptions to usual and customary care (i.e., for determining which cases should be sent to peer review); the reporting formats used by providers to report treatment plans, by peer reviewers to document reviews, and by third parties to report the results of the review process; manuals and other educational and informational materials intended for the use of reviewers, providers, patients, and third-party payers; policies and procedures guiding the actual review process; peer reviewer rosters and educational programs for reviewers; a computer-based management information system; and staff. During the early years of the peer review program, administrative time was devoted almost exclusively to the development of the structure. Currently, responsibilities for structure are significantly less time-consuming. The requirement is to maintain the structure while working with association governance groups, CHAMPUS, and other third-party payers to coordinate changes and revisions.

Process

Most of the staff time of the offices is devoted to the process of peer review. The responsibilities in this regard include monitoring the paper flow among third-party payers, peer reviewers, and the associations; monitoring the work of the peer reviewers and paying them for their reviews; monitoring the peer-review-related activities of the third parties; generating reports summarizing the performance of the peer reviewers, the third parties, and the associations; and providing information and consultation to providers, reviewers, and third parties, and the public. Our primary responsibility with regard to the process of review is to see that all parties conduct their peer-review activities according to the principles of the contracts (e.g., satisfying time lines and maintaining confidentiality) and consistent with the law and association policy.

Outcome: Administrative

As administrators, we need to be able to demonstrate that our offices are achieving certain standards of performance. We use our computerized management information systems to collect data, which are then used in the evaluation of our activities. The areas of focus include the performance of three groups: peer reviewers, third-party payers, and association staff. For example, the data gathered on peer reviewers include demographic and other characteristics; case volume and case turnaround time; patterns of recommendations and the extent to which they provide justification for their recommendations; and the amounts they are paid for review. These data form the basis for generating summary reports on reviewer performance, which are used by association governance groups and by CHAMPUS in their monitoring of the programs. The data are also the basis of routine feedback to reviewers about their performance. We also aggregate data on third parties' turnaround time, on their use of peer reviewers (i.e., whether they use proper procedures for the selection of reviewers for specific cases), and on their level of agreement with peer-reviewer recommendations. We also collect information on our own reviewer use, on our turnaround times for case processing and for payment of reviewers, and on the frequency of our feedback to reviewers.

This ongoing accountability process provides critical information about our primary concern: the effectiveness of our administration. The data allow us to assess the outcome of our efforts. There is a second aspect of outcome that, although subsumed under administration, significantly affects the directors' abilities to be effective proponents of the peer review system. We refer to the impact of peer review on the therapeutic relationship and to the related questions of the reliability and validity of the document-based process.

Outcome: Impact, Reliability, and Validity

The relative lack of data on outcome (when compared to the data available on the administrative process) has been significant to us to the extent that it has affected our ability to appear to be responsive to inquiries and criticisms, and to provide a rationale for the review process that has been chosen by the associations. Consequently, we have initiated investigations of outcome through our offices and have encouraged association governance to do likewise, and we have supported external studies to the extent that our resources have allowed.

We have been most successful in investigating certain questions of impact. We have done this via surveys of practitioners who have been subject to the peer review program. We have also conducted approximately a dozen small-sample surveys of peer reviewers and practitioners regarding their attitudes about the peer review process and about related professional issues. Information on the effect of the process on the treatment plan or on the therapeutic relationship remains almost exclusively anecdotal, however, primarily because the relevant patient data need to through the cooperation of providers and third-party payers. Both groups have been reluctant to cooperate in gathering such data.

We have been able to conduct or assist in conducting a number of investigations of the reliability of the peer-reviewer decision-making process (Cohen & Pizzirusso, 1982; Dall & Claiborn, 1982). Cohen and his colleagues, for example, are responsibale for a rather comprehensive program of research focusing on the effects of the theoretical orientation of reviewer and provider on the recommendations and judgments made by reviewers.

Significant unanswered questions about document-based review in mental health relate to the validity of the process, that is, the extent to which the treatment report accurately describes what is going on in therapy. Despite a long tradition of this type of review in medical care, data are scarce. Our lack of success in answering these questions relates both to the difficulty and expense involved in gathering data and to the providers' fear of disrupting the therapeutic process if, for example, the patient were asked questions about the process or if an independent observer were asked to make an assessment of the treatment.

Although third-party payers have all cooperated in limited ways with survey research conducted by the two associations, their attitude has been that they are buying the service of peer review and have little interest in answering what they perceive to be scientific questions. Furthermore, they have raised serious doubts about the ethics of their participating in any studies if the patient would need to be involved directly. The responsibility, therefore, has fallen to the associations, and to date, neither has made an active commitment to expanding the scope of investigations of the peer review process.

EDUCATION FOR PEER REVIEW

From the beginning, the associations were committed to the principle that the peer review systems would be considered developmental for a significant period of time following implementation. It was anticipated that, because the programs were significantly different from previous efforts, many difficulties would need to be resolved throughout the early years. It was also acknowledged that making changes in such a large-scale system, involving as it did many complex organizations (professional association, third-party payers, and the federal government), would necessarily take a long time. In addition, as practitioners lacked knowledge about utilization review in general and this innovative review program in particular, the association realized that education would be an important part of program implementation.

The directors knew, then, that the primary responsibilities of our offices would be to coordinate and facilitate change that would inevitably occur slowly, at the same time developing and coordinating educational programs focusing on the review process, its rationale, and the political realities. Little did we realize how slow the process of change would be as the controversy increased and the associations found themselves in a defensive posture. Little did we realize, also, how the educational needs would increase, given the unexpectedly slow pace of developing and implementing change, and given what we were to discover about the average practitioner's ignorance of both peer review and the third-party system. Indeed, the educational task would soon appear almost impossible and would become one that severely taxed the resources originally thought to be adequate to accomplish it.

IN THE EYE OF THE STORM

The administrators of the peer review program were in the unique position of being able to recognize the problems that either resulted from actual implementation of the system or that had been caused by errors in judgment made during early developmental phases of the programs. We were able to detect the "bugs"—those aspects of the process that did not work as planned and hence caused problems. For example, the review process (with the exception of the review of traditional psychoanalysis) was designed to be "orientation-free." That is, problems, progress, and goals were to be expressed in terms of the patient's functioning, an approach minimizing the need for the treatment plan to be written in the terminology of the provider's particular theoretical orientation. More important, this approach was to obviate the need to assign reviewers to providers on the basis of orientation. (An explicit assumption was that there is a universal language of behavior that could be understood by all experienced therapists.) Peer reviewers were also to adopt an orientation-free perspective in peforming their reviews. Difficulties arose when some peer reviewers demonstrated an inability to perform without injecting their own orientation biases into their comments. This was felt to be a greater problem for psychology, and that association instituted training programs for reviewers in an attempt to eliminate those reviewers' tendencies. In later years, the APA also stimulated a move to "match" provider and reviewer on the basis of orientation.

We were also able to detect those procedures that *did* work as planned and, in doing so, caused problems that had been unforeseen. For example, CHAMPUS had given us instructions not to communicate directly with the claims processors when problems arose. Because the FIs were responsible as contractors to the government, and not to the associations, all communication was to go through CHAMPUS. We soon discovered that, although the communications were conveyed reasonably well and CHAMPUS became aware of all the problems, the problems were being intensified by the slowness of the process. With this knowledge and with CHAMPUS's tacit approval, we started shortcutting the system, developing direct lines of communication with FI personnel.

The administrators were also in a position to diagnose the mood of the practitioner and to determine that not all their problems and unhappiness were related to the peer review system (even though many of them seemed to attribute all third-party problems to peer review). It was our responsibility to respond to all letters and calls from providers and patients. It soon became apparent that our role would be not only to help resolve the immediate problems of peer review and to convey to association governance the difficulties that practitioners were having with the system, but also to help our professions to develop increased sophistication about the practices and principles of third-party payment and to deal effectively with these systems.

In our attempts to educate providers in this way, we found ourselves in the unique position of being potentially caught in the middle among all constituent groups. While attempting to explain third-party policies to psychologists, psychiatrists, and patients, we were in danger of being perceived as apologists for insurance companies. In fact, what we tried to do was to give the providers the information they would need to do three things: stay within the law and professional ethical principles (e.g., by explaining the legal consequences of submitting inaccurate diagnoses, billing for missed appointments, or routinely forgiving copayments); know their and their patients' rights with respect to the policy; and be appropriately assertive with the third party while not alienating the personnel with whom they had to communicate.

Finally, the directors were in a position to observe how the implementation of the review process caused renewed controversy with respect to some long-standing intraprofessional differences. Many traditionally long-term therapists, for example, perceived the review program as a mandate for them to adopt shorter term strategies. Similarly, some psychiatrists with antimedication preferences felt pressure to adopt what they perceived to be palliative "quick cures" of medication. Perhaps because of the reimbursement factor, the peer review program, like few other professional movements, raised practitioners' level of anxiety about these and similar issues. Peer review was seen as a serious threat to professional autonomy. Whether or not the threat was real, the controversy was intensified by the existence of these long-standing and strongly held differences among practitioners in both professions.

CONCLUSION

It is, of course, true that implementing peer review in the context of an already-complicated and little-understood third-party system has made life more difficult for all parties, especially psychiatrists and psychologists providing mental health services. We acknowledge also that the current program of review has its imperfections. We are well aware, however, that pressures for accountability will only increase in the foreseeable future as solutions are sought to problems in financing health care, and that health care professions will have to live, if not with peer review, then with possibly less desirable alternate strategies for allocating resources. Through efforts such as the peer review programs, the American Psychiatric Association and the American Psychological Association are attempting to stay in the forefront of the groups dealing with these problems and to ensure their roles as participants in structuring solutions. Neither group believes that it has the luxury of withdrawing from the arena, lest a totally unacceptable strategy be chosen. In this case, it may be true that "The devil we know is better than the devil we don't know."

In our role as administrators of the mental-health peer-review programs, and within the limits of our responsibilities, we have endeavored to make the system work as well as it can and, at the same time, to help our professions learn to cope with peer review and with the larger context of health care financing. In the tradition of the mental health professions, we attempt to be responsive to the needs of our constituent groups and at the same time to give psychologists and psychiatrists the knowledge that they need to deal with reality, which in this case is the peer review program.

REFERENCES

Asher, J. (1981). *Assuring quality mental health services: The CHAMPUS experience.* (ADM81-1099). Washington, DC: U.S. Government Printing Office.

Chodoff, P. (1972). The effect of third party payment on the practice of psychotherapy. *American Journal of Psychiatry, 129,* 540–545.

Cohen, L., & Pizzirusso, D. (1982). Peer review of psychodynamic psychotherapy: Experimental studies of the APA/CHAMPUS program. *Evaluation and the Health Professions, 5,* 415–436.

Dall, O., & Claiborn, W. L. (1982). An evaluation of the Aetna pilot peer review project. *Psychotherapy: Theory, research, and practice, 19,* 3–8.

Donabedian, A. (1982). *Criteria and standards of quality. Quality assessment and monitoring* (Vol. 2). Ann Arbor, MI: Health Administration Press.

Harrington, B. S. (1984). Psychiatric practice must change to survive floundering economy. *Psychiatric News, 19*(6), 1, 22–23.

Lassen, C. L. (1982). The Colorado Medicare Study: Perspectives of the peer review committees. *Professional Psychology, 13,* 105–111.

Morton, S. I. (1982). Peer review: A view from within. *Professional Psychology, 13,* 141–144.

Newman, D. E. (1974). Peer review: A California model. *Psychiatric Annals, 4,* 75–85.

Rodriguez, A. R. (1983). Psychological and psychiatric peer review at CHAMPUS. *American Psychologist, 38,* 941–947.

Sechrest, L., & Hoffman, P. O. (1982). Philosophical underpinnings of peer review. *Professional Psychology, 13,* 14–18.

Sharfstein, S. S., Muszynski, S., & Meyers, E. (1984). *Health insurance and psychiatric care: Update and appraisal.* Washington, DC: American Psychiatric Press.

Shueman, S. A., & Troy, W. G. (1982). Education for peer review. *Professional Psychology, 13,* 14–18.

Tryon, G. S. (1983). Pleasures and displeasures of full-time private practice. *The Clinical Psychologist, 36*(4), 45–48.

Willens, J., & DeLeon, P. H. (1982). Political aspects of peer review. *Professional Psychology, 13,* 23–26.

Wilson, S. (1982). Peer review in California: Summary findings in forty cases. *Professional Psychology, 13,* 517–521.

21

Peninsula Hospital Community Mental Health Center Quality Assurance System

LORRAINE L. LUFT AND DONALD E. NEWMAN

INTRODUCTION

The quality assurance system at Peninsula Hospital Community Mental Health Center has evolved over time in response to the particular characteristics and requirements, both internal and external, of this organization. Thus, the current mix of the components that comprise the quality assurance system has been added at different times to satisfy different needs. In quality assurance, as in service delivery, the administrative philosophy of the center includes ongoing evaluation and monitoring to determine how adequately each program is functioning, as well as to identify unmet needs. It is important to remember, therefore, that the range of quality assurance activities described in this chapter was not planned or implemented in one thrust; as the center has grown, as the financial climate has changed, and as external requirements have multiplied, the center's quality assurance program has grown.

STRUCTURE OF THE CENTER

The model of quality assurance at Peninsula developed out of its unique history and its private practice model of service delivery. Because a full account of the early history of the center is documented elsewhere (Vaughan, Newman, Levy, & Marty, 1973), a brief summary here will suffice. In 1964, the hospital opened an inpatient mental-health unit, the psychiatrists on the hospital's professional staff began to meet as a separate section of the department of medicine, and federal funding became available for community mental health centers. By 1969, a comprehensive center had been established, and the professional staff had organized into a department of psychiatry, which allowed affiliate membership for licensed clinical social workers and psychologists in private practice.

LORRAINE L. LUFT AND DONALD E. NEWMAN • Peninsula Hospital Community Mental Health Center, 1783 El Camino Real, Burlingame, California 94010.

The center provides a full range of mental health services to its four-city suburban catchment area of 95,000 people. For the majority of center patients, who are unable to afford the full costs of treatment, funding is available under a contract from San Mateo County, which designates Peninsula responsible for one of the county's four regions. Each patient is treated by a privately practicing psychiatrist, psychologist, or social worker who maintains primary therapeutic responsibility even when the patient is hospitalized or participates in the various partial-care programs staffed by Mental Health Center employees. The center has a salaried psychiatrist as its director and a staff of about 85 (full-time equivalent) employees who run the center's inpatient units; partial care, crisis, intake and referral, and child and adolescent programs; community services; program evaluation; in-service education; and administrative services. Staff members are carefully screened—by the hospital's personnel department and the team with which they will work, as well as by their supervisor—before being hired. Written performance evaluations are completed and discussed with each employee annually, and if there are problems with an individual, more frequent counseling and documentation are provided.

DEPARTMENT OF PSYCHIATRY STRUCTURE

The 83-member attending staff who comprise the interdisciplinary department of psychiatry annually elect its chairperson, who then appoints members to chair various committees, such as proctorship, audit, continuing education, and utilization review. An advisory committee is composed of all officers of the department and all chairpersons of standing committees. The responsibility for quality assurance is shared by the department of psychiatry and the center staff, just as treatment is provided with collaboration between the two groups. With varying degrees of input from the center staff as appropriate, the department of psychiatry maintains quality assurance functions through credentials review for membership, monitoring of privileges through review and proctorship, clinical audits, review of suicides and complications, three peer-review committees, and recently an experimental high-user review committee.

CREDENTIALS REVIEW

As part of the application process to join the staff of Peninsula Hospital, the hospital bylaws require a review of the education and training of the applicant. As a first step, the department of psychiatry requires licensure and/or graduation from an approved school and an interview to determine the therapist's strengths and weaknesses. All of the qualified professionals must comply with the procedures for privileges and proctorship before they are allowed therapeutic responsibility for any cases.

PRIVILEGES AND PROCTORSHIP

Each member of the department of psychiatry must request specific therapeutic privileges: inpatient, partial hospitalization (adult, adolescent, and child), electroshock therapy, and special somatic therapies; outpatient individual (adult, adolescent, and child), group (adult, adolescent, and child), conjoint family, and couple therapy, and biofeedback; mental health consultation (adult, adolescent, and child); and psychological testing (adult, adolescent, and child). Each privilege has its own training requirements, and therapists must document their training and expertise before that privilege can be *provisionally* granted and proctors assigned.

In contrast to supervision or consultation, proctorship is the monitoring and evaluation of the applicant's clinical work by two members of the department. Detailed guidelines and evaluation forms are provided in the department of psychiatry proctorship policies, which clarify the requirements and provide a format for the proctors' reports. Proctorship must be of sufficient depth and extent so that the proctor has an opportunity to evaluate the skills of the applicant in treating several patients; typically, it lasts over a year and may even be extended beyond 2 years. On the basis of information provided by the proctors, the three-member proctorship committee recommends to the advisory committee that the applicant be granted the privilege without proctorship, that proctorship be continued, or that the privilege be cancelled and the proctorship terminated. As part of its recommendations, the proctorship committee may recommend further training, therapy, and/or analysis for the applicant. Even after all of these requirements are met, a therapist may be called into question because privileges must be renewed annually. Privileges can be revoked, and for serious difficulties or infractions of professional integrity, a therapist may be asked to leave the department.

Thus, before a patient is assigned a therapist for treatment, that therapist must demonstrate training, experience, and ability in the specific therapeutic modality. Thus, the first assurance of quality of service rests with the privileges and proctorship procedures of the department of psychiatry. The committee maintains its own minutes and individual files on each department member, and all reports remain confidential.

Audit Committee

This committee is composed of representatives of professionals from the department of psychiatry (with one appointed chairperson), Community Mental Health Center nursing and administrative staff, and a member of the hospital's medical records department. The committee, with the aid of program evaluation staff, generally conducts three or four audits each year. First, a consensus is reached by the committee members on a topic for review. This is usually based on identified management, procedural, or placement problems; a desire to confirm that appropriate care is being delivered; or a need to establish accepted levels of practice in specific situations. An audit may be targeted to examine particular diagnostic categories, nursing practices, management issues, or outcomes of care. The committee routinely reviews all incident reports to identify problems that may be beneficial to audit. Topics that have recently been audited include clinical and administrative problems in the discharge process, medication errors and their connection with a new medication-reporting system, profiles of violence-prone patients, the incidence of physical injuries (particularly falls and athletic accidents) and their relationship to particular medications, and an elopement risk study. Audit results are presented to the full department of psychiatry at its monthly meetings and are also disseminated as appropriate to Mental Health Center staff for corrective action. In conjunction with audit activities, educational programs have been established for nursing staff on proper clinical record-keeping.

SUICIDE AND COMPLICATIONS REVIEW COMMITTEE

This committee is composed of five members of the department and head nurses from the center's relevant treatment units. Another mechanism for quality control, this committee meets monthly to review all suicides, attempted suicides, suicidal gestures, and other complications, such as elopement and assaultive behavior. Whereas the audit committee gathers data and prepares reports and recommendations on substantial numbers of such cases, this committee reviews individual cases in depth with the attending therapist present for a more educative approach. An example of a case recently reviewed involved a lengthy hospitalization with some suicidal gesturing during that period. The committee identified a need for more limit-setting by the inpatient staff and the attending therapist.

PEER REVIEW COMMITTEES

Since 1971, the center has had an active peer review system that authorizes all treatment of publicly funded patients. There are four peer review committees meeting weekly: the adult outpatient committee, the child outpatient committee, and two inpatient–partial-care committees. These committees meet with the treating therapist to work out a treatment plan appropriate to the patient's history, diagnosis, and goals for treatment. Such an authorization specifies the particular mode of treatment and length of time the patient will receive it during a given episode of care.

Each of the outpatient review committees is made up of three professionals from the department of psychiatry. Two are appointed by the chairperson of the department, and the third, who is both a member of the department and an employee of the Mental Health Center, is appointed by the center director. The reviewers assigned to the child outpatient committee are all specialists in child and adolescent treatment. The two inpatient–partial-care committees have identical functions but meet at different times of the week, so that inpatients can be reviewed within no more than 4 days of their admission to the hospital. Two attending professionals are appointed to each of these committees by the chairperson of the department of psychiatry; the center's representatives are the head nurse and the psychiatrist director of the team to which the patient is assigned.

The outpatient committees meet weekly for 2 hours, and the inpatient–partial-care committees each meet for 1½ hours weekly. Each case discussion is generally limited to 30 minutes. When more than one professional is providing treatment to a family, all the therapists involved are asked to participate when a family member's case is reviewed to ensure that treatment will not be fragmented. The joint meeting also allows for a recommendation of family therapy with the consensus of the therapists working with various members of the family.

The criteria for selecting cases for review are explicitly enumerated for each type of care. For outpatients, six visits are initially allowed for diagnostic evaluation, and then, if the patient is to continue in treatment, the case must be presented to the appropriate review committee for authorization. Inpatients are to be scheduled for review within 48 hours of admission; if this is impossible because the peer review schedule is completely booked, the earliest available time is scheduled, and an interim review is done by the unit director. For partial-care patients, the scheduling of the review depends on the intensity of the particular program. Adult and adolescent patients in their respective hospital-based day-care programs must be reviewed after four months in the program; patients in the community day-care or activity group-therapy programs must be reviewed after 6 months; and those in the community activity program (chronic patients) are reviewed after 1 year. Patients who have been presented to any of the review committees and who require treatment beyond that authorized at the first review must be presented again to the appropriate committee in order to continue therapy.

The objectives of each peer review are to address each of these issues: (a) the appropriateness of the admission; (b) the appropriateness of the treatment plan, that is, services and medication; (c) the length of stay; (d) discharge plans; and (e) the authorized treatment. The reviewers are provided with each patient's clinical chart, minutes from any previous reviews of the case, and a summary record of service utilization and the medications prescribed. The peer review process essentially follows a case review format with a brief presentation by the treating therapist explaining the rationale for the current treatment plan and then open discussion between the review committee members and the therapist. Minutes are dictated on each case that document a summary of the diagnostic evaluation and focus primarily on goals, direction, treatment modality, and duration. The minutes are treated as confidential material and are filed separately from the patient's medical record. No copies are distributed, but the treatment therapist, the program, and the business office are each sent a memorandum outlining the authorized treatment.

No therapist is paid for treatment unless there is peer review authorization for the services, so that the authority of the peer review committees is backed up by financial control. The review process, as it was initially established in 1971, was primarily a peer utilization procedure. With fixed limits on the county contract to care for publicly funded patients, the need to carefully ration the funds available for treatment in all of the modalities became clear in the 1st year of operation with the private practice model (Levy, 1974; Newman, 1974). However, over the years, the value of the peer review system as a quality assurance mechanism became obvious. It became clear that quality review could not be separated from utilization review, and now both aspects are an integral part of the system.

Although maintaining the authority for allocating treatment resources, the actual committee review resembles a clinical case conference, which fosters education more than control (Newman & Luft, 1974). In fact, an evaluation study confirmed that most therapists felt they received helpful consultation from the peer review committee, and more than 90% found peer review to be an educational experience that promoted acceptance of the peer review process (Luft & Newman, 1977; Luft, Sampson, & Newman, 1976).

EXPERIMENTAL HIGH-USER COMMITTEE

After reviewing thousands of cases with the peer review system described above, the center realized that this approach did not allow for a longitudinal review of the care of some of the most long-term and difficult patients, who utilize a disproportionate share of the center's services and budget. Clearly, with the fixed budgets for treating low-income patients, the center has strong financial incentives to provide the most cost-effective care possible to its patients. Out of these concerns, the center added the special experimental high-user peer-review committee to deal with the high-utilizing population.

Recognizing that the task of this committee was a particularly difficult one, with much material to review, the committee members decided to schedule the cases for a full hour's review instead of the usual half hour scheduled by the other peer review committees. This was a face-to-face clinical review with the therapist that focused on finding new treatment approaches that would be more helpful to the patient and that would also hopefully reduce utilization of services that were not beneficial to the patient. The reviewers were instructed to dictate minutes of each review, which included both the presenting therapist's formulation of the case and the reviewers' critique.

From the start of the high-user committee in 1978, the center was also interested in studying its impact. A pilot study comparing the first group of patients reviewed by the committee with the highest users during a previous period suggested that the committee could substantially reduce the utilization of services. Later, a randomized controlled study was used to evaluate the impact of this peer review on high-cost community-mental-health patients (Luft, 1983). The subjects were low-income, publicly funded patients who used at least 16 days of inpatient care within 1 year and had total billings of at least $5,000 for the same period. Fifteen pairs of subjects were selected by matching the number of inpatient days, total billings, and major diagnosis.

One subject of each pair was randomly assigned to an experimental group whose cases were reviewed by the high-user peer-review committee, and the other was assigned to a control group whose cases were not reviewed. Patients, therapists, and reviewers were blind to the existence of the study, which was conducted by the center's program evaluation staff. The aim of the study was to compare the use of services, the expenditures, and the clinical outcomes for the experimental group after high-user peer review with the comparable data for the control group.

Contrary to hypotheses and the pilot study, compared to the control group the experimental group did not have greater reductions in use or billings during 6 months following review. Overall, both groups experienced reductions of 60%–65% in billings, representing reductions of about $3,500 per patient over 6 months. The clinical outcomes, as measured by therapists' ratings on the Brief Hopkins Psychiatric Rating Scale (Derogatis, 1977) and patients' self-ratings on the Profile of Adaptation to Life–Clinical Form (1978), were at least as good for the experimental group and perhaps somewhat better than for the control group on level of functioning 6 months after the review.

The difference in findings between this evaluation study and the earlier pilot may be partially attributed to differences in study design between the controlled and uncontrolled studies. There were also differences in the functioning of the committee during the two periods, which may explain the divergent results. Instead of one consistent committee, during the latter study there were six teams, each with two reviewers, perhaps resulting in a more variable and less standard intervention. In addition, the committee function in the latter period was described as an exploratory endeavor, and instructions to committee reviewers and presenting therapists did not mention the specific goal of reducing utilization. The exploratory nature of the committee, rather than a limited-setting approach, was also reflected in the recommendations made by the committee: in more than half of the cases reviewed, the committee recommended no changes in treatment. When the pilot study was conducted in 1977, the review committee had just been implemented, and there may have been different expectations on the part of those presenting cases to the committee as well as those serving as reviewers. Engaged in a new endeavor, all of the participants may have been more zealous in their efforts and may have had higher expectations of the committee work, perhaps resulting in a more powerful intervention than was true in subsequent years.

Nonetheless, in reading the proceedings of this committee, it is abundantly clear that these were indeed very challenging patients with chronic and severe disturbances. Even while often offering the therapist no clear-cut alterations in treatment plans, it appeared that the reviewers did provide support and validation to the therapists working with such difficult cases. The results and conclusions from this evaluation study were presented to the cochairpersons of the high-user committee, as well as to the chairperson of the department of psychiatry, so that they could use these findings in addressing whether to continue, modify, or disband this experimental committee. Currently, the committee has discontinued reviewing cases, pending further study and discussion.

PROGRAM EVALUATION

As just illustrated, the program evaluation unit of the Mental Health Center also plays an integral role in quality assurance activities. The objectives of this unit are to identify and measure community mental health needs; to provide process evaluations, including the collection and integration of utilization data on clients, staff, services, and funding; to conduct outcome evaluations that assess the impact and cost-effectiveness of the direct and indirect services provided; and to respond to other requests from the center and hospital administrations, the department of psychiatry, the citizens' advisory board, the catchment area residents, and the external funding sources (e.g., county, state and federal) regarding accountability, program planning, and measures of effort, effectiveness, adequacy, efficiency, and the comparative value of program operations.

The director of program evaluation reports directly to the center director and is an active member of the administrative team of the center. She participates in regular center administrative meetings and management decision-making, both providing information from evaluation work completed and receiving suggestions for future evaluation projects. The other evaluation staff (a program evaluation analyst and half-time statistical clerk), working under the supervision of the director, assist on both ongoing types of program monitoring and special outcome studies.

The evaluation unit is responsible for monitoring the peer review activities, as well as the utilization of services. Thus, quarterly reports are prepared that summarize each committee's activities, including the number of cases reviewed and the amount of treatment authorized. These summaries are given to the center director, who then reports the results to the peer reviewers and to the full department of psychiatry at their monthly meetings, so that all the practitioners can be aware of the trends in authorization and treatment patterns and their impact on the center's community budgets.

This type of information is used in conjunction with other utilization and budgetary data to provide for ongoing monitoring of the public budgets. Such careful review allows for funds to be spent prudently, obviating a need to restrict care and thereby potentially to compromise the quality of treatment at the end of each contract year. Each month, the program evaluation staff uses revenue and expenditure data to project levels of utilization by service mode for the remainder of the year. This projection enables management to make rational decisions concerning the allocation of center resources and staff to the various service modes.

The program evaluation unit also conducts clinical care studies that are reported to the center administration and the department of psychiatry. In addition to the previously cited evaluation studies pertaining to the peer review system, the center has published reports of therapists', patients', and inpatient staff's views of treatment modes and outcomes (Luft, Smith, & Kace, 1978) and a comparative cost-effectiveness study of hospital-based and community-based partial-care programs (Luft & Fakhouri, 1979). Other studies of our clinical programs have included evaluation of the various treatment environments using the Ward Atmosphere Scale (Moos, 1974b) for the inpatient units and the Community-Oriented Programs Environment Scale (Moos, 1974a) for the partial care programs. In addition, patient evaluation questionnaires are routinely completed at termination by all adult patients and the parents of children in treatment regarding satisfaction with the treatment received and the acceptability of services, as well as changes in the patients' symptoms and levels of functioning. Similarly, client satisfaction questionnaires are completed by participants in all community services programs, classes, and parent discussion groups to indicate the quality and effectiveness of indirect services, and to provide vital input concerning community mental-health needs. The staff then uses this information in formulating goals and priorities for the coming year.

FEEDBACK BETWEEN CENTER AND DEPARTMENT OF PSYCHIATRY

The assurance of the quality of care in this setting can be achieved only with an ongoing, comprehensive dialogue between the center staff and the department of psychiatry. As is probably already evident, such feedback takes place through informal dialogue at many levels and through various formal structures, both providing for a system of checks and balances. The center director and the chairperson of the department meet weekly over lunch in order to keep communication current. Similarly, the directors of the various inpatient and partial care programs, as well as the rest of the Mental Health Center staff, are in constant touch with the therapists privately treating their patients, so that treatment can be coordinated and the possibility of conflict or confusion can be reduced.

Of course, problems arise, and each group can help serve in a watchful capacity with the other group, creating a system of checks and balances. When there might be a problem with a staff nurse, for example, often it will be one of the attending therapists who will bring it up for discussion; likewise, if one of the attending therapists is having difficulty with a case or in getting along with the staff, it may well be one of the paraprofessional mental-health workers or a nurse who brings it to the attention of the director. In contrast to other settings, where close staff loyalties might preclude problems' coming to the surface, this system of using outside attendings provides a useful tool that frequently prevents blind spots and countertransference problems.

As already mentioned, there are close ties between the program evaluation staff and the department of psychiatry in quality assurance activities. In addition to monitoring the peer review activities of the department, the evaluation staff routinely communicate the results of all relevant work to the department, as well as to the full center staff, again so that coordinated changes may be made to enhance the quality of care. On occasion, members of the department also initiate studies, and program evaluation staff assist or collaborate on study design, data collection, analysis, or interpretation of results.

Center staff and department members also collaborate to improve the quality of care through formal committee structures. As described earlier, center representatives participate in peer review, suicide and complication review, and audit committees. Similarly, center staff join department members on program committees, such as the child and adolescent committee, which oversee existing programs and help to develop new ones. For example, the child and adolescent committee recently began an exploration of the possibility of opening an adolescent inpatient unit. In the course of such discussion, the directors of both adolescent services and adult inpatient services were intimately involved in discussion with the private attendings, and the program evaluation staff provided data on the utilization of the adult inpatient units by adolescents. In this way, more informed decisions can be made to ensure appropriate, quality care.

EXTERNAL REQUIREMENTS

Most of the quality assurance elements described above were developed to meet the internal needs of the Mental Health Center with its use of private practitioners and limited county contracts. As noted earlier, the peer review system, which serves as the core, was implemented in 1971, before many mental-health facilities were concerned about quality assurance. Since that time, however, there has been a proliferation of external requirements relating to utilization review and/ or quality assurance. In fact, most of the external pressures have more to do with limiting utilization rather than necessarily improving quality because much of the motivation has been generated by fiscal constraints.

In particular, the utilization review requirements of the state of California have had a major impact on quality assurance proceedings at Peninsula. In addition to the formal peer-review system described above, the center has added a number of in-house utilization-review components in compliance with the state requirements. For inpatients, the center conducts a separate review on the appropriateness of admission. At the time of admission, the head nurse of the unit screens the patient for one of the following criteria: evidence of danger to self, others, or property; evidence of seriously disordered behavior and/or impaired reality testing; or evidence of a need for planned medication evaluation or special treatment requiring hospitalization. If the admission is appropriate, the head nurse authorizes admission and a stated length of stay. This initial length of stay may not exceed predetermined guidelines by diagnosis. If the head nurse reviewer has a question about the appropriateness of the admission, she or he confers with the unit director. This psychiatrist then reviews the admission and discusses the case with the attending therapist if there is a question about the appropriateness of admission. If the attending therapist chooses to appeal the unit director's decision regarding the authorized stay, she or he may refer the review to the peer-review-committee chairperson if the concern is a clinical one, or to the center director if the concern is administrative, for a final decision.

Similarly, partial care patients are scheduled for review by the inhouse interdisciplinary review team before the 12th visit to the program. Authorizations may not exceed 24 program visits. Patients considered appropriate for partial hospitalization treatment are those patients who would otherwise need acute hospitalization if partial care services were not available. In addition, the peer review committee hears all cases before 4 months of treatment have elapsed and every 6 months thereafter, as well as any cases of particular concern referred by the inhouse review team or the unit director.

Partial care programs for the chronically ill patients who would need long-term institutionalization if the partial care program were not available have a different review system. These cases are reviewed by an in-house interdisciplinary team before the first 30 calendar days of the treatment program have elapsed, and the team may authorize treatment only up to 90 calendar days at a time. This in-house team also refers cases of concern to the peer review committee, which routinely reviews community day-care patients at least once every 6 months and community activity-program patients at least once every 12 months.

For outpatients, peer review was already required within six initial visits; however, the new MediCal procedures require a utilization review of the therapist's progress notes before every 16th visit. These progress notes document the same types of information submitted to peer review: diagnosis (if different), clinical symptomatology, treatment plan, therapeutic goals, prognosis, and any significant change in drug regimen.

For all state-funded patients, there are also requirements for medication monitoring. At each peer-review session, there is now a medication-monitoring check-list with the following items: appropriateness of medication(s) and dosage(s), evidence of patient's informed consent, critical adverse reaction, adherence to treatment regimen, patient's ability to manage own medication(s), and recommendations for any correction needed. Program evaluation staff prepare statistical summaries of these checklists and submit the quarterly reports required by the state. Quarterly reports are also submitted to the audit committee for review and recommendation of education programs if appropriate.

DISCUSSION

In order to meet the new state requirements, Peninsula had to add a number of additional reviews to the existing system. How much these new components contribute to the assurance of quality care is open to question. What is clear is that these new requirements add considerably to the paperwork, bureacracy, and cost involved in the care of patients. The treatment of the publicly funded patient, with so many checkpoints and requirements for documentation, thus becomes a more onerous and stressful task for the caregivers. These requirements increase the staff's level of frustration and tension and thereby diminish the professional quality of work. Furthermore, the veritable explosion of paperwork has necessitated the hiring of "paper people," which further increases the cost of patient care.

The relationship between the quantity and quality of documentation and the quality of care remains to be proved. Although the therapists find the face-to-face peer-review session generally educative and helpful, they resent the burgeoning paperwork that, at best, seems neutral vis-à-vis quality and, at worst, detracts from the quality of treatment. Any quality-assurance system that moves away from an examination of the clinical issues inherent in the individual case and toward forced conformity to statistical norms or arbitrary limits does not have quality as its primary goal. It appears that the various funding sources often impose their limits under the guise of "quality assurance," and this development ironically adds to the costs while diminishing the quality of care.

Clinical systems will be hard-pressed to preserve their quality and true quality-assurance mechanisms in the face of this new challenge. It is indeed unfortunate that such externally imposed systems are generally not flexible enough to be adapted to individual settings. In the absence of internally designed quality-assurance and utilization-review systems, such state-mandated procedures might be both necessary and cost-effective. However, in a setting like Peninsula, they tend to have a detrimental impact. Thus, the Peninsula model, which from the start had strong internal incentives to provide the most cost-effective treatment possible, is being eroded by the arbitrary imposition of external requirements. It would appear that funding sources might more wisely pass on their budgetary constraints in the form of fixed budgets to those local providers that are capable of making the administrative and clinical decisions necessary to offer judicious, high-quality treatment.

REFERENCES

Derogatis, L. R. (1977). *The psychopathology rating scale series: A brief description*. Unpublished manuscript, Johns Hopkins University.

Levy, A. (1974). Private peer review for fiscal control of publicly funded programs. *Hospital and Community Psychiatry, 25,* 235–238.

Luft, L. L. (1983). *A clinical and fiscal evaluation of peer review for high-cost mental health patients*. Unpublished dissertation, Pacific Graduate School of Psychology, Menlo Park, CA.

Luft, L. L., & Fakhouri, J. (1979). A model for a comparative cost-effectiveness evaluation of two mental health partial care programs. *Evaluation and Program Planning, 2,* 33–40.

Luft, L. L., & Newman, D. E. (1977). Therapists' acceptance of peer review in a community mental health center. *Hospital and Community Psychiatry, 28,* 889–894.

Luft, L. L., Sampson, L. M., & Newman, D. E. (1976). Effects of peer review on outpatient psychotherapy: Therapist and patient followup survey. *American Journal of Psychiatry, 133,* 891–895.

Luft, L. L., Smith, K., & Kace, M. (1978) Therapists', patients', and inpatient staff's views of treatment modes and outcomes. *Hospital and Community Psychiatry, 29,* 505–511.

Moos, R. (1974a). *Community-oriented programs environment scale manual*. Palo Alto, CA: Consulting Psychologists Press.

Moos, R. (1974b). *Ward atmosphere scale manual*. Palo Alto, CA: Consulting Psychologists Press.

Newman, D. E. (1974). Peer review: A California model. *Psychiatric Annals, 4,* 75–85.

Newman, D. E., & Luft, L. L. (1974). The peer review process: Education versus control. *American Journal of Psychiatry, 131,* 1363–1366.

PAL-C Scale: Profile of adaptation of life clinical scale manual. (1978). Roanoke, VA: Institute for Program Evaluation.

Vaughan, W. T., Newman, D. E., Levy, A., & Marty, S. (1973). The private practice of community psychiatry. *American Journal of Psychiatry, 130,* 24–27.

22

Developing a Quality Assurance Program in a University Counseling Center

There Is Always More Than Meets the Eye

INTRODUCTION

As the 1960s drew to a close, those involved in higher education in the United States began to witness the first signs of a significant lessening of resources, which, for many years, had been flowing at an increasing rate into colleges and universities. Those responsible for the training of psychological service providers and the delivery of mental health services in the public sector could discern a similar pattern.

As state governments' budgets and federal grant support for higher education shrank in the next decade, the 1970s inevitably saw, within colleges and universities, increasingly fierce interprogram competition for internal funding. This exacerbated the historic vulnerability of "academic support" programs (such as student health and mental health, career development, and placement services) vis-à-vis instructional programs. Thus it was that student services sought increasingly to represent and justify their mission to their sanctioners—college and university administrators, students, clients, and the community at large—in nonreactive ways.

The kinds of responses made by student counseling centers to what was, in essence, a paradoxical situation were varied and interesting. Faced with spending resources to attract, or even retain, scarce resources, these agencies made increasing use of social ecological principles as manifest in outreach and primary prevention, "off-site" consultation with campus constituents, facilitation in mutual support groups, and the distribution of brochures and newsletters designed to increase counseling center visibility and, therefore, the number and appropriateness of referrals.

There were, however, two other ways in which counseling services units at-

---

WARWICK G. TROY • California School of Professional Psychology, 2235 Beverly Boulevard, Los Angeles, California 90057.

tempted to anticipate and meet the increasingly onerous challenge that budget reduction posed to service delivery goals. One general approach involved changes in the system (structure and function) of service delivery. Examples include the use of structured psychoeducational groups; self-management techniques, including specifically focused skill-development workshops; the development of "packaged" alternatives to formal therapy involving group and workshop learning experiences (made replicable across other personnel and leaders by formal handbooks and manuals); introduction of fee-for-service and limitations on numbers of sessions for which clients would be eligible; and the identification, training, and use of non-professionals in service delivery.

Another general response by which counseling centers attempted to accommodate to straitened circumstances was the use of formal accountability procedures. Thus, statistics on clients' service contacts were maintained and in-house evaluative research on the efficacy and cost-effectiveness of programs was conducted. In short, the *laissez-faire* attitudes that had been the tradition in many of these agencies (e.g., intake therapists had little reason not to assign every client to relatively long-term or open-ended treatment) came to be seen as not invariably appropriate. Many agencies began making decisions regarding which clients would be entitled to what services. Furthermore, those delivering the services were asked to modify their ways, often by adopting more highly structured, time limited strategies of intervention. The concepts, if not the terminology of *utilization review* and *resource allocation* (as well as the important related concept of *quality assurance*) began to be increasingly explored. As resources further dwindled, agency administrators who were responsible for justifying annual requests for the funds looked for ways to demonstrate both that they were providing cost-effective services and that those services were targeted to the individuals and groups with the greatest need.

In the late 1970s the author became the director of a psychological services center in a medium-sized (approximately 8,000 students) urban state university near Baltimore, Maryland. Primarily as a result of a reorganization that preceded his hiring, the staff of the center had significantly decreased while the demand for services had increased. At the same time, over the preceding years the clinic had developed a reputation, especially among faculty, as being a center whose services were of questionable merit and whose staff qualifications and training were insubstantial. Consequently, because of deficiencies in image, the probability of faculty or staff referring a student whose academic progress was being impeded by personal problems (the most appropriate referral for the center) was felt to be unacceptably low.

To deal with these problems and to develop a mechanism for the collection of service delivery data that might help justify the existence and continued funding of the center at a level no lower than that currently obtaining, the decision was made to develop and implement a formal system of accountability. The development of the system rested on three main assumptions: that it had to work within the constraints of existing agency resources; that a unitary model of accountability would be inadequate, and hence, that a multifaceted approach needed to be adopted in which the interdependency of components would be explicitly recognized; and that certain components should be developed before others.

What was finally chosen was an accountability model consisting of three components, implemented by phases. *Phase 1* took the form of a campuswide outreach program aimed at all campus sanctioners, direct and indirect, of center services. *Phase 2* consisted of the implementation of a formal mechanism for monitoring agency service and delivery and staff time outlay. *Phase 3* involved the development of a peer-review-based quality assurance approach to the case management of individual counseling and psychological services. This general sequential model clearly presupposed that, without concerted and protracted "grassroots" campus outreach and consultation efforts (Phase 1), procedures for keeping close track of clients across programs and concomitant staff activities (Phase 2) would have limited impact on sanctioners of counseling center services.

This chapter has as its primary focus Phase 3 of the agency accountability system outlined above. This somewhat technical aspect is emphasized because a model of quality assurance that uses the collective judgments of professional colleagues, or "peers," is uncommon in counseling centers and poses some interesting problems in implementation. Nonetheless, for this counseling center to have decided to ensure the quality of direct services only (Phase 3) without having earlier developed strong community outreach (Phase 1) and service statistics and staff activity analysis (Phase 2) would, in the author's judgment, have been maladroit if not frivolous. For an agency such as a university counseling center to be accountable, it has to be *seen* as being accountable. That is why ongoing efforts directed toward those who are likely to be the consumers of center services, referral agents, and administrative sanctioners are so important. In fact, in this particular counseling center, by far the majority of accountability efforts by the staff were directed to Phases 1 and 2. Only in the latter part of the author's tenure at the center did development of the peer review model of quality assurance for direct services begin in earnest. It was in the context, then, of the first two components of the general accountability system—already on a reasonable footing—that the counseling center eventually was able to devote energy and resources to the complex task of creating a workable system of quality assurance for individual counseling and psychotherapy.

Toward the end of Phase 2, approximately 3 years after the beginning of the drive toward accountability, a picture of the key pieces that were to be the framework of the quality assurance system began to emerge. Thus, the groundwork was being laid for movement toward Phase 3, which eventually was to consist of a time-limited (12-session) model of service delivery, a formal case management and documentation system, and a peer review process that would allow evaluation and consultation for ongoing care and would enable staff to make decisions about which clients were to be given extended services (more than 12 sessions). By the time the author left the center, 5 years after arriving, the quality assurance system was in the early stages of operation. The years in between involved arduous planning, educating, convincing, and cajoling to create among staff, university administrators, and the university community an acceptance of, if not a commitment to, the principles and the process of accountability.

The remainder of this chapter describes developments in the counseling center as the quality assurance program was planned and began to be implemented. Within the limitations of the participant-observer status of the author, the emphasis is on the ways in which agency staff were affected by and affected both the planning and the actual implementation of the program. The process to be depicted is the diffusion of an innovation as staff and director sought to grapple with its reality. Therefore, the account that follows necessarily reflects the apprehensions, anxieties, and conflicts of the center staff, as well as collegial problem-solving, far-sighted planning, and professionalism. In order that this process may be better understood, the next section begins with an analysis of the potential costs and benefits of innovations such as quality assurance programs. A detailed description of the rationale and the elements of the quality assurance program itself then follows.

CHOOSING A PARTICULAR MODEL OF ACCOUNTABILITY

There are obviously many ways to do quality assurance. The challenge for any agency, however, is to select a model that affords balance between what the agency administrator wants to accomplish, on the one hand, and the availability of agency *resources* (staff competence, attitudes, and types of clients), together with the *expectations* that the constituent groups have developed for the unit, on the other. In essence the likely benefits and the probable costs of a number of potential models must be weighed carefully.

OBJECTIVES

The ultimate justification for the changes described in this chapter was survival. Given the diminished resources and the rather poor image of the agency, some explicit demonstration of accountability was imperative. It was hoped that the promulgation of a formal accountability program would improve the image of the center by demonstrating the commitment of both director and staff to high-quality services and to reasonable and responsible utilization of resources. Given the problems commonly encountered in implementing these types of programs, however (Luke & Boss, 1981; Newman & Luft, 1974), this hope alone was not sufficient justification. It was hoped that other, more concrete benefits would accrue.

The primary potential benefit of an accountability system is that the data generated through the program can be used to provide sanctioners (in this case, university administrators) with specific empirical evidence of the value to the community of the services delivered by the agency. Because these administrators were not mental health professionals, it was important to generate some "hard" data that these particular sanctioners would find meaningful. The system that was eventually designed for the center yielded data on utilization (Phase 2) as well as on treatment efficacy. To the delight of several research-oriented staff members, the same data base would facilitate the conduct of evaluation research studies, something that had never before been done at the center.

It was hoped that the peer review program of quality assurance would help ensure both the high quality of the services and equity in the distribution of center resources. Even more important, however, was the hope that it would provide the agency with a mechanism that would assist in making the difficult decisions about which clients would (and would not) get longer term care. It is a well-known fact that the clients who generally seek services in university clinics or counseling centers are the types of people therapists tend to consider appropriate for longer-term therapy (i.e., they are middle class, intelligent, and verbal). In addition, they tend to be attractive (i.e., rewarding) clients. Hence, implementation of an arbitrary time limit could be antithetical to both the staff's predilections and their professional values. The peer review process was intended to provide the center staff with a reasonable, professionally acceptable mechanism for implementing the time-limited mandate.

Finally, the new system was seen as a potentially valuable in-service training mechanism for staff and trainees. It was hoped that staff members would benefit both from the activities necessary to create the system and from the ongoing case-review process. The center interns and practicum students would be formally exposed to a critical introduction to accountability and its mechanisms—ideas that would be novel to most of them, but which would most likely remain a significant factor in their professional lives.

THE PRICE TO BE PAID

The director was aware that the potential advantages to an agency of being formally accountable are sometimes outweighed by the problems that routinely accompany the development and implementation of such systems. For various reasons, staff members may not enthusiastically embrace innovations that require of them increased personal accountability for the type, duration, and intensiveness of their professional activities. Their expectations of autonomy, their faith in their own judgment and traditional ways of doing things, and their valuing of organizational stability may result in attempts to maintain the status quo (Luke & Boss, 1981). Hence, the decision to increase accountability in this formal way was made with full knowledge that clinic staff (at least initially) would very likely be something other than willing partners in the endeavor, and that their conscious or unconscious resistance would influence greatly the pattern of adoption of the system.

Even with the full cooperation of all staff members, developing an effective accountability system is very time-consuming. Development and implementation requirements reduce the amount of time that can be devoted to service delivery—a fact that rarely escapes staff members who are trained to do direct service delivery, like to do it, and probably do it better than they do accountability. In the case of the center, the process did affect service delivery and, to a greater extent, campus outreach and consultation efforts. This effect was of significant concern to some staff and to representatives of student government and other organizations, who were displeased that the center staff did not remain as available for *ad hoc* consultation.

A large portion of accountability time is necessarily devoted to instrumentation, its development, its implementation, and its monitoring. Time is required to gain adequate consensus on both the form and the content of instruments. In addition, staff and clients must cooperate in the often tedious process of piloting measures. Following the pilot phase, the instruments continue to be time-consuming. They are the one component of the system guaranteed to affect all staff members and often all clients. They easily become the symbol of an intrusive system and a focus for staff hostility.

If higher level administrators are naive about psychological services, there is always the possibility that the development of any metasystem will be viewed with skepticism and seen as trivial in comparison with those activities more obviously related to the direct service needs of students. For this particular campus, it was felt that the history of the previous 3 years of deliberate outreach and consultation in academic administration and governance would be sufficient to sustain the positive image of the center. In any event, the strategy called for ongoing consultation with the administration leadership in academic and student affairs during the formative stages of the quality assurance program.

Finally, the director's previous experience with accountability programs (Shueman & Troy, 1982; Troy & Magoon, 1979a) had given him a heightened sensitivity to the potential problems and prompted his decision to "hasten slowly" in the development and implementation of the system. The vehicle chosen was team building. It was hoped that, in the process of formulating the policy and procedures, staff would develop a valuing of the process and, concomitantly, a sense of ownership of the program. Because the development plan also addressed directly the need to incorporate an intensive review of research, concepts, and mechanisms in the area of quality assurance in the center's own staff-development program, the team-building approach to the development of the innovation would be synchronous with the professional growth of agency staff.

CONTINGENCIES

If an agency carefully weighs the costs and benefits of being formally accountable, it is easy to conclude that implementing an accountability system is not worth the cost: the costs are too high. This conclusion may be arrived at even in the absence of direct *external* pressure on an agency for accountability. As has already been observed, although external administrative pressure was not directly discernible, there was a sense of threat—longer term, to be sure, but threat nonetheless. The center staff all felt this unease, and it seemed to be linked to perceptions surrounding the ultimate survival of the center. The decision, however, to justify anticipated costs by pressing ahead with the quality assurance program was based as much on an explicit value stance relating to professionalism in service delivery and evaluation as it was on a response to a perceived threat.

In choosing a model, however, it is obvious that professional values and judgments need to be tempered with consideration of the requirements of the community being served by the agency (in this case, the university community, including students, staff, and faculty), the needs of the agency sanctioners (in this case, the university administrators), and the needs of the agency staff. Questions such as how time-consuming the processes will be, how much accompanying effort and psychological cost will be demanded of clients and staff, and what kinds of data will be meaningful and useful to the sanctioners—all help to determine the form of the model for the particular agency. And the agency must be able to justify the allocation of resources to this type of endeavor.

A major axiom in the choice of a suitable model is the prescription: *Start where you are, with what you've got.* Most publicly funded agencies have little or no control over their clientele, their staff, or their fiscal base. Furthermore, it is still necessary to do business as usual before, during, and after the development and implementation of the system. Such constraints determine the characteristics of the system and how much time and effort can be devoted to its development and maintenance.

If, historically, the mission of the agency has not included accountability, it is necessary to consider the inevitable effects of agency inertia. Change is never easy, and the introduction of an accountability model, which may be perceived from within as the imposition of unwelcome scrutiny or "belt tightening," is likely to cause more resistance than most other types of change.

Finally, change must be initiated with full knowledge of the prevailing institutional values: What do the various constituent groups value, which of these groups must be pleased, and which are in a position to provide effective support for the agency in its efforts to change?

CAPITALIZING ON ORGANIZATIONAL CLIMATE

As stated above, the university community—particularly the faculty—harbored vaguely negative feelings about the clinic. It was a negativism born of imperfect knowledge of the clinic's history and ignorance of current staff and agency policies. More important, it was related to relatively strong negative feelings among faculty toward the office of student affairs, the administrative arm of the campus that had the organizational responsibility for the counseling center. This particular campus had a strong academic (research) orientation. A large majority of the faculty had received their graduate training at Ivy League or other prestigious universities, had been recruited for their research achievements or promise, and saw, or wished to see, their campus as academically elite. Because the prevailing value was academic excellence, many faculty and departmental chairpersons viewed the student affairs office as competing with them for precious resources while being unworthy of those resources that the unit already had.

So that we could capitalize on this persvasive attitude, goals for the agency were couched in terms that, it was hoped, would be compelling to faculty concerned with excellence, in the context of the agency's attempts to attain service excellence. Specifically, the implementation of the system was presented as one step in attaining formal accreditation for the clinic through the Internationl Association of Counseling Services. It was felt that the depiction of agency service delivery in the context of a formalized quality-assurance program would provide a unique and valuable emphasis on case-by-case management, rather than the more traditional accountability-by-professional-qualifications approach. That individual case management by the case contractor (staff counselor) would be mediated by collegial review by small teams of professionals (peers) was, while not unique to counseling centers (H. Korn, personal communication, August, 1978), certainly very uncommon.

There were three additional ways in which the academic bias of the faculty was exploited. First, as part of the center's general campus outreach program (Phase 1), the agency mission and goal statements that were circulated to the academic departments emphasized the importance of early intervention by the center staff in mediating successful academic outcomes for students who might be emotionally and academically at risk. By underscoring a clear role for faculty as appropriate and timely referral agents, and by providing ongoing consultation to departments on referral strategies, a process was presented for the amelioration of a problem of deep concern to the academic community: a troubling student dropout rate. And because, in publicly funded institutions of higher education, reductions in student numbers directly affect the fiscal base of the academic departments, the essential communality between student affairs and academic affairs objectives was highlighted, and the student-development and mental-health goals were concomitantly advanced.

Second, the director began a concerted (and ultimately successful) effort to have himself and several staff members (who had doctorates in psychology) given adjunct faculty appointments. This appeared to be a reasonable goal because those for whom the requests were made had already attained relatively impressive scholarly and professional records. The goal was to have the clinic staff thought of as *professional* (service) *faculty* rather than as *counselors,* and to blur the distinction between service faculty and academic faculty. The final strategy was to encourage the staff to conduct research as an accountability activity and to publicize the fact that the clinic had become—among other things—a center for research.

The director's attempt to legitimize the agency by making it more congruent with the intensely academic environment also met the needs of the university administration. A previous regional-accreditation site-visit team had questioned the university's support of its student affairs component. The center's seeking of formal accreditation and the increased campus support that (it was hoped) would accompany the process, would underscore the commitment of the campus administration to the remediation of the deficiencies identified by the representatives of the accrediting body.

To summarize, the counseling center elected to proceed with its quality assurance (peer review) program in direct service delivery because we felt we must (survival), because we could (resources and expertise), because we felt we should (professional values), because there were certain advantages in doing so (campus credibility and agency accreditation), and because we felt the benefits outweighed the costs.

THE MODEL

Three primary factors affected the adoption of the model finally implemented at the center. The first was the director's commitment to a process embodying a particular set of professional values, in this case those in the American Psychological Association's *Standards for Providers of Psychological Services* (APA, 1977). These standards dictated that the system be based on a written treatment plan that could be used for "establishing accountability, obtaining informed consent, and providing a mechanism for subsequent peer review" (p. 8). The director's hope was that the system would help ensure a high quality of professional services in the clinic. This was the *quality assurance* goal of the program.

The second factor was the obvious need for a mechanism that would help us to achieve a more cost-effective use of resources. This was the *utilization review* (UR) goal. More and more students were descending on the clinic in need of services, and it seemed to the staff that the mean level of distress or disturbance of self- and other referred clients coming through the door had increased greatly in as short a time as three years. The center had at the time neither a waiting list nor a session limit, and the staff were feeling increasingly out of control in their agency and at the mercy of the multitude of prospective clients. Because a waiting list was, in principle, acceptable to no one, it was clear that some kind of session limit would be necessary. Parenthetically, because it is often said that any constraints (e.g., a time limit) imposed on the provider of services are potential threats to service quality (Mattson, 1984), the documentation system would need to lend itself to data collection and evaluation efforts that could help us to assess the efficacy of services thus provided.

The third factor was the need for a system that could transcend clinicians' disciplines and theoretical orientations. In other words, it needed to be *generic*. The clinical staff consisted of doctoral-level psychologists and licensed clinical social workers with differing orientations and levels of training and education. In addition, there were significant numbers of graduate students involved in training experiences in the agency. The documentation system had to be something that one service provider could complete (whether staff or trainee) and another could read and understand, regardless of professional bias, level of training, theoretical orientation, or status.

ELEMENTS OF THE MODEL

The primary component of the accountability system was a panel review of individual cases. It was explicitly intended *not* to reflect the form of a traditional case conference. Rather, the focus was on *case management*, that is, on service planning decisions addressing such issues as the necessity for and the appropriateness of care. The document base for the review process was an *individual service plan* (ISP), which provided for information on the client's presenting problems, service goals, intervention strategies, and the expected length of treatment (see Figure 1, pp. 477–480). The ISP was similar to the problem-oriented case-record systems commonly used in public and private agencies and in the national peer-review programs conducted by the American Psychological Association and the American Psychiatric Association (Stricker, 1983). With its focus on client functioning and observable behavior, the system was designed not only to transcend disciplines and theoretical orientations, but also to be amenable to content analysis for the purposes of data collection. The ISP was developed by the director, in consultation with the staff, based on previous experience with accountability systems and knowledge of the available technology.

The limit to be imposed on the length of treatment was determined by the entire staff. The center was fortunate to have on its staff a psychologist who was coauthoring a book describing a systematic research program on time-limited psychotherapy. Included in the research for the book was an exhaustive review of the literature on the effectiveness of this type of therapy, given various limits. The center staff participated in several in-service sessions during which this resident expert discussed the research. The goal was for the staff to make a group decision about the particular limit that would apply at the clinic.

The research seemed to indicate that either 8 or 12 sessions was the optimum length for time-limited work in a setting such as the counseling center. Therefore, for reasons both theoretical and utilitarian (e.g., the length of the academic calendar), the 12-session limit was adopted. As certain staff members, however, were therapy-oriented and valued longer term work, and because everyone agreed on the professional development value of longer term experience, the decision was made also to allow each senior staff counselor and trainee to carry two longer term clients in his or her caseload.

The intake process was not changed significantly with the new system, except for the requirement that the intake counselor complete Sections A and B (see Figure 1) of the ISP. Clients were assigned to staff primarily as a function of availability and a rotation model was used. The counselor who was assigned the client wrote the ISP, which was subsequently used as a basis for peer review.

Criteria for peer review

Cases were chosen for peer review according to four criteria:

1. The case contractor (staff counselor) believed that the client would require more than 12 sessions.
2. The client had completed 6 sessions postintake.
3. The client was receiving "extended" services (more than 12 sessions), and the number of sessions was a multiple of 6.
4. The case contractor requested consultation on a case.

In other words, clients were to be reviewed automatically at 6 sessions and, if they were seen past the 12-session limit, at each subsequent sixth session. In addition, therapists could request review if they wanted to assign a client to an extended service contract, or if they wished consultation on a case.

Staff members were assigned to three-person case-management review teams that were intended to meet twice monthly to review the cases selected according to the criteria. The purposes were to evaluate and make recommendations about the effectiveness of treatment strategies and to act as arbitrators in cases where a therapist wished to contract for extended services with a particular client. As it was crucial to maximize the efficacy of the case review meetings, the staff determined that (a) comments had to be constructive and the tone collegial; (b) the focus had to be primarily on postintake dispositional considerations and not on protracted discussions of symptom etiology; (c) the case contractor was required to "hear out" any comments from his or her colleagues but was not required to incorporate any such suggestions into the client's ISP; and (d) if, however, the case contractor elected not to modify the service plan of the client in accordance with the recommendations of his or her peers, a formal justification had to be presented to the

INDIVIDUALIZED SERVICE PLAN (ISP): CASE MANAGEMENT RECORD

Confidential   Confidential

Client's Name: _____ S.S.N.: _____

Address: _____ Phone(s): _____(H)

_____ _____(W)

_____(Dorm)

Date Initial Contact: _____   ISP File No: _____

A. INITIAL INTERVIEW REPORT

1. Intake Interviewer: 2. **Appointment:**   emergency ____

   _____   (check one)   reg. sched. ____

                                                             walk-in ____

3. **Projected Service(s):**   Primary _____   Collateral/Adjunctive _____

   (circle one) Category: P-S   E-V   Acad.

   (circle one) Mode: Indiv.   Gp.   Conj.   W/S

4. **Estimated Duration:**

   (circle one)   0   1-3   4-6   7-9   10-12   12+

5. **Tests Administered:**

   (circle one)   SCII   NSVCS   SDS   ACL   OTHER _____

- -

6. **Disposition:**

   (check as   ____ Terminate on intake   ____ Offer cnslg. contract

   appropriate)   ____ Terminate & refer out   ____ C.C. staff counselor

                    _____   _____

                    _____   _____

Assessed Degree Client Dysfunction (to be completed by ISP contractor after case assignment):

1 = no/min. dysfunction	Job/School .... ( )	Rel. w/Other(s) .... ( )
2 = some " "	Home/Family .... ( )	Bodily Functions .... ( )
3 = moderate "	Protect/n Self/Other(s) .... ( )	Personal Comfort .... ( )
4 = severe	Other .... ( )	
5 = no basis to judge	(specify) _____	

FIGURE 1. Individualized service plan (ISP): Case management record.
(*Continued* on pp. 478–480.)

review group. Thus, in the formative stages of this particular model, peer feedback was to be *advisory* only—a precaution taken in view of the literature on peer review, which attests to the importance of reducing insecurity and defensiveness through collegial guidance rather than through inflexibly mandating compliance (Shueman & Troy, 1982).

B. PRESENTING SITUATION ON INTAKE

1. Context: 2. Problem(s):

C. SERVICE PLAN

1. Objectives:

Date: _____   Negotiated by: _____ (clt)

_____ (cnslr)

D. SERVICE PLAN FOR EXTENDED CONTRACT

1. (Revised) Objectives: Beginning Session # _____

Date: _____

THE DYNAMICS OF IMPLEMENTATION

Because of the likely resistance to such a quality assurance program, it is crucial that a person with vested authority take responsibility for its development and maintenance within the agency. In an effort to ensure that the staff would be accountable for their own participation in the program, the director assumed overt responsibility for the program and its associated activities: initial team-building, ongoing in-service development activities, and the maintenance of the peer review process. All staff were asked to report directly to him with respect to the peer review activities.

There was dissatisfaction with two specific aspects of the system: filling out forms and the utilization review responsibility. There was also a somewhat

3. Projected Goals, Interventions, Comments:	E. COLLATERAL CONTACTS
2. Procedures:	3. Progress: (include est. of # sessions to come)
2. (Revised) Procedures:	3. Progress on Extended Contract:

unexpected problem related to the referral of students who needed more service than could be provided by the clinic.

Completing Forms

Even though guidelines had been developed and training sessions had been held to assist the staff in learning to complete the ISP, there were common difficulties that, on more than one occasion, were interpreted by the director as being due to negative feelings about the documentation requirements. For example, forms frequently were left uncompleted. On other occasions, lack of timeliness was the problem.

F. FINAL PROGRESS REPORT

Counselor: _____ Date Submitted: _____

Date ISP Began: _____ Date ISP Ended: _____ # of Sessions/Hrs.: _____

Description of Progress at Termination of ISP:	Interview Dates (indicate no shows)

Interview Dates (indicate no shows)

1. _____   13. _____
2. _____   14. _____
3. _____   15. _____
4. _____   16. _____
5. _____   17. _____
6. _____   18. _____
7. _____   19. _____
8. _____   20. _____
9. _____   21. _____
10. _____ 22. _____
11. _____ 23. _____
12. _____ 24. _____

Final Disposition: (Describe any plans for external referral, subsequent client self-referral, follow-up, etc.)

G. EVALUATION OF ISP

1. Counselor Estimate of Degree of Client Dysfunction at Termination of ISP:

2. Clt. Assess. Severity Pres. Problem(s):

Job/School .... ( )   Rel. w/Other(s) ...... ( )   a. Pre-intake (PIF):   1   2   3   4

Home/Family .... ( )   Bodily Functions .... ( )   b. Term. ISP (F/U # 1):   1   2   3   4

Protect/n Self/   Personal Comfort .... ( )   c. 6 mo. post ISP: 1   2   3   4
Other(s) .... ( )   (F/U # 2):

Other (specify) .... ( )

3. Overall Assess./Progress Toward ISP Goals:

	Cnslr.	Clt.
a. Termin. ISP: (F/U # 1)	1 2 3 4	1 2 3 4
b. 6 mo. post-ISP: (F/U # 2)		1 2 3 4

H. PEER REVIEW

Initial: Date _____   Foll. Sess. # _____

Subsequent: Date _____

Foll. Sess. # _____

Case Management Team: (A) _____ (B) _____

CMT Members: _____

CC Form 2/82 —

A second common problem was related to specific documentation requirements. This was staff members' repeated failure to document their intervention strategies adequately. Whereas an adequate response to this section of the treatment plan ("Procedures" of the ISP) might be "relationship building" or "inventory of client resources," a common response was the generic and uninformative "psychotherapy." Staff counselors had little difficulty in expressing problems, stating goals, and even evaluating progress toward goals as a part of the peer review process. The problems arose when they were asked to state clearly what they were doing to help the client attain the goals, that is, the precise format, intensity, and duration of direct and/or collateral interventions. It should be added that this difficulty in delineating what might be called the topology of treatment is one that occurs frequently in the context of functional case-management documentation, if one is to take as evidence the reports submitted under the national peer-review program conducted by the American Psychological Association (S. A. Shueman, personal communication, August 1984).

Utilization Review Responsibilities

What was soon to become apparent to all center staff was that making decisions about who would get less treatment was a difficult and unpleasant task. Previously, the director had been the person who worried about resource allocation. With the implementation of the new system, however, staff members were given the responsibility for making the decisions. Although willingly engaging in the informal discussion of cases, staff members tended to resist making the difficult decisions about termination of treatment or referral outside the agency. They also disliked the innovation that accompanied the implementation of the UR system: placing prospective educational or vocational clients in a holding group. Students routinely failed to show up for these groups—not a surprise to most staff—and staff were left with the feeling that they had failed to meet the needs of these students and that they were wasting time allocating resources to group interventions.

Case Management Review Meetings

Although the discussion in program-planning meetings seemed to indicate initial general acceptance of the peer review process by staff, significant problems arose. That they so soon emerged is interesting, given that an inordinate amount of time had already been spent in laborious discussions of "safety" issues in the peer review process. Staff members selected their own teams, and it quickly became clear that the teams functioned in very different ways. One team (shorter term orientation) tended to meet haphazardly, and when it did meet, attendence was poor. A propensity of this team was "process" discussion on how members felt about clients, problems with the center's utilization review system, and the peer review process itself. Formal case-management tasks barely got dealt with, and case documentation was inconsistent or missing. (The director was a member of this team!) Another (longer term orientation) team met very regularly indeed but operated in the traditional case-conference format—a model that, for a number of years, had been greatly valued for professional development purposes by some team members.

Clearly, agency inertia was present: in one instance, the team elected to ignore, or deflect, the task; in the other, the task was formally observed, but the format of a previously used quality-assurance mechanism was preserved. Such eventualities may have been dismaying to the director, but such resistance tends to be the rule rather than the exception, at least in the formative stages of peer-review-based quality-assurance programs (Newman & Luft, 1974; Shueman & Troy, 1982). How to reduce such resistance in a noncoercive fashion while preserving the educational function of peer review is a significant challenge to mental health program administrators.

Trainees

Interestingly, most of the difficulties experienced by the staff over the new system were not experienced by the students in training—neither interns nor practicum students. These students did not have the history of their more experienced colleagues and had not had the opportunity to establish habits that would interfere with acceptance of the new ways of doing things. They were being initiated into a system in which hard decisions had to be made and were made routinely. They did not know what it was like to have the options that the staff members had had until that time in their professional lives and in their own training. To be "raised" in an environment of accountability effectively eliminates many of the alternatives provided by traditional training, alternatives that tend to assume unlimited access to service resources for clients. This fact argues persuasively for early exposure of the graduate trainee to the principles and practices of accountability (R. J. Bent, personal communication, January 1983; Shueman & Troy, 1982).

Referral Issues

The imposition of time limits required the center staff, for the first time, to make a systematic and careful evaluation of external referral sources in the community at large. The center was being forced to rely more than ever before on external treatment alternatives for students who could not be adequately served under the time-limited model. Teams of staff members visited agencies to assess possibilities for referrals to those agencies. The variables that affected their assessment included the apparent quality of the services provided, the costs and possible financing arrangements, and accessibility considerations (i.e., Was the student in the agency's catchment area?). The staff also investigated the extent to which they could consult on a case or receive feedback from the agency following the referral.

What was found did little to make the staff feel sanguine about the new system. The public care system was overcommitted and understaffed (both in number of staff and in staff qualifications), and the private sector was too costly for many students. This information could, however, hardly be called a surprise. Rather, it was a reminder of the harsh realities of the mental-health delivery system for those who are in need but who have few financial resources and little or no access to third-party coverage.

Because of the potential difficulties of referral, it became apparent that the six-session peer-review point would need to focus seriously on both the client's need for additional treatment and the possibilities of an appropriate referral. And so it was that staff members found themselves needing to anticipate these issues early in their work with clients in order to responsibly discharge their role as case contractors.

DISCUSSION

In the process of developing this utilization review and quality assurance program, center staff learned what many agencies before them had learned: many factors militate against the success of such a program even if everyone involved in its development believes in its necessity. The general learning was that being held to formal accountability mechanisms is difficult and may well appear to be inconsistent with the existing professional values of the staff. The "intellectual" aspects of the process (specifically, the case conference or, at least, the notion of case review) were acceptable to the staff members; but filling out forms, making decisions about referral or termination, and reviewing the work of colleagues and, in turn, being reviewed were seen by some as anathema to their roles as caregivers and autonomous professionals. Consequently, in order to survive, a system needs to be an integral part of an agency rather than merely a mechanistic appendage. It is a truism that innovations lacking the critical mass of collective ownership can be subverted by the active or passive resistance of staff members.

In the case of this agency, the system was embedded in a program of staff development from its conceptualization through its implementation. This factor contributed to one of the signal accomplishments of system implementation: it resulted in staff members' reflecting seriously on accountability issues such as accessibility to care, the importance of keeping good records, and the significance of being perceived as accountable. No longer did *any* staff members say that these issues were not important or that they were concerns for the director only.

The price of development was 2 years in which significant amounts of time were taken from service delivery, with a consequent diminution of the center's image among some sanctioners. Those involved in both student government and residence-hall programming were less impressed by the accountability efforts than were other sectors of the university community. They saw the agency-staff as "playing psychologists" at the expense of campus outreach and service on demand. One of the most noticeable effects of this effort, which was ultimately aimed at better service to the campus community, was a significant reduction in the amount of time spent in outreach.

But the return on the investment was, as hoped, a better organized center, better clinical work resulting from discussions of case management, better record-keeping, and a better research and evaluation data base. It was the director's observation that the staff perceived a higher level of professionalism in their agency. They were more sophisticated about accountability, which was rapidly becoming one of the major issues in mental health service delivery. The staff as a unit exhibited more cohesion and an improved capacity for group problem-solving.

The agency sanctioners (the university administrators) also perceived a higher level of professionalism in the agency's protracted efforts to "put its house in order." They were also pleased with the successful bid by the center for accreditation. Concomitantly, there was evidence that the faculty regarded the clinic with increased respect, as manifested in much more frequent consultation with center staff on issues both student-related and personal. Clearly, the hard work put into the center's faculty liaison program was seen to be justified, even if some among the counseling center's own colleagues in the student affairs division chose to see in the center's revivified academic image signs that it had deserted its erstwhile comrades for the insubstantial lure of academic respectability.

POSTSCRIPT

As the title of this chapter suggests, there was more to implementing this quality assurance program than meets the eye. Just as there were real, anticipated costs of program development, there were negative consequences that were not foreseen. For example, the center never really directly confronted what may be thought of as the "primary prevention paradox," that is, the phenomenon that successful efforts to identify early target groups particularly at risk tend to overpower the capacity of the service unit to provide appropriate care. To be sure, the results of primary prevention *may* be that clients who receive care are less impaired, but triage has to be used if the resources of the agency remain fixed.

In the case of the center, one of the effects of Phase 1 was to increase significantly the number of self- and, especially, other-referred clients at the very time when a session limit was instituted. This may not have been serious had the center been able to provide all such care in group or workshop formats with the emphasis on information, guidance, and self-management skill development. What the center found, however, was that a significant proportion of the new referrals had significant psychological impairment requiring intensive or longer term care, as well as collateral (psychiatric) services. (The increasing severity of problems seen at the center was seen as a consequence, in part at least, of increasingly open university admissions, of deinstitutionalization, and of a reduction in public mental-health services.)

Another unanticipated negative consequence was the reduction in the quality and accessibility of the center's educational–vocational counseling (E–V) program. Simply put, it was accorded lower priority because of the inordinate focus on the quality assurance for "clinical" services. This effect was particularly distressing to the director and some of the staff because the E–V program had traditionally been well-received and utilized, and was seen to be one of the center's strengths. This is an instance in which utilization review standards for the E–V program—as opposed to clinical services—were compromised.

Also serious was the fact that Phase 3, the peer-review quality-assurance program, simply did not work as well as had been hoped. This impressive-looking mechanism was not wholly accepted by staff, was inconsistently managed, and was seen by certain colleagues as pedantic elitism. It even appeared to reduce accessibility to care!

In retrospect, two explanations seem to advance themselves. First, by the end of the formative stages of the development of the program, the schism that had appeared to divide staff long before the innovation was developed had by no means been bridged. Thus, in a staff whose commitments were characterized by a clinical, longer term, psychotherapeutic orientation, on the one hand, and a developmental, outreach, short-term counseling orientation, on the other, powerful differences in professional values served to attenuate the impact of Phase 3 work. It is conceivable that the director's efforts to mediate between the groups may have exacerbated the schism by freeing each group from the responsibility of their being accountable to the consequences of their stances.

The explanation that seems more compelling relates to the particular role played by the director. In his zeal to facilitate program development through team-building and staff professional development, he played the role of peacemaker from his self-appointed position at the hub of the wheel. This action permitted noncompliance because it permitted noncommunication across the schism. In his anxiety to ensure not only staff ownership of the quality assurance program, but *harmonious* ownership, the director chose to wink at—if not quite ignore—staff noncompliance and the pernicious effects of an unbridged schism.

Finally, the director was not scrupulous enough in ensuring the development of written, procedural guidelines in the form of program implementation manuals. Guidelines were promulgated, to be sure, but a formal policy and procedures manual never appeared. Without such a document, and in the absence of clearly articulated staff responsibilities, the staff were free to rely on selective recall of precedents affecting program operation.

In retrospect, it seems impossible to conclude other than that, in light of the director's knowledge of the schism and its developmental vagaries, he should have managed, rather than facilitated. He should have established clearer procedures and held the staff more accountable.

In his need to be seen as an educator, and to have the center accepted by the wider academic community, the director may have exacerbated the inherent vulnerability of all innovations in their formative stages. As defined by the realization of objectives, programmatic achievements were real enough; nonetheless, due to the admixture of his normative position and management style, his stewardship of the quality assurance program was to some extent compromised.

To say that a pattern of such events in a service delivery setting is by no means unique is likely true, but less than helpful. That this chapter is an attempt, through narrative, to bridge a gap between what happens and what actually tends to be reported in these matters may be more helpful.

REFERENCES

American Psychological Association. (1977). *Standards for providers of psychological services.* Washington, DC: Author.

Luke, R., & Boss, R. (1981). Barriers limiting the implementation of quality assurance programs. *Health Services Research, 16,* 305–314.

Mattson, M. R. (1984). Quality assurance: A literature review of a changing field. *Hospital and Community Psychiatry, 35,* 605–616.

Newman, D. E., & Luft, L. L. (1974). Peer review process: Education versus control. *American Journal of Psychiatry, 132,* 1363–1366.

Shueman, S. A., & Troy, W. G. (1982). Education and peer review. *Professional Psychology, 13,* 58–65.

Stricker, G. (1983). Peer review systems in psychology. In B. D. Sales (Ed.), *The professional psychologist's handbook.* New York: Plenum Press.

Troy, W. G., & Magoon, T. M. (1979). Activity analysis in a university counseling center: Daily time recording or time estimates? *Journal of Counseling Psychology, 26,* 58–63.

23

Psychology Peer Review in an IPA-Model HMO

BETH EGAN O'KEEFE AND JOSEPH N. CRESS

Health maintenance organizations (HMOs) are not a new phenomenon. Kaiser Aluminum Company created the Kaiser-Permanente Plan in California in 1942 to provide health care services for its employees in areas with few such services. Kaiser-Permanente was one of the first HMOs in the country and is still one of the biggest of the "closed panel" type of HMOs.

There has been a proliferation of new HMOs since 1942. It has even been predicted that there could be as many as 450 HMOs serving as many as 20 million people by the end of the 1980s. Blue Cross and Blue Shield of Iowa has even discussed forming a statewide HMO involving all physicians in the state.

HMOs are formed by community groups with the intention of providing a broad range of services to consumers, by the providers themselves, or by industry with the intention of providing the best services possible within budgetary constraints. Quality assurance is an integral part of the HMO philosophy, ensuring the provision of the best, most cost-efficient services possible.

An HMO is not an insurance company. It is a health-care delivery system. Subscribers (or their employers) are charged a flat monthly premium that covers a wide range of routine and catastrophic medical care. Preventive medicine and outpatient care are usually stressed.

The first HMOs, such as Kaiser-Permanente in California, delivered care out of their own clinics. Physicians, psychologists, and other health-care providers were employees of the organization, on salary. Such HMOs often had their own hospitals. These "closed panel" organizations are still a prevalent form of the health maintenance organization.

Psychologists have frequently been involved in health maintenance organizations as staff psychologists providing adjunct psychological care to the subscribers of the HMO. Nicholas Cummings and his colleagues showed long ago that psychologists were valuable team members in a closed panel HMO, actually reducing the total cost to the HMO by providing appropriate psychological care rather

BETH EGAN O'KEEFE • 2100 52nd Avenue, Moline, Illinois 61265.   JOSEPH N. CRESS • 4645 Brady Street, Davenport, Iowa 52806.

than expensive, but in some cases unnecessary, medical care (Cummings & Follette, 1979a,b; Follette & Cummings, 1979). Closed panel HMOs frequently welcome psychologists and are comfortable havens for them, like the VA hospitals of the 1960s.

The independent practice association (IPA) form of the HMO can be seen as an HMO without walls. This form is really a hybrid, combining the traditional HMO feature of prepaid health care with the traditional private-practice feature of service delivery in the practitioner's office on a fee-for-service basis. Psychologists, physicians, and other health-care providers contract with the HMO and are not limited to seeing only HMO subscribers. This form is called an *independent practice association model HMO* because the providers usually belong to one or more independent practice associations that represent the providers with the HMO. Another name for an HMO without walls is an *open panel IPA*.

The IPA-model HMO tends to be more like its cousin, the preferred provider organization (PPO), than it is like a closed panel HMO. In a PPO, a group of providers band together to contract with another group, an insurance carrier or a company, to serve their members. For example, Health Care Group A would sign a contract to treat Company Y's employees for X dollars. Often, Company Y would let the contract out for bidding in the competitive marketplace before signing with Health Care Group A.

IPA-model HMOs, unlike some closed panel HMOs, do not always welcome psychologists with open arms. IPA-model HMOs are essentially the province of private practicing physicians, many of whom have never worked closely with psychologists. Some IPA-model HMOs, including the Bay State Health Care Foundation in Massachusetts and the Quad-City Health Plan in Iowa/Illionois, were reluctant at first to contract with psychologist providers but finally did so.

IPA-model HMOs frequently use peer review mechanisms to assure efficient care, depending on the reviewers to challenge unnecessary and expensive procedures and hospitalizations. The quality assurance procedures of the Quad-City Health Plan, an IPA-model HMO that does contract with psychologist providers, are reviewed by a committee of psychologist practitioners, a sensible but seemingly relatively rare occurrence in HMOs. Most quality assurance now affecting psychologists in IPA-model HMOs appears to be being done primarily by medical committees. Thus, the Quad-City Health Plan is a pioneer in psychology HMO peer-review procedures.

The Quad-City Health Plan (QCHP) is located in the Quad-City area, encompassing parts of Iowa and parts of Illinois, on the Mississippi River. The biggest cities are Davenport, Iowa, and Rock Island and Moline, Illinois. The total metropolitan population of the Quad-City area is over a third of a million people. The area is a center for farm implement manufacturing; there is also a large federal arsenal in the area.

The QCHP was started by a local consortium of industries in 1979. Its birth was not an easy one: there was considerable physician opposition in the Illinois Quad-City area at the start. Finally, three physician IPAs (one in Illinois, one in Iowa, and one for osteopathic physicians) were formed and contracted to provide medical services to the members of the HMO. A bistate psychological IPA was formed later.

Illinois corporate law mandated a separate IPA for psychologists, as members of different professions cannot be in the same corporation. Thus psychologists could not join the medical IPAs.

HMO subscribers were at first only employees of the community's largest company, and the HMO's principal sponsor, John Deere. Now, however, subscribers include employees of several large industries and some small employee groups. The Quad-City Health Plan is now a powerful force in the community, with approximately 70,000 local residents depending on it for their health care needs. Members, on joining, pick a family physician. They are not limited to that physician, but they cannot go to a non-HMO physician without a prior written referral from an HMO physician. Over 95% of local Iowa and Illinois physicians are now affiliated with the HMO.

As part of a compromise to gain affiliation with the Quad-City Health Plan, HMO subscribers needing to see participating psychologists must be referred by a plan physician. It was thought at first that the necessity for physician referral to psychologists might be onerous. This has certainly not proved to be the case; the practice has brought the primary-care physician and the psychologist into much closer contact and, in many cases, has created new, effective treatment teams.

It was important to the QCHP that members be provided quality, cost-effective services. Those goals were also very important to the psychologists affiliated with the plan. Early in the psychologists' negotiations with the HMO, a peer review committee was formed, and a peer review manual, with practice standards for local psychologists, was written and approved by the members of the psychology IPA. Table 1 reproduces selected items from the peer review manual. Psychologists wanting to be affiliated with the HMO were required to practice by the standards in the manual and to agree to cooperate with the peer review process.

For several reasons, there was probably more cooperation with the peer review process in the Quad-City psychology community than is usual. Most important, psychologists wanted inclusion in the local HMO, and the local HMO mandated peer review. If they wanted to be affiliated with the local HMO, psychologists had to agree to actively cooperate with the peer review process. Second, there was an understanding between the psychologists and the HMO that psychologists would

TABLE 1. Selected Items from Psychologists' Outpatient Peer Review Manual

---

2. Individual psychotherapy sessiions should have a duration of 60 minutes or less, except in unusual circumstances wherein the provider should note a brief justification for the extended session. In any event, the extended session should not exceed 120 minutes.

7. Group psychotherapy sessions should be not less than 60 nor more than 120 minutes in duration.

8. There should be no more than 10 persons enrolled in a group per psychotherapist. Exceptions should be explained.

11. Participants in psychotherapy should be at least 4 years of age.

17. Psychological assessment should not be provided routinely, but should be provided when it has been determined either medically or psychologically that it is indicated.

---

eventually go "at risk" like the physicians. In an "at-risk" situation, it is in psychologists' best interests financially to control unnecessary services. Third, there was strong support among leading local psychologists for peer review and quality assurance, and a strong peer review committee was appointed by the president of the psychology IPA, including psychologists with quality assurance and peer review experience. For example, one of the committee members has been a Civilian Health and Medical Program of the Uniformed Services (CHAMPUS) peer reviewer since that program's inception. Another was responsible for starting and supervising a complicated, comprehensive management information system (MIS) at a large local mental-health center. The third member of the original committee had served on the quality assurance committee of a local mental-health center for several years. Two of the three committee members were also officers of the psychology IPA.

The mandate of the QCHP peer-review committee is the provision of quality care, offering services that are valid, called for, and appropriate. This mandate ensures the protection of the consumer and also aids the cost efficiency of the HMO. To do this, the peer review committee developed a form to be completed by the psychologist provider. The three-page form itself, which can be seen in the Appendix, was an adaptation of other peer-review forms already in use, principally the from us used by CHAMPUS and by the federal Blue Cross and Blue Shield programs at that time.

Currently, psychology quality assurance in the QCHP is rudimentary process evaluation. Outcome evaluation is neither politically nor technically feasible at this time. The QCHP peer-review committee does hope eventually to establish minimal standards for competent and cost-effective psychological care. A related goal is the development of regional norms of care, including base rates for particular diagnostic categories, treatment modalities, lengths of treatment, and so on. Finally, the committee hopes that the peer review process will assist independent practitioners by giving the kind of case feedback that most stopped receiving when they left agency work and entered the private arena.

The QCHP psychology-peer-review committee fully recognizes that appropriately completing the "Outpatient Psychological Treatment Profile" does not guarantee competent, high-quality, cost-effective care. Nevertheless, failing to complete the form satisfactorily suggests carelessness and insufficient accountability and documentation at the least, and possible substandard care at the worst. As can be seen in the Appendix, the review form consists of over 60 variables that are grouped into six major categories. "Treatment Status" includes demographic data, basic numerical data with respect to the evaluation and intervention episode, and diagnosis. "Degree of Impairment" determines the severity of symptoms over six different areas of functioning. "Degree of Subjective Distress" assesses the extent of the patient's vocalized discomfort. "Symptoms" is a section of the form that specifies 17 different areas of symptomatology and then allows for open-ended responses if necessary. The section on "Treatment Goals" asks the therapist to specify the plan of treatment and its relationship to the presenting problems. The final section, entitled "Specific Components of Treatment," asks the provider to indicate the nature, frequency, and session length of treatment. In addition, information is requested with respect to collateral contracts and consultations with the referring physician and other professionals.

On a case-by-case basis, the committee, now consisting of five members, determines whether the diagnostic impression appears to be appropriate and warranted with respect to the presenting symptoms. In addition, the committee carefully scrutinizes the treatment goals articulated by the provider to make sure that they touch on all major presenting problems. With respect to certain presenting problems, the committee determines whether appropriate collateral and conjoint consultation has been pursued. For example, local practice standards dictate that disorders among latency-age children require at least parallel parental contact if not family psychotherapy. In addition, indications of anorexia and bulimia among adolescents alert the peer review committee about the necessity for family involvement in the treatment process. In addition, the committee believes that it is highly important that the providers remain in contact with the referring physician to keep him or her abreast of the psychological status of the patient.

With respect to the general practice of an individual provider, the committee also reviews summary printouts generated by the HMO's management-information system. With these data, the review committee is able to compare an individual's practice norms for the provider group. For example, local psychology norms do not advocate routine formal psychological assessment using a standardized battery for every patient. The relative frequency of such evaluations for each provider is compared to the group norm. In general, the committee attempts to determine whether the provider's practice habits or individual therapeutic procedures fall within the shibboleth of "usual, customary, and reasonable."

Length of treatment is another variable studied by the peer review committee. The QCHP authorizes $1,500 per family member for psychological outpatient care per year. At $75 per hour of care, which has been about the local average, 20 hours of care are provided. Thus, the QCHP obviously does not intend to pay for psychoanalysis. Yet, QCHP staff has tended also to view with suspicion the client who needs only one to five treatment sessions. There is no intention to pay for treatment of minor psychological upsets. The intent is to pay for the treatment of acute psychological disturbances that are severe enough to interfere with the client's functioning in some way. The average number of treatment sessions per episode of care for each provider is reviewed, and variance with local provider patterns is noted.

Cases can be brought to the committee for review in any one of a number of ways. The vast majority come from a formalized sampling procedure using the HMO's management-information system. After the provider joins the psychology IPA, she or he is reviewed on her or his first six patients at the end of eight treatment sessions or at termination, whichever comes first. Thereafter, a 10% random sampling procedure has been used. Also, all episodes of care totaling more than 20 sessions are automatically reviewed. In addition, the HMO staff can ask the committee to review any case in which they feel there may be some questionable or unreasonable claims or practices. Furthermore, the provider can request a review in anticipation that a particular case may trigger a review because of practices that may seem questionable or inappropriate and may need clarification. Finally, a consumer may initiate a review by complaining to the HMO.

Once a case has been selected for review, the Quad-City Health Plan liaison person, who interfaces with the psychology IPA, sends a treatment profile form to the provider. No identifying data are on the form itself for either the provider or the patient; instead, code numbers are used for both. The provider is required to complete the form within 2 weeks. Copies of the treatment form are made and then sent to each peer reviewer. Before the peer-review committee meeting, the members examine each of the forms and make notes with respect to completeness, extent of accountability, appropriateness of care, and so on.

When consumers enroll in the HMO, they automatically authorize quality care assessment. They realize and consent that the care they receive may very well be reviewed. They also understand that they will be anonymous to the reviewers and will be identified only by a code. Only the HMO has a file correlating the code with the actual name. In this sense, patient confidentiality is tightly preserved. With respect to the provider, however, although confidentiality is discussed in theory, it cannot be adhered to in practice. In actually reviewing the form, the committee members see only a code number identifying the provider. For two reasons, however, most of the committee members are eventually able to associate a name with the provider. First of all, because of the limited number of providers, the reviewers are quite familiar with the practitioners' theoretical orientations and preferred modes of intervention. Second, whenever the committee determines that further clinical information is necessary or that corrective feedback is appropriate, the clinician is actually approached directly, face-to-face. Although such an approach is not feasible in a large peer-review system, it is the committee's belief that a more personal and less formal approach is beneficial for both committee members and psychology providers. The committee members cannot be cavalier in their approach and must be willing to meet face-to-face with the provider to give him or her direct corrective feedback. The provider knows that he or she is not being dealt with by an unseen and invisible force; rather, the contact is more on the basis of a consultation than on the basis of an authoritative fiat.

Approximately every other month, depending on the number of cases to be reviewed, the peer review committee meets for approximately 2 hours, together with liaison personnel from the HMO. The president of the psychology IPA is ex officio a member of the committee and often attends. Between six and eight people are usually present. In the meeting, the QCHP liaison personnel present any relevant issues that they feel need to be brought to the attention of the committee. After that business is completed, each form is reviewed individually. If any concerns are generated, a committee member is designated to give feedback to the service provider after the meeting. During the meeting, the committee, together with the liaison personnel, also review the summary printouts and discuss whether a psychologist's practice pattern appears to be at variance with the practices of other providers. If any indications of variance emerge, a committee member is again designated to discuss the variance issues with the psychologist. After the liaison personnel have left, the clinical members of the committee sometimes convene briefly to determine how each of the psychologists in question should be approached.

After the meeting, each designated committee member contacts the psychologist whose profile report was open to question. At present, the contacts are made either over the phone or face-to-face. The questions or issues are pointed out to the psychologist and brought to his or her attention for review and consideration. During the first feedback episode, the committee usually does not explicitly ask for a concrete behavioral change. A summary letter is sent to the psychologist. At the next peer review meeting, the member reports back to the committee the results of the session with the individual provider. From that point on, the committee carefully reviews the reports and summary data of that psychologist to determine whether the psychologist is changing in the direction suggested by the committee. If the questionable behavior continues, the psychologist is then issued a formal warning stating that if the recommended concrete changes are not observed within a specified time framework, the psychologist will be subject to possible suspension. If the behavioral change is not forthcoming, the peer review committee recommends to the IPA that the psychologist be suspended for a specified length of time. During the period of suspension, the psychologist is not reimbursed by the HMO for treating participants. In addition, the psychologist loses all IPA voting privileges during the period of suspension. If the questionable behavior occurs again after the psychologist has been reinstated after suspension, the committee recommends to the IPA group that the psychologist be terminated. Thus far, the action of the peer review committee has been entirely educative and informative; no formal warnings or sanctions have been necessary.

No case reviews have been initiated by consumers, and none have been proposed by providers themselves. All have come either from the selection process or from the HMO itself. It may be instructive to summarize a few of the cases in which corrective feedback has been given to psychology providers.

In one case, a provider appeared to be routinely administering psychological tests to his patients. In fact, at the outset, more than half this provider's patients were billed for psychological assessment. The member was told that his practice was at considerable variance with the practice norms of other members of the provider group. He argued the case for psychological screening, but in the end, he acquiesced and agreed to limit the frequency of routine assessment.

Another psychologist used some rather severe diagnostic categories for young adolescents without sufficient documentation. After feedback, the psychologist agreed that the diagnoses were perhaps too harsh and consented to be more circumspect in the future.

In another case, a psychologist who to the committee's knowledge was not especially trained in the area of neuropsychology submitted a claim for a neuropsychological evaluation. The psychologist fully admitted that he had no specialized training in neuropsychology but said that he had billed the evaluation as neuropsychological assessment because he had administered one test that is often used in neuropsychological assessment batteries.

In another instance, the HMO requested the committee to examine what appeared to be rather unusual billing practice. The HMO had received a bill for approximately 10 group psychotherapy sessions over a 3-month period. The liaison personnel were concerned that something irregular was occurring, because they had requested monthly billings. On consultation with the provider, it was apparent that it was this particular clinic's policy to bill time-limited group psychotherapy at the end of the entire episode of care and not on a monthly basis.

In its first 22 months of operation, the committee reviewed 126 treatment reports. Only six reports required discussion with the psychology providers. At the quarterly IPA meetings, the peer review committee reports on general trends and concerns and shares with the members any issues raised by the HMO. On one occasion, for example, the committee members noticed a number of apparent errors in coding psychological assessments. The HMO provides for three different categories of psychological assessment with different rates and different limits. A trend among a number of psychologists was to code an evaluation category that was actually reimbursed at a lower rate than that to which they were entitled. In this case, the feedback was warmly received!

Ideally, the process of quality assurance has at least three distinctive components. The first is a mechanism that guarantees that the members of the provider pool will meet minimal standards of competence. The second is a format for assessing the process of psychotherapy itself, determining that usual, customary, and standard procedures are used. The third consists of outcome evaluation. In the current format, only the first two are feasible.

Candidates for entrance into the IPA must have a background commensurate with the criteria listed in the *National Register of Health Service Providers in Psychology*. By having certain high standards for admission to the psychology IPA, it can be reasonably ensured that all psychology providers will have at least minimal competence.

The assessment of process, the second component of quality assurance in psychological treatment, is admittedly subjective, arbitrary, and intuitive. It focuses only on what the clinician does and not on the results. It asks the reviewers to formalize the ideal—or at least adequate—criteria for treatment, and then to compare with these criteria the reported behaviors of a given provider in a given therapeutic relationship. Despite these shortcomings, process measurement can be used as an index of cost-effective care.

At present, the third phase of quality assurance, outcome evaluation, is not well represented in our current peer-review scheme. The reasons for the omission are entirely pragmatic. The IPA has neither sufficient time nor sufficient resources to establish an adequate, proven outcome evaluation. Furthermore, the HMO itself is interested primarily in process and not outcome. Therefore, it is not willing, at least at present, to underwrite the expense of establishing an outcome evaluation program.

Another shortcoming of the current peer-review program is that the management information system that generates the summary data was not designed for psychology input. Instead, it was designed primarily for physicians, particularly for psychiatrists. A number of the IPA members have experience in developing and maintaining management information systems, and there may be an opportunity for significant input in the future.

A final limitation of the current procedure is that an unscrupulous provider could easily misrepresent data, especially symptoms, to argue for the appropriateness and reasonableness of a certain kind of care.

The peer review process is now in its 7th year. Several substantial changes have been or will be made in the near future. In the beginning of the peer review program, initial feedback was not documented in writing. Feedback was conveyed either over the phone or face-to-face. Beginning in January 1984, the verbal feedback has been summarized in a letter to the provider. This particular change was proposed by the HMO in view of its own experience with some recalcitrant physicians. Accordingly, the change was not precipitated by any difficulties in dealing with psychologists *per se.* The committee plans to improve the tracking of individual cases from one point of review to another. Many cases are actually reviewed more than once. Initially, they may have been reviewed because of the random sampling process. Recently, the 10% random sampling process was discontinued because of the large burden placed on the providers and reviewers and because of the success of the educative process with the providers. Providers are taking seriously issues of quality of care, and this seriousness is reflected in the excellent quality of the treatment forms now being done by providers. However, all cases are automatically reviewed after 20 treatment sessions. Accordingly, approximately 10% of those cases reviewed because of treatment lasting 20 or more sessions may also have been reviewed earlier through random process. Treatment forms completed by the psychologist at the two different times are compared for a determination of whether the psychologist is following the treatment plans, making progress, and reaching the stated goals.

The current treatment form was never intended to be final. Rather, it was seen as a form that would evolve over time. This committee plans to incorporate some of the features of the recently revised "Outpatient Mental Health Report" suggested by the American Psychological Association. In particular, the committee is considering the differentiation of intermediate and final treatment goals. Furthermore, it is considering asking for some indicators of progress in therapy at the time the report is written. In those cases where the number of treatment sessions has exceeded 20, the committee may ask the therapist to document progress and changes, including any modifications of treatment goals and procedures. In all fairness, however, because of the relatively high rate of review, a revised form will not require significantly more time to complete than the current form. This brevity is important for the committee as well, because members are not reimbursed for the time they spend on peer review activities.

In considering the inclusion of some measures of outcome, the committee does not plan to evaluate therapists on a case-by-case basis. All therapists are very familiar with those cases in which all the right things were done and the outcome was minimal, and with those cases in which all the wrong things were done and the outcome was fantastic. However, the committee is considering reviewing each provider's outcome results with respect to diagnostic groups, age groups, and so on.

The committee also plans to update and to modify the peer review manual. Regional norms have changed slightly over the past few years, and these changes will be reflected in the new manual.

In order to establish greater reliability and validity on the review form, some of the nonclinical data on the form will be compared to data contained within the HMO management information system. This change will allow more accurate assessment of the status of treatment, whether it is continuing or terminated, the number of sessions to date, the actual numbers of hours of assessment billed, the diagnoses, and so on. In the opinion of the committee, an accurate status assessment will discourage an unscrupulous provider from falsifying a treatment form.

At the present time, the peer review process has no quality-control mechanism to ensure the validity of the data submitted by the provider. Although there is no benefit in misrepresenting more severe cases as less severe, there is clearly an advantage in misrepresenting less severe cases as more severe. However, as a result, the HMO has raised the possibility of outpatient chart review. Some HMO administrators have hinted that, to address the issue of quality care and cost containment aggressively, reviewers need to have access to the actual clinical file to determine the validity of the diagnosis, the type and length of treatment, whether or not patient rights are being protected, and so on. Clearly, this process is expensive, and the efforts to obtain consensual validation of appropriate criteria would be exceedingly laborious. If chart review is mandated, it is the committee's plan to select charts on a random basis. If, during this sampling procedure, clear patterns of concern emerge, then sampling may be done more frequently from among cases that appear to have a higher probability of deficient quality or excess cost. Although a chart review process would afford greater quality control, the committee will resist this process as much as possible and will rely on the good faith of its members until such faith is betrayed.

Profile analysis is a long-term goal of the peer review process. Because of the small group and the limited number of patients currently treated, data need to be gathered over a relatively long period of time in order to make up a sufficiently large data base. Once the large data base is established, patterns of use, deviations from regional norms, and emerging trends in both care and cost can all be useful. These profiles will be determined according to a number of variables, including age, diagnosis, patient characteristics, individual therapist characteristics, and the characteristics of outpatient psychotherapy clinics.

At the present time, no formal appeal system exists within the peer review process. Recommendations by reviewers can be contested in only two ways. An individual clinician can request to meet with the peer review committee and openly discuss the rationale behind the particular recommendation that was found by the provider to be unacceptable. If the provider is still not satisfied, he or she then can bring the issue before the entire group of directors of the IPA. Conceivably, a contested case could be appealed at the state level. The peer review process supported by the American Psychological Association would be a final course of appeal in case the issue cannot be resolved at the regional level.

The QCHP psychology peer review process has now been in place almost 7 years. The cooperation of psychologist providers has been excellent, and the working relationship with the HMO staff has been cordial and businesslike. Although the system could certainly be improved, and there are plans to improve it in the not-too-distant future, the peer review committee feels comfortable with the results and system to date. Although there are changes in the wings for the Quad-City Health Plan and psychologists, peer review will no doubt continue to play an important role.

APPENDIX

QUAD CITY HEALTH PLAN: OUTPATIENT PSYCHOLOGICAL TREATMENT PROFILE

                                                    Provider ID #

OUTPATIENT PSYCHOLOGICAL TREATMENT PROFILE

1. TREATMENT STATUS

    QCHP Patient ID #: _____

    Patient's D.O.B.: ___/___/___   Sex: ___ M ___ F

    Date of this report: ___/___/___   Date treatment began: ___/___/___

    Continuing: _____   Terminated: _____   Date: _____

    # of sessions to date: _____
    # of hours testing: _____
    # of hours additional assessment: _____
    Expected total # of additional sessions: _____
    List relationship of other family members currently in treatment with you:

    Conjointly   _____

    Separately   _____

    With other mental health providers: _____

    Diagnosis: 1. _____
               2. _____

2. DEGREE OF IMPAIRMENT: (Circle appropriate descriptions)

    Employment or School   (a) Unimpaired (b) Mild (c) Moderate (d) Severe
    Home/family   (a) Unimpaired (b) Mild (c) Moderate (d) Severe
    Social   (a) Unimpaired (b) Mild (c) Moderate (d) Severe
    Bodily Function   (a) Unimpaired (b) Mild (c) Moderate (d) Severe
    Interpersonal Relations   (a) Unimpaired (b) Mild (c) Moderate (d) Severe
    Protection of Self & Others (a) Unimpaired (b) Mild (c) Moderate (d) Severe

---

IDENTIFYING INFORMATION   QCHP Patient ID # _____

Patient's Name: _____ Provider's ID # _____
Date Submitted: ___/___/___

Patient's Diagnosis: 1. _____
                     2. _____

I certify that to the best of my knowledge and belief the information on this
Treatment Report is complete and correct.

                    _____
                    (Provider's Signature)

TREATMENT PROGRAM - 2

3. <u>DEGREE OF SUBJECTIVE DISTRESS</u>: (Circle appropriate description)

   (a) None   (b) Mild   (c) Moderate   (d) Severe

4. <u>SYMPTOMS</u>: (Circle appropriate descriptions)

   (1) Psychophysiological, specify: _____

   (2) ANXIETY   (9) CONVERSION/DISSOCIATIONS
       a. moderate
       b. severe   (10) OBSESSIONS/COMPULSIONS
       c. incapacitating
                                        (11) IMPULSIVE BEHAVIOR
   (3) PHOBIA
       a. moderate   (12) DRUG ABUSE
       b. severe
                                        (13) ALCOHOL ABUSE
   (4) DEPRESSION
       a. moderate   (14) MARITAL DISTURBANCES
       b. severe
                                        (15) AGGRESSIVE BEHAVIOR, Specify:
   (5) SELF-PUNITIVE BEHAVIOR
       a. moderate   _____
       b. severe
       c. suicidal   (16) CHILDHOOD DISTURBANCE, Specify:

   (6) THOUGHT DISORDERS   _____
       a. delusions
       b. hallucinations   (17) INTERPERSONAL DISTURBANCES, Specify:
       c. paranoid thinking
                                             _____
   (7) IMPAIRED BRAIN FUNCTIONING
       a. confusion   (18) OTHERS _____
       b. memory loss
       c. disorientation   _____
       d. impaired judgment
                                             _____
   (8) MANIC BEHAVIOR
       a. hypomanic   _____
       b. severe
                                             _____

5. <u>TREATMENT GOALS</u>, Specify:

TREATMENT PROGRAM - 3

6. <u>SPECIFIC COMPONENTS OF TREATMENT</u>:

		<u>Therapy to Date</u>	<u>Planned for Future</u>
A.	<u>Primary Therapy</u>		
	1. Type	_____	_____
	2. Frequency	___ per wk mo (circle one)	___ per wk mo (circle one)
	3. Sessions length	___ minutes	___ minutes
B.	<u>Additional Therapy</u>		
	1. Type	_____	_____
	2. Frequency	___ per wk mo (circle one)	___ per wk mo (circle one)
	3. Sessions length	___ minutes	___ minutes
C.	<u>Collateral Contacts</u>		
	1. Type (specify with whom; e.g. family members, significant others)	_____	_____
	2. Frequency	___ per wk mo (circle one)	___ per wk mo (circle one)
D.	<u>Consultations</u>		
	1. Type (specify with whom; e.g. physician, minister)	_____	_____
	2. Frequency	___ per wk mo (circle one)	___ per wk mo (circle one)

REFERENCES

Cummings, A., & Follette, T. (1979a). Brief psychotherapy and medical utilization. In A. Kiesler, A. Cummings, & R. VandenBos (Eds.), *Psychology and national health insurance: A sourcebook.* Washington, DC: American Psychological Association.

Cummings, A., & Follette, T. (1979b). Psychiatric services and medical utilization in a prepaid health plan setting: 2. In A. Kiesler, A. Cummings, & G. R. VandenBos (Eds.), *Psychology and national health insurance: A sourcebook.* Washington, DC: American Psychological Association.

Follette, T., & Cummings, A. (1979). Psychiatric service and medical utilization in a prepaid health plan setting. In C. A. Kiesler, A. Cummings, & G. R. VandenBos (Eds.), *Psychology and national health insurance: A sourcebook.* Washington, DC: American Psychological Association.

Index